Lecture Notes in Computer Science   14135

Founding Editors

Gerhard Goos
Juris Hartmanis

The series Lecture Notes in Computer Science (LNCS), including its subseries Lecture Notes in Artificial Intelligence (LNAI) and Lecture Notes in Bioinformatics (LNBI), has established itself as a medium for the publication of new developments in computer science and information technology research, teaching, and education.

LNCS enjoys close cooperation with the computer science R & D community, the series counts many renowned academics among its volume editors and paper authors, and collaborates with prestigious societies. Its mission is to serve this international community by providing an invaluable service, mainly focused on the publication of conference and workshop proceedings and postproceedings. LNCS commenced publication in 1973.

Ignacio Rojas · Gonzalo Joya · Andreu Catala
Editors

Advances in Computational Intelligence

17th International Work-Conference
on Artificial Neural Networks, IWANN 2023
Ponta Delgada, Portugal, June 19–21, 2023
Proceedings, Part II

 Springer

Editors
Ignacio Rojas 🅾
University of Granada
Granada, Spain

Gonzalo Joya
University of Malaga
Málaga, Spain

Andreu Catala
Polytechnic University of Catalonia
Vilanova i la Geltrú, Spain

ISSN 0302-9743 ISSN 1611-3349 (electronic)
Lecture Notes in Computer Science
ISBN 978-3-031-43077-0 ISBN 978-3-031-43078-7 (eBook)
https://doi.org/10.1007/978-3-031-43078-7

This Springer imprint is published by the registered company Springer Nature Switzerland AG
The registered company address is: Gewerbestrasse 11, 6330 Cham, Switzerland

Paper in this product is recyclable.

Preface

We are proud to present the set of final accepted papers for the 17th edition of the IWANN conference - the International Work-Conference on Artificial Neural Networks - held in Ponta Delgada, São Miguel, Azores Islands, (Portugal) during June 19–21, 2023.

IWANN is a biennial conference that seeks to provide a discussion forum for scientists, engineers, educators, and students about the latest ideas and realizations in the foundations, theory, models, and applications of hybrid systems inspired by nature (neural networks, fuzzy logic, and evolutionary systems) as well as in emerging areas related to these topics. As in previous editions of IWANN, it also aims to create a friendly environment that could lead to the establishment of scientific collaborations and exchanges among attendees. The proceedings include the communications presented at the conference.

Since the first edition in Granada (LNCS 540, 1991), the conference has evolved and matured. The list of topics in the successive Call for Papers has also evolved, resulting in the following list for the 2023 edition:

1. *Deep Learning*
2. *Learning and adaptation*
3. *Emulation of cognitive functions*
4. *Bio-inspired systems and neuro-engineering*
5. *Advanced topics in computational intelligence*
6. *Agent-based models*
7. *Time series forecasting*
8. *Robotics and cognitive systems*
9. *Interactive systems and BCI*
10. *Machine Learning for Industry 4.0 solutions*
11. *AI Health*
12. *AI in 5G technology*
13. *Social and Ethical aspects of AI*
14. *General applications of AI*

At the end of the submission process, and after a careful peer review and evaluation process (each submission was reviewed by at least 2, and on the average 2.7, program committee members or additional reviewers), 108 papers were accepted for oral or poster presentation, according to the reviewers' recommendations.

During IWANN 2023 several special sessions were held. Special sessions are a very useful tool for complementing the regular program with new and emerging topics of particular interest for the participating community. Special sessions that emphasize multi-disciplinary and transversal aspects, as well as cutting-edge topics are especially encouraged and welcome, and in this edition of IWANN comprised the following:

- **SS01: Ordinal Classification**
  Organized by: Victor M. Vargas, David Guijo-Rubio, Pedro A. Gutiérrez
- **SS02: Machine Learning in Mental Health**
  Organized by: Pepijn van de Ven
- **SS03: Interaction with Neural Systems in both Health and Disease**
  Organized by: Pablo Martínez Cañada, Jesus Minguillón Campos
- **SS04: Deep Learning applied to Computer Vision and Robotics**
  Organized by: Enrique Dominguez, José García-Rodríguez, Ramon Moreno Jiménez
- **SS05: Applications of Machine Learning in Biomedicine and Healthcare**
  Organized by: Miri Weiss Cohen, Daniele Regazzoni, Catalin Stoean
- **SS06: Neural Networks in Chemistry and Material Characterization**
  Organized by: Ruxandra Stoean, Patricio García Báez, Carmen Paz Suárez Araujo
- **SS07: Real-World Applications of BCI Systems**
  Organized by: Ivan Volosyak
- **SS08: Spiking Neuron Networks: Applications and Algorithms**
  Organized by: Elisa Guerrero Vázquez, Fernando M. Quintana Velázquez
- **SS09: Deep Learning and Time Series Forecasting: Methods and Applications**
  Organized by: Francisco Martínez Álvarez, Verónica Bolón-Canedo, David Camacho
- **SS10: ANN HW-Accelerators**
  Organized by: Mario Porrmann, Ulrich Rückert

In this edition of IWANN, we were honored to have the presence of the following invited speakers:

1. Alberto Bosio, Full Professor at the INL – École Centrale de Lyon, France. *Title of the presentation: Reliable and Efficient Hardware for Trustworthy Deep Neural Networks*
2. Amaury Lendasse, Department Chair Information and Logistics Technology Faculty, University of Houston, USA. *Title of the presentation: Metric Learning with Missing Data*

It is important to note that for the sake of consistency and readability of the book, the presented papers are not organized as they were presented in the IWANN 2023 sessions, but classified under 14 chapters. The organization of the papers is in two volumes arranged basically following the topics list included in the call for papers. The first volume (LNCS 14134), entitled "Advances in Computational Intelligence. IWANN 2023. Part I", is divided into five main parts and includes contributions on:

1. Advanced Topics in Computational Intelligence
2. Advances in Artificial Neural Networks
3. ANN HW-Accelerators
4. Applications of Machine Learning in Biomedicine and Healthcare
5. Applications of Machine Learning in Time Series Analysis

In the second volume (LNCS 14135), entitled "Advances in Computational Intelligence. IWANN 2023. Part II", is divided into nine main parts and includes contributions on:

1. Deep Learning and Applications
2. Deep Learning Applied to Computer Vision and Robotics
3. General Applications of Artificial Intelligence
4. Interaction with Neural Systems in Both Health and Disease
5. Machine Learning for Industry 4.0 Solutions
6. Neural Networks in Chemistry and Material Characterization
7. Ordinal Classification
8. Real-World Applications of BCI Systems
9. Spiking Neural Networks: Applications and Algorithms

The 17th edition of the IWANN conference was organized by the University of Granada, University of Malaga, and Polytechnical University of Catalonia.

We would like to express our gratitude to the members of the different committees for their support, collaboration and good work. We specially thank our Honorary Chairs (Joan Cabestany, Alberto Prieto and Francisco Sandoval), the Technical Program Chairs (Miguel Atencia, Francisco García-Lagos, Luis Javier Herrera and Fernando Rojas), the Program Committee, the Reviewers, Invited Speakers, and Special Session Organizers. Finally, we want to thank Springer LNCS for their continuous support and cooperation.

June 2023

Ignacio Rojas
Gonzalo Joya
Andreu Catala

Organization

Steering Committee

Davide Anguita	Università degli Studi di Genova, Italia
Andreu Catalá	Universitat Politècnica de Catalunya, Spain
Marie Cottrell	Université Paris 1 Panthéon-Sorbonne, France
Gonzalo Joya	University of Málaga, Spain
Kurosh Madani	Université Paris-Est Créteil, France
Madalina Olteanu	Université Paris Dauphine – PSL, France
Ignacio Rojas	University of Granada, Spain
Ulrich Rückert	Universität Bielefeld, Germany

Program Committee

Kouzou Abdellah	Djelfa University, Algeria
Vanessa Aguiar-Pulido	University of Miami, USA
Arnulfo Alanis	Instituto Tecnológico de Tijuana, Mexico
Ali Alkaya	Marmara University, Turkey
Amparo Alonso-Betanzos	University of A Coruña, Spain
Gabriela Andrejkova	.
Davide Anguita	University of Genoa, Italy
Cecilio Angulo	Universitat Politècnica de Catalunya, Spain
Javier Antich Tobaruela	University of the Balearic Islands, Spain
Alfonso Ariza	University of Málaga, Spain
Corneliu Arsene	SC IPA SA, Romania
Miguel Atencia	University of Málaga, Spain
Jorge Azorín-López	University of Alicante, Spain
Halima Bahi	Badji Mokhtar – Annaba University, Algeria
Juan Pedro Bandera Rubio	University of Málaga, Spain
Oresti Banos	University of Granada, Spain
Bruno Baruque	University of Burgos, Spain
Lluís Belanche	Universitat Politècnica de Catalunya, Spain
Francisco Bonnín	University of the Balearic Isles, Spain
Julio Brito	University of la Laguna, Spain
Pablo C. Cañizares	Universidad Complutense de Madrid, Spain
Joan Cabestany	Universitat Politècnica de Catalunya, Spain
Eldon Glen Caldwell	University of Costa Rica, Costa Rica

Jose Luis Calvo-Rolle	University of A Coruña, Spain
Azahara Camacho	WATA Factory, Spain
Hoang-Long Cao	Vrije Universiteit Brussel, Belgium
Jaime Cardoso	University of Porto, Portugal
Carlos Carrascosa	GTI-IA DSIC Universidad Politécnica de Valencia, Spain
Francisco Carrillo Pérez	University of Granada, Spain
Pedro Castillo	University of Granada, Spain
Andreu Català	Universitat Politècnica de Catalunya, Spain
Ana Rosa Cavalli	Télécom SudParis, France
Valentina Colla	Scuola Superiore S. Anna, Italy
Raúl Cruz-Barbosa	Universidad Tecnológica de la Mixteca, Mexico
Miguel Damas	University of Granada, Spain
Daniela Danciu	University of Craiova, Romania
Enrique Dominguez	University of Málaga, Spain
Grzegorz Dudek	Częstochowa University of Technology, Poland
Gregorio Díaz	University of Castilla - La Mancha, Spain
Marcos Faundez-Zanuy	Escola Superior Politècnica Tecnocampus, Spain
Enrique Fernandez-Blanco	University of A Coruña, Spain
Manuel Fernandez-Carmona	University of Málaga, Spain
Carlos Fernandez-Lozano	University da Coruña, Spain
Antonio Fernández-Caballero	University of Castilla-La Mancha, Spain
Jose Manuel Ferrandez	Univ. Politecnica Cartagena, Spain
Oscar Fontenla-Romero	University of A Coruña, Spain
Leonardo Franco	University of Málaga, Spain
Emilio Garcia-Fidalgo	University of the Balearic Islands, Spain
Francisco Garcia-Lagos	University of Málaga, Spain
Jose Garcia-Rodriguez	University of Alicante, Spain
Patricio García Báez	University of La Laguna, Spain
Pablo García Sánchez	University of Granada, Spain
Rodolfo García-Bermúdez	Universidad Técnica de Manabíl, Ecuador
Patrick Garda	Sorbonne Université, France
Peter Glösekötter	Münster University of Applied Sciences, Germany
Juan Gomez Romero	University of Granada, Spain
Manuel Graña	University of the Basque Country, Spain
Elisa Guerrero	University of Cadiz, Spain
Jose Guerrero	Universitat de les Illes Balears, Spain
Bertha Guijarro-Berdiñas	University of A Coruña, Spain
David Guijo-Rubio	University of Córdoba, Spain
Alberto Guillen	University of Granada, Spain
Pedro Antonio Gutierrez	University of Córdoba, Spain

Luis Herrera	University of Granada, Spain
Cesar Hervas	.
Wei-Chiang Hong	Asia Eastern University of Science and Technology, Taiwan
Petr Hurtik	University of Ostrava, Czech Republic
M. Dolores Jimenez-Lopez	Rovira i Virgili University, Spain
Juan Luis Jiménez Laredo	Université Le Havre Normandie, France
Gonzalo Joya	University of Málaga, Spain
Vicente Julian	Universitat Politècnica de València, Spain
Otoniel Lopez Granado	Miguel Hernández University de Elche, Spain
Rafael Marcos Luque Baena	University of Málaga, Spain
Ezequiel López-Rubio	University of Málaga, Spain
Kurosh Madani	Lissi/Université Paris-Est Créteil, France
Bonifacio Martin Del Brio	University of Zaragoza, Spain
Luis Martí	Inria Chile Research Centre, Chile
Pablo Martínez Cañada	University of Granada, Spain
Francisco Martínez Estudillo	Universidad Loyola Andalucía, Spain
Francisco Martínez-Álvarez	Universidad Pablo de Olavide, Spain
Montserrat Mateos	Universidad Pontificia de Salamanca, Spain
Jesús Medina	University of Cádiz, Spain
Salem Mohammed	Mustapha Stambouli University, Algeria
Jose M. Molina	Universidad Carlos III de Madrid, Spain
Miguel A. Molina-Cabello	University of Málaga, Spain
Juan Moreno Garcia	University of Castilla-La Mancha, Spain
John Nelson	University of Limerick, Ireland
Alberto Núñez	Universidad Complutense de Madrid, Spain
Madalina Olteanu	SAMM, Université Paris Dauphine - PSL, France
Alfonso Ortega	Universidad Autónoma de Madriod, Spain
Alberto Ortiz	Universitat de les Illes Balears, Spain
Osvaldo Pacheco	University of Aveiro, Portugal
Esteban José Palomo	University of Málaga, Spain
Massimo Panella	University of Rome "La Sapienza", Italy
Miguel Angel Patricio	Universidad Carlos III de Madrid, Spain
Jose Manuel Perez Lorenzo	University of Jaen, Spain
Irina Perfilieva	University of Ostrava, Czech Republic
Vincenzo Piuri	University of Milan, Italy
Hector Pomares	University of Granada, Spain
Mario Porrmann	Osnabrück University, Germany
Alberto Prieto	University of Granada, Spain
Alexandra Psarrou	University of Westminster, UK
Fernando M. Quintana	University of Cádiz, Spain
Pablo Rabanal	Universidad Complutense de Madrid, Spain

Juan Rabuñal	University of A Coruña, Spain
Md. Ahsanur Rahman	North South University, Bangladesh
Sivarama Krishnan Rajaraman	National Library of Medicine, USA
Ismael Rodriguez	Universidad Complutense de Madrid, Spain
Ignacio Rojas	University of Granada, Spain
Ricardo Ron-Angevin	University of Málaga, Spain
Antonello Rosato	"Sapienza" University of Rome, Italy
Fernando Rubio	Universidad Complutense de Madrid, Spain
Ulrich Rueckert	Bielefeld University, Germany
Addisson Salazar	Universitat Politècnica de València, Spain
Roberto Sanchez Reolid	University of Castilla-La Mancha, Spain
Noelia Sanchez-Maroño	University of A Coruña, Spain
Francisco Sandoval	University of Málaga, Spain
Jorge Santos	ISEP, Portugal
Jose Santos	University of A Coruña, Spain
Prem Singh	GITAM University-Visakhapatnam, India
Catalin Stoean	University of Craiova, Romania
Ruxandra Stoean	University of Craiova, Romania
Carmen Paz Suárez-Araujo	Universdad de Las Palmas de Gran Canaria, Spain
Javier Sánchez-Monedero	Universidad Loyola Andalucía, Spain
Claude Touzet	Aix-Marseille University, France
Daniel Urda	University of Burgos, Spain
Pepijn van de Ven	University of Limerick, Ireland
Víctor Manuel Vargas	University of Córdoba, Spain
Francisco Velasco-Alvarez	University of Málaga, Spain
Alfredo Vellido	Universitat Politècnica de Catalunya, Spain
Francisco J. Veredas	University of Málaga, Spain
Michel Verleysen	Université catholique de Louvain, Belgium
Ivan Volosyak	Rhine-Waal University of Applied Sciences, Germany
Miri Weiss Cohen	Braude College of Engineering, Israel
Mauricio Zamora	University of Costa Rica, Costa Rica

Contents – Part II

Deep Learning and Applications

Deep Learning Applied to Computer Vision and Robotics

General Applications of Artificial Intelligence

Interaction with Neural Systems in Both Health and Disease

Machine Learning for 4.0 Industry Solutions

Neural Networks in Chemistry and Material Characterization

Ordinal Classification

Real World Applications of BCI Systems

Spiking Neural Networks: Applications and Algorithms

Contents – Part I

Advances in Artificial Neural Networks

ANN HW-Accelerators

Applications of Machine Learning in Biomedicine and Healthcare

Deep Learning and Applications

Predicting Wildfires in the Caribbean Using Multi-source Satellite Data and Deep Learning

J. F. Torres[1]([⊠]), S. Valencia[2,3], F. Martínez-Álvarez[1], and N. Hoyos[4]

[1] Data Science and Big Data Lab, Universidad Pablo de Olavide, Seville, Spain
{jftormal,fmaralv}@upo.es
[2] GIGA, Escuela Ambiental, Facultad de Ingeniería, Universidad de Antioquia, Medellín, Colombia
[3] Grupo de investigación en Ecología Aplicada, Escuela Ambiental, Facultad de Ingeniería, Universidad de Antioquia, Medellín, Colombia
santiago.valencia8@udea.edu.co
[4] Departamento de Historia y Ciencias Sociales, Universidad del Norte, Barranquilla, Colombia
nbotero@uninorte.edu.co

Abstract. Wildfires pose a significant threat to the environment and local communities, and predicting their occurrence is crucial for effective management and prevention. The Caribbean region is particularly susceptible to wildfires due to factors such as human activities, climate change, and natural causes. In this article, we propose a comprehensive methodology that combines data from multi-source satellite data and applies a range of predictive models. The results demonstrate the potential of deep learning techniques for identifying high-risk areas and developing effective fire management strategies. They also highlight the importance of continued research and investment in this area to improve the accuracy of predictive models and ultimately ensure the safety of communities and the environment. The findings have important implications for policymakers and stakeholders in the Caribbean region, who can use this information to develop more effective fire management strategies to minimize the impact of wildfires on the environment and local communities. By identifying high-risk areas, preventative measures such as controlled burns and improved fire management strategies can be implemented.

Keywords: Fire detection · Forecasting · Deep learning · Satellite data

1 Introduction

The occurrence of wildfires in the Caribbean region is a growing concern due to the devastating impact they can have on the environment and local communities. Predicting the likelihood of a wildfire can help mitigate its effects, minimize the loss of life and property, and allow for a more effective response from emergency services.

© The Author(s), under exclusive license to Springer Nature Switzerland AG 2023
I. Rojas et al. (Eds.): IWANN 2023, LNCS 14135, pp. 3–14, 2023.
https://doi.org/10.1007/978-3-031-43078-7_1

Wildfires in the Caribbean region can be caused by a variety of factors, including human activities such as agriculture and land-use changes, as well as natural causes like lightning strikes. With climate change leading to hotter and drier conditions, the risk of wildfires is increasing, making it essential to predict their occurrence. Predictive modeling can help identify high-risk areas, which can be targeted for preventative measures such as controlled burns or improved fire management strategies.

Predicting the occurrence of wildfires in the Caribbean region is essential for effective wildfire management and prevention. With data science techniques such as machine learning algorithms and geospatial analysis, accurate predictive models can be built to help identify high-risk areas and develop effective fire management strategies [1]. It is crucial to continue research and investment in this area to ensure the safety of the environment and local communities.

To this end, this article proposes the use of various data sources, including sensors such as MODIS, Terra, SRTM, IGAC, GPWv4.11, Terraclimate and CGIAR-CSI, to address the problem of wildfire detection in a wide region of the Caribbean. The collected data had to be meticulously analyzed, merged, and preprocessed to generate a comprehensive master dataset that can be used to train and evaluate predictive models.

In this study, six predictive models with different characteristics were trained and evaluated to analyze their performance. These models are Support Vector Machines, Logistic Regression, K-Nearest Neighbors, Decision Tree, Artificial Neural Network and Convolutional Neural Network. These models were chosen due to their proven effectiveness in previous studies in predicting the likelihood of wildfires in different regions.

Overall, the results of this study can help inform policymakers and stakeholders in the Caribbean region to develop more effective fire management strategies to minimize the impact of wildfires on the environment and local communities. The study highlights the importance of continued research and investment in this area to improve the accuracy of predictive models and ultimately ensure the safety of communities and the environment.

The rest of the article is structured as follows. Section 2 presents a literature review, analyzing the latest works related to the objective of this article. Section 3 describes the methodology proposed. Section 4 reports and discusses the results. Finally, Sect. 5 draws the conclusions from this work.

2   Related Works

There are several techniques used in predictive modeling for wildfire occurrence in the Caribbean region. One approach is to use machine learning algorithms that can analyze large datasets of environmental and climate variables such as temperature, humidity, wind speed, and vegetation cover. These algorithms can then build models that can predict the likelihood of a wildfire occurring based on these variables [10, 19].

A concise review on forest fire prediction techniques was published in 2022 [8], including machine and deep learning and image processing approaches. When compared to alternative techniques, these methods produced significantly better outcomes.

The geospatial analysis involves the use of geographic information systems to map and analyze data related to wildfires. This can include data on land cover, topography, and fire history, which can help identify high-risk areas and inform fire management strategies [21]. Additionally, remote sensing data can be used to monitor changes in vegetation cover and detect early signs of wildfires.

In 2019, Tehrany et al. introduced an ensemble model for the spatial prediction of tropical forest fire susceptibility machine learning and multi-source geospatial data [20]. The approach was tested on data from Vietnam the Logit-Boost classifier reached the most accurate results.

Later, Kalantar et al. proposed a forest fire susceptibility prediction based on remote sensing and machine learning with resampling algorithms [11]. The authors tested their approach on data from Iran and outperformed several well-established methods, reaching an Area Under the ROC Curve (AUC) value of 0.91.

Another approach for the prediction of forest fire using multispectral satellite measurements was introduced in [15]. Data from Nepal villages were used to validate the accuracy of the approach, showing a strong correlation between forest fire, temperature and precipitation.

Also in 2021, Rashkovetsky et al. [17] proposed a wildfire detection method from satellite imagery using U-Net architecture. The results showed that the fusion of Sentinel-2 and Sentinel-3 data provided the best detection rate in clear conditions, whereas the fusion of Sentinel-1 and Sentinel-2 data showed a significant benefit in cloudy weather.

The authors in [7] predicted potential wildfire severity across Southern Europe with satellite imagery and geospatial data available at the planetary scale. The work aimed at becoming a benchmark for progress in the prediction of fire danger and sets the basis for the design of pre-burn management actions.

A wildfire growth prediction and evaluation approach using Landsat and MODIS data can be found in [16]. Data from Croatia were used to validate their approach, in which Random Forest with input Landsat 8 spectral bands and indices resulted in the highest classification accuracy.

3 Methodology

The methodology presented in this study consists of three main stages for predicting the likelihood of wildfire occurrences in the Caribbean region, as illustrated in Fig. 1. Subsequent sections describe each of the blocks comprising the methodology.

3.1 Study Area

The Caribbean region of Colombia covers approximately 139,300 km^2 between the border with Panama to the west, and the border with Venezuela to the east.

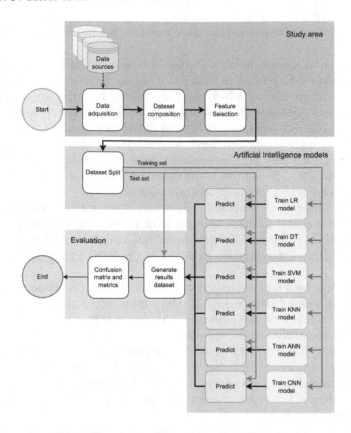

Fig. 1. Flowchart of the proposed methodology.

Most of the region is composed of lowlands with elevations that do not exceed 200 m above sea level (MASL), except for the high coastal massif of the Sierra Nevada de Santa Marta ($\approx$ 5700 MASL), and smaller massifs with elevations up to $\approx$500 MASL. Climate ranges from arid in the northern Guajira Peninsula, where annual precipitation is <500 mm, to very humid near Panama's border with annual values up to 3000 mm.

Intra-annual precipitation variability is modulated by the latitudinal migration of the Intertropical Convergence Zone (ITCZ) [14] with dry and windy conditions during December-March when the ITCZ is located to the south, and wet conditions during September-November when the ITCZ moves to northward. Accordingly, the region's biomes include tropical desert in the north (i.e. Guajira Peninsula), tropical dry forest throughout most of the area, and tropical rainforest near the Andean foothills and the border with Panama. Other sources of rainfall variability include the Chocó and San Andrés jet streams [4,13]; the former is associated with intense precipitation in the Western Andean Cordillera, while the later is related with windy conditions during December-February and June-August [4]. Interannual rainfall variability is associated with ocean-atmosphere oscillations including El Niño Southern Oscillation (ENSO), the Quasi-biennal oscillation (QBO), the Tropical North Atlantic index (TNA), and their interactions [5,12,18]. Relative humidity ranges from 65% in the Guajira Peninsula, to 90% in the southwest near Panama [9]. Mean monthly temperature in the lowlands ranges between 26°–30°C with little variation throughout the year [9], while maximum monthly temperature in the lowlands ranges between 30°–35°C.

The region has undergone significant land cover changes, particularly as a result of the rapid expansion of pasture starting the late 1800s-early 1900s [2, 6]. As a result, by the 1970s the Caribbean region already had overall high values of human footprint, indicative of a long history of human intervention [3]. Accordingly, current land cover in the lowlands is dominated by pasture for cattle and crops. Forest and secondary growth is mostly limited to the highlands, while there are extensive wetlands along the main rivers and coastal wetlands, particularly the Magdalena River, and the Santa Marta coastal lagoon.

After obtaining data from different sources, a matching process was performed by utilizing the temporal component of the data. This matching process involved identifying and aligning similar data points from various sources based on their timestamps. Once the data was matched, it was then analyzed for correlation between different attributes.

In order to identify the most relevant attributes for the study, a process of attribute selection was carried out. This involved analyzing the correlation between all the available attributes and selecting only the most relevant ones for the study. This selection process helped to simplify the dataset and eliminate any unnecessary variables, making it easier to draw meaningful insights from the data. As such, the final dataset consists of 7576 samples and 33 features, with one of the columns (fire) serving as the class label for performing binary classification.

3.2   Artificial Intelligence Models

This article presents the results of training various artificial intelligence models to predict the occurrence of wildfires in the Caribbean, utilizing the dataset generated in the preceding section. Specifically, six different algorithms were applied, including Support Vector Machines (SVM), Logistic Regression (LR), K-Nearest Neighbors (KNN), Decision Trees (DT), Artificial Neural Networks (ANN), and Convolutional Neural Networks (CNN). The performance of each model was evaluated and compared in terms of accuracy to assess their suitability for wildfire prediction in the Caribbean region. These models represent a significant advancement in wildfire prediction and have the potential to inform decision-making for wildfire management and prevention efforts. To further explain the methodology used in training the artificial intelligence models, the dataset was split into two sets: a training set and a test set. The training set comprised 70% of the total dataset, while the remaining 30% was reserved for test.

Support Vector Machines (SVM) is a popular algorithm for classification problems in machine learning. SVMs work by finding a hyperplane that optimally separates the training data into different classes. This is achieved by maximizing the margin, which is the distance between the hyperplane and the closest data points from each class. SVMs are particularly effective when dealing with high-dimensional feature spaces, where they can capture complex decision boundaries that other algorithms may struggle with. SVMs can also handle non-linearly separable data by projecting the input data into a higher-dimensional space using a kernel function. In this higher-dimensional space, SVMs can find a linearly separable hyperplane that corresponds to a non-linear decision boundary in the original feature space. The main disadvantage of SVMs is that they can be sensitive to the choice of hyperparameters, such as the kernel function and the regularization parameter. Nonetheless, SVMs remain a popular choice for classification tasks due to their strong theoretical foundations and empirical performance.

Logistic Regression (LR) is a widely-used algorithm for binary classification in machine learning. Unlike Support Vector Machines, which aim to find a hyperplane that optimally separates the training data into different classes, Logistic Regression models the probability of each class given the input features. Specifically, Logistic Regression models the log-odds of the positive class as a linear function of the input features, which is then transformed to a probability using the sigmoid function. During training, the model's parameters are learned by maximizing the likelihood of the training data. Logistic Regression is a simple yet effective algorithm that can handle linearly separable and non-linearly separable data, although its performance can be limited by the complexity of the decision boundary. One of the key advantages of Logistic Regression is that it provides interpretable coefficients that can be used to understand the relative importance of each feature in predicting the class label. Overall, Logistic Regression is a popular choice for binary classification tasks, particularly when interpretability is important.

K-Nearest Neighbors (KNN) is a non-parametric algorithm used for classification problems in machine learning. The algorithm works by assigning a new data point to the class that is most common among its k-nearest neighbors in the training set. The value of k is a hyperparameter that can be tuned based on the complexity of the problem and the size of the training data. One of the main advantages of KNN is its simplicity and flexibility, as it can handle non-linear decision boundaries and can be used with any distance metric. However, KNN can be computationally expensive for large datasets and requires careful preprocessing of the input data to ensure that the distance metric is meaningful. KNN is also sensitive to the choice of k, which can affect the model's bias-variance tradeoff. Overall, KNN is a popular choice for classification problems when interpretability and flexibility are important considerations.

Decision Tree DT. Its fundamental principle is to create a tree-like model of decisions and their possible consequences based on a set of training data. The algorithm starts by selecting the most significant feature from the input data to create the root node of the tree. Then, the algorithm recursively splits the data into smaller subsets based on the values of the selected feature. This process continues until a stopping criterion is met, such as a specified depth of the tree or a minimum number of samples required to make a decision. The final result is a tree-like model of decisions that can be used to classify new data points based on their features. The decision tree algorithm is straightforward to interpret and can handle both categorical and numerical data, making it a popular choice for many classification tasks

Artificial Neural Networks (ANN) are a class of algorithms used for classification problems in machine learning. ANNs are inspired by the structure and function of the human brain and consist of multiple layers of interconnected nodes or neurons. During training, the model's parameters, which are the weights and biases of the neurons, are adjusted to minimize the difference between the model's predictions and the true labels of the training data. ANNs can handle non-linear decision boundaries and can be used with various loss functions and activation functions, depending on the problem and the data. One of the main advantages of ANNs is their ability to capture complex patterns in the data, although this comes at the cost of increased model complexity and a greater risk of overfitting. ANNs require large amounts of training data and can be computationally expensive to train, particularly for deep architectures with many layers. Despite these challenges, ANNs have achieved state-of-the-art performance in many classification tasks, including image recognition, natural language processing, and speech recognition.

Convolutional Neural Networks (CNN) are a class of deep learning algorithms used primarily for image classification tasks. CNNs are similar to ANNs in that they consist of layers of neurons, but they are designed to capture spatial patterns in the input data by applying convolutional filters to the input images. During training, the model's weights are learned by backpropagation, a gradient-based optimization technique that adjusts the model's parameters to minimize the difference between the predicted and actual labels of the training data. CNNs can handle large and complex image datasets, and are able to learn hierarchical representations of the input data. This is achieved by using pooling layers to reduce the dimensionality of the features, and by stacking multiple convolutional layers to learn increasingly complex and abstract features. CNNs have achieved state-of-the-art performance on many image recognition tasks, including object detection, segmentation, and classification. Despite their success, CNNs are computationally expensive and require large amounts of training data to achieve high accuracy.

3.3   Evaluation

Evaluating the performance of a model is crucial to determine its effectiveness in solving a particular problem. Evaluation metrics are used to measure how well a model performs on a given dataset. In classification problems, commonly used metrics include accuracy, precision, sensitivity, and specificity, among others. In this section, these metrics and their interpretation are discussed.

– Accuracy measures the percentage of correctly classified instances out of the total number of instances.

$$Acc. = \frac{TP + TN}{TP + TN + FP + FN} * 100 \tag{1}$$

– Sensitivity measures the ability of a classification model to correctly identify positive instances.

$$Sen. = \frac{TP}{TP + FN} * 100 \tag{2}$$

– Specificity measures the ability of a classification model to correctly identify negative instances.

$$Spe. = \frac{TN}{TN + FP} * 100 \tag{3}$$

– Positive Predictive Value (PPV) measures the percentage of true positive predictions out of the total positive predictions made by the model.

$$PPV = \frac{TP}{TP + FP} * 100 \tag{4}$$

– Negative Predictive Value (NPV) measures the percentage of true negative predictions out of the total negative predictions made by the model.

$$NPV = \frac{TN}{TN + FN} * 100 \tag{5}$$

– Matthews Correlation Coefficient (MCC) measures the quality of binary classifications, ranging from -1 to 1, where 1 represents a perfect classification, 0 represents a random classification, and -1 represents a total disagreement between the prediction and the actual values.

$$MCC = \frac{TP * TN - FP * FN}{\sqrt{(TP + FP)(TP + FN)(TN + FP)(TN + FN)}} \tag{6}$$

where TP means true positive, which represents the number of instances that were correctly classified as positive by the model. TN denotes true negative, which represents the number of instances that were correctly classified as negative by the model. FP refers to false positive, which represents the number of instances that were classified as positive but were actually negative. Finally, FN connotes false negative, which represents the number of instances that were classified as negative but were actually positive.

4 Results

The results obtained in this study provide valuable insights into the effectiveness of different methods for classifying the dataset. The classification accuracy achieved by each method was evaluated and compared in order to identify the most suitable approach for this particular task. The findings of this study have

Table 1. Confusion matrix of the models.

	SVM	LR	KNN	ANN	DT	CNN
TP	18783	31051	29344	29786	30369	30329
FP	12657	369	2096	1674	1071	1111
FN	652	7188	758	306	800	347
TN	4854	318	6748	7200	6706	7159

important implications for the field and can guide future research in the development of improved classification models. In this discussion section, we will delve into the results obtained and analyze the implications of these findings for the field.

Table 1 presents the confusion matrices for the six used models. Among the models, the CNN exhibits the highest TP value (30329) and the lowest FN value (347), indicating that it correctly classified the majority of positive instances while making fewer false negative predictions. On the other hand, LR has the highest FP value (369), indicating that it made the most false positive predictions, followed by KNN (2096) and ANN (1674). SVM has the highest FN value (2652), indicating that it made the most false negative predictions, followed by LR (7188) and DT (800).

Overall, the DT model has the highest TN value (6706), indicating that it correctly classified the majority of negative instances while making fewer false positive predictions. SVM has the lowest TN value (4854), indicating that it made the most false positive predictions, followed by LR (318) and KNN (6748).

Table 2. Performance of the models.

Model	Acc. (%)	Sen. (%)	Spe. (%)	PPV (%)	NPV (%)	MCC
SVM	60.69	64.67	59.74	27.72	87.63	0.19
LR	80.54	04.24	98.76	44.98	81.20	0.09
KNN	92.67	89.90	93.33	76.30	97.48	0.78
ANN	94.92	95.92	94.68	81.14	98.98	0.85
DT	95.20	89.34	96.59	86.23	97.43	0.85
CNN	96.26	95.38	96.47	86.57	98.87	0.89

Table 2 presents the performance of various models regarding the metrics discussed in Sect. 3.3. CNN leads the models with the highest accuracy (96.26%), followed by DT (95.20%) and ANN (94.92%). In contrast, SVM, LR and KNN exhibit lower accuracy, with SVM having the lowest performance (60.69%).

SVM displays relatively higher sensitivity (64.67%) than specificity (59.74%), while LR shows the opposite, with higher specificity (98.76%) and lower sensitivity (4.24%). KNN and ANN demonstrate balanced sensitivity and specificity, whereas DT and CNN exhibit high specificity (96.59% and 96.47%, respectively) and comparatively lower sensitivity (89.34% and 95.38%, respectively).

Regarding PPV and NPV, CNN shows the highest values (86.57% and 98.87%, respectively), indicating better accuracy in predicting true positives and true negatives. DT and ANN also show high PPV and NPV, whereas SVM, LR and KNN exhibit lower values.

A breakdown of the predictions grouped by each month can be seen in Fig. 2. As shown, the SVM algorithm performed the worst in all months. However, despite the fact that the CNN-based model offered the best overall results, it was not the best in every month of the year. Specifically, it can be observed that in the months of October and November, it was surpassed by simpler methods like KNN. This may be due to class imbalance and a lack of fire examples for that time period.

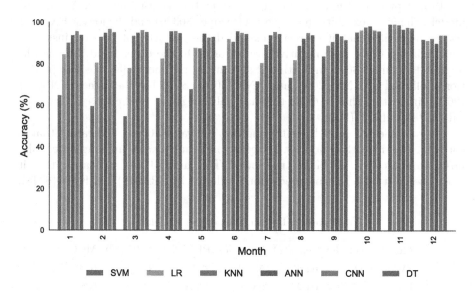

Fig. 2. Models accuracy by month.

The dataset is imbalanced, which may cause the models to be biased towards the majority class and perform poorly on the minority class. Therefore, it is vital to consider additional evaluation metrics, such as MCC, to assess the model's performance accurately. MCC measures the correlation between the predicted and actual values, with values closer to 1 indicating better performance. CNN, DT, and ANN exhibit the highest MCC (0.89, 0.85, and 0.85, respectively), while SVM, LR and KNN have lower values. In conclusion, the CNN model shows the highest performance regarding the accuracy, PPV, NPV and MCC metrics.

5   Conclusions

The use of multi-source satellite data and deep learning techniques can aid in predicting the likelihood of wildfires in the Caribbean region. This study employed six different predictive models, including Support Vector Machines, Logistic Regression, K-Nearest Neighbors, Decision Tree, Artificial Neural Network, and Convolutional Neural Network, to analyze the effectiveness of each model. The results indicate that the Convolutional Neural Network model has the highest true positive value and the lowest false negative value, indicating that it correctly classified the majority of positive instances while making fewer false negative predictions. On the other hand, Logistic Regression has the highest false positive value, making the most false positive predictions. In conclusion, the study demonstrates the potential of deep learning techniques for identifying high-risk areas and developing effective fire management strategies. The use of multi-source satellite data and deep learning algorithms can improve the accuracy of predictive models, thereby enabling policymakers to implement better measures for wildfire prevention and management.

Acknowledgements. The authors would like to thank the Spanish Ministry of Science and Innovation for the support under the projects PID2020-117954RB and TED2021-131311B, and the European Regional Development Fund and Junta de Andalucía for projects PY20-00870 and UPO-138516. This work has also been funded by the *Becas Iberoamérica: Santander Investigación 2021.*

References

1. Ahamad, M.: Arima-based forecasting of the effects of wildfire on the increasing tree cover trend and recurrence interval of woody encroachment in grazing land. Resourc. Environ. Sustain. **10**, 100091 (2022)
2. Ausdal, S.V.: Potreros, ganancias y poder. Una historia ambiental de la ganadería en Colombia, 1850–1950. Historia Crítica (39E), pp. 126–149 (2009)
3. Ayram, C.A., Etter, A., Díaz-Timoté, J., Buriticá, S.R., Ramírez, W., Corzo, G.: Spatiotemporal evaluation of the human footprint in Colombia: Four decades of anthropic impact in highly biodiverse ecosystems. Ecol. Ind. **117**, 106630 (2020)
4. Bernal, G., Poveda, G., Roldán, P., Andrade, C.: Patrones de variabilidad de las temperaturas superficiales del mar en la costa caribe colombiana. Revista de la Academia Colombiana de Ciencias Exactas, Físicas y Naturales **30**(115), 195–208 (2006)
5. Enfield, D.B., Alfaro, E.J.: The dependence of Caribbean rainfall on the interaction of the tropical Atlantic and pacific oceans. J. Clim. **12**(7), 2093–2103 (1999)
6. Etter, A., McAlpine, C., Possingham, H.: Historical patterns and drivers of landscape change in Colombia since 1500: a regionalized spatial approach. Ann. Assoc. Am. Geogr. **98**(1), 2–23 (2008)
7. Fernández-García, V., Beltrán-Marcos, D., Fernández-Guisuraga, J.M., Marcos, E., Calvo, L.: Predicting potential wildfire severity across Southern Europe with global data sources. Sci. Total Environ. **829**, 154729 (2022)

8. Gaikwad, A., Bhuta, N., Jadhav, T., Jangale, P., Shinde, S.: A review on forest fire prediction techniques. In: Proceedings of the IEEE International Conference On Computing, Communication, Control And Automation, pp. 31–35 (2022)

9. IDEAM Instituto de hidrología, meteorología y estudios ambientales: Atlas climatológico de Colombia (2015)

10. Kadir, E.A., Kung, H.T., Rosa, S.L., Sabot, A., Othman, M., Ting, M.: Forecasting of fires hotspot in tropical region using LSTM algorithm based on satellite data. In: Proceedings of the IEEE Region 10 Symposium, pp. 1–7 (2022)

11. Kalantar, B., Ueda, N., Idrees, M.O., Janizadeh, S., Ahmadi, K., Shabani, F.: Forest fire susceptibility prediction based on machine learning models with resampling algorithms on remote sensing data. Remote Sens. **12**, 3682 (2020)

12. Poveda, G.: La hidroclimatología de Colombia: una síntesis desde la escala interdecadal hasta la escala diurna. Revista de la Academia Colombiana de Ciencias Exactas, Físicas y Naturales **28**(107), 201–222 (2004)

13. Poveda, G., Mesa, O.J.: On the existence of lloró (the rainiest locality on earth): Enhanced ocean-land-atmosphere interaction by a low-level jet. Geophys. Res. Lett. **27**(11), 1675–1678 (2000)

14. Poveda, G., Waylen, P.R., Pulwarty, R.S.: Annual and inter-annual variability of the present climate in northern south America and Southern Mesoamerica. Palaeogeogr. Palaeoclimatol. Palaeoecol. **234**(1), 3–27 (2006). Late Quaternary climates of tropical America and adjacent seas

15. Qadir, A., Talukdar, N.R., Uddin, M.M., Ahmad, F., Goparaju, L.: Predicting forest fire using multispectral satellite measurements in Nepal. Remote Sens. Appl. Soc. Environ. **23**, 100539 (2021)

16. Radocaj, D., Jurisic, M., Gasparovic, M.: A wildfire growth prediction and evaluation approach using Landsat and MODIS data. J. Environ. Manag. **304**, 114351 (2022)

17. Rashkovetsky, D., Mauracher, F., Langer, M., Schmitt, M.: Wildfire detection from multisensor satellite imagery using deep semantic segmentation. IEEE J. Sel. Top. Appl. Earth Observ. Remote Sens. **14**, 7001–7016 (2021)

18. Restrepo, J.C., et al.: Freshwater discharge into the Caribbean Sea from the rivers of Northwestern South America (Colombia): magnitude, variability and recent changes. J. Hydrol. **509**, 266–281 (2014)

19. Rim, C., Om, K., Ren, G., Kim, S., Kim, H., Kang-Chol, O.: Establishment of a wildfire forecasting system based on coupled weather-wildfire modeling. Appl. Geogr. **90**, 224–228 (2018)

20. Tehrany, M.S., Jones, S., Shabani, F., Martínez-Álvarez, F., Bui, D.T.: A novel ensemble modeling approach for the spatial prediction of tropical forest fire susceptibility using LogitBoost machine learning classifier and multi-source geospatial data. Theoret. Appl. Climatol. **137**, 637–653 (2019)

21. Xie, W., He, M., Tang, B.: Data-enabled correlation analysis between wildfire and climate using GIS. In: Proceedings of the 3rd International Conference on Information and Computer Technologies, pp. 31–35 (2020)

Embedded Temporal Feature Selection for Time Series Forecasting Using Deep Learning

M. J. Jiménez-Navarro[1]([⊠]) [iD], M. Martínez-Ballesteros[1] [iD],
F. Martínez-Álvarez[2] [iD], and G. Asencio-Cortés[2] [iD]

[1] Department of Computer Science, University of Seville, 41012 Seville, Spain
{mjimenez3,mariamartinez}@us.es
[2] Data Science and Big Data Lab, Pablo de Olavide University, 41013 Seville, Spain
{fmaralv,guaasecor}@upo.es

Abstract. Traditional time series forecasting models often use all available variables, including potentially irrelevant or noisy features, which can lead to overfitting and poor performance. Feature selection can help address this issue by selecting the most informative variables in the temporal and feature dimensions. However, selecting the right features can be challenging for time series models. Embedded feature selection has been a popular approach, but many techniques do not include it in their design, including deep learning methods, which can lead to less efficient and effective feature selection. This paper presents a deep learning-based method for time series forecasting that incorporates feature selection to improve model efficacy and interpretability. The proposed method uses a multidimensional layer to remove irrelevant features along the temporal dimension. The resulting model is compared to several feature selection methods and experimental results demonstrate that the proposed approach can improve forecasting accuracy while reducing model complexity.

Keywords: feature selection · embedded · neural network · time series · forecasting

1 Introduction

Embedded feature selection has become the preferred approach for feature selection due to its combination of simplicity, efficiency, and remarkable results. However, just some techniques include feature selection in its design, which force to rely on less efficient or effective methods. Deep learning is one example of technique which does not embed an effective feature selection, which is one of the causes for obtaining poor results for tabular data in spite of the remarkable results in areas like artificial vision and natural language processing.

Time series forecasting [7] is one of the most common tabular data in the industry and a critical area of research that aims to predict future values of a variable based on its past information. While traditional time series forecasting

I. Rojas et al. (Eds.): IWANN 2023, LNCS 14135, pp. 15–26, 2023.
https://doi.org/10.1007/978-3-031-43078-7_2

models often use all available variables, including potentially irrelevant or noisy features, this can lead to overfitting and poor performance. Feature selection can help address this issue by selecting the most informative variables in the temporal and feature dimensions. However, selecting the right features and the right moments from the past information can be challenging for time series models.

In this paper, we introduce a novel method for time series forecasting called Time Selection Layer (TSL)[1], which extends the FADL framework proposed by Jimenez Navarro et al. [6]. A deep learning-based method is proposed for time series forecasting that incorporates feature selection to improve the efficacy and interpretability of the model. The proposed method includes an additional layer at the top of the model which selects the relevant features and time steps. This layer is trained during the backpropagation process, which, once the training process has finished, removes irrelevant features without additional training steps nor complex mechanisms. The previous methodology selected the features and all the past moments, which was incompatible with time series forecasting scenarios where a univariate time series or all the input features were used in the prediction.

The resulting model is compared to a baseline model that uses all available features and three filter-based methods because they do not require extra computation, and experimental results demonstrate that the proposed approach can improve forecast accuracy while reducing model complexity in all scenarios. Additionally, the selected features provide insight into the underlying patterns and drivers of the time series, aiding in the interpretation of the forecast results.

The main contributions of this paper are as follows:

- Simple and effective solution for feature selection in time series with minimal/no parametrization.
- General-purpose approach, which may be applied to almost any type of time series independently of its nature or task.
- Efficient interpretability approach which requires no extra computation.
- Comparison between 9 different time series forecasting datasets and 4 baseline approaches with a remarkable improvement in efficacy.

The rest of the article is organized as follows. In Sect. 2, we provide a brief overview of related work on time series forecasting and feature selection. In Sect. 3, we describe the proposed method in detail. Section 4 presents the experimental setting used to compare the different datasets and feature selection approximations. In Sect. 5, the experimental results and an analysis of the selected features of the proposed method between all datasets and feature selection methods are reported. Finally, Sect. 6 discusses the main conclusions of this work.

---

[1] Python implementation and experimentation have been included in the following repository https://github.com/manjimnav/TSSLayer.

2   Related Works

This section conducts a literature review of embedded feature selection methods applied to neural networks. We will review the advantages and limitations of each technique and discuss its applicability to different types of datasets and problems.

Yuan et al. [10] propose an embedded feature selection method for neural networks using the point-centered convolutional neural network. The study is applied to moldy peanuts identification problem using hyperspectral images. The authors propose using the weights in the first convolutional layer as an estimator of the relevance of the neural network for each band. The method provided competitive accuracy compared to conventional approaches. In our work, the goal is to filter irrelevant features by removing or keeping the input with the same value. In the work proposed by the authors, it is possible to assign a weight with a continuous value, which may difficult to determine if a feature is relevant.

Zhang et al. [11] propose to use the Group Lasso penalty to embed the feature selection in a neural network. The authors apply their method to a set of well-known baseline datasets compared with other regularization approaches. The groups in the Group Lasso penalty represent all the weights in the first layer connected to a layer. Each group has a regularization using a smoothing approximation to the Lasso penalty to make it fully differentiable. However, the proposal described may be only applied to feed-forward layers, which limits the applicability to other models. In our work, we propose a general purpose method for any type of neural networks.

Cancelan et al. [2] propose the E2E-FS method to embed feature selection in neural networks. The authors apply this method to different microarray challenges and artificially modified datasets for feature selection. The E2E-FS includes an additional loss function method in order to filter a fixed number of features and remove the influence of irrelevant features. In our work, the method automatically selects the amount of features, which usually is a subset of the total features. In addition, no change to the loss function is needed to filter the features as in our approach, as we approximate the Heaviside function to make it differentiable.

Borisov et al. [1] propose a general-purpose layer called CancelOut to filter irrelevant features. The method is applied to three baseline datasets compared to other embedded and not embedded feature selection methods. The methodology proposed by the authors consists of the use of an element-wise multiplication between the inputs and a set of filters. The filters are built by applying a sigmoid function to a set of weights, and the goal is to remove the influence of the irrelevant input features. However, for feature selection, a threshold is needed, which may have a great influence in the results. In our work, no threshold must be parameterized, which makes our method more generalizable.

3   Methodology

This section is divided into three subsections to describe the main methodology. In Sect. 3.1, the nomenclature used during the figures, formulas, and explanations is provided. In Sect. 3.2, the methodology is described in detail, focusing on the changes from the previous work and its implications. In Sect. 3.3 the weight initialization and regularization strategies for the TSL are described.

3.1   Nomenclature

In this section, the main elements used in the description of the TSL are reported.

- D represents the number of features used for the neural network. Note that in univariate time series $D = 1$.
- M represents the number of past moments used in the input, also called the window.
- X represents the input matrix with size MxD.
- W_L represents the TSL weights with size MxD.
- $\hat{H}$ represents the Heaviside function approximation.
- $\circ$ represents the Hadamard product.
- H represents the Heaviside function.
- σ represents the hard sigmoid function.
- δ represents the "detach" function which ignores the propagation of the gradient in the backpropagation algorithm.

3.2   Description

Figure 1 summarizes the suggested approach which involves the introduction of a new layer, referred to as the TSL, just after the input layer. Specifically, this layer operates on a per-element basis by linking each input to a distinct neuron within the TSL. These neurons act as a filter for the input, enabling to remove the irrelevant features from the network when required. Initially, all inputs are selected. Then, the TSL weights and the rest of the weights in the network are optimized together via backpropagation, with positive and negative weights assigned to each feature. Positive weights preserve the influence of the feature, while negative weights nullify the influence by setting it to zero. Due to the weights in TSL are optimized during the backpropagation process, the filter must collaborate to minimize the loss function embedding the feature selection process into the neural network design.

   The filter in TSL is implemented by applying a Heaviside approximation to the matrix W_L setting the weights to zero or one. The resulting binarized weights are defined as the mask, which is multiplied by the input X using the Hadamard product, which removes the influence of the filtered inputs.

   Therefore, the operation performed by the TSL is defined as follows:

$$TSL(X^{MxD}) = \hat{H}(W_L^{1xD}) \circ X^{MxD} . \tag{1}$$

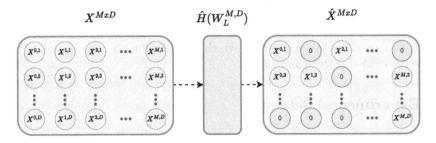

Fig. 1. TSL layer applied to an input matrix X with M past moments for D features. Note that the output $\hat{X}$ has the same dimensions, but some moments were set to zero. Finally, the output is used as input for a neural network.

where the function H applies the following operation to the weight $W_L^{i,j}, i \in M, j \in D$:

$$H(W_L^{i,j}) = \begin{cases} 1, & if\ W_L^{i,j} >= 0, \\ 0, & if\ W_L^{i,j} < 0, \end{cases} \quad\quad \begin{array}{c}(2)\\(3)\end{array}$$

Note that if a feature mask contains a zero weight for a feature at a specific moment, it will affect all the remaining layers and reduce the influence of the removed features. This fact may lead to a higher variance in training time compared to a neural network without TSL.

As the Heaviside function is not differentiable, an approximation was implemented for this function with a differentiable version. The operation was performed for each weight $W_L^{i,j}$ in W_L as follows:

$$\hat{H}(W_L^{i,j}) = \sigma(W_L^{i,j}) - \delta(W_L^{i,j} - H(W_L^{i,j})). \quad\quad (4)$$

Using this approximation, the gradients can propagate through the TSL using the Heaviside function without requiring extra parameterization or regularization.

3.3   Weight Initialization and Regularization

The weight initialization and regularization are essential to achieve a good selection. Note that this initialization and regularization are applied to the TSL independently of the rest of the neural network.

As mentioned in the previous section, initially all features and moments are selected, and during the training the features are selected. In this work, we may consider that selection uses a backward approximation for feature selection. For this reason, the weights are initialized as a positive value. However, a large positive value would introduce a bias to keep the features, as reducing the weights would require more iterations than near-zero values. Thus, the weights are initialized as a near-zero positive value of 0.01.

Regularization is responsible for penalizing the amount of selected inputs. In this case, we considered the same penalization independently of the feature or moment to avoid any inductive bias, which will introduce a penalization of 0.01 for every selected feature and moment.

4   Experimental Setting

This section aims to provide a comprehensive understanding of the experimental design and implementation, allowing for reproducibility and evaluation of the results divided into five subsections. Section 4.1 outlines the datasets used in the study, including their sources, sizes, and characteristics. Section 4.2 describes the various transformations applied to the datasets, such as normalization and division. Section 4.3 provides a detailed explanation of the hyperparameters used in the study, including their values and how they were chosen. Section 4.4 outlines the different feature selection methods used in the study, including any modifications or adaptations made. Finally, Sect. 4.5 describes the performance metrics used to evaluate the results of the study.

4.1   Datasets

The experiments were performed by selecting eight data sets consisting of different time series from various sources and fields, such as maintenance of the power transformer, consumption of electricity, pollution, etc. In Table 1, each dataset will be described in detail, including the number of features, sample frequency, and forecasting strategy. The forecasting strategy can be either many-to-many or many-to-one. Many-to-many strategy uses multiple features as input and outputs multiple features that may be the same as the input or a subset. Many-to-one strategy uses multiple features as input and only one feature as output, usually one of the input features.

Table 1. Summary of datasets used in the experiment

Dataset	Instances	Time Span	Features	Frequency	Strategy
ETT	17420	2016–2018	7	1 h	Many-to-one
Electricity	26304	2011–2014	320	15 min	Many-to-many
ExchangeRate	7588	1990–2016	8	1 day	Many-to-many
$TorneoCO$	96409	2005–2015	4	10 min	Many-to-one
$TorneoNO_2$	96409	2005–2015	4	10 min	Many-to-one
$TorneoO_3$	96409	2005–2015	4	10 min	Many-to-one
$TorneoPM_{10}$	96409	2005–2015	4	10 min	Many-to-one
Traffic	7544	2015–2016	862	1 h	Many-to-many

The Electricity Transformer Temperature (ETT) [12] datasets consist of insulating oil temperature samples obtained from two power transformer in China.

The dataset has 17,420 records collected from 2016 to 2018, with data sampled every minute and hour. In this work, the hourly version from both transformers was used, called ETTh1 and ETTh2. The datasets consist of a multivariate time series containing five features: High UseFul Load (HUFL), High UseLess Load (HULL), Middle UseFul Load (MUFL), Middle UseLess Load (MULL), Low UseFul Load (LUFL), Low UseLess Load (LULL), and Oil Temperature (OT), which is the target. Therefore, the forecasting strategy used is the many-to-one approach, where the seven features are used as input, and only the oil temperature is the output. In this work, single-horizon forecasting is used for all datasets.

The Electricity [4] dataset contains electricity consumption measurements obtained from 370 clients in Portugal. In this work, the grouped version [8] is used, containing 320 time series collected from 2011 to 2014, with 26,304 records for each time series sampled every 15 min. The dataset consists of 320 consumption time series as features. The target is to predict all the time series; therefore, the forecasting strategy used is the many-to-many approach.

The ExchangeRate [4] dataset contains exchange rate measurements obtained from eight countries. The dataset has 7,488 records sampled every day from 1990 to 2016. The features of the data set consist of daily exchange rates from the eight countries. Australia, United Kingdom, Canada, Switzerland, China, Japan, New Zealand, and Singapore. The forecasting strategy used in this case is the many-to-many approach, which means that the eight features will be used as output.

The Torneo [5] dataset contains pollution and meteorological measurements obtained in Seville, Spain. The dataset has 96,409 records sampled every 10 min from 2005 to 2015. The features in the dataset consist of four pollutants (CO, NO_2, O_3, and PM_{10}) and three meteorological variables (Temperature, Wind direction, and Wind speed). The dataset was divided into four different datasets, each using a many-to-one forecasting strategy for each pollutant: TorneoCO, TorneoNO_2, TorneoO_3, and TorneoPM_{10}. As the number of target features is considerably small, this division was made to study the relevant features based on just one pollutant.

The Traffic [3] dataset contains the road occupancy rate of the California Department of Transportation. The dataset has 17,544 records collected from 862 sensors placed in California during 2015 and 2016, sampled every hour. The features consist of 862 sensors, using the many-to-many forecasting strategy.

4.2 Preprocessing

The data must be processed before using it in the experiment. First, the data are standardized to ensure that each feature has a mean of zero and a standard deviation of one. Then, a windowing process is applied to the time series, which divides the time series into fixed size windows with contain the information and is fed into the neural network. The size of the window is an important parameter that needs to be optimized based on the specific problem and data characteristics. The dataset is then divided into training, validation, and testing sets using the 70%, 10%, and 20% of the data, respectively. Each split preserves the temporality between the instances and between themselves.

4.3 Hyperparameters Definitions

The hyperparemeters configure the training process and have a great impact on its behavior. The hyperparameters have been divided into three groups: data, model, and feature selection hyperparameters.

The primary hyperparameter of the data is the window size. In this study, we used a different sequence size for each dataset. For ETT and Torneo datasets the window size is 12 or 24, in Electricity is 4 or 8, in ExchangeRate is 2 or 7 and Traffic 3 or 6. Furthermore, during the windowing process, we subsampled the datasets by building a window for all X records, as defined by the shift which is 24 h for ETT, Electricity and Torneo datasets while a shift of 3 instances where selected for ExchangeRate and Traffic. For the forecast step, only one value was selected.

Regarding the model, the hyperparameters were consistent across all datasets and methods. A simple neural network with two hidden layers was selected, where the number of neurons in the first hidden layer was equal to half of the input features, while a quarter of the input features was employed in the second hidden layer. During training, we utilized the Adam optimizer for 100 epochs. In addition, we used the early stop technique to stop the training process once the training loss did not decrease for at least 10 epochs.

For feature selection hyperparameters, two types exist depending on the approach. For filter methods, a threshold is employed as a hyperparameter, which can contain values ranging from 50% to 100% of total relevance, in increments of 5%. For the TSL, the single hyperparameter employed is the regularization term, which can be set to $1e^{-3}$, $1e^{-4}$, or $1e^{-5}$.

A grid search is used for all possible combinations of hyperparameters in order to find the best set of hyperparameters for each feature selection method and model. The search uses the training data to optimize the model with the set of hyperparameters and the metrics calculated over the validation set are used as the quality measure.

4.4 Methods

In this section, the main methods used for the comparison are detailed. First, the neural network is evaluated without feature selection as the baseline method. Then, three filter-based feature selections are employed: Linear, Correlation (Corr) and Mutual Information (MI). For the filter methods, a threshold must be optimized, as mentioned in Sect. 4.3. This threshold will determine the amount of features which represents the thr total relevance, being $thr \in [50, 100]$ the threshold percentage.

Linear selection uses the weights of a linear model with L1 regularization to determine the importance of features. Corr selection uses the Pearson correlation with the target variable to determine the importance of the characteristic. MI selection uses the Mutual Information [9] between the inputs and the target variable for determining the feature importance.

4.5 Metrics

To evaluate the performance of the model, three different metrics will be used. To evaluate the efficacy of the different methods, the mean squared error (MSE) will be employed. The number of selected features will be used to relate the method with the best efficacy with the number of relevant features needed. The formula is as follows: $MSE(Y_{pred}^n, Y_{true}^n) = \frac{1}{N} \sum_{n=1}^{N} (Y_{pred}^n - Y_{true}^n)^2$. Finally, the total time used for each method including the hyperparameter optimization process will be detailed.

5 Results and Discussion

In this section, the results obtained for each selection method and dataset are reported. The discussion is divided into three sections. First, the best efficacy results are analyzed for each dataset and method. Then the best hyperparameters obtained based on the best configuration are commented. Finally, the number of features of the best methods in each dataset.

5.1 Efficacy

Table 2 presents the MSE obtained for each selection method and dataset, as well as the corresponding improvement in MSE compared to the NS case (no selection), which is reported in parentheses.

Table 2. Mean squared error obtained for each selection method and dataset. Note that the improvement (between 0 and 1) with respect to the nonselection is reported in parentheses.

Dataset	NS	Corr	Linear	MI	TSL
ETTh1	0,399	0,157 (0,61)	0,058 (0,85)	0,073 (0,82)	**0,053 (0,87)**
ETTh2	0,590	0,135 (0,77)	**0,060 (0,9)**	0,166 (0,72)	0,111 (0,81)
Electricity	2,366	2,366 (0,00)	1,963 (0,17)	2,366 (0,00)	**0,869 (0,63)**
ExchangeRate	0,645	0,645 (0,00)	0,408 (0,37)	0,645 (0,00)	**0,387 (0,40)**
TorneoCO	0,512	0,253 (0,51)	0,228 (0,55)	0,243 (0,53)	**0,188 (0,63)**
TorneoNO2	0,802	0,422 (0,47)	0,415 (0,48)	0,494 (0,38)	**0,276 (0,66)**
TorneoO3	0,919	0,420 (0,54)	**0,251 (0,73)**	0,373 (0,59)	0,278 (0,70)
TorneoPM10	0,715	0,507 (0,29)	0,440 (0,38)	0,538 (0,25)	**0,415 (0,42)**
Traffic	0,299	0,213 (0,29)	0,220 (0,26)	0,227 (0,24)	**0,186 (0,38)**

For the NS method, the MSE is consistently higher or equal to that of the other methods, indicating that feature selection generally enhances prediction performance in the context of time series forecasting.

The Corr and MI methods demonstrate a significant improvement in MSE compared to NS, except for the Electricity and ExchangeRate datasets, where they perform similarly to NS. This suggests that Corr and MI are not always reliable indicators of feature relevance.

The TSL method produces the best results in seven out of nine datasets, followed by Linear, which yields the best results in the remaining two datasets. It seems that there are datasets with a linear relationship between the target and the features, which justifies the remarkable results.

In conclusion, TSL appears to exhibit the most consistent and optimal performance overall, which seems to indicate that our method is more generalizable to other problems without losing efficacy.

5.2   Best Hyperparameters

This section reports the hyperparameters found for the selection methods with the best efficacy. The goal is to identify general patterns for the different selection methods.

Table 3. Best hyperparameters found for each selection algorithm and dataset based on the efficacy.

Dataset	Window size					Threshold			Regularization
	NS	Corr	Linear	MI	TSL	Corr	Linear	MI	TSL
ETTh1	24	12	24	12	12	0.60	0.60	0.60	5e−3
ETTh2	12	12	24	12	12	0.80	0.55	0.65	5e−4
Electricity	8	8	8	8	8	0.60	0.60	0.50	1e−2
ExchangeRate	3	3	7	3	3	0.50	0.85	0.50	5e−5
TorneoCO	24	24	24	24	24	0.50	0.50	0.70	1e−2
TorneoNO2	12	24	12	24	24	0.75	0.50	0.70	1e−2
TorneoO3	12	12	12	12	24	0.55	0.55	0.90	1e−2
TorneoPM10	12	24	12	24	24	0.95	0.60	0.85	5e−3
Traffic	6	6	6	6	6	0.60	0.75	0.85	5e−4

Table 3 displays the best hyperparameters identified for each dataset and selection method, including the window size for the input time series, the threshold for the feature selection methods, and the regularization term used in our proposed approach.

Regarding window size, no consistent pattern emerges between methods. In some cases, the optimal window size is the maximum allowed, while in others, a smaller size is preferred.

For the threshold, no consistent pattern is observed for each feature selection method. The values range from 50% to 95%, making it challenging to optimize this parameter as no universally applicable value can be identified.

Concerning regularization in TSL, the optimal values typically fall within the range of 5e−3 to 1e−2. Thus, a regularization term between these two values may be a suitable initial value.

5.3   Selected Features

In this section, the number of features selected for each feature selection method with optimal hyperparameters is analyzed.

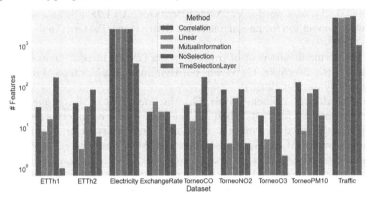

Fig. 2. Features obtained for each method using the best hyperparameters. Note the logaritmic scale.

Figure 2 displays the number of features obtained by each selection method. It is important to note that the number of features selected is not related to efficacy.

Overall, we observe that TSL tends to select fewer features than other methods. We hypothesize that the excellent efficacy results are a consequence of this selection strategy. Methods such as Corr or MI may not effectively filter out irrelevant features due to the high degree of interdependence among the features and moments. Linear appears to filter adequately when the feature space is small, but it may encounter difficulties when the number of original features is large.

6   Conclusions and Future Works

In this study, we have introduced a simple embedded feature selection approach for deep learning by adding a new layer. This layer acts as a filter that eliminates the impact of features with a temporal dimension. Our findings demonstrate, our proposed TSL outperforms other methods in most of the tested datasets with a straightforward parametrization and less information.

As future work, we intend to enhance the layer to consider the locality principle of selected moments. This involves increasing the likelihood of selecting moments around the chosen moments.

Acknowledgements. The authors would like to thank the Spanish Ministry of Science and Innovation for the support under the projects PID2020-117954RB and TED2021-131311B, and the European Regional Development Fund and Junta de Andalucía for projects PY20-00870, PYC20 RE 078 USE and UPO-138516.

References

1. Borisov, V., Haug, J., Kasneci, G.: CancelOut: a layer for feature selection in deep neural networks. In: Proceedings of 28th International Conference on Artificial Neural Networks. Artificial Neural Networks and Machine Learning - ICANN 2019: Deep Learning, pp. 72–83 (2019)
2. Cancela, B., Bolón-Canedo, V., Alonso-Betanzos, A.: E2E-FS: an end-to-end feature selection method for neural networks. IEEE Trans. Pattern Anal. Mach. Intell. pp. 1–12 (2020)
3. CDT: California department of transportation (2015). https://pems.dot.ca.gov/
4. Godahewa, R., Bergmeir, C., Webb, G., Hyndman, R., Montero-Manso, P.: Electricity hourly dataset (2020)
5. Gómez-Losada, A., Asencio-Cortés, G., Martínez-Álvarez, F., Riquelme, J.C.: A novel approach to forecast urban surface-level ozone considering heterogeneous locations and limited information. Environ. Model. Softw. **110**, 52–61 (2018)
6. Jiménez-Navarro, M.J., Martínez-Ballesteros, M., Sousa, I.S., Martínez-Álvarez, F., Asencio-Cortés, G.: Feature-aware drop layer (FADL): a nonparametric neural network layer for feature selection. In: Proceedings of 17th International Conference on Soft Computing Models in Industrial and Environmental Applications (SOCO 2022), pp. 557–566 (2023)
7. Jiménez-Navarro, M.J., Martínez-Ballesteros, M., Martínez-Álvarez, F., Asencio-Cortés, G.: PHILNet: a novel efficient approach for time series forecasting using deep learning. Inf. Sci. **632**, 815–832 (2023)
8. Lai, G., Chang, W., Yang, Y., Liu, H.: Modeling long- and short-term temporal patterns with deep neural networks. ACM, pp. 95–104 (2018)
9. Shannon, C.E.: A mathematical theory of communication. ACM SIGMOBILE Mob. Comput. Commun. Rev. **5**(1), 3–55 (2001)
10. Yuan, D., Jiang, J., Gong, Z., Nie, C., Sun, Y.: Moldy peanuts identification based on hyperspectral images and point-centered convolutional neural network combined with embedded feature selection. Comput. Electron. Agric. **197**, 106963 (2022)
11. Zhang, H., Wang, J., Sun, Z., Zurada, J.M., Pal, N.R.: Feature selection for neural networks using group lasso regularization. IEEE Trans. Knowl. Data Eng. **32**(4), 659–673 (2020)
12. Zhou, H., et al.: Informer: beyond efficient transformer for long sequence time-series forecasting. In: Proceedings of the AAAI Conference on Artificial Intelligence, vol. 35, pp. 11106–11115 (2021)

CauSim: A Causal Learning Framework for Fine-Grained Image Similarity

Hichem Debbi[✉]

Department of Computer Science, University of M'sila, M'sila, Algeria
`hichem.debbi@univ-msila.dz`

Abstract. Learning image similarity is useful in many computer vision applications. In fine-grained visual classification (FGVC), learning image similarity is more challenging due to the subtle inter-class differences. This paper proposes CauSim: a framework for deep learning image similarity based on causality. CauSim applies counterfactual reasoning on Convolution Neural Networks (CNNs) to identify significant filters responding to important regions, then it measures the similarity distance based on the counterfactual information learned with respect to each filter. We have verified the effectiveness of the method on the ImageNet dataset, in addition to four fine-grained datasets. Moreover, comprehensive experiments conducted on fine-grained datasets showed that CauSim can enhance the accuracy of existing FGVC architectures. The results can be reproduced using the code available in the GitHub repository https://github.com/HichemDebbi/CauSim.

Keywords: CNN · Attention · FGVC · Image similarity · Causality · Counterfactuals

1 Introduction

The attention mechanism is useful in many machine learning tasks due to its similarity with us as humans. This ability of identifying important regions led to implement this mechanism into CNNs for identifying the most discriminative regions, which could be helpful in many tasks. However, when it comes to challenging tasks such as Fine-grained visual classification (FGVC), attention and identifying discriminitive regions will be more challenging.

Fine-grained visual classification (FGVC) aims to recognize objects from subcategories with subtle inter-class differences. While classical CNN architectures had promising results on many datasets such as ImageNet [14], when it comes to FGVC, they fail to achieve good results, especially since input images usually are subjected to factors such as pose, viewpoint, or location of the object in the image, and thus the classification can be rationally affected. So, in order to recognize hundreds of subcategories under the same category, such as birds [18], dogs [9], flowers [13] and cars [10], we need more sophisticated architectures.

As more sophisticated techniques, we have discriminative feature learning techniques that apply different attentions [5,20,22,26]. These attention techniques have shown promising results, but they still suffer from the inability to

I. Rojas et al. (Eds.): IWANN 2023, LNCS 14135, pp. 27–38, 2023.
https://doi.org/10.1007/978-3-031-43078-7_3

find multiple discriminative regions at once [24]. We have also metric learning, which is a kind of learning which aims at identifying similarities between different pairs. Due to the importance of metric learning in computer vision, we had many applications in face verification [3], classification [8], and product search [2], either by using Siamese or triplet.

The causal hierarchy as described by Pearl consists of three main levels. At the first level comes association, which is used mainly to express statistical relationships, and then comes interventions, which is based on changing what we observe. Then at the top level of this hierarchy we find counterfactuals, which combine both association and intervention. So it would help to answer the questions, was it X that caused Y ? What if I had acted differently?

Recently Beckers and Halpern [1] addressed the problem of abstracting causal models, they develop a high-level macro causal model that describes for instance beliefs, which is a faithful abstraction of a low-level "micro" model that describes the neuronal level.

In this paper, we propose CauSim: a causal learning framework for fine-grained similarity. In CNNs, it is evident that the main building blocks that lead to such a decision are the filters. So, we consider filters as actual causes. Once the causal learning process is complete, for each outcome we obtain some filters that have more contribution for the classification than others. For each filter, we assign a value representing the difference in prediction probability when this filter is removed or simply set to zero. This value refers to the importance of the filter with respect to the related class, i.e. the higher value the filter gets, the most important is for the predicted class. We refer to these values by *Counterfactual information (CI)*.

Since filters are identified at the level of each convolutional layer, and since convolutional layers in CNNs have a hierarchical form, we can say that the filters at low levels have a causal effect on the output of the last layer. So CNNs actually express causality abstraction, and thus we consider only the last convolutional layer of every CNN architecture. While counterfactual causality has been widely used in the context of CNNs with different tasks such as explanation [25], to our knowledge, causal reasoning has not been addressed before for learning image similarity and FGVC.

We will show that our approach is very effective for capturing intra-class variance through many experiments. First, we will show that our method could identify different categories among different classes. This is proved by identifying the category of Dogs (120 breeds) in the 1k Imagenet dataset, in addition to identifying the birds' category of 59 species. We also run experiments on fine-grained datasets in order to challenge it against classes in the same category. Our method was very effective in identifying similar classes that most human might confuse due to their highest visual similarities. Finally, we challenge CauSim as a possible FGVC technique, which resulted in very promising results. By integrating CauSim with Binlinear (BCNN) [11], we report promising results for identifying fine-grained features on four fine-grained data sets: Oxford flowers [13], CUB-200-2011 [18], Stanford Cars [10] and Stanford Dogs [9].

We describe in the following the main contributions of this paper:

– We propose CauSim, a causal learning framework based on CNNs, which can identify the subtle similar features between different classes.
– CauSim employs a distance metric on the filters' Counterfactual Information (CI) learned to identify to which extent classes are similar. Classes whose CI values are highly correlated are more likely to be similar.
– For the FGVC task, CauSim can complement existing architectures, where it does not need any annotations or bounding boxes, neither in the training nor in the testing phase, and no retraining is needed.

2   Related Work

Causality in CNNs: The causal-based methods suggest to rely on the cause-effect principle. By employing either statistical, intervention or counterfactual approaches, many causal approaches have been proposed in the context of CNNs for different tasks. The main task in which causality has been largely used is explanation [12, 15].

While most causality approaches were proposed for the aim of explanation, recently some new computer vision tasks have also been addressed. Yang et al. [21] has addressed the do-calculus for proposing a causal attention (CATT) mechanism that can be used in the context of Image Captioning (IC) and Visual Question Answering (VQA). They used causality interventions to model the cause-effect between appearing objects in the image scene, by replacing trivial correlation between objects, and thus mitigating the bias caused by the hidden confounders in the dataset.

Metric Learning and FGVC. Metric learning has been firstly proposed in Siamese network that tries to minimize distance between positive pairs. Another approach considers a triplet of two positive pairs and negative one from three instances. Given a query image, once would like to find most similar images, which is helpful in modern search engines. Many works later tried to enhance this technique [16, 19]. Similar approach [17] for metric learning has been applied recently on FGVC by taking object's parts instead of the whole image, and then compute the correlation between these parts.

While [26] uses a mechanism of multi-attention to compute the features of two different parts, a recent work considers the correlation between different channels to extract discriminitive features [6]. They introduced a contrastive channel interaction (CCI) with a contrastive loss to model the channel-wise correlations between two images. Compared with existing methods, it requires training in one stage, and requires two images for computing the correlated information. Such an operation is performed by substracting weights information of two images.

This approach of considering channels interaction for the aim of FGVC has been addressed before in CGNL [23] and TSAN [27]. These two methods comparing to the previous one considers positive channels interaction in contrast to the previous one, which considers negative channels interactions. Similarly to CCI, Trilinear Attention Sampling Network (TASN) provides an attention module which generates attentions through modeling inter-channel relationship, but it requires hundreds of part proposals to be learned and fed to the network. However, the architecture consists of many other modules and steps such as knowledge distilling, which helps to extract the learned details.

Unfortunately, although these techniques made considerable advances, they still suffer from identifying the same regions for different classes, since they detect object parts in isolation without considering their inherent correlations. In our paper we will show how addressing correlation between learned features based on counterfactual causality would allow to disentangle subtle details between different classes in the same category, thus most similar classes are easily identified.

Similarity on ImageNet

ImageNet is useful for studying image similarity, since it has a large number of classes, and the images are in a high resolution. In addition, they are verified to contain the relevant concepts. One similar work to ours in visual similarity has been also applied on the level of ImageNet dataset [4], but without employing any deep learning technique. They studied whether images are from the same basic-level category, which will reinforce our intuition about semantic similarity. Similarly to this work, we evaluated CauSim on Imagenet dataset, by addressing large subcategories, which are the dogs category consisting of 120 breeds, and the birds category of 59 birds, this is among 1k classes. For performing this task, [4] consider a prototype representing every category, which is similar to our technique. However, we identify a prototype for every class through the CNN architecture itself, which is identified in a high accuracy among the instances of this class. Based on our experiments, we arrive to a conclusion that semantic similarity implies visual similarity, which has been already proven by most visual recognition approaches [4].

3   Causal Learning for Fine-Grained Similarity

It is well known that the most important layers of CNNs are the convolution layers, which include filters. Actually, a filter represents the basic element of the network that gives activations for different regions in the input image. Several research directions on CNNs have emerged in the aim of understanding the effect of filters. In this section, we will also show how addressing the causal effect of filters could be beneficial in many ways. While many works investigated counterfactuals on input images, in our technique we let both the input image as well as the CNN architecture intact, without any modification and retraining.

3.1   Filters as Actual Causes

Due to the importance of counterfactuals, Halpern and Pearl have extended the definition of counterfactuals by Lewis to build a rigorous mathematical model of causation, which they refer to as structural equations [7]. We can adapt this

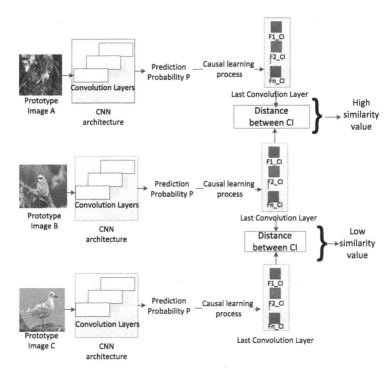

Fig. 1. An overview of CauSim: The approach consists of three main steps. After identifying a prototype image for each class through the CNN architecture itself, it is passed to causal learning process, where we measure the impact of each filter on the prediction probability. The final step is to measure the distance between the vectors of filters' CIs of each class, thus obtaining the final similarity.

definition in CNN by considering filters as actual causes for a predicted class φ. However, since the predicted class is returned with a probability P, φ should not be just a Boolean formula that refers to the class predicted. Therefore, we should define counterfactuals in terms of the prediction probability. To do so, we should ask the following question: what would be the probability of φ, when removing a filter f at a specific layer l, while keeping all the filters intact ?

In our architecture, we consider a prototype image for each class. This prototype is identified by the architecture itself, which is identified among the existing instances by the CNN architecture as the one with the highest confidence P. The intuition about this, is that a candidate prototype includes most of the important features in higher visibility, which enables the CNN architecture to identify it in an accurate way.

Now we can introduce the definition of an actual cause in CNNs.

Actual Cause. Let us consider a filter at a specific layer l denoted $f_l \in F$, where F refers to all filters. We say that f_l is causal for a prediction φ, if its own removal, where all the filters are kept the same, decreases the prediction probability P into P'. We refer to the difference $P - P'$ for a filter f_l by the counterfactual information, and it is denoted by CI_{f_l}. A filter is not causal if its $CI_{f_l} <= 0$.

According to the definition of causality [7], the set of filters F can be partitioned into two sets F_W^φ and F_Z^φ, where the set F_W^φ refers to filters causing the prediction φ, i.e. their removal will decrease the prediction probability P, whereas F_Z^φ are not causal, in way that the removal of a filter in F_Z^φ does not affect P, or otherwise increases it. When the removal of a filter leads to increasing P, this means that this filter has a negative effect on the decision, which means that it is responsible for increasing the prediction probability of another class, not the current class.

Abstraction of Causality

Beckers and Halpern [1] investigated the problem of abstracting causal models, when micro variables have causal effect on macro variables at higher levels. Actually this represents an essential key to our framework of causality. A CNN consists of many convolution layers that consist of many filters. The last layers are able to learn complex features, since their filters are applied after many convolutions on the input image. In other words, the last convolution layer can abstract visible features, since it lays in the top hierarchy of the causal network. Therefore, its filters can be considered as macro-variables in the highest level, which capture the effect of filters in the lowest-level layers (micro-variables).

This intuition behind abstraction of causal models leads us to consider only the filters of the last convolution layer, since they represent the top macro-variables that contribute to the final predicted class.

3.2 Images Similarity

When the causal learning process is complete, for each class we will have all the causal and non-causal filters, which refer to related and non-related features to this class respectively. We benefit from the counterfactual information for measuring the similarity between different classes by evaluating the differences of the CIs of their filters. To measure the similarity, we consider the CIs as vectors, and then we apply on them the Pearson correlation coefficient technique, to measure to which extent these vectors share similar values. Pearson correlation is a statistical technique for measuring the linear correlation between two sets of data. It can deal with both positive and negative values, and its result is normalized between -1 and 1. Figure 1 describes the CauSim architecture, and as we see, two classes are considered very similar if their CI' vectors are highly correlated.

For identifying similarity between different classes, this approach has been tested on many datasets: ImageNet [14], and four FGVC datasets: CUB-200-2011 birds, Stanford Dogs, Stanford Cars and Flowers 102.

ImageNet. ImageNet consists of 1k classes of different categories, thus it is very challenging for our approach. We made an experiment on ImageNet, by considering the VGG16 architecture. We consider only the last convolution layer "conv5_4" for computing the CIs of filters, which consists of 512 filters. All vectors of the 1k classes are stocked in a dictionary for similarity computation.

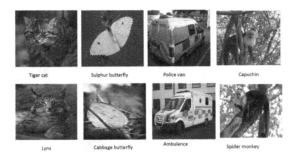

Fig. 2. Some CauSim results with VGG16 on ImageNet (Top-1 Similar classes are vertically aligned)

The experiment consists of challenging our approach for identifying categories of similar classes of the ImageNet dataset. We choose the largest categories in ImageNet, dogs category with 120 breeds, and birds with 59 species. We conducted quantitative experiments by identifying top 1 and top 3 similar classes to each dog/bird. Then we compute the top-1 error; i.e. top-1 similar class which is not a dog/bird, and top-3 error, i.e. there is no dog/bird in top-3 similar classes. The quantitative results are shown in the Table 1 using the Pearson correlation technique. As we see from the results, CauSim is very efficient in identifying categories for a given dataset, knowing that ImageNet is a very challenging dataset, due to its large number of classes, and thus many unrelated classes could share some similar features.

Figure 2 shows some similar samples. As we see, it is evident that classes in the same category such as monkeys and butterflies are very similar, which means that visual similarity implies semantic similarity, which has been proved by most visual recognition approaches [4].

Table 1. Quantitative similarity results on ImageNet using Pearson correlation

Category	Number	Top 1-error	Top 3-error
Dogs	120	35.00	14.16
Birds	59	42.37	25.42

Fine-Grained Datasets. Actually, similarity between different classes would be very helpful and more challenging when dealing with classes within the same category, such as birds and dogs. Therefore, we challenge CauSim on four fine-grained datasets: Flowers 102 [13], Caltech-UCSD Birds-200-2011 [18], Stanford

Fig. 3. Qualitative results of CauSim on CUB-200-2011, Stanford dogs, Stanford cars and Flowers 102

dogs [9] and Stanford cars [10]. As there is no exact mean for evaluating this experiment quantitatively, we report some qualitative results in Fig. 3, which shows some similar samples. For these datastes, we rely on the pre-trained architecture VGG19 for Flowers 102, and the FGVC architecture BCNN for the rest of datasets. BCNN has the VGG16 as a backbone, so similarly to the ImageNet dataset, the causal learning process is performed on the last convolutional layer "conv5_4" for every class.

As we see in the figure, our technique is able to capture every subtle detail between classes in an efficient way, where classes which are identified very similar by CauSim, are those mostly confused by humans.

4   CauSim for FGVC

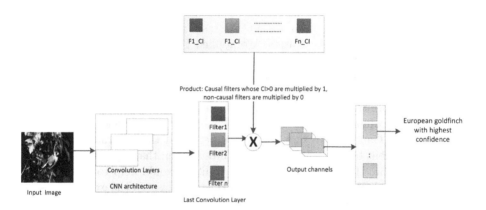

Fig. 4. An overview of CauSim for FGVC: Every filter of the last convolution layer is either multiplied by one (active) if its $CI > 0$, otherwise is multiplied by zero (inactive). Thus we are weighting the attention of CNN architecture, and the class having the most compatible causal vector with the CNN attentions would be returned.

Table 2. Accuracy results on four FGVC datasets.

Model	Dataset	Model's accuracy	CauSim mean-acc
VGG19	Flowers 102	95.42%	**96.82%**
BCNN-VGG16	Stanford Dogs	84.21%	**99.94%**
BCNN-VGG16	Cub200-2011	77.85%	**92.42%**
BCNN-VGG16	Stanford cars	75.91%	**97.25%**

As we saw in the previous qualitative results, CauSim is able to identify subtle similar details between classes in FGVC datasets. This led us to test its effect on the FGVC task. The process of adapting CauSim for FGVC is depicted in Fig. 4. After learning the counterfactual information of each filter on the last convolutional layer, we propose to use this information as a vector of features that should be incorporated in the classification process. The process is as follows: the final activations of the last layer are weighed by making a product of these activations with respect to CI of each filter. As we saw in the previous section, the filters are either causal or no-causal, so the product is performed on the filters of the last convolutional layer with a vector of binary values, when a filter f_l is causal (i.e. $CI_{f_l} > 0$), is multiplied by 1 (active), and zero (0) otherwise. Actually, we try to reinforce the attention of the architecture on an input image with respect to every CI vector, and then the final prediction is identified as the one having more features present in the input image. This has to be done for every class, which is known to prior, so the aim here is to test CauSim ability to identify all the features relative to a specific class, and thus classifying the given image with high accuracy.

To show its effectiveness for the FGVC task, we conducted several experiments on four fine-grained datasets. We applied CauSim on different architectures as follows: VGG19 on Flowers 102, and BCNN [11] with VGG16 as a backbone on CUB-200-2011, Stanford dogs and Stanford cars respectively. The results of the experiments are presented in Table 2. All the experiments have been done on pre-trained networks without any need to annotations or bounding boxes. The column acc-mean refers to the mean of all accuracies of every class when applying CauSim. We should mention here that the accuracy is obtained in this experiment by providing the causal filters given that we know the truth-label in prior. So, the aim here is to see if CauSim would help to reinforce the attention of the FGVC architecture toward the true class. In the experiments we found that for all samples in all datasets, the class is not just correctly predicted, but also is predicted with at least 0.9 confidence

As we see in Table 2, CauSim is able to increase the classification accuracy of VGG19 on Flowers, and BCNN on the rest of datasets. For BCNN having VGG16 as a backbone, the increasing of accuracy is very promising from 84.21% to 99.94%. As BCNN employs two VGG16 networks, CauSim is executed on both of them, thus increasing their effectiveness and enhancing the total accuracy.

One advantage of CauSim is its simplicity, and it does not need any modification on the input image, nor the architecture. Furthermore, no annotations or bounding boxes are needed. More interesting thing about CauSim: it can be applied on any CNN architecture, in which the last convolutional layer induces the important features leading to the final decision. Another important feature regarding CauSim, is that it can be also used to enhance the accuracy of FGVC architectures themselves, particularly those having CNN architectures as a backbone.

5   Conclusion and Future Work

In this paper we proposed CauSim, a causal learning framework based on CNNs for identifying fine-grained similarity between images. CauSim makes use of counterfactual reasoning and causality abstraction to identify filters corresponding to the features discriminating a class from another. Using similarity distances on the counterfactual information learned, resulted in identifying the most similar classes in the large dataset ImageNet, in addition to four fine-grained datasets.

CauSim shows preliminary promising results for the FGVC task when combined with the BCNN network. As future work, we aim extend it as a fully FGVC architecture. In addition, since CauSim shows good results in identifying the most important features, we aim to test its effectiveness for the task of Weakly Supervised Object Localization (WSOL).

References

1. Beckers, S., Halpern, J.Y.: Abstracting causal models. In: AAAI (2017)
2. Bell, S., Bala, K.: Learning visual similarity for product design with convolutional neural networks. ACM Trans. Graph. **34**(4) (2015)
3. Chopra, S., Hadsell, R., LeCun, Y.: Learning a similarity metric discriminatively, with application to face verification. In: 2005 IEEE Computer Society Conference on Computer Vision and Pattern Recognition (CVPR'05), pp. 539–546 (2005)
4. Deselaers, T., Ferrari, V.: Visual and semantic similarity in imagenet. In: CVPR 2011, pp. 1777–1784 (2011). https://doi.org/10.1109/CVPR.2011.5995474
5. Fu, J., Zheng, H., Mei, T.: Look closer to see better: Recurrent attention convolutional neural network for fine-grained image recognition. In: 2017 IEEE Conference on Computer Vision and Pattern Recognition (CVPR), pp. 4476–4484 (2017). https://doi.org/10.1109/CVPR.2017.476
6. Gao, Y., Han, X., Wang, X., Huang, W., Scott, M.R.: Channel interaction networks for fine-grained image categorization. In: IEEE Conference on Computer Vision and Pattern Recognition, pp. 10818–10825 (2020)
7. Halpern, J., Pearl, J.: Causes and explanations: a structural-model approach part i: causes. In: Proceedings of the 17th UAI, pp. 194–202 (2001)
8. Hoffer, E., Ailon, N.: Deep metric learning using triplet network. In: Feragen, A., Pelillo, M., Loog, M. (eds.) Similarity-Based Pattern Recognition, pp. 84–92 (2015)
9. Khosla, A., Jayadevaprakash, N., Yao, B., Li, F.F.: Novel dataset for FGVC: Stanford dogs. In: CVPR Workshop (2011)
10. Krause, J., Stark, M., Deng, J., Fei-Fei, L.: 3D object representations for fine-grained categorization. In: 2013 IEEE International Conference on Computer Vision Workshops, pp. 554–561 (2013)
11. Lin, T.Y., RoyChowdhury, A., Maji, S.: Bilinear CNN models for fine-grained visual recognition. In: 2015 IEEE International Conference on Computer Vision (ICCV), pp. 1449–1457 (2015). https://doi.org/10.1109/ICCV.2015.170
12. Narendra, T., Sankaran, A., Vijaykeerthy, D., Mani, S.: Explaining deep learning models using causal inference. arXiv:1811.04376 (2018)
13. Nilsback, M.E., Zisserman, A.: Automated flower classification over a large number of classes. In: ICVGIP, pp. 722–729 (2008)
14. Russakovsky, O., et al.: Imagenet large scale visual recognition challenge. Int. J. Comput. Vision **115**(3), 211–252 (2015)
15. Schwab, P., Karlen, W.: Cxplain: causal explanations for model interpretation under uncertainty. NeurIPS, pp. 10220–10230 (2019)

16. Sohn, K.: Improved deep metric learning with multi-class n-pair loss objective. In: Proceedings of the 30th International Conference on Neural Information Processing Systems. NIPS'16, pp. 1857–1865 (2016)

17. Sun, M., Yuan, Y., Zhou, F., Ding, E.: Multi-attention multi-class constraint for fine-grained image recognition. In: Ferrari, V., Hebert, M., Sminchisescu, C., Weiss, Y. (eds.) ECCV 2018. LNCS, vol. 11220, pp. 834–850. Springer, Cham (2018). https://doi.org/10.1007/978-3-030-01270-0_49

18. Wah, C., Branson, S., Welinder, P., Perona, P., Belongie, S.: The caltechucsd birds-200-2011 dataset. Technical report, California Institute of Technology (2011)

19. Wang, J., Zhou, F., Wen, S., Liu, X., Lin, Y.: Deep metric learning with angular loss. In: 2017 IEEE International Conference on Computer Vision (ICCV), pp. 2612–2620 (2017). https://doi.org/10.1109/ICCV.2017.283

20. Xiao, T., Xu, Y., Yang, K., Zhang, J., Peng, Y., Zhang, Z.: The application of two-level attention models in deep convolutional neural network for fine-grained image classification. In: 2015 IEEE Conference on Computer Vision and Pattern Recognition (CVPR), pp. 842–850 (2015)

21. Yang, X., Zhang, H., Qi, G., Cai, J.: Causal attention for vision-language tasks. In: CVPR 2021, pp. 9847–9857 (2021)

22. Yang, Z., Luo, T., Wang, D., Hu, Z., Gao, J., Wang, L.: Learning to navigate for fine-grained classification. In: Ferrari, V., Hebert, M., Sminchisescu, C., Weiss, Y. (eds.) Computer Vision – ECCV 2018. LNCS, vol. 11218, pp. 438–454. Springer, Cham (2018). https://doi.org/10.1007/978-3-030-01264-9_26

23. Yu, C., Zhao, X., Zheng, Q., Zhang, P., You, X.: Hierarchical bilinear pooling for fine-grained visual recognition. In: Ferrari, V., Hebert, M., Sminchisescu, C., Weiss, Y. (eds.) ECCV 2018. LNCS, vol. 11220, pp. 595–610. Springer, Cham (2018). https://doi.org/10.1007/978-3-030-01270-0_35

24. Zhao, B., Wu, X., Feng, J., Peng, Q., Yan, S.: Diversified visual attention networks for fine-grained object classification. IEEE Trans. Multimedia **19**(6), 1245–1256 (2017)

25. Zhao, W., Oyama, S., Kurihara, M.: Generating natural counterfactual visual explanations. In: IJCAI 2020, pp. 5204–5205 (2011)

26. Zheng, H., Fu, J., Mei, T., Luo, J.: Learning multi-attention convolutional neural network for fine-grained image recognition. In: 2017 IEEE International Conference on Computer Vision (ICCV), pp. 5219–5227 (2017)

27. Zheng, H., Fu, J., Zha, Z.J., Luo, J.: Looking for the devil in the details: learning trilinear attention sampling network for fine-grained image recognition. In: Proceedings of the IEEE Conference on Computer Vision and Pattern Recognition, pp. 5012–5021 (2019)

NOSpcimen: A First Approach to Unsupervised Discarding of Empty Photo Trap Images

David de la Rosa[1(✉)], Antón Álvarez[2], Ramón Pérez[2], Germán Garrote[3],
Antonio J. Rivera[1]📵, María J. del Jesus[1]📵, and Francisco Charte[1]📵

[1] Department of Computer Science, University of Jaén, Jaén, Spain
{drrosa,arivera,mjjesus,fcharte}@ujaen.es
[2] WWF Spain, Madrid, Spain
{aalvarez,rapayal}@wwf.es
[3] Consejería de Medio Ambiente y Agua. Junta de Andalucía, Seville, Spain
German.garrote.alonso@juntadeandalucia.es

Abstract. A key tool in wildlife conservation is the observation and monitoring of wildlife using photo-trapping cameras. Every year, thousands of cameras around the world take millions of images. A large proportion of these are empty – they do not show any animals. Sorting out these blank images requires considerable effort from biologists, who spend hours on the task. It is therefore of particular interest to automate this task. So far, systems have been proposed which are based on the use of supervised learning models. In order to learn, these systems require the annotation of images to indicate where animals are located within them. NOSpcimen (NOn-SuPervised disCardIng of eMpty images based on autoENcoders) system takes a different approach. It relies on unsupervised learning mechanisms. Thus, no prior annotation work is required to automate the process of discarding empty images.

Keywords: Unsupervised learning · deep learning · neural networks · clustering · robust autoencoder

1 Introduction

Monitoring the natural environment is vital to the preservation of species, but it is not a trivial task. It requires professionals capable to recognize the presence of animals and time to understand their behavior [1].

The use of photo-trapping cameras [2] is usually one of the best options to solve this problem. A photo-trapping camera is a tool for the surveillance and study of the wild environment from a fixed installation. These devices provide data on the location, population, size and interaction of species. The camera has a motion sensor connected to it, so that when an animal moves close to the sensor it causes the camera to fire, recording a burst of images without interfering with the animal's behavior.

© The Author(s), under exclusive license to Springer Nature Switzerland AG 2023
I. Rojas et al. (Eds.): IWANN 2023, LNCS 14135, pp. 39–51, 2023.
https://doi.org/10.1007/978-3-031-43078-7_4

One of the major disadvantages of this technique is the problem of blank images. A large proportion of the captured images are empty, they do not show any animals. Many situations can cause this problem, such as the motion sensor detecting the movements of an inanimate object, the animal passing by too quickly to be captured by the camera or the negative impact of high temperatures on motion sensors. For this reason, the use of software that automatically filters out empty images is a useful tool for many professionals in this field.

Machine learning (ML), a subfield of artificial intelligence, focuses on the creation of systems able to learn and adapt to the problem on the basis of the training data provided. Regarding to this problem, the use of ML implies the creation of a training dataset of images to develop models for the detection of empty and non-empty images.

Most of the ML proposed systems follow a supervised approach, that require the annotation of images in the training phase (which is expensive to manufacture since it needs human actions to identify, categorize and annotate the data) to indicate where animals are located within them. Then, the model compares the prediction and the actual value and use that information to infer knowledge that will allow them to recognize patterns in non-labelled images.

NOSpcimen takes an unsupervised approach that avoids the required image annotation work, apart from a previous separation of empty and non-empty images. The idea is to group the set of images according to their features, train an autoencoder (AE) for each cluster to reconstruct empty images and train a Multilayer Perceptron neural network (MLP) to classify images as empty or non-empty by taking the reconstruction errors.

The rest of this paper is structured as follows. Starting with preliminaries in Sect. 2, NOSpcimen's details are explained in Sect. 3. Section 4 focuses on the experimentation carried out in order to optimize the models and on the results achieved. Lastly, Sect. 5 explains the conclusions reached after the completion of the project.

2   Preliminaries

Many solutions can be found in the literature to solve the empty images filtering problem. They are grouped into two main areas.

On the one hand, solutions based on computer vision. The study [3] assumes that when the camera trap detects activity, it takes photographs with the same background. The only difference would be found in the movement of the animal. If it were a set of blank images, there would be no difference between them.

On the other hand, deep learning-based solutions, where a neural network is trained to differentiate between empty and non-empty images. For example, the study [4] trained a convolutional neural network using the ResNet-18 architecture and more than three million images for identification of wildlife in camera trap images, achieving an accuracy between 82 and 98%, depending on the test set. Also, [5] follows a similar scheme, training very deep convolutional neural network for automatic species classification.

Another supervised example is MegaDetector [6], a free tool to classify images as it contains animals, vehicles, persons or if it is empty. The latest version, the fifth, incorporates the YOLOv5 network, allowing users to classify their own images with high processing speed (up to three times faster than the previous version that used the Faster-RCNN architecture). This model is trained on bounding boxes from a variety of ecosystems with both private and public data.

As previously described, these studies follow supervised approaches. To the best of our knowledge, there are not unsupervised approaches for the empty images filtering problem on photo trap images.

Apart of deep learning models such as those described, it is important to define the concept of AE. An AE [7] (see Fig. 1) is a type of neural network that aims to reconstruct the input of the network onto the output, creating constraints so that the output is not a direct copy of the input. This is achieved by creating a coding layer, which represents the encoding of the input given by the encoder layers. Then, the decoder layers reconstruct the input based on its latent representation in the encoding.

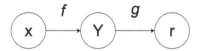

Fig. 1. Structure of an AE. x represents the input, f the encoder, Y the encoding, g the decoder and r the output

Several types of AE can be found in order to improve the results of a basic structure for different problems [7]. For example, denoising or robust autoencoders (RAE) are more tolerant to noisy inputs, sparse autoencoders include a penalty in the learning process and convolutional autoencoders use convolutional layers, which works properly with images as input.

Focusing on RAE models [8], the main difference with a classic AE resides in the loss function used. Correntropy based loss function [9] allows RAE to be more tolerant to noisy inputs. The loss function is presented in Eqs. 1 and 2, where y_{pred} represents the RAE prediction, y_{true} the actual value and σ a constant that represents the kernel size.

$$Loss = - \sum kernel(y_{pred} - y_{true}) \tag{1}$$

$$kernel(\alpha) = \frac{1}{\sqrt{2\pi}\sigma} \exp(-\frac{\alpha^2}{2\sigma^2}) \tag{2}$$

The presented AE architectures are used in many different problems, such as data compression [10], dimensionality reduction and data visualization [11], data denoising [12] or even as a module to solve an anomaly detection problem, as will be shown in Sect. 3.

3 Our Proposal: NOSpcimen System

The empty images filtering problem fits well as a classification problem. Nevertheless, NOSpcimen aims to transform the classification task into an anomaly detection problem where images containing some animals are considered as anomalies. The survey in [13] presents an overview of research methods in deep learning-based anomaly detection. Focusing on unsupervised deep anomaly detection, AE are the most common used architecture.

In this line, our working hypothesis is as follows. An AE is able to reconstruct the input onto the output. Therefore, if an AE is trained to only reconstruct empty images, the reconstruction error will be higher when passing a non-empty image as input to the model. Evaluating the reconstruction error, e.g. using a neural network, we may be able to distinguish if an image contains animals.

Nonetheless, the set of images are highly variable depending on the time of the day when the photography was taken (daytime or night), the brightness of the environment or the characteristics of the terrain (with heavy vegetation or in dry environments). This results in an AE having to learn many different image patterns, which causes lower accuracy in the reconstruction process. For this reason, we propose separating the dataset into several groups of images with similar features, so that one AE would be trained for each group, specializing in the characteristics of that group.

By joining all these blocks together, NOSpcimen is proposed as an unsupervised solution for the empty images filtering problem. Its diagram is presented in Fig. 2. Details are provided in the following subsections.

3.1 Image Segmentation Using K-Means Clustering

A clustering algorithm was trained by using empty images, creating N groups of images based on the image histogram, a representation of the number of pixels in an image as a function of their intensity.

Specifically, we have implemented Lloyd K-means algorithm [14]. Each image is assigned to a cluster by measuring the distance to the center of each one, the centroids, and choosing the closest group.

3.2 Robust Autoencoders for Image Reconstruction

As described in Sect. 2, RAEs are more tolerant to noisy inputs so that it has been decided to use RAEs instead of vanilla AEs.

Once the image clusters have been determined, the next step is to create the RAE models, one for each group of images trained only with his own empty images in order to optimize their reconstruction.

Following this logic, the reconstruction of images where animals appear should be worse, so measuring the reconstruction error we can determine the presence of animals in each image.

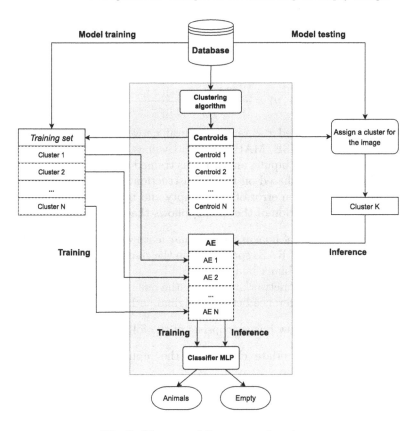

Fig. 2. Diagram of the proposed system

3.3 Multilayer-Perceptron Neural Network for Image Classification Based on Reconstruction Error

The last step is to develop a method to classify images based on the difference between original and reconstructed images given by RAEs. To measure that difference, three metrics have been used: mean squared error (MSE) and mean absolute error (MAE), where m is the image width, n the image height, Y the original image and $\hat{Y}$ the reconstructed image, and structural similarity (SSIM), where x is the original image, y the reconstructed image μ is the average of the image, σ^2 is the variance of the image, σ the covariance of the image, $c_1 = (k_1 L)^2$ and $c_2 = (k_2 L)^2$, where L is the dynamic range of the pixel values, $k_1 = 0.01$ and $k_2 = 0.03$ by default (see Eqs. 3, 4 and 5).

$$MSE = \frac{1}{m * n} \sum_{i=0}^{m-1} \sum_{j=0}^{n-1} [Y(i,j) - \hat{Y}(i,j)]^2 \qquad (3)$$

$$MAE = \frac{1}{m * n} \sum_{i=0}^{m-1} \sum_{j=0}^{n-1} |Y(i,j) - \hat{Y}(i,j)| \tag{4}$$

$$SSIM(x,y) = \frac{(2\mu_x\mu_y + c_1)(2\sigma_{xy} + c_2)}{(\mu_x^2 + \mu_y^2 + c_1)(\sigma_x^2 + \sigma_y^2 + c_2)} \tag{5}$$

In this line, we decided to create a neural network model capable to do that. Concretely, using MSE, MAE and SSIM values and the cluster identifier associated to the image as inputs, an MLP was trained to determine the presence of animals in each image based on the reconstruction errors. This network was trained with reconstruction error of both empty and non-empty images.

As summary, the creation of the model follows the steps below:

1. Creation of N groups of images with similar features.
2. Development of a set of RAEs specialized in the characteristics of each cluster trained to reconstruct blank images.
3. Development of a MLP network to classify the images as empty or non-empty taking as input the difference between original and reconstructed images.

The classification of new images consists of the following steps:

(a) Determine the appropriate cluster for the input image using centroids obtained in step 1.
(b) Reconstruct the image using the RAE associated with the cluster.
(c) Calculate the reconstruction error between the reconstructed and original image and predict the presence of animals using the MLP network.

4   Experimentation

In this section how NOSpcimen has been empirically tested is explained. Section 4.1 introduces the dataset and the preprocessing of the images. After that, the training process of every model will be analyzed, focusing on clustering in Sect. 4.2, on RAE models in Sect. 4.3 and ending up with MLP classifier in Sect. 4.4. Finally, obtained results are presented in Sect. 4.5.

4.1   Dataset

The photo trap images we will work with belong to WWF Spain. We have a total of 92 573 images, of which 37 503 are empty and 55 070 contains some animals.

To prepare data for model input, preprocessing is necessary. Original images are 2560 pixels wide and 1920 pixels high, with a depth of three bits per pixels, since they are RGB images. Those dimensions are excessive for deep learning models for several reasons, such as limited RAM memory or execution time. Moreover, large dimensions do not imply better results. Hence size reduction is necessary. After a visual analysis of a random subset of images applying a scaling in the image width to 512, 384, 256 and 192 pixels, we decided to scale the images to 384 pixels wide and 288 pixels high, maintaining the same depth.

In addition, all the camera trap images have timestamps at the bottom that include the date and time the photograph was taken. Removing this data reduces the input noise of the neural network.

Finally, the train test validation split is performed by randomly choosing images for empty and non-empty classes following the proportion 60-20-20, 60% images for training, 20% for validation and 20% for testing.

4.2   Clustering Algorithm

Two different implementations of the K-means algorithm were chosen: Scikit-learn version [15] and NLTK version [16]. The main difference resides in the distance metric: while the former only allows euclidean distance, the latter provides many other metrics, as cosine, Manhattan or correlation distances.

As stated in Sect. 3.1, clustering algorithm will be based on the use of the histogram of the images, which provides a good representation of the inner properties of the images. The number of bins for the histograms will be $(85 \times 85 \times 85)$, grouping the pixels values into bins of size three.

Different parameters were analyzed to get the best combination: color model (RGB and HSV), distance metric (euclidean, cosine, Manhattan and correlation distance), and number of clusters that will be formed (between 1 and 12). Results were measured using intra-cluster distance (ICD) obtained.

Table 1. Color model experiments.

Color model	ICD
RGB	770.05
HSV	1 157.47

Table 2. Distance metric experiments.

Distance metric	ICD
Euclidean	770.05
Canberra	820.86
Cosine	856.57
Manhattan	902.32

As shown in Tables 1 and 2, RGB color model contributes less ICD than HSV. Euclidean distance is also the best distance metric in terms of ICD.

Concerning the number of clusters, Fig. 3 shows a steeper slope between six and seven clusters, from which the slope is much smoother.

Because of presented results, we decided to use RGB color model, euclidean distance and seven clusters of images.

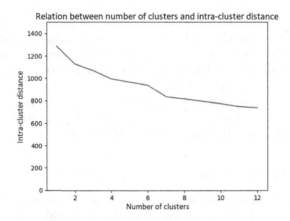

Fig. 3. Relation between number of clusters and intra-cluster distance.

4.3 Robust Autoencoders

RAE models decisions are focused in two aspects: model architecture and model parameters.

As to model architecture, after some testing, the chosen one is shown in Fig. 4. Two tests were created in order to verify that it gives optimal results. On the one hand, reducing the size of the network worsens the results as it does not have the capacity to process the information provided by the images. On the other hand, increasing the size of the model by adding more layers and reducing the encoding layer's size also worsen the results. If the encoding layer is excessively small, the decoder network will not be able to reconstruct the image properly.

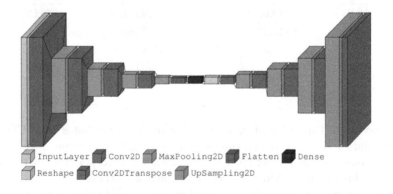

Fig. 4. Architecture of RAE chosen.

Table 3 resumes the model parameters, settled based on loss and MSE functions obtained during the model training (see Fig. 5). The figures show a stagnation from 70 epochs onwards in both loss and MSE functions, which proves that choosing more than 70 epochs is a bad option. Some testing was made in order to determine if a lower number of epochs is better. Nevertheless, these tests gave slightly worse results, so we decided to stick with the initial option. Optimizer and learning rate parameters have default values.

All RAEs maintain the same architecture and parameters.

Table 3. RAE chose parameters

Parameter	Value
Epochs	70
Batch size	16
Optimizer	Adam
Learning rate	0.001

Table 4. MLP chosen parameters.

Parameter	Value
Epochs	120
Batch size	16
Optimizer	Adam
Learning rate	0.001
Input	MSE, MAE, SSIM, CID

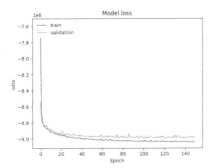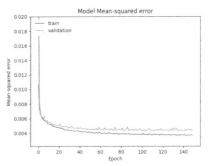

Fig. 5. Loss and MSE functions evolution during model training.

4.4 Multilayer-Perceptron Classifier

Following the same scheme of RAE models, the MLP experimentation will focus in model architecture and model parameters.

The input of the MLP network is simpler than RAEs, as it will only be composed of three different error metrics between reconstructed and original image and one integer representing the cluster identifier. Therefore, the model architecture will be also simpler. It is formed by two densely connected layers of 20 neurons and a softmax output layer that allows to classify the input as an animal or empty observation.

Model parameters are presented in Table 4. The network was trained based on cross-entropy loss between predicted and true labels, which calculates a score that summarizes the average difference between the actual and predicted probability distribution for each input. A test was performed in order to determine whether including the cluster identifier (CID) in the input layer improved the results. As shown in Fig. 6, accuracy is slightly improved when including CID. Also, accuracy value stagnates in 0.88–0.89 from 120 epochs onwards.

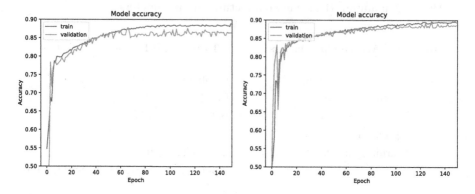

Fig. 6. MLP accuracy comparison including cluster identifier (right) or not (left)

4.5   Results

This subsection covers the results achieved by NOSpcimen on the test set. Figure 7 presents the confusion matrix and the ROC curve. Moreover, Table 5 summarises the classification performance. The confusion matrix shows a majority of correct predictions (top left corner and bottom right corner) compared with incorrect ones. This is also endorsed by the 0.89 F1-score. AUC is also a good metric to prove the proper performance of the system, with a value of 0.955, near its maximum possible value.

Overall, given its predictive performance, we can state that the developed software fulfill the proposed objectives.

Finally, a comparison between Megadetector v5a and NOSpecimen using the same image test set is also presented in Table 5. Megadetector confidence threshold was set at 0.2.

As can be seen, Megadetector's results are slightly better than those of NOSpcimen in terms of F1-score and AUC. Nevertheless, it is remarkable that Megadetector is a supervised approach, trained with much more data than NOSpcimen (various millions for the former against just 40 000 images for the latter). This fact greatly simplify the training phase, as in a supervised approach it is necessary to dedicate a time to annotate the data while this is avoided in a non-supervised approach.

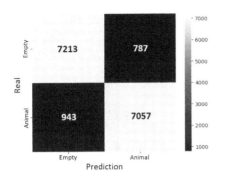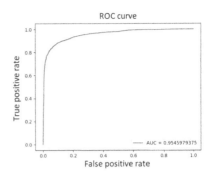

Fig. 7. Confusion matrix and ROC curve achieved.

Table 5. Precision, recall, F1 score, area under ROC curve and accuracy values

Class	Precision	Recall	F1-score	AUC	Accuracy
Empty	0.884	0.902	0.892	–	–
Animals	0.900	0.882	0.890	–	–
Global	0.892	0.892	0.891	0.955	0.892
Megadetector	0.955	0.955	0.960	0.991	0.960

5 Conclusions

With the results presented in the previous section, we can state that the proposed objectives have been achieved. The effort to empty images filtering task could be notably reduced by using NOSpcimen, allowing the professionals to focus only on non-trivial images, those that contain animals, instead of dedicate valuable time to the manual filtering.

In that line, a web application has been developed in order to make an usable prototype of this software. Figure 8 shows the main screen of the application (since the application will be used by Spanish conservationists, the user interface is in Spanish). Some parameters can be modified in order to adjust the software to the users' needs, as if they want to copy or move the images, if they want to store dubious images, etc. In addition, a threshold value can be seen in the image, which can take values between [0.5, 1]. The model set a probability for each image to be empty or not, in a range [0,1]. In case that the user do not need to store dubious images, each image will be assigned to the more probable class. On the other hand, if dubious images are stored, only images that exceed the threshold value will be assigned to a class. The rest will be assigned to *dubious* class. Following this reasoning, a high threshold implies that many images will be classified as dubious but with a low false positive or false negative rate, while a low threshold implies fewer images will be classified as dubious but with more false positives or false negatives.

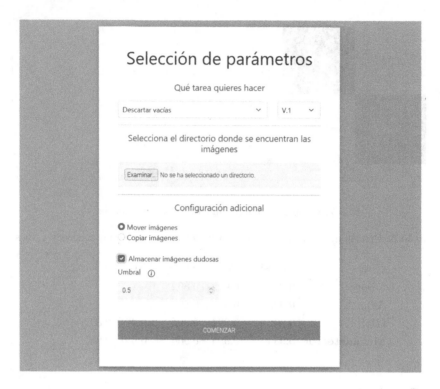

Fig. 8. Main page of the application.

The conservationists of WWF Spain are currently testing and validating this prototype with their own datasets. They plan to integrate the stable product as a component in their workflow being developed to optimize the monitoring of the Iberian lynx using camera trapping.

Lastly, as seen in Subsect. 4.5, considering that this is a first approach to unsupervised discarding of empty photo-trapping images, our results are close to those of other supervised and state-of-the-art approaches, proving that unsupervised learning have the potential to provide valuable insights into biodiversity conservation problems. We are currently working to introduce some changes to improve this prototype. One of our objectives is to use other types of autoencoders, such as masked autoencoders [17], to increase the distance between reconstruction errors of empty and non-empty images. A clustering algorithm improvement is also planned to get more homogeneous groups of similar size by trying other distance metrics.

Acknowledgements. The research carried out in this study is part of the project "ToSmartEADs: Towards intelligent, explainable and precise extraction of knowledge in complex problems of Data Science" financed by the Ministry of Science, Innovation and Universities with code PID2019-107793GB-I00/AEI/10.13039/501100011033. Also, this work was partly enabled by Antón Alvarez's participation in the CV4Ecology Summer Workshop, supported by the Caltech Resnick Sustainability Institute.

References

1. Tuia, D., et al.: Perspectives in machine learning for wildlife conservation. Nat. Commun. **13**(1), 792 (2022)
2. De Bondi, N., White, J.G., Stevens, M., Cooke, R.: A comparison of the effectiveness of camera trapping and live trapping for sampling terrestrial small-mammal communities. Wildlife Research **37**(6), 456–465 (2010)
3. Wei, W., Luo, G., Ran, J., Li, J.: Zilong: a tool to identify empty images in camera-trap data. Eco. Inform. **55**, 101021 (2020)
4. Tabak, M.A., et al.: Machine learning to classify animal species in camera trap images: applications in ecology. Methods Ecol. Evol. **10**(4), 585–590 (2019)
5. Villa, A.G., Salazar, A., Vargas, F.: Towards automatic wild animal monitoring: identification of animal species in camera-trap images using very deep convolutional neural networks. Ecol. Inform. **41**, 24–32 (2017)
6. Beery, S., Morris, D., Yang, S., Simon, M., Norouzzadeh, A., Joshi, N.: Efficient pipeline for automating species id in new camera trap projects. Biodiversity Inf. Sci. Stand. **3**, e37222 (2019)
7. Charte, D., Charte, F., García, S., del Jesus, M.J., Herrera, F.: A practical tutorial on autoencoders for nonlinear feature fusion: taxonomy, models, software and guidelines. Inf. Fusion **44**, 78–96 (2018)
8. Qi, Y., Wang, Y., Zheng, X., Wu, Z.: Robust feature learning by stacked autoencoder with maximum correntropy criterion. In: 2014 IEEE International Conference on Acoustics, Speech and Signal Processing (ICASSP), pp. 6716–6720. IEEE (2014)
9. Liu, W., Pokharel, P.P., Principe, J.C.: Correntropy: a localized similarity measure. In: The 2006 IEEE International Joint Conference on Neural Network Proceedings, pp. 4919–4924. IEEE (2006)
10. Theis, L., Shi, W., Cunningham, A., Huszár, F.: Lossy image compression with compressive autoencoders. arXiv preprint arXiv:1703.00395 (2017)
11. Hinton, G.E., Salakhutdinov, R.R.: Reducing the dimensionality of data with neural networks. Science **313**(5786), 504–507 (2006)
12. Xie, J., Linli, X., Chen, E.: Image denoising and inpainting with deep neural networks. Adv. Neural. Inf. Process. Syst. **25**, 341–349 (2012)
13. Chalapathy, R., Chawla, S.: Deep learning for anomaly detection: a survey. arXiv preprint arXiv:1901.03407 (2019)
14. Lloyd, S.: Least squares quantization in PCM. IEEE Trans. Inf. Theory **28**(2), 129–137 (1982)
15. Pedregosa, F., et al.: Scikit-learn: machine learning in python. J. Mach. Learn. Res. **12**, 2825–2830 (2011)
16. Bird, S., Klein, E., Loper, E.: Natural Language Processing with Python: Analyzing Text with the Natural Language Toolkit. O'Reilly Media Inc, Sebastopol (2009)
17. He, K., Chen, X., Xie, S., Li, Y., Dollár, P., Girshick, R.: Masked autoencoders are scalable vision learners. In: Proceedings of the IEEE/CVF Conference on Computer Vision and Pattern Recognition (CVPR), pp. 16000–16009, June 2022

Efficient Transformer for Video Summarization

Tatiana Kolmakova[1] and Ilya Makarov[2,3(✉)]

[1] HSE University, Moscow, Russia
tskolm@ya.ru
[2] Artificial Intelligence Research Institute (AIRI), Moscow, Russia
[3] AI Center, NUST MISiS, Moscow, Russia
iamakarov@misis.ru

Abstract. The amount of user-generated content is increasing daily. That is especially true for video content that became popular with social media like TikTok. Other internet sources keep up and easier the way for video sharing. That is why automatic tools for finding core information of content but decreasing its volume are essential. Video summarization is aimed to help with it. In this work, we propose a transformer-based approach to supervised video summarization. Previous applications of attention architectures either used lighter versions or loaded models with RNN modules, that slower computations. Our proposed framework uses all advantages of transformers. Extensive evaluation on two benchmark datasets showed that the introduced model outperform existed approaches on the SumMe dataset by 3% and shows comparable results on the TVSum dataset.

Keywords: Video Summarization · Deep Learning · Transformers

1 Introduction

Day by day consumption of video content is growing. According to Cisco Global Networking Trends Report [8], by 2022, Internet video will represent 82% of all business Internet traffic. Video summarization aimed to provide a faster way of understanding what happened in the video and save precious time. As for surveillance cameras, a shorter video may include only suspicious events that occurred in the area. Educational videos might be summarized to present core ideas from lectures [1,40]. And video from sports events might be shortened to the most intriguing moments.

The main goal of video summarization is to provide an essential core of the original video. There are two main types of resulting videos. The first is selecting key-frames and composing a storyboard [27,38,39]. The second is finding the most important video segments and collecting them in chronological order into

The work of Ilya Makarov was made in the framework of the strategic project "Digital Business" within the Strategic Academic Leadership Program "Priority 2030" at NUST MISiS.

a video skim. The second way is preferable because it includes motions and may include sounds, which helps to make the summary look more natural. But storyboards also have their pros, such as it represents a static sequence of keyframes, so there is no limitation in time and no restriction on data organization.

Since work describing transformers has been published [22,33,48], this type of architecture captures new areas. The power of this architecture is processing elements of the input sequence simultaneously. In unsupervised approaches [2, 18,25,26,32] authors implemented an attention module but also used RNN nets, that slowed the total architecture.

The contributions of this work as follows:

- We propose a transformer-based neural network architecture for supervised video summarization, that is not slowed by RNN.
- We provide a study of the importance of positional encoder in the video summarization.
- We conduct an experiment on two benchmark datasets that shows a performance of a proposed model.

The remainder of the paper is organized as follows. In Sect. 2, we present a study of the existed approaches to video summarization. In Sect. 3, we explain the proposed architecture. In Sect. 4, benchmark datasets are shown and provided an experimental setup. In Sect. 5, we display the results of the experiment, and in Sect. 6, we discuss the results of the model, in Sect. 7, we provide a conclusion.

2   Related Work

Since inventing typography, the amount of information has constantly increased. The task of shrinking the volume but keeping the core meaning of the subject started for the text content in the late 19501950ss from work [28]. Thus far, many methods have been developed. Since [7] basic approach of text summarization field was a sequence to sequence models. This method consists of two neural nets – encoder and decoder, both of which are recurrent neural nets (RNNs). At the start, the input sentence is tokenized and, each token is converted into a vector. The encoder accepts sentence token by token and produces a hidden vector for each input token [15]. The last hidden vector goes as an input to the decoder. Each decoder's cell receives a hidden vector from the previous cell and produces an output token and hidden vector. The decoder finishes work when its cell produces an empty token. In the context of videos, each frame is presented in vector form with the help of pre-trained neural nets, such as GoogleNet [46], AlexNet [24], Inception V3 [47], etc. The first work that used the sequence to sequence approach for video summarization was [35,52]. Researchers used bidirectional LSTM for modeling temporal video dependencies. Each backward and forward hidden vector of the LSTM cell afterward goes into a multilayer perceptron. Model enhanced by the determinantal point process – a probabilistic approach that helps to choose more diverse frames. The followed work of [53] also with sequence to sequence approach. The authors used a hierarchical encoder with two layers: for the frame and shot level vectors. On top of the decoder, another

LSTM encoder is used. The loss function compares the output of the encoder over summary with the output of the initial two-layers encoder. In this way, the authors ensure that the summary is close to the original video. In [54] authors gave us an idea of using multiple layers of RNN. They were inspired by CNN architecture and proposed two layers of RNN net. In their work [55] authors use bi-LSTM for short detecting and another bi-LSTM for selecting key shots.

In classical sequence to sequence models, the decision about each frame is based only on the output of the last RNN unit. But some information might not be found by a fixed-size vector. Especially it is a problem for long input sequences. The idea of creating a context vector that collects all hidden values and passes them into the decoder was presented in [5]. With the help of this vector, the model can find a more sophisticated relationship between input and target values. Pioneered this method in the video summarization field [19]. Their framework has a bidirectional LSTM network as an encoder and decoder with an attention mechanism. A similar architecture is described in [6]. The principal idea of work is the attention mechanism – it uses a prior distribution of web videos learned by VAE. In [20] researchers used an LSTM with self-attention as an encoder and a decoder with an attention mechanism. Their model was taught to predict the distribution of the importance scores, for this purpose, KL divergence was used. KL divergence loss is also used in the article [9]. Authors provide their own CNN architecture to convert pictures into vectors. In addition, depth information [23,30,31,34] can be used for selecting key-frames in addition to RGB information [56]. Pure attention mechanism was used in [11]. The authors claim that sequence to sequence models are very demanding and propose a computationally less expensive network. In [25] use multi-head and multi-layer attention mechanism, inspired by transformers [45,48]. The raw video goes throughout the sequence to sequence net, which is learned by reconstruction loss. The encoder of this sequence to sequence model shares weights with the other LSTM encoder that acted over attention layers. And the result of those LSTMs should be comparable. Highlights predicted by classifier above attended features. The idea of transformers also used in [44] article. This work provides a bidirectional GRU network with a multi-attention module above. To increase the receptive field of LSTM, [21] proposed to attribute each input frame to different chunks and strides. Each of the streams goes separately throughout bidirectional LSTM. Attention is based on vectors of differences. [50] provided a solution to the softmax bottleneck problem. They use a mixture of attention over bidirectional LSTM output and 3D CNN output.

To intensify the attention mechanism proposed in [49] a multi-layer architecture. Each layer consists of an LSTM and memory units. Each LSTM cell could have direct access to the memory unit from its layer and, each memory unit has a connection to the memory unit from the next layer. That configuration helps to increase the receptive field of LSTM. In [12] authors go deeper and use the global attention mechanism.

Datasets of videos where every video frame is evaluated with an importance score are very expensive. That is why the number of works tries to summarize the video in an unsupervised manner. To simplify, a framework of such methods consists of two nets: summarizer and generator. The summarizer neural net chooses keyframes and, the generator net should restore the whole video sequence based on provided frames. Usually, GAN [14] is used for recovering video by selected frames. To make keyframes diverse, researchers use an additional loss function that calculates the measure of variety frames at the summary.

The first who used this approach to video summarization was in [29]. They implemented the VAE-GAN model consists of four LSTM networks: sLSTM produces a subset of frames, eLSTM transforms frames into a fixed-size vector, dLSTM reconstructs the whole video by an encoded vector, and cLSTM distinguishes what type of video is given as input – reconstructed or the original. A similar idea introduced in [13,51], researchers want to maximize the mutual information between the summary and the original video. After the selection step, the authors use two GANs: forward, which tries to generate the whole video by selected frames, and backward that tries to get keyframes by the original video. Sparse loss is used to make sure that the selector network will not choose all frames. In [4] took architecture of [29] and improved its performance by implementing a stepwise, label-based approach for the adversarial part of the model. In their next work [2], the authors extended original architecture and added an attention mechanism. In [18] generator and discriminator consist of two layers: multi-head attention and BiLSTM. The discriminator tries to distinguish input features and features produced by the generator. Based on GAN as well, the model constructed in [26] but is formulated as supervised. At the first stage, it uses shot-level GAN, which aims to reconstruct shots by keyframes. At the second stage, the multi-attention mechanism is used to select keyframes from previous level keyframes.

3 Proposed Method

3.1 Model Architecture

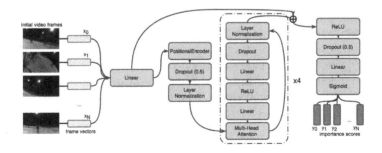

Fig. 1. Model inputs frames embeddings and outputs importance scores.

As shown in Fig. 1 the first layer is linear. That should help the model find a more appropriate projection of the feature vectors. As we converted frames into vectors and thus, the network has a lower amount of information, this layer should add flexibility to architecture.

We work with a sequence where dependencies between frames are essential because major events are not happening in one frame, thus we should not miss any placement details. To this matter, we use a positional encoder that shifts the input vector to the vector to determine its position in the input sequence. Finding a unique vector for every place in the input sequence is accomplished by collecting D different sinus and cosine frequencies depending on position numbers. After processing the sequence into an internal representation, we use multi-head attention with a feed-forward network. Multihead attention uses vanilla scaled dot-product attention and the feed-forward network is a two-layer net with ReLU activation. The last block of the model is a ReLU activation with a linear layer and sigmoid that projects values into $(0, 1)$. All over the model, we use a lot of dropout layers to avoid overfitting and layer normalization layers to avoid gradient decay.

4   Experiment

In this section, we showing benchmark datasets (Sect. 4.1), discuss an evaluation metric (Sect. 4.2), describing the evaluation setup (Sect. 4.3), and provide details of model implementation (Sect. 4.4).

4.1   Datasets

We evaluate the method's performance based on datasets used in the video summarization field:

- **TVSum** [43] consists of 50 videos from YouTube, each video length between 1 and 10 min. Videos are divided into 10 categories. Each video has at least 20 annotations from different users.
- **SumMe** [16] consists of 25 videos. These videos are from diverse categories like events, sports, cooking, etc. The duration of the videos is less than 7 min. Each video was annotated at least by 15 people. Human annotators made videos individually by selecting segments from original videos.
- **OVP** [10] consists of 50 video, each video length from 1 min to 4 min. Annotation provided in form of key-frames and produces by 5 users.
- **YouTube** [10] consists of 50 videos, length between 1 and 10 min. Annotation provided the same way as for the OVP dataset.

Following the established procedure of features gathering since [52], processing starts from downsampling video to 2 fps. Each video frame after that goes to GoogleNet [46], and vector from layer *pool5* becomes a feature descriptor for the frame.

Datasets provide the ground truth data in different formats. To make a fair comparison with other methods possible, we need a summary based on the key shots for both test and train data. For the SumMe dataset, such information is provided in the raw dataset, but in the TVSum we have a target value presented as an importance score for each frame. Following the procedure, that described in [52] and used in future works [11,25] we process each video by algorithm:

1. Segment video on disjoint intervals by KTS algorithm [36].
2. For each interval calculate the average importance score
3. Each frame in the same interval gets an importance score equals to the interval importance score.
4. Use a knapsack dynamic programming task to choose key shorts with the limit to total time. Usually, the length of the summary is limited to 15% of the original video length.

Datasets OVP and YouTube are used for augmented learning settings.

4.2   Evaluation Metric

As a metric to compare the agreement between ground truth summary and generated summary we use a key-shot F-score. Such evaluation criterion appeared in [52] and, afterward, papers follow this protocol to make a comparison.

Let A be a generated summary and B be ground truth, then

$$precision = \frac{overlapped\ frames\ of\ A\ and\ B}{A}$$

$$recall = \frac{overlapped\ frames\ of\ A\ and\ B}{B}$$

And the final score is calculated as

$$F = 2 \times \frac{precision \times recall}{precision + recall} \times 100$$

But between datasets exists a different approach to measure the score. For the TVSum dataset, each generated summary is compared to all user summaries. As a result, we get an F-measure that is averaged throughout those pairs. As for the SumMe dataset, for each machine-generated summary, the closest user-generated summary was obtained, so F-measure is maximized. This concept is used according to [17].

4.3   Evaluation Protocol

Following evaluation protocol in [52] we evaluate two different settings.

- **Canonical** It is a classical supervised learning approach where the dataset is divided into three parts: train, test, validation. But, as we have only a small amount of data, we will follow the protocol from [11]. This article suggests using K-fold validation for checking the resulting measure. Authors used 5-fold in each 80% of data became train and, the rest 20% is test.
- **Augmented** For this method, we use other datasets to expand our data. We chose 80% of the dataset and united it with others (OVP, YouTube), thus collecting a wide training set. The rest 20% remains for the test. Idea is that despite differences in the datasets they could be beneficial in the learning process.

4.4   Implementation Details

For converting frame into feature vector we use a GoogleNet pre-trained on ImageNet and get an output from pool5. The resulting dimension is D=1024. The first linear layer is a matrix from $\mathbf{R}^{D \times D}$, each dropout layer is taken with a rate of 0.5. Multi-Head Attention module consists of h=4 heads and, the dimensions for internal subspaces are $d_k = d_v = 64$. We use four such layers enhanced with a feed-forward network (FFN). Linear layers in FFN are $\mathbf{R}^{D \times 2D}$ and $\mathbf{R}^{2D \times D}$. The last linear layer is $\mathbf{R}^{D \times 1}$. We used Adam optimizer and MSELoss, the learning rate was 1e-4. The summary is produced by the knapsack algorithm.

5   Results

We provide a comparison of state-of-the-art models against ours on two benchmark datasets TVSum and SumMe in Table 1. Results for models are taken from the papers. As could be seen, our method produces competitive results and outperform models on the SumMe dataset. It is a 3.5% improvement to the highest previous score. From the table also could be seen that models with a higher f1-score are produced by the attention mechanism [11,19,25,26].

Following [3,11], we compare scores of all models to scores calculated for randomly generated and user-generated summaries as shown in Table 2. The randomly generated summary is created by sampling from a uniform distribution. And scores for user-generated summaries calculated 1) between all users and averaged, 2) ground truth summary is generated based on user frame importance scores and, afterward, each user summary is compared to ground truth.

Big difference in the results of two approaches for calculating F-score of user-generated summaries authors [11] explained as ground truth summaries are measured for key-shots, but between users, F-score measured at the frame level. And as key-shots are longer, than they are more likely to overlap.

Let us start the analysis from the TVSum dataset. As we could see, scores of models without attention are close to the scores of random summaries. As for the models with attention, they perform like a human and even slightly better

Table 1. Comparison with existing approaches on TVSum and SumMe datasets on two different learning settings. Number in the bold indicates best result.

Model	TVSum		SumMe	
	C	A	C	A
vsLSTM [52]	54.2	57.9	37.6	41.6
dppLSTM [52]	54.7	59.6	38.6	42.9
SUM-GAN$_{sup}$ [29]	56.3	61.2	41.7	43.6
SUM-FCN [37]	56.8	59.2	47.5	51.1
DR-DSN$_{sup}$ [57]	58.1	59.8	42.1	43.9
H-MAN [26]	60.4	61.0	51.8	52.5
M-AVS [19]	61.0	61.8	44.4	46.1
VASNet [11]	61.42	62.37	49.71	51.09
MC-VSA [25]	**63.7**	**64.0**	51.6	53.0
Ours	63.12	63.41	**53.5**	**53.72**

Table 2. F-score calculated in three ways: 1) between all users and averaged 2) between each user score and ground truth score 3) for random generated summary

F-score calculation method	TVSum	SumMe
between users comparison [11]	53.8	31.1
comparison to ground truth [11]	63.7	64.2
random summary [3]	54.4	40.2

on the augmented setting. Models at the TVSum dataset at peak quality, but as for SumMe, there is a way for improvement. We consider that two important parts of our model are the first linear layer and the positional encoder, which have not been in the previous architectures. To check this idea we evaluate the model without those elements and compare the contribution of these units (see Table 3).

Table 3. Comparing the contribution of architecture units

Model	TVSum	SumMe
full	63.12	53.5
without positional encoder	62.8	51.5
without linear	61.65	51.32

We see that the positional encoder improves the model's performance. That means temporal information is valuable for summarization. And, as the linear layer makes an impact, the ability to tune features for the model's needs also significant.

6   Discussion

To demonstrate the quality of generated summaries, we will look at the videos for which the model predicts importance scores that have the highest and the lowest F-measure compared to ground truth. The lowest score on K-fold validation has video #38 with an F1 score of 43.6%. We can see the prediction of importance scores of this video in Fig. 2. If we have a closer look at this prediction, we notice that the model guesses points of change well. It has a lack of confidence, but as we use the knapsack algorithm above the results of the prediction it is not so important. In Fig. 3 we see a generated summary, where the blue line shows the model's prediction and the orange line shows averaged user's predictions. We may notice that users are not united in this summary. The measure of user consistency is 57.87 for this video and it is the lowest value for F-measure between users and ground truth.

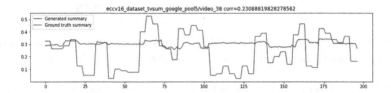

Fig. 2. Prediction of importance scores on video #38 TVSum from the test sample. Importance scores predicted for key-shots.

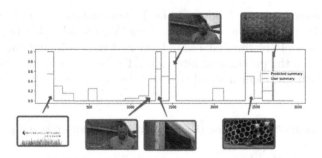

Fig. 3. Generated summary for video #38 TVSum from the test sample. In the blue frames are chosen by the model and in orange averaged user summaries. (Color figure online)

Figure 4 shows model's prediction with the highest F-measure - 72.1%. In this video model more confident but also cannot catch peaks. The measure of user confidence with ground truth is 72.08%. Generated summary compared to user-based summaries is shown in Fig. 5. As could be seen, our framework highlights all significant peaks in the importance scores of the video.

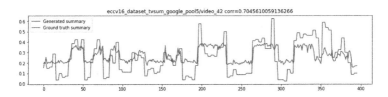

Fig. 4. Prediction of importance scores on video #42 TVSum from the test sample. Importance scores predicted for key-shots.

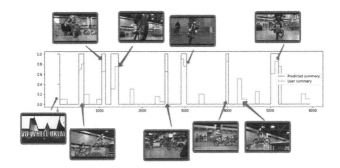

Fig. 5. Generated summary for video #42 TVSum from the test sample. In the blue frames are chosen by the model and in orange averaged user summaries. (Color figure online)

7   Conclusion

Video content becomes more and more popular. It is produced by people all over the world daily and shared on the Internet. With such an amount of data, easy to miss important information [41, 42]. Video summarization is used to highlight major moments in the video, which helps to decrease its volume and save content.

We proposed a deep neural framework for supervised video summarization. The central module in our framework is transformer-like multi-layer multi-head attention. And this was enhanced by the projection layer and by a feed-forward network that produces importance scores. We showed that using a positional encoder increase quality of the model because it helps to use temporal information. The proposed architecture was evaluated under two benchmark datasets. We compared the performance of the model with existed methods, random generators, and humans. On the SumMe dataset, our model outperforms the competing methods by 3%, but not human scores, and on the TVSum dataset, the model shows comparable to state-of-the-art and human scores.

References

1. Abdrahimov, A., Savchenko, A.V.: Summarization of videos from online events based on multimodal emotion recognition. In: Proceedings of International Russian Automation Conference (RusAutoCon), pp. 436–441. IEEE (2022)
2. Apostolidis, E., Adamantidou, E., Metsai, A.I., Mezaris, V., Patras, I.: Unsupervised video summarization via attention-driven adversarial learning. In: Ro, Y.M., Cheng, W.-H., Kim, J., Chu, W.-T., Cui, P., Choi, J.-W., Hu, M.-C., De Neve, W. (eds.) MMM 2020. LNCS, vol. 11961, pp. 492–504. Springer, Cham (2020). https://doi.org/10.1007/978-3-030-37731-1_40
3. Apostolidis, E., Adamantidou, E., Metsai, A.I., Mezaris, V., Patras, I.: Video summarization using deep neural networks: a survey. arXiv preprint arXiv:2101.06072 (2021)
4. Apostolidis, E., Metsai, A.I., Adamantidou, E., Mezaris, V., Patras, I.: A stepwise, label-based approach for improving the adversarial training in unsupervised video summarization. In: Proceedings of the 1st International Workshop on AI for Smart TV Content Production, Access and Delivery, pp. 17–25 (2019)
5. Bahdanau, D., Cho, K., Bengio, Y.: Neural machine translation by jointly learning to align and translate. arXiv preprint arXiv:1409.0473 (2014)
6. Cai, S., Zuo, W., Davis, L.S., Zhang, L.: Weakly-supervised video summarization using variational encoder-decoder and web prior. In: Ferrari, V., Hebert, M., Sminchisescu, C., Weiss, Y. (eds.) Computer Vision – ECCV 2018. LNCS, vol. 11218, pp. 193–210. Springer, Cham (2018). https://doi.org/10.1007/978-3-030-01264-9_12
7. Cho, K., et al.: Learning phrase representations using RNN encoder-decoder for statistical machine translation. arXiv preprint arXiv:1406.1078 (2014)
8. Cisco: Global networking trends report (2020). https://www.cisco.com/c/dam/m/en_us/solutions/enterprise-networks/networking-report/files/GLBL-ENG_NB-06_0_NA_RPT_PDF_MOFU-no-NetworkingTrendsReport-NB_rpten018612_5.pdf
9. Datt, M., Mukhopadhyay, J.: Content based video summarization: finding interesting temporal sequences of frames. In: 2018 25th IEEE International Conference on Image Processing (ICIP), pp. 1268–1272. IEEE (2018)
10. De Avila, S.E.F., Lopes, A.P.B., da Luz Jr, A., de Albuquerque Araújo, A.: Vsumm: a mechanism designed to produce static video summaries and a novel evaluation method. Pattern Recogn. Lett. **32**(1), 56–68 (2011)
11. Fajtl, J., Sokeh, H.S., Argyriou, V., Monekosso, D., Remagnino, P.: Summarizing videos with attention. In: Carneiro, G., You, S. (eds.) ACCV 2018. LNCS, vol. 11367, pp. 39–54. Springer, Cham (2019). https://doi.org/10.1007/978-3-030-21074-8_4
12. Feng, L., Li, Z., Kuang, Z., Zhang, W.: Extractive video summarizer with memory augmented neural networks. In: Proceedings of the 26th ACM International Conference on Multimedia, pp. 976–983 (2018)
13. Feygina, A., Ignatov, D.I., Makarov, I.: Realistic post-processing of rendered 3d scenes. In: Proceedings of the 45th ACM International Conference SIGGRAPH (SIGGRAPH'18), pp. 1–2. ACM, New York, USA, 12–16 August 2018. https://doi.org/10.1145/3230744.3230764
14. Goodfellow, I.J., et al.: Generative adversarial networks. arXiv preprint arXiv:1406.2661 (2014)
15. Grachev, A.M., Ignatov, D.I., Savchenko, A.V.: Neural networks compression for language modeling. In: Shankar, B.U., Ghosh, K., Mandal, D.P., Ray, S.S., Zhang, D., Pal, S.K. (eds.) PReMI 2017. LNCS, vol. 10597, pp. 351–357. Springer, Cham (2017). https://doi.org/10.1007/978-3-319-69900-4_44

16. Gygli, M., Grabner, H., Riemenschneider, H., Van Gool, L.: Creating summaries from user videos. In: Fleet, D., Pajdla, T., Schiele, B., Tuytelaars, T. (eds.) ECCV 2014. LNCS, vol. 8695, pp. 505–520. Springer, Cham (2014). https://doi.org/10.1007/978-3-319-10584-0_33

17. Gygli, M., Grabner, H., Van Gool, L.: Video summarization by learning submodular mixtures of objectives. In: Proceedings of the IEEE Conference on Computer Vision and Pattern Recognition, pp. 3090–3098 (2015)

18. He, X., et al.: Unsupervised video summarization with attentive conditional generative adversarial networks. In: ACM Multimedia, pp. 2296–2304 (2019)

19. Ji, Z., Xiong, K., Pang, Y., Li, X.: Video summarization with attention-based encoder-decoder networks. IEEE Trans. Circuits Syst. Video Technol. **30**(6), 1709–1717 (2019)

20. Ji, Z., Zhao, Y., Pang, Y., Li, X., Han, J.: Deep attentive video summarization with distribution consistency learning. IEEE Trans. Neural Networks Learn. Syst. (2020)

21. Jung, Y., Cho, D., Kim, D., Woo, S., Kweon, I.S.: Discriminative feature learning for unsupervised video summarization. In: Proceedings of the AAAI Conference on Artificial Intelligence, vol. 33 (01), pp. 8537–8544 (2019)

22. Karpov, A., Makarov, I.: Exploring efficiency of vision transformers for self-supervised monocular depth estimation. In: 2022 IEEE International Symposium on Mixed and Augmented Reality (ISMAR), pp. 711–719. IEEE (2022)

23. Korinevskaya, A., Makarov, I.: Fast depth map super-resolution using deep neural network. In: Proceedings of the 17th IEEE International Symposium on Mixed and Augmented Reality (ISMAR'18), pp. 117–122. TU Munich, IEEE, New York, USA, October 16–20 2018. https://doi.org/10.1109/ISMAR-Adjunct.2018.00047

24. Krizhevsky, A., Sutskever, I., Hinton, G.E.: Imagenet classification with deep convolutional neural networks. In: Advances in Neural Information Processing Systems, vol. 25, pp. 1097–1105 (2012)

25. Liu, Y.T., Li, Y.J., Wang, Y.C.F.: Transforming multi-concept attention into video summarization. In: Proceedings of ACCV (2020)

26. Liu, Y.T., Li, Y.J., Yang, F.E., Chen, S.F., Wang, Y.C.F.: Learning hierarchical self-attention for video summarization. In: 2019 IEEE International Conference on Image Processing (ICIP), pp. 3377–3381. IEEE (2019)

27. Lomotin, K., Makarov, I.: Automated image and video quality assessment for computational video editing. In: van der Aalst, W.M.P., Batagelj, V., Ignatov, D.I., Khachay, M., Koltsova, O., Kutuzov, A., Kuznetsov, S.O., Lomazova, I.A., Loukachevitch, N., Napoli, A., Panchenko, A., Pardalos, P.M., Pelillo, M., Savchenko, A.V., Tutubalina, E. (eds.) AIST 2020. LNCS, vol. 12602, pp. 243–256. Springer, Cham (2021). https://doi.org/10.1007/978-3-030-72610-2_18

28. Luhn, H.P.: The automatic creation of literature abstracts. IBM J. Res. Dev. **2**(2), 159–165 (1958)

29. Mahasseni, B., Lam, M., Todorovic, S.: Unsupervised video summarization with adversarial lstm networks. In: Proceedings of the IEEE Conference on Computer Vision and Pattern Recognition, pp. 202–211 (2017)

30. Makarov, I., Aliev, V., Gerasimova, O.: Semi-dense depth interpolation using deep convolutional neural networks. In: Proceedings of the 25th ACM International Conference on Multimedia (MM'17), pp. 1407–1415. FX Palo Alto Laboratory. ACM, New York, 23–27 October 2017. https://doi.org/10.1145/3123266.3123360

31. Makarov, I., Aliev, V., Gerasimova, O., Polyakov, P.: Depth map interpolation using perceptual loss. In: Proceedings of the IEEE International Symposium on

Mixed and Augmented Reality (ISMAR'17). pp. 93–94. Ecole Centrale de Nantes, France, IEEE, New York, USA, 09–13 October 2017. https://doi.org/10.1109/ISMAR-Adjunct.2017.39

32. Makarov, I., Bakhanova, M., Nikolenko, S., Gerasimova, O.: Self-supervised recurrent depth estimation with attention mechanisms. PeerJ Comput. Sci. **8**(e865), 1–25 (2022). https://doi.org/10.7717/peerj-cs.865

33. Makarov, I., Borisenko, G.: Depth inpainting via vision transformer. In: Proceedings of the 19th IEEE International Symposium on Mixed and Augmented Reality (ISMAR'21), pp. 286–291. INSA/IRISA, IEEE, New York, USA, 04–08 October 2021. https://doi.org/10.1109/ISMAR-Adjunct54149.2021.00065

34. Makarov, I., et al.: On reproducing semi-dense depth map reconstruction using deep convolutional neural networks with perceptual loss. In: Proceedings of the 27th ACM International Conference on Multimedia (MM'19), pp. 1080–1084. CNRS-IRISA, ACM, New York, USA, 21–25 October 2019. https://doi.org/10.1145/3343031.3351167

35. Maslov, D., Makarov, I.: Online supervised attention-based recurrent depth estimation from monocular video. PeerJ Comput. Sci. **6**(e317), 1–22 (2020). https://doi.org/10.7717/peerj-cs.317

36. Potapov, D., Douze, M., Harchaoui, Z., Schmid, C.: Category-specific video summarization. In: Fleet, D., Pajdla, T., Schiele, B., Tuytelaars, T. (eds.) ECCV 2014. LNCS, vol. 8694, pp. 540–555. Springer, Cham (2014). https://doi.org/10.1007/978-3-319-10599-4_35

37. Rochan, M., Ye, L., Wang, Y.: Video summarization using fully convolutional sequence networks. In: Ferrari, V., Hebert, M., Sminchisescu, C., Weiss, Y. (eds.) ECCV 2018. LNCS, vol. 11216, pp. 358–374. Springer, Cham (2018). https://doi.org/10.1007/978-3-030-01258-8_22

38. Savchenko, A.V., Belova, N.S.: Statistical testing of segment homogeneity in classification of piecewise-regular objects. Int. J. Appl. Math. Comput. Sci. **25**(4), 915–925 (2015)

39. Savchenko, A.V., Savchenko, L.V.: Towards the creation of reliable voice control system based on a fuzzy approach. Pattern Recogn. Lett. **65**, 145–151 (2015)

40. Savchenko, A.V., Savchenko, L.V., Makarov, I.: Classifying emotions and engagement in online learning based on a single facial expression recognition neural network. IEEE Trans. Affect. Comput. **13**(4), 2132–2143 (2022)

41. Savchenko, A., Khokhlova, Y.I.: About neural-network algorithms application in viseme classification problem with face video in audiovisual speech recognition systems. Optical Memory Neural Networks **23**(1), 34–42 (2014)

42. Sokolova, A.D., Kharchevnikova, A.S., Savchenko, A.V.: Organizing multimedia data in video surveillance systems based on face verification with convolutional neural networks. In: van der Aalst, W.M.P., et al. (eds.) AIST 2017. LNCS, vol. 10716, pp. 223–230. Springer, Cham (2018). https://doi.org/10.1007/978-3-319-73013-4_20

43. Song, Y., Vallmitjana, J., Stent, A., Jaimes, A.: Tvsum: summarizing web videos using titles. In: Proceedings of the IEEE Conference on Computer Vision and Pattern Recognition, pp. 5179–5187 (2015)

44. Sung, Y.L., Hong, C.Y., Hsu, Y.C., Liu, T.L.: Video summarization with anchors and multi-head attention. In: 2020 IEEE International Conference on Image Processing (ICIP), pp. 2396–2400. IEEE (2020)

45. Svidovskii, K., Semenkov, I., Makarov, I.: Attention-based models in self-supervised monocular depth estimation. In: 2022 IEEE 20th Jubilee International Symposium on Intelligent Systems and Informatics (SISY), pp. 000443–000448. IEEE (2022)

46. Szegedy, C., et al.: Going deeper with convolutions. In: Proceedings of the IEEE Conference on Computer Vision and Pattern Recognition, pp. 1–9 (2015)
47. Szegedy, C., Vanhoucke, V., Ioffe, S., Shlens, J., Wojna, Z.: Rethinking the inception architecture for computer vision. In: Proceedings of the IEEE Conference on Computer Vision and Pattern Recognition, pp. 2818–2826 (2016)
48. Vaswani, A., et al.: Attention is all you need. arXiv preprint arXiv:1706.03762 (2017)
49. Wang, J., Wang, W., Wang, Z., Wang, L., Feng, D., Tan, T.: Stacked memory network for video summarization. In: Proceedings of the 27th ACM International Conference on Multimedia, pp. 836–844 (2019)
50. Wang, J., et al.: Query twice: Dual mixture attention meta learning for video summarization. In: Proceedings of the 28th ACM International Conference on Multimedia, pp. 4023–4031 (2020)
51. Yuan, L., Tay, F.E., Li, P., Zhou, L., Feng, J.: Cycle-sum: cycle-consistent adversarial LSTM networks for unsupervised video summarization. In: Proceedings of the AAAI Conference on Artificial Intelligence, vol. 33 (01), pp. 9143–9150 (2019)
52. Zhang, K., Chao, W.-L., Sha, F., Grauman, K.: Video summarization with long short-term memory. In: Leibe, B., Matas, J., Sebe, N., Welling, M. (eds.) ECCV 2016. LNCS, vol. 9911, pp. 766–782. Springer, Cham (2016). https://doi.org/10.1007/978-3-319-46478-7_47
53. Zhang, K., Grauman, K., Sha, F.: Retrospective encoders for video summarization. In: Ferrari, V., Hebert, M., Sminchisescu, C., Weiss, Y. (eds.) ECCV 2018. LNCS, vol. 11212, pp. 391–408. Springer, Cham (2018). https://doi.org/10.1007/978-3-030-01237-3_24
54. Zhao, B., Li, X., Lu, X.: Hierarchical recurrent neural network for video summarization. In: Proceedings of the 25th ACM International Conference on Multimedia, pp. 863–871 (2017)
55. Zhao, B., Li, X., Lu, X.: HSA-RNN: hierarchical structure-adaptive RNN for video summarization. In: Proceedings of the IEEE Conference on Computer Vision and Pattern Recognition, pp. 7405–7414 (2018)
56. Zhingalov, K., Karpov, A., Makarov, I.: Multi-modal RGBD attention fusion for dense depth estimation. In: 2022 IEEE 20th Jubilee International Symposium on Intelligent Systems and Informatics (SISY), pp. 000109–000114. IEEE (2022)
57. Zhou, K., Qiao, Y., Xiang, T.: Deep reinforcement learning for unsupervised video summarization with diversity-representativeness reward. In: Proceedings of the AAAI Conference on Artificial Intelligence, vol. 32 (1) (2018)

Driver's Condition Detection System Using Multimodal Imaging and Machine Learning Algorithms

Paulina Leszczełowska[✉], Maria Bollin, Karol Lempkowski, Mateusz Żak, and Jacek Rumiński

Faculty of Electronics Telecommunications and Informatics, Gdansk University of Technology, Gdansk, Poland
paulinaleszczelowska@gmail.com, jacek.ruminski@pg.edu.pl

Abstract. To this day, driver fatigue remains one of the most significant causes of road accidents. In this paper, a novel way of detecting and monitoring a driver's physical state has been proposed. The goal of the system was to make use of multimodal imaging from RGB and thermal cameras working simultaneously to monitor the driver's current condition. A custom dataset was created consisting of thermal and RGB video samples. Acquired data was further processed and used for the extraction of necessary metrics pertaining to the state of the eyes and mouth, such as the eye aspect ratio (EAR) and mouth aspect ratio (MAR), respectively. Breath characteristics were also measured. A customized residual neural network was chosen as the final prediction model for the entire system. The results achieved by the proposed model validate the chosen approach to fatigue detection by achieving an average accuracy of 75% on test data.

Keywords: Machine Learning · Neural Networks · Fatigue · Drowsiness · Driver · Thermal · RGB

1 Introduction

The National Highway Traffic Safety Administration (NHTSA) reports that approximately 91,000 accidents are caused by drowsy driving every year. Approximately 50,000 people are injured and 800 are killed as a result of these accidents [24]. Since many of these accidents are not reported, the number of car accidents caused by drowsy driving is likely grossly underestimated. According to the European Commission, 10 to 25% of all accidents were due to driver fatigue [29]. As a result, a growing number of companies are implementing driver-state monitoring systems as a safety measure. This number will increase annually as a result of laws and regulations mandating the installation of such systems. Due to this, driver fatigue detection has become a topic of great interest in the scientific community. Existing solutions can be divided into three categories: monitoring

I. Rojas et al. (Eds.): IWANN 2023, LNCS 14135, pp. 66–78, 2023.
https://doi.org/10.1007/978-3-031-43078-7_6

the status of the vehicle's equipment, measuring the physiological parameters of the driver, and monitoring the driver's behavior [15].

This paper presents a system that simultaneously uses video data captured by thermal and RGB cameras to determine the state of the driver. This multimodal approach enables the extraction of driver-specific features.

The main contributions of this work are summarized as follows:

1. Creation of a unique dataset containing over 200 videos of nearly 20 distinct participants captured by RGB and thermal cameras and manually annotated.
2. Adaptation of methods for extraction of feature associated with fatigue symptoms.
3. Proposition of the RNN-based classification model for a drowsy driver detection.

The following section is a discussion of the various approaches utilized by other researchers in the field of fatigue detection. Section 3 describes the data acquisition process, obtained dataset, alongside the proposed solution architecture. Further implementation details and results are described. Sections 4, 5 and 6 discuss the obtained results and draw conclusions.

2   Related Works

A systematic literature review identified state-of-the-art driver state detection solutions, including conditions, datasets, features, and classification models.

2.1   Conditions of Data Collection

Conditions of data collection are one of the most influential factors in determining the final results of prediction models. Real, simulated, and regular contexts have been identified. The authors of [27] discuss the risks associated with driver drowsiness detection field experiments conducted under actual traffic conditions. Nevertheless, they highlight the model efficiency benefits of such an approach.

As stated by the authors of [4], the vast majority of driver drowsiness detection datasets used by researchers are collected under simulated driving conditions. This method eliminates the risks associated with recording actual drivers. Numerous researchers (e.g., [10,18]) have utilized data collected under conditions unrelated to driving. They portrayed a person in front of the camera who was either drowsy or awake or feigned fatigue by yawning or closing their eyes.

2.2   Datasets

In numerous cases, researchers created their own datasets to meet their needs. For instance, the authors of [27] created an experimental dataset of drowsy and non-drowsy faces extracted from datasets such as Bing Search API and Kaggle.

Twelve distinct public datasets were found, three of which were directly related to the drowsiness detection and the remaining nine were used as a feature extraction aid. The Driver Drowsiness Detection Dataset collected by the NTHU Computer Vision Lab was the most popular of these three datasets [32]. Another dataset, INVEDRIFAC, contained driver data recorded while driving a vehicle [2]. The University of Texas at Arlington Real-Life Drowsiness Dataset (UTA-RDD) [13] is the last dataset found that is directly associated with drowsiness detection and consists of videos captured in everyday settings.

Other datasets were utilized for intermediate system components such as face detection, eye state detection, and yawn detection. There were four separate face detection datasets: WIDER FACE (e.g. [20]), MTFL [17], FER2013 [6], and Celeba [8]. For eye state detection, five datasets were utilized: CEW (e.g. [15]), MRL (e.g. [26]), ZJU [15], and two datasets downloaded from Kaggle [26,28]. Two datasets used for yawn detection were YawnDD (e.g. [4]) and one of the previously mentioned Kaggle datasets [4].

2.3 Features

Various image-derived characteristics have been employed by researchers to assess the condition of drivers. The majority decided to extract eye- and mouth-state-related features. The authors of [28] calculated Eye Aspect Ratio (EAR) and Mouth Aspect Ratio (MAR) as the primary characteristics derived from the video data. Another article [30] has introduced the measures of Percentage Eye Closure (PERCLOS), which can be defined as the ratio of closed eyes to the number of open eyes, and Frequency of Mouth (FOM), whose calculation is very similar to that of PERCLOS. Some authors decided to use less popular indicators of drowsiness such as head movements (e.g. [27]), eyebrow furrowing [7], speed of facial muscle movement [34], and temperature of the face [23].

2.4 Machine Learning Algorithms

Non-deep Learning. The authors of [21] evaluated four machine learning models, k-nearest neighbors (KNN), centroid displacement-based KNN (CDNN), support vector machine (SVM), and random forest (RF), for the classification of different states, such as fatigue, yawning, etc., from features extracted before the softmax layer in the gamma fatigue detection network (GFDN) model. This method did not improve the classification accuracy of their deep learning model.

The authors of [3,10], and [18] utilized SVM and histogram-oriented gradient (HOG) to detect faces and facial landmarks in video frames. These algorithms were also used to classify the state of drivers in [3] The authors of [10,18] also utilized non-deep learning classification algorithms, such as Fisher's linear discriminant analysis (FLDA), the Bayesian Classifier, and SVM.

Deep Learning. The vast majority of solutions utilized Convolutional Neural Networks (CNN) for either feature extraction or decision-making. For example, in [31], researchers used a stacked CNN model to extract features and a CNN classifier with a Softmax layer to determine whether a driver is sleepy or not.

CNNs were used in both stages of implementation in [28], and the final classifier achieved one of the highest levels of accuracy among existing models (over 0.96).

Conv-LSTM is a classification method introduced by [33] consisting of two submodels: CNN for feature extraction and the Long Short-Term Memory (LSTM) model to interpret the extracted features across consecutive frames with 97.2% accuracy. Researchers in [6,14], and [16] utilized LSTM models as well. Many researchers also applied VGG architectures (e.g., [25,28]). VGG-FaceNet was used alongside AlexNet and FlowNet for feature representation learning in [25]. In [28], however, researchers implemented VGG16 transfer learning concepts for continuous monitoring of the driver's state and achieved over 96% accuracy.

3   Method

3.1   Experimental Setup

To ensure a diverse data set, a suitable study population was identified, and it was decided to include 18-to-60-year-old individuals of either gender with a valid driver's license. Additionally, the subjects were prohibited from consuming caffeinated beverages on the day of the recording. For the recording and processing of participants' faces, the appropriate legal documentation and consents were acquired. The final volunteer group consisted of 5 women and 14 men with the mean age of 26 years.

Using a commercially available steering wheel and gas and brake pedals, volunteers controlled simulated vehicles. The simulation was set in a city, had no particular goal, and was displayed on an ultrawide monitor, ensuring a more immersive experience. Certain crowd densities and weather conditions were chosen to create a stress-free environment capable of inducing drowsiness.

The aforementioned RGB and Thermal cameras, in conjunction with Google Coral, were used to capture the participants. Both cameras and the Coral device were encased in custom 3D-printed cases. A Vernier Respiration Monitoring Belt was incorporated into the simulator setup in order to establish a benchmark for breath metrics. Using openCV, it was possible to record simultaneously from both cameras in a specified loop for a given session. Following a predetermined number of recording sessions, data from both devices was aggregated further. The RGB camera recorded from a top-down perspective, allowing for accurate facial feature capture, whereas the thermal camera recorded from the level of the steering wheel. This placement of the thermal camera enabled the capture of participants' nostrils, allowing for the analysis of breath metrics.

Each participant followed the same procedure. Before participating in the project, a team member would explain the procedure, have them sign the required agreements, and conduct a brief survey to determine whether they were fatigued and had consumed caffeine earlier in the day. The participant was then given up to ten minutes to familiarize themselves with the simulator's configuration. After confirming readiness, participants proceeded to drive in a predetermined environment. Participants were asked in a post-recording survey whether they experienced any fatigue during the procedure. The form also included questions regarding the user's opinion of the proposed system (Fig. 1).

Fig. 1. Simulator

3.2 Dataset

As a result of conducted sessions 244 RGB and thermal recordings from 19 different volunteers were obtained. Thermal videos have a resolution of 160 × 120 pixels and a frame rate of 8–9 fps. RGB recordings have a framerate of 30 fps and a resolution of 640 × 480 pixels. The average duration of thermal and RGB recording was 30 s. Based on participant survey responses, every recording was annotated. Label 1 means a recorded subject is fatigued, while label 0 means the opposite. In addition, RGB videos were also annotated frame-by-frame. These annotations noted when a subject had an open mouth or blinked.

3.3 System Architecture

In Fig. 2 the overall data processing pipeline for the proposed system is presented. The system is based on simultaneous recording of the driver's face by both thermal and RGB cameras. Each recording is being processed in order to acquire facial metrics connected with eye and mouth states from RGB images and breath metrics from thermal ones. Data is then aggregated and given to the pre-trained RNN model which determines the current physical state of the driver.

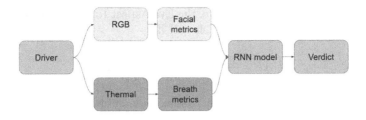

Fig. 2. Proposed system architecture

3.4 RGB Images Pipeline

Two of the most important attributes in assessing a driver's condition are the eye state and mouth state [5]. In this study, the RGB frames were used to extract blinking and yawning frequencies. The detection of the face and its distinguishing features in each frame of recording is the first stage of RGB video processing. Firstly, the OpenCv library's implementation of the Haar cascade is used to detect faces. The facial points on a cropped person's face are then mapped using the pre-trained facial landmarks detector from the dlib package, which is able to estimate the placement of 68 coordinates as shown in Fig. 3.

Fig. 3. Facial landmarks [1]

In order to determine the driver's mouth state, six of the detected points are used to calculate the Mouth Aspect Ratio (MAR) [12]. It is defined by the average of three distances between points on the upper and lower lips.

$$MAR = \frac{\|p_{62} - p_{68}\| + \|p_{63} - p_{67}\| + \|p_{64} - p_{66}\|}{3} \qquad (1)$$

For the purpose of evaluating the state of the eye, two methods are employed. The first is founded on the same principle as mouth-state determination. In order to determine the degree of eye closure, the Eye Aspect Ratio (EAR) is calculated using twelve facial points (six for each eye) [9].

$$EAR = \frac{\|p_{38} - p_{42}\| + \|p_{39} - p_{41}\|}{2 \cdot \|p_{37} - p_{40}\|} \tag{2}$$

A second approach is used to classify eyes as open or closed. In order to achieve this the VGG16-based classifier was adapted and initialized with the weights of the VGGFace model [22]. In order to ensure better generalization and prevent overfitting, the model was trained with the independent Closed Eyes in the Wild (CEW) dataset and only tested on our self-gathered data.

3.5   Thermal Images Pipeline

The information obtained from thermal imaging is used to calculate the rate of the driver's respiratory rate. Due to the low resolution of thermal imaging, facial features may be indistinguishable; thus, histogram equalization, which normalizes the intensity distribution for a given range, was used to enhance visual contrast. Because respiration changes the temperature around the nostril area, it is important to analyze only this small area. To find the region of interest, a model trained on thermal imaging was used [19]. After determining the region of interest, the mean pixel value was calculated. The resulting values are then used as a signal representing the time-averaged pixel value. The pixel values around the nose should change during respiration due to temperature differences.

The final steps aim to reduce noise and filter the signal to produce a waveform for calculating the respiration rate. The signal is refined using the asymmetric least squares smoothing algorithm [11] with $\lambda = 100$ and p = 0.1. For the purpose of the study, the average respiration rate was calculated. The duration of single respiration was calculated by dividing the number of frames between local minima in the processed signal by the frame rate of the Lepton camera. The local minima were found using scipy's find_peaks function [1]. The average respiratory rate was obtained by normalizing the average duration of respiration.

3.6   Classification Model

The final decision model was based on Recurrent Neural Network (RNN) to which video segments were fed. Each video was separated into 10-second-long windows. This window was then shifted by one second's worth of frames, yielding a total of twenty windows per video. Then, five features were calculated for each window extracted from a video using the metrics described in earlier sections. These features included the mean EAR, the percentage of frames with closed eyes, the mean MAR, the maximum MAR, and the mean respiration rate. The model consists of one GRU unit and two linear layers. It returns a single value between 0 and 1 that indicates the likelihood that the driver is experiencing drowsiness. A decision was made in accordance with a chosen threshold of 0.5. A diagram of the neural network is shown in Fig. 4.

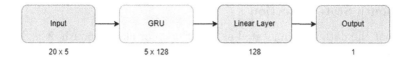

Fig. 4. Classification neural network diagram

4   Results

4.1   Eye State Detection

To estimate the average state of both eyes simultaneously, the mean of the EAR metrics was calculated separately for each eye. As can be deduced from Table 1 presenting the comparison of both approaches for eyes open classification, the trained VGG16 model performed exceptionally well on our dataset. The distribution of classes was unbalanced due to the fact that, under realistic conditions, people's eyes are open for much longer than a typical blink, which lasts only a fraction of a second.

Table 1. Metrics for eyes open classification

Metric	VGG16	EAR threshold
Accuracy	**0.9019**	0.8669
Balanced Accuracy	**0.8561**	0.81
Sensitivity	**0.9101**	0.8771
Specificity	**0.8021**	0.7424

The classification outcomes were used to calculate the number of eye blinks per video. The results were evaluated using Mean Absolute Error (MAE) and Root Mean Squared Error (RMSE). Table 2 displays the error values corresponding to the comparison of classification by simple EAR threshold. As can be seen, the CNN model provides significantly more accurate results than the EAR.

Table 2. Error measures for number of eye blinks per video

Metric	VGG16	EAR threshold
MAE	**3.55**	25.38
RMSE	**5.41**	32.7

4.2 Mouth State Detection

Given that the data was annotated as either open or closed mouth, it is difficult to estimate the mouth state accuracy of the calculated metric. A threshold of 5.5 was applied to the MAR value for quality control purposes, and any value above this threshold was considered an open mouth.

The average accuracy, Area Under Receiver Operating Characteristic Curve (AUROC), and F1 score were calculated for sample recordings based on the binary classification of a mouth state. The classification of the mouth state yielded great results, as presented in Table 3. This ensured that the MAR values calculated for particular frames accurately reflected the condition of the mouth.

Table 3. Metrics for mouth state classification

Metric	MAR threshold
Accuracy	0.915
AUROC	0.895
F1 score	0.787

4.3 Respiratory Rate Detection

For each subject the estimated signal form thermal imaging was obtained. Average rate for all subjects is 16.3 with standard deviation of 2.37. Finally the estimated respiratory rate was used as one of the features. Figure 5 illustrates the average respiratory rate for the participants, however some had to be omitted due to insufficient data.

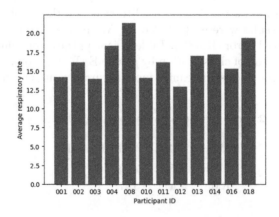

Fig. 5. Average respiratory rate for participants

4.4   Drowsiness Detection

The adapted RNN model was trained using the Adam optimizer with early stopping based on the value of balanced accuracy score on the evaluation set with 15 epochs of patience. The model was able to achieve a balanced accuracy of approximately 81.5% at its peak. On the test set the model managed to achieve a 75% which is promising considering the small size of the dataset. Metrics achieved on the test set are depicted in Table 4.

Table 4. Metrics for drowsiness detection

Metric	Proposed model
Accuracy	0.8
Balanced Accuracy	0.75
F1 Score	0.86

5   Discussion

In this paper, a model for drowsiness detection using two data sources is presented. This approach is novel; consequently, it was impossible to locate data that met all the requirements of the study. The scalability of the proposed dataset is a massive undertaking involving numerous factors, each of which affects the process of data acquisition differently. One of the limitations present in the current version of the dataset is the length of the recordings (30 s). Acquiring larger amount of longer videos from various age groups would potentially allow for a better generalization of the system and also be more accurate to the real driving environment. Good example of such extensive dataset is YawnDD which is commonly used in fatigue detection [4,29]. Nonetheless, the dataset presented in this paper enabled the achievement of reliable results with room for improvement in future research.

Incorporating new parameters into the system, in addition to the respiratory rate mean value presented in this paper, may have the potential to significantly improve the system's accuracy. Examples of such parameters are the regularity of respiratory cycles, the alteration of respiratory patterns, etc.

Possible improvements may include also intermediate models tasked with feature extraction. The result presented in the preceding section demonstrated that blink detection by neural network model is more precise than EAR threshold alone. Due to this, it would be advisable to incorporate a deep learning-based method for yawning detection, as it was successfully done in [6].

6   Conclusion

In the course of this study, a system for detecting driver drowsiness was developed. It is based on a multimodal approach and employs both RGB and thermal cameras. Therefore, it was possible to acquire the driver's physiological characteristics non-invasively. With data regarding a participant's eyes, mouth, and breathing rate, the system was able to achieve a test accuracy of 0.75. This result demonstrates that this particular type of system is reliable, and it is possible that it could be enhanced with additional data samples. It is plausible that the addition of vehicle state variables commonly employed in the automotive industry could further improve the prediction and increase road safety as a whole.

Acknowledgement. We wanted to issue a special thank you to Gdańsk University of Technology and AI Tech programme for supporting us in the process of developing this system and giving us access to the university's resources.

References

1. 68 facial landmark coordinates from the ibug 300-w dataset. Accessed 15 Mar 2023
2. Invedrifac (2019)
3. Akshara, R., Karthik, J., Reddy, E.S.C., Nayak, R.G.: An efficient fatigue detection system by inspecting facial behavioral aspects. In: 2021 Third International Conference on Inventive Research in Computing Applications (ICIRCA) (2021)
4. Bai, J.: Two-stream spatial-temporal graph convolutional networks for driver drowsiness detection. IEEE Trans. Cybern. **52**(12), 13821–13833 (2022)
5. Bajaj, J.S., Kumar, N., Kaushal, R.K.: Comparative study to detect driver drowsiness. In: 2021 International Conference on Advance Computing and Innovative Technologies in Engineering (ICACITE), pp. 678–683 (2021)
6. Chen, S., Wang, Z., Chen, W.: Driver drowsiness estimation based on factorized bilinear feature fusion and a long-short-term recurrent convolutional network. Information **12**(1), 3 (2020)
7. Chmielińska, J., Jakubowski, J.: Detection of driver fatigue symptoms using transfer learning. Bull. Polish Acad. Sci. Tech. Sci. **66**(6), 869–874 (2018)
8. Cui, Z., Sun, H.-M., Yin, R.-N., Gao, L., Sun, H.-B., Jia, R.-S.: Real-time detection method of driver fatigue state based on deep learning of face video. Multimedia Tools Appl. (2021)
9. Dewi, C., Chen, R.-C., Chang, C.-W., Wu, S.-H., Jiang, X., Yu, H.: Eye aspect ratio for real-time drowsiness detection to improve driver safety. Electronics **11**(19), 3183 (2022)
10. Dey, S., Chowdhury, S.A., Sultana, S., Hossain, M.A., Dey, M., Das, S.K.: Real time driver fatigue detection based on facial behaviour along with machine learning approaches. In: 2019 IEEE International Conference on Signal Processing, Information, Communication & Systems (SPICSCON) (2019)
11. Eilers, P., Boelens, H.: Baseline correction with asymmetric least squares smoothing. Unpubl. Manuscr (2005)
12. Fear-The-Lord. Fear-the-lord/drowsiness-detection github. Accessed 26 Feb 2023

13. Ghoddoosian, R., Galib, M., Athitsos, V.: A realistic dataset and baseline temporal model for early drowsiness detection. In: 2019 IEEE/CVF Conference on Computer Vision and Pattern Recognition Workshops (CVPRW) (2019)
14. Guo, J.-M., Markoni, H.: Driver drowsiness detection using hybrid convolutional neural network and long short-term memory. Multimedia Tools Appl. **78**(20), 29059–29087 (2018)
15. Huang, R., Wang, Y., Guo, L.: P-fdcn based eye state analysis for fatigue detection. In: 2018 IEEE 18th International Conference on Communication Technology (ICCT) (2018)
16. Jamshidi, S., Azmi, R., Sharghi, M., Soryani, M.: Hierarchical deep neural networks to detect driver drowsiness. Multimedia Tools Appl. **80**(10), 16045–16058 (2021)
17. Jia, H., Xiao, Z., Ji, P.: Fatigue driving detection based on deep learning and multi-index fusion. IEEE Access **9**, 147054–147062 (2021)
18. Kumar, A., Patra, R.: Driver drowsiness monitoring system using visual behaviour and machine learning. In: 2018 IEEE Symposium on Computer Applications & Industrial Electronics (ISCAIE) (2018)
19. Kuzdeuov, A., Koishigarina, D., Aubakirova, D., Abushakimova, S., Varol, H.A.: Sf-tl54: a thermal facial landmark dataset with visual pairs. In: 2022 IEEE/SICE International Symposium on System Integration (SII), pp. 748–753 (2022)
20. Li, K., Gong, Y., Ren, Z.: A fatigue driving detection algorithm based on facial multi-feature fusion. IEEE Access **8**, 101244–101259 (2020)
21. Liu, W., Qian, J., Yao, Z., Jiao, X., Pan, J.: Convolutional two-stream network using multi-facial feature fusion for driver fatigue detection. Future Internet **11**(5), 115 (2019)
22. Parkhi, O.M., Vedaldi, A., Zisserman, A.: Deep face recognition (2015)
23. Moazen, I., Nahvi, A.: Implementation of a low-cost driver drowsiness evaluation system using a thermal camera. SAE Technical Paper Series (2021)
24. NHTSA. Drowsy driving. Accessed 15 Jan 2023
25. Park, S., Pan, F., Kang, S., Yoo, C.D.: Driver drowsiness detection system based on feature representation learning using various deep networks. In: Chen, C.-S., Lu, J., Ma, K.-K. (eds.) ACCV 2016. LNCS, vol. 10118, pp. 154–164. Springer, Cham (2017). https://doi.org/10.1007/978-3-319-54526-4_12
26. Pawar, R., Wamburkar, S., Deshmukh, R., Awalkar, N.: Driver drowsiness detection using deep learning. In: 2021 2nd Global Conference for Advancement in Technology (GCAT) (2021)
27. Phan, A.-C., Nguyen, N.-H.-Q., Trieu, T.-N., Phan, T.-C.: An efficient approach for detecting driver drowsiness based on deep learning. Appl. Sci. **11**(18), 8441 (2021)
28. Sri, B.R., Akanksha, Y., Puthali, R., Anuradha, T.: Early driver drowsiness detection using convolution neural networks. In: 2021 Second International Conference on Electronics and Sustainable Communication Systems (ICESC) (2021)
29. Mobility & Transport Road Safety. Fatigue and crash risk. Accessed 15 Jan 2023
30. Savas, B.K., Becerikli, Y.: Real time driver fatigue detection system based on multi-task connn. IEEE Access **8**, 12491–12498 (2020)
31. Suresh, Y., Khandelwal, R., Nikitha, M., Fayaz, M., Soudhri, V.: Driver drowsiness detection using deep learning. In: 2021 2nd International Conference on Smart Electronics and Communication (ICOSEC) (2021)
32. Weng, C.-H., Lai, Y.-H., Lai, S.-H.: Driver drowsiness detection via a hierarchical temporal deep belief network. In: Chen, C.-S., Lu, J., Ma, K.-K. (eds.) ACCV 2016. LNCS, vol. 10118, pp. 117–133. Springer, Cham (2017). https://doi.org/10.1007/978-3-319-54526-4_9

33. Yarlagadda, V., Koolagudi, S.G., Kumar, M.V.M., Donepudi, S.: Driver drowsiness detection using facial parameters and rnns with lstm. In: 2020 IEEE 17th India Council International Conference (INDICON) (2020)

34. Zhao, L., Wang, Z., Zhang, G., Gao, H.: Driver drowsiness recognition via transferred deep 3d convolutional network and state probability vector. Multimedia Tools Appl. **79**(35–36), 26683–26701 (2020)

On-Line Authenticity Verification of a Biometric Signature Using Dynamic Time Warping Method and Neural Networks

Krzysztof Walentukiewicz$^{(\boxtimes)}$ ⓘ, Albert Masiak ⓘ, Aleksandra Gałka ⓘ, Justyna Jelińska ⓘ, and Michał Lech ⓘ

Gdansk University of Technology, 80-233 Gdansk, Poland
krzysztof.walentukiewicz2@gmail.com, albert.masiak99@gmail.com, ola@galka.pl, justynaj99r@gmail.com, mlech.ksm@gmail.com

Abstract. To ensure proper authentication, e.g. in banking systems, multimodal verification are becoming more prevalent. In this paper an on-line signature based on dynamic time warping (DTW) coupled with neural networks has been proposed. The goal of this research was to test if combining neural networks with DTW improves the effectiveness of verification of a handwritten signature, compared to the classifier based on fixed thresholds. The DTW algorithm was used as a feature extraction method and a similarity measure. Two neural network architectures were tested: multilayer perceptron (MLP) and one with convolutional neural network (CNN). A dataset containing model, verification and forged signatures gathered from a research group using a biometric pen has been created. The research has proved that the DTW coupled with neural networks perform significantly better than the baseline method - DTW model based on constant thresholds. The results are presented and discussed in this paper.

Keywords: dynamic time warping · handwritten signature verification · neural networks · CNN

1 Introduction

In recent years, there has been an increasing need for secure and reliable methods of authentication. Handwritten signatures are a flawed and not reliable way of authenticating oneself, as they can easily be forged. Due to the development of technologies allowing the analysis of biological traits, which are unique identifying characteristics, such as fingerprints or a scan of a retina, there has been an increasing number of biological based authentication services used throughout the industry.

Based on the data acquisition method, signature's biometric verification methods are typically classified into two categories: static (off-line) and dynamic

© The Author(s), under exclusive license to Springer Nature Switzerland AG 2023
I. Rojas et al. (Eds.): IWANN 2023, LNCS 14135, pp. 79–90, 2023.
https://doi.org/10.1007/978-3-031-43078-7_7

(on-line) signature verification. Off-line signatures solely contain information on the shape of the signature, while on-line signatures also contain details about dynamic features changing as the signature is produced, such as pen velocity, acceleration, angular position and pressure. Off-line signature verification involves scanning a document that contains the signature to get a digital image of it. On the other hand, on-line signature verification makes use of specialized equipment like a digitizing tablet or a pressure-sensitive pen to record the writing motions [3,8].

On-line signature authentication using a biometric pen allows for a less vulnerable to fraud way of analyzing signatures, by dynamically collecting samples from various sensors during the signature verification process, such as pen pressure and acceleration, which can later be used to measure similarities of a model signature and currently examined one.

One of the methods used for on-line signature verification is Dynamic Time Warping (DTW), which is an algorithm used to measure similarities between two signals that can be of different lengths. More detailed description of this method is included in a later part of this paper. Previous research has proposed a verification method based on fixed DTW thresholds determined experimentally on a training set [12]. Our goal was to test a hypothesis that using neural networks to verify the authenticity of a handwritten signature, parameterized using the Dynamic Time Warping method (DTW), improves the effectiveness of verification compared to a classifier based on fixed thresholds.

In this paper, we propose a method for verifying the authenticity of a handwritten signature using a combination of dynamic time warping (DTW) and neural networks. The DTW method is used to compare the similarity of two biometric signatures and to parametrize them. In next step the neural network is trained to classify the signatures as belonging to a person trying to authenticate themselves or as a forgery attempt based on the DTW similarity score. We performed a statistical analysis of the models performance.

2   Related Works

A Systematic Literature Review (SLR) has been conducted, which revealed a research gap in the topic of the combination of neural networks and DTW. The research of the related works showed a wide range of different approaches used recently to solve the on-line signature verification problem. However, the solution proposed in this article - combining DTW algorithm with neural networks turned out to be less common than it was presumed. Only about 20% of all methods used in this field used a combination of DTW and Machine Learning (ML) algorithms. The most popular ML techniques involve Support Vector Machine and Recurrent Neural Networks, but Feed-forward Neural Networks are also among the most explored ways of dealing with the authentication problem.

During the SLR process it has been discovered that over 46% of the articles found used only the DTW method for signature verification. It was expected that many would use this method, because it was widely used in a similar field - word recognition in the late 1970's [16] and early 1980's [14]. Despite that, one of the first usage of a DTW method for curve matching and signature verification was presented in 1999 [13].

A threshold based DTW signature identification method using signature envelope was presented in [4]. The scheme used basic features such as X, Y coordinates of given signature and was tested on a Japanese handwritten signature dataset. In this approach, the authors developed personalized models, that created a decision boundary based on the maximum and minimum variations of the X and Y signals after DTW method was applied. Although the approach was simple, it managed to outperform previously proposed approaches by achieving an accuracy score of almost 80% and a False Acceptance Rate (FAR) and False Rejection Rate (FRR) scores of 27.35% and 15.18%, respectively.

A method that used the whole DTW matrix in combination with the DTW scores derived from comparing two signatures was proposed in [18]. Until the work of Sharma et. al. prior works utilized only the DTW scores to authenticate a signature. It has been shown in this paper, that using the fusion of the DTW score and the whole DTW matrix can improve the accuracy of detecting authentic signatures.

Different feature extraction methods have been used in combination with neural networks. A discrete wavelet transform (DWT) was one of the successful ones [5]. From X and Y coordinates, the features of pen movement angles were calculated. Afterwards, every signal was independently transformed by the DWT and combined in a signature feature vector. Then a neural network decided whether or not the signature was authentic. It has been shown that using the DWT with neural networks can lead to 90% success rate, which shows that such combination can be a very powerful tool.

The concept of connecting DTW with neural networks was proposed among others in [21]. A Deep Dynamic Time Warping combines a Siamese Network that extracts a feature sequence from each of the signature signals with a DTW block that aligns the sequences of two inputs. It has been shown that such approach can achieve lower Equal Error Rate (EER) than using only DTW or only a Siamese network.

As an improvement of the previous work the authors took advantage of features that DTW extracts and added Siamese Network to it [22]. The Siamese network was incorporated directly into the DTW algorithm, leading to a novel method called Prewarping Siamese Network. The optimization was done using a local embedding loss. For training the model, four datasets were used: MCYT-100 [15], BiosecurID SONOF [6], and SUSIG visual and blind sub-corpuses [10]. This novel approach resulted in the EER value at around 2.11%.

In [2] the authors decided to fuse the scores of 3 classifiers - Deep Bidirectional Long Short-Term Memory (BiLSTM), Support Vector Machines (SVM) with DTW and SVM with different comparator, proposed in the paper. The signature was recognized as genuine when the sum of the scores of 3 classifiers for genuine signature was higher than for forged. While the separated scores of those classifiers were rather weak, the fusion of them resulted in EER lower than 1% on both SVC2004 [23] and MCYT-100 datasets.

During the SLR process 35 of the 81 found articles did not include DTW at all. One of such approaches was a method described in [11]. The authors were inspired by the latest progress on Recurrent Neural Networks (RNN) and tried to implement it into the problem of signature authentication. However, it requires a relatively large training set and significant amount of computational power. On the other hand the results are promising, getting EER at around 2.37% on SVC-2004 dataset.

Another approach that involved RNN was [20]. Authors tried to combine RNN with a Siamese architecture trained on the BiosecurID dataset. Different training scenarios of authentication problem were considered: skilled forgeries, random forgeries and combination of skilled plus random forgeries. The results were as follows: 5.50% EER for skilled forgeries and 3.00% for random forgeries.

3   Dynamic Time Warping

For the purpose of comparing two signatures made using a biometric pen it was necessary to use an algorithm that could recognize them as belonging to the same person even if the signatures varied in length. Dynamic Time Warping (DTW) is one such algorithm. The result of running a DTW algorithm is a matrix with minimal alignment costs between samples. The path divergence from the diagonal with its costs is distinctive for a given case and because of that it can be used to differentiate between a forgery attempt and authentic signature. The DTW algorithm formula is described below.

Let's assume two signatures F and G:

$$F = f_1, f_2, f_3, ..., f_n \tag{1}$$

$$G = g_1, g_2, g_3, ..., g_m \tag{2}$$

The distance between them can be described as follows:

$$d(f_i, g_j) = |f_i - g_j| \tag{3}$$

The cells in matrix are computed as a cost function:

$$\gamma_{i,j} = d(f_i, g_j) + min(\gamma_{i-1,j-1}; \gamma_{i-1,j}; \gamma_{i,j-1}) \tag{4}$$

The matrix as a whole can be used as a representation of the comparison between two signals. However, it is possible to get similar information from the optimal cost path in a matrix. To get the path the first thing is to find the last element in the matrix and move back to the first element by applying the equations below:

$$w' = \{w_k, w_{k-1}, ..., w(0)\} \quad max(m; n) \le k < m + n + 1 \tag{5}$$

$$w'_l = \begin{cases} (i-1, j-1) & \gamma_{i+1,j+1} = min(\gamma_{i-1,j-1}; \gamma_{i-1,j}; \gamma_{i,j-1}) \\ (i-1, j) & \gamma_{i+1,j} = min(\gamma_{i-1,j-1}; \gamma_{i-1,j}; \gamma_{i,j-1}) \\ (i, j-1) & \gamma_{i,j+1} = min(\gamma_{i-1,j-1}; \gamma_{i-1,j}; \gamma_{i,j-1}) \end{cases} \tag{6}$$

One of the methods to find out if both signatures are similar, proposed in previous research [12] uses the comparison of the result from DTW algorithm p_s' with given threshold p_{THR}. Thanks to this measurement it is possible to get a degree of similarity in range 0 to 1. It may be done using the equation below:

$$p = \begin{cases} 1, & p_s' < p_{THR} \\ \frac{p_{THR}}{p_s'}, & p_s' > p_{THR} \end{cases} \tag{7}$$

4   Dataset

The first part of our research was to collect proper data. To do so, 33 people were gathered. The participants were 16–30 years old, representing both genders. In the process of gathering data the study group was divided into smaller subsets of 2 to 5 people. Each person from the sub-group submitted 5 model and 5 verification signatures. Then, every other member of the same sub-group tried to forge the person signature 5 times, firstly just after they saw the model signature (random forgery) and secondly after practicing singing for someone else (skilled forgery).

Each such signature consists of 13 different signals which are:

- acceleration measured in all three dimensions
- angular position in all three dimensions
- pressure of the pen
- angle and acceleration with respect to gravity in three dimensions

Moreover, some signatures were removed and not used in the dataset due to biometric pen defect that paused recording the signals in the middle of signature. Finally the data consist of 322 model signatures, 328 verification signatures and 537 forged signatures.

4.1   Data Processing

Each verification and forged signature was compared to corresponding model signature using DTW method. For such pairs of signatures *fastdtw* library [17] was used to calculate accumulated distance and DTW path. We decided to use the independent DTW algorithim [19], therefore calculations were made for each of 13 signals separately.

Instead of calculating the whole matrix at once the *fastdtw* library uses the divide and conquer method to make approximation about the DTW matrix and then performs more detailed calculations in the smaller parts. This implementation of DTW method was compared with one made by authors and with another from *dtw* library [7]. It was the fastest of them in terms of time of execution, while producing fairly accurate results.

Based on the samples of the DTW path we calculated distances between the model and the compared signal samples, which resulted in creation of new feature used in the dataset called *pairwise cost*. Therefore the basic dataset consists of *accumulated DTW distance, DTW path and pairwise cost* for every signal in every model and compared signature pair. All of the features were stored in the .json files.

From that data a labeled dataset has been created. The processed output of DTW method was divided into two classes. Verification and model signatures were labeled as 1 whereas output from forged and model as 0. Subsequently the training set contains 1071 elements of class 0 and 640 elements of class 1.

4.2 Test Set

For each individual one randomly selected verification signature was set aside into test set. Then we generated DTW output of that verification signature combined with every of 5 model individual's signature. Finally, there were five true samples in the test set for each of the 33 test subjects.

The number of falsified samples depended on the type of forgery carried out for the individual. For each person, there was at least one forger (test groups ranged from two to four people) who performed five random forgery attempts, or five random forgeries and five skilled forgery attempts. Therefore, in the test set for some examinees there are five samples of forgeries, and for some there are ten. Since there was a need to represent both random and skilled forgeries in the test dataset, we opted to include both types of forgeries for some participants. In the end, the test set contains 383 samples, 223 of class 0 and 160 of class 1.

4.3 MLP Models Datasets

Using the pairwise cost data the dataset for MLP models has been created. For the first MLP model, average and standard deviation of pairwise cost have been calculated, whereas the average of pairwise cost and the averaged area under the DTW path has been used in the second one.

4.4 CNN Model Dataset

The input of the CNN model is the DTW cost matrix calculated between corresponding sensors from model signatures. All the matrices has been resized to the their average size in dataset. Resizing was necessary due to fixed size of model input to standarize DTW matrices, which were of various sizes, depending on length of signals gathered. We have decided to use linear interpolation to upscale an image or area interpolation to downscale an image. In order to input this to model all matrices are merged into tensor of size $13 \times 270 \times 300$.

5 Models

When creating and training the models the main concern was relative small size of the dataset. Therefore, architectures were rather shallow and simple to train them properly and achieve satisfactory results. In total three models were created: two MLP models with different datasets and a CNN model.

5.1 MLP Models

The first model architecture was based on Multilayer Perceptron (MLP). It consisted of two hidden layers: 64 and 32 neurons respectively with the addition of dropout with probability 30% to prevent overfitting. As an activation function ReLU function was chosen. Model was trained for 100 epochs using Binary Cross Entropy as a loss function and batch size of 64. For optimization ADAM was chosen with the following parameters:

$$lr = 0.001, \beta_1 = 0.9, \beta_2 = 0.999, \epsilon = 1 * 10^{-7}$$

The training process was stopped earlier when in 10 last epochs loss haven't improved. This architecture was trained with two different datasets resulting in two distinct models. When it comes to the first model, dataset with mean and standard deviation of optimal path in DTW matrix, for the second the mean and area under optimal path.

5.2 CNN Model

The second developed model architecture was based on Convolutional Neural Networks. The idea was to treat cost matrices produced by DTW as images and input them to the network to predict if given signature is forged or not. Convolutional stage of the model consists of three convolutional layers with 2 dimensional filters in number 16, 32, 64. For all of them padding was set to *same*, kernel size 3 × 3 and activation function was *ReLU*. After each convolutional layer were pooling layer, performing max pooling operation in 2 dimensions with window size 2 × 2 and padding same. Next stage is consisting of 2 dense layers with sizes 128 and 32. For both of them ReLU was used as an activation function and dropout with probability 20%. The training was performed for 100 epochs with a loss function *Binary Cross Entropy*. Batch size was 32 and the optimizer was chosen to be ADAM with parameters :

$$lr = 0.001, \beta_1 = 0.9, \beta_2 = 0.999, \epsilon = 1 * 10^{-7}$$

During training early stopping was used after 10 epochs without improvement in loss value.

6 Experiments and Statistical Analysis

Given the biometric nature of the solutions proposed in this and prior works [12], and their potential applications in industries such as banking, more informative metrics were elected. Specifically, we evaluated the False Acceptance Rate (FAR), defined as the ratio of the number of forged signatures accepted by the system to the total number of forgeries, and the False Rejection Rate (FRR), defined as the ratio of the number of authentic signatures rejected by the system to the total number of authentic signatures.

Ideally, both of these metrics would be equal to zero, but they are dependent on one another - as one metric's value gets lower the second's tends to get higher. During the evaluation of the results, we have given more attention to the FAR metric, as it carries more weight in applications such as the banking industry, where the goal is to minimize the ratio of forgeries accepted by the system, even at the expense of a higher number of incorrectly rejected authentic samples.

6.1   Model Evaluation and Comparison

The proposed models have been implemented and evaluated on a test set described in paragraph Sect. 4.2. The results of the evaluation, along with the specified metrics, are shown in Table 1.

Table 1. Metrics for the models

	constant threshold	CNN	MLP (std)	MLP (area)
FAR	25.07%	**1.57%**	6.01%	4.96%
FRR	18.80%	**13.58%**	27.15%	24.54%

The results of the evaluation indicate that each of the models proposed in this study significantly outperforms the model based on fixed thresholds in terms of the False Acceptance Rate (FAR) metric. It is noteworthy that the convolutional model, achieves the lowest FAR value of 1.57%. This represents a notable improvement compared to the baseline model. On the other hand, the Multilayer Perceptron (MLP) models, while achieving lower FAR values, exhibit higher False Rejection Rate (FRR) values, which is an undesirable behavior.

In addition to evaluation on the whole test set, the performance of the proposed models was assessed on only random forgery samples and only skilled forgery samples, contained in the test set. The results of this comparison are presented in Table 2:

Table 2. FAR values for random and skilled forgeries

	constant threshold	CNN	MLP (std)	MLP (area)
Random forgeries	20.00%	**0.00%**	0.83%	0.83%
Skilled forgeries	29.17%	**1.67%**	8.33%	4.17%

The assesment results of FAR values indicate that all proposed models outperform the baseline significantly. For random forgeries all FAR values are below 1%, which indicates that all models are secure, reliable and resilient to a forger, who has only seen given signature for a brief moment. As for the skilled forgeries the proposed models still outperform the baseline, but a greater variation can be observed. A noteworthy result is that of a CNN model which FAR value is below 2% even for skilled forgeries.

6.2   Statistical Analysis of the Results

To evaluate the statistical significance of the results of this study, the Cochran's Q test was employed. This non-parametric statistical test is used to determine whether k treatments have identical effects [1]. In this case, the treatments were the performance of the proposed models.

The following null and alternative hypotheses were used:

- null hypothesis (H_0): The performance of all the models is equally effective - the proportion of correct predictions is the same between all models.
- alternative hypothesis (H_1): There is a difference in performance between the models - the proportion of correct predictions in at least one of the models is different.

If the p-value associated with the test statistic is less than a certain significance level (for the purpose of this comparison $\alpha = 0.05$ has been chosen), the null hypothesis can be rejected and it can be concluded that there is sufficient evidence to say the proportion of correct predictions is different for at least one of the models. Cochran's Q test statistic and p-value has been calculated and presented in Table 3.

Table 3. Cochran's Q test results

χ^2	88.32
p-value	$5.03 * 10^{-19}$

There is sufficient evidence to reject the null hypothesis and conclude that there is a difference in performance between the models.

The McNemar post-hoc test has been conducted to examine the statistical significance of differences between pairs of models. The McNemar test is a well-known statistical test for analyzing the statistical significance of differences in classifier performance [9]. This test is also a χ^2 test for goodness of fit that compares the distribution of counts expected under the null hypothesis with the observed counts. The following null and alternative hypotheses were used:

- null hypothesis (H_0): The performance of the two analyzed models is equally effective - the proportion of correct predictions is the same.
- alternative hypothesis (H_1): There is a difference in performance between the models - the proportion of correct predictions is different.

In the same way as in the Cochran's Q test, if the p-value associated with the test statistic is less than a certain significance level (for the purpose of this comparison $\alpha = 0.05$ has been chosen), the null hypothesis can be rejected and it can be concluded that there is sufficient evidence to say the proportion of correct predictions is different for the models.

The McNemar test statistic (χ^2) and p-value have been calculated and the results are presented in Table 4.

Table 4. McNemmar's tests results versus baseline

	vs. CNN	vs. MLP (std)	vs. MLP (area)
χ^2	69.54	8.28	15.67
p-value	$7.48 * 10^{-17}$	$4 * 10^{-3}$	$7.52 * 10^{-5}$

There is sufficient evidence to reject the null hypothesis for the comparison between the baseline model and the models proposed in this paper. It can be concluded that there is a difference in performance between the models.

Table 5. McNemmar's tests within models

	CNN vs. MLP (std)	CNN vs. MLP (area)	MLP vs. MLP (area) vs. (std)
χ^2	44.50	30.56	2.97
p-value	$2.54 * 10^{-11}$	3.24^{-8}	0.08

As can be seen in the Table 5 the only pair of models for which the null hypothesis can not be rejected is the pair of MLP models, so it cannot be concluded with sufficient confidence, that there is a difference between the performance of these models.

7   Conclusion

Combination of neural networks with DTW algorithm can be effective in biometric signature verification and significantly outperforms model based on fixed thresholds. Statistical tests showed significant differences between proposed models and the one based on fixed thresholds. It is worth noting that there is no statistically significant difference between two approaches to MLP model, yet area MLP was slightly better in performance. From three developed models the CNN model displayed the highest accuracy as well as low FAR which is crucial in authentication systems. An increase in performance will require significantly more data to train the model, which will result in better generalization and more robust model. Further works could additionally include gathering better trained forgeries and training the models with them to increase the security level.

References

1. Cochran, W.G.: The comparison of percentages in matched samples. Biometrika **37**(3–4), 256–266 (1950). https://doi.org/10.1093/biomet/37.3-4.256
2. Dhieb, T., Boubaker, H., Njah, S., Ben Ayed, M., Alimi, A.M.: A novel biometric system for signature verification based on score level fusion approach. Multimedia Tools Appl. **81**(6), 7817–7845 (2022). https://doi.org/10.1007/s11042-022-12140-7

3. Doroz, R., Porwik, P., Orczyk, T.: Dynamic signature verification method based on association of features with similarity measures. Neurocomputing **171**, 921–931 (2016). https://doi.org/10.1016/j.neucom.2015.07.026. https://www.sciencedirect.com/science/article/pii/S0925231215010036
4. Durrani, M.Y., Khan, S., Khalid, S.: Versig: a new approach for online signature verification. Cluster Comput. **22**, 7229–7239 (2019)
5. Fahmy, M.M.: Online handwritten signature verification system based on dwt features extraction and neural network classification. Ain Shams Eng. J. **1**(1), 59–70 (2010). https://doi.org/10.1016/j.asej.2010.09.007. https://www.sciencedirect.com/science/article/pii/S2090447910000080
6. Galbally, J., Diaz-Cabrera, M., Ferrer, M.A., Gomez-Barrero, M., Morales, A., Fierrez, J.: On-line signature recognition through the combination of real dynamic data and synthetically generated static data. Pattern Recogn. **48**(9), 2921–2934 (2015)
7. Giorgino, T.: Computing and visualizing dynamic time warping alignments in r: the dtw package. J. Stat. Softw. **31**, 1–24 (2009)
8. Jain, A.K., Griess, F.D., Connell, S.D.: On-line signature verification. Pattern Recogn. **35**(12), 2963–2972 (2002). https://doi.org/10.1016/S0031-3203(01)00240-0. https://www.sciencedirect.com/science/article/pii/S0031320301002400
9. Kavzoglu, T.: Chapter 33 - object-oriented random forest for high resolution land cover mapping using quickbird-2 imagery. In: Samui, P., Sekhar, S., Balas, V.E. (eds.) Handbook of Neural Computation, pp. 607–619. Academic Press (2017). https://doi.org/10.1016/B978-0-12-811318-9.00033-8. https://www.sciencedirect.com/science/article/pii/B9780128113189000338
10. Kholmatov, A., Yanikoglu, B.: Susig: an on-line signature database, associated protocols and benchmark results. Pattern Anal. Appl. **12**, 227–236 (2009)
11. Lai, S., Jin, L., Yang, W.: Online signature verification using recurrent neural network and length-normalized path signature descriptor. In: 2017 14th IAPR International Conference on Document Analysis and Recognition (ICDAR), vol. 1, pp. 400–405. IEEE (2017)
12. Lech, M., Czyżewski, A.: Handwritten signature verification system employing wireless biometric pen. In: Bembenik, R., Skonieczny, Ł, Protaziuk, G., Kryszkiewicz, M., Rybinski, H. (eds.) Intelligent Methods and Big Data in Industrial Applications. SBD, vol. 40, pp. 307–319. Springer, Cham (2019). https://doi.org/10.1007/978-3-319-77604-0_22
13. Munich, M., Perona, P.: Continuous dynamic time warping for translation-invariant curve alignment with applications to signature verification. In: Proceedings of the Seventh IEEE International Conference on Computer Vision, vol. 1, pp. 108–115 (1999). https://doi.org/10.1109/ICCV.1999.791205
14. Myers, C., Rabiner, L., Rosenberg, A.: Performance tradeoffs in dynamic time warping algorithms for isolated word recognition. IEEE Trans. Acoust. Speech Signal Process. **28**(6), 623–635 (1980). https://doi.org/10.1109/TASSP.1980.1163491
15. Ortega-Garcia, J., et al.: Mcyt baseline corpus: a bimodal biometric database. IEE Proc. Vision Image Signal Process. **150**(6), 395–401 (2003)
16. Sakoe, H., Chiba, S.: Dynamic programming algorithm optimization for spoken word recognition. IEEE Trans. Acoust. Speech Signal Process. **26**(1), 43–49 (1978)
17. Salvador, S., Chan, P.: Toward accurate dynamic time warping in linear time and space. Intell. Data Anal. **11**(5), 561–580 (2007)
18. Sharma, A., Sundaram, S.: On the exploration of information from the dtw cost matrix for online signature verification. IEEE Trans. Cybern. **48**(2), 611–624 (2018). https://doi.org/10.1109/TCYB.2017.2647826

19. Shokoohi-Yekta, M., Hu, B., Jin, H., Wang, J., Keogh, E.: Generalizing dynamic time warping to the multi-dimensional case requires an adaptive approach. Citeseer (2015)
20. Tolosana, R., Vera-Rodriguez, R., Fierrez, J., Ortega-Garcia, J.: Exploring recurrent neural networks for on-line handwritten signature biometrics. IEEE Access **6**, 5128–5138 (2018)
21. Wu, X., Kimura, A., Iwana, B.K., Uchida, S., Kashino, K.: Deep dynamic time warping: end-to-end local representation learning for online signature verification. In: 2019 International Conference on Document Analysis and Recognition (ICDAR), pp. 1103–1110 (2019). https://doi.org/10.1109/ICDAR.2019.00179
22. Wu, X., Kimura, A., Uchida, S., Kashino, K.: Prewarping siamese network: learning local representations for online signature verification. In: ICASSP 2019–2019 IEEE International Conference on Acoustics, Speech and Signal Processing (ICASSP), pp. 2467–2471. IEEE (2019)
23. Yeung, D.-Y., et al.: SVC2004: first international signature verification competition. In: Zhang, D., Jain, A.K. (eds.) ICBA 2004. LNCS, vol. 3072, pp. 16–22. Springer, Heidelberg (2004). https://doi.org/10.1007/978-3-540-25948-0_3

SlideGCN: Slightly Deep Graph Convolutional Network for Multilingual Sentiment Analysis

El Mahdi Mercha[1,2]($\boxtimes$), Houda Benbrahim[1], and Mohammed Erradi[1]

[1] ENSIAS, Mohammed V University in Rabat, Rabat, Morocco
elmahdi.mercha@um5s.net.ma
[2] HENCEFORTH, Rabat, Morocco

Abstract. Multilingual sentiment analysis refers to the process of sentiment scoring while gathering insights from data in different languages. Many research studies have been conducted to perform multilingual sentiment analysis. However, most of these studies focus on the short-distance semantics which consists in modeling local consecutive word sequences. In this work, we consider the global word co-occurrence in the whole corpus, which capture both short- and long-distance semantics, to convey more meaningful insights for the analysis. We propose an approach called MSA-GCN (Multilingual Sentiment Analysis based on Graph Convolutional Network) while supporting both short- and long-distance semantics. We build a single heterogeneous text graph for a multilingual corpus based on sequential, semantic, and statistical information. Then, a slightly deep graph convolutional network learns embeddings for all nodes in a semi-supervised manner. Extensive experiments are carried out on various datasets, and the results demonstrate the effectiveness of the proposed approach.

Keywords: Multilingual sentiment analysis · Graph convolutional network · Deep learning · Natural language processing

1 Introduction

Sentiment analysis is a typical problem of text classification in natural language processing (NLP). Several studies have been conducted for sentiment analysis, due to its significance in retrieving insightful information for decision making. Even if the majority of the early studies focused on developing single-language systems, attention has recently shifted to creating multilingual systems. Both classic machine learning and deep learning methods have been widely explored in multilingual sentiment analysis (MSA) [2,23]. Previously, classic machine learning methods investigated a range of traditional classifiers with various feature extraction methods, such as bag-of-words. Recently, with the development of language modeling techniques, namely word embeddings, deep learning became a dominant paradigm in sentiment analysis. Several word embedding methods

© The Author(s), under exclusive license to Springer Nature Switzerland AG 2023
I. Rojas et al. (Eds.): IWANN 2023, LNCS 14135, pp. 91–103, 2023.
https://doi.org/10.1007/978-3-031-43078-7_8

have been developed to encode semantic and syntactic information and other linguistic patterns, such as Word2Vec [18] and fastText [8]. Some dominating architectures, particularly long short-term memory (LSTM) [22] and convolutional neural network (CNN) [10], made extensive use of such methods to learn text representations. However, these architectures capture the short-distance semantics in local consecutive word sequences, but they miss global word co-occurrence in the whole corpus, which conveys the long-distance semantics [19].

Over the last few years, a new approach known as graph neural network (GNN) have gained considerable attention because of the great expressive power of graphs [4]. Motivated by the success of the methods developed for representation learning [18], graph representation learning methods have been invented to overcome the limitations of the traditional approaches. These methods aim to combine information from the graph topologies and attributes into low-dimension embeddings to serve different tasks.

In this work, we propose an approach for Multilingual Sentiment Analysis using Graph Convolutional Network (MSA-GCN). Inspired by the success of TextGCN [24], we construct a single heterogeneous text graph to model the entire data of a multilingual corpus. The words and documents serve as nodes in the constructed graph. The relationships between these nodes are built based on semantic, sequential, and statistical information. We propose SlideGCN (SLIghtly DEep Graph Convolutional Network), which is a variant of the vanilla graph convolutional network (GCN) [12], to model the graph. The task of MSA is then considered as a node classification problem. The proposed approach can learn insightful representations for documents, and it achieves strong classification performance without using external sentiment information. Furthermore, the experiments show the prominent results of the proposed approach in a variety of language combinations, indicating its robustness against language variation. To the best of our knowledge, this is the first comprehensive study which explore graph neural network for multilingual sentiment analysis. The main contributions are summarized as follows:

1. This work proposes a novel method to automatically represent the whole multilingual corpus with a single heterogeneous text graph, based on semantic, sequential and statistical information.
2. It proposes a new model, named SlideGCN, to model the heterogeneous graph and learn predictive representations.
3. Extensive experiments are conducted on a variety of language combinations to reveal the efficiency and the robustness of the general MSA-GCN approach against language variation.

2   Related Works

Multilingual sentiment analysis has recently received a lot of interest, and various studies have been conducted [17]. These studies focused on performing language-independent and translation-free systems capable of learning meaningful information directly from multilingual data.

Intensive work have been proposed based on machine learning and deep learning with different word representations. Abudawood et al. [2] investigated different classical machine learning classifiers with various hand-crafted features, such as bag-of-words and n-grams. Medrouk et al. [16] investigated CNN and LSTM in the multilingual problem without focusing on special features. Shakeel et al. [20] suggested a hybrid architecture based on the CNN and LSTM models to learn both long-term dependencies and n-gram features. Jaballi et al. [7] proposed a bidirectional LSTM network preceded by a preprocessing stage which encapsulate contextual information from multilingual feature sequences.

Although word-based representations are the extensively utilized approaches for handling MSA, some studies have suggested character-based approaches. Almost all character-based approaches rely on CNN to create efficient architectures that learn insightful information from character features [23].

GNNs can be used to model relationships between words or documents, which can lead to improved performance on tasks such as text classification [12,24], and sentiment analysis [13]. There are several graph representation learning methods that have been developed for NLP tasks, including GCN [12], and graph attention networks [21], among others. These methods vary in their architecture and the way they propagate information across the graph, but they all aim to learn meaningful representations of graph-structured data.

3   The Proposed Method

3.1   Heterogeneous Text Graph Construction

We construct a single heterogeneous text graph to model the entire multilingual corpus by considering words and documents as nodes. Two types of edges are built between nodes: word-document edges and word-word edges. On one hand, the word-document edges are built based on statistical information. The weight of the edge between word node and document node is the *term frequency-inverse document frequency*, which reflects the importance of a word to a document in a corpus. On the other hand, we define the word-word edges based on two variants of language properties of information: sequential and semantic information. To create the sequential relations, we use the sliding window strategy to collect statistics on word co-occurrence throughout the whole corpus. The weight between two word nodes is then determined based on *positive pointwise mutual information* (PPMI). Formally, the weight of a word pair w_i, w_j is computed as follows:

$$PPMI(w_i, w_j) = max\left(log\frac{p(w_i, w_j)}{p(w_i)p(w_j)}, 0\right) \tag{1}$$

$$p(w_i, w_j) = \frac{\#W(w_i, w_j)}{\#W} \tag{2}$$

$$p(w_i) = \frac{\#W(w_i)}{\#W} \tag{3}$$

where $\#W(w_i, w_j)$ is the number of sliding windows in which both words w_i and w_j co-occurred, $\#W(w_i)$ is the number of sliding windows containing the word w_i, and $\#W$ is the total number of the sliding windows in the whole corpus.

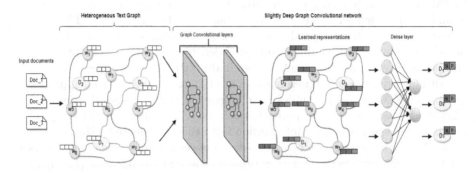

Fig. 1. The architecture of our proposed MSA-GCN approach. A heterogeneous text graph is constructed based on sequential, semantic, and statistical information to model the whole corpus. Then, SlideGCN learns effective graph node representation to identify the sentiment class of each document.

Building the word-word relations using just sequential information may results in a disconnected graph. Therefore, we adopt semantic information to bridge the gap across languages and build a single linked graph. The semantic edge weights are obtained based on the similarity of multilingual word embeddings aligned in a single vector space. Therefore, we use the fastText embeddings that have been aligned in a single space [5]. However, preliminary experiments reveal that merging directly the sequential and semantic edge weights lead to worst results, due to that the edge weights come from different distributions. So, instead of directly using the cosine similarity, we propose to use the exponential of a scaled cosine similarity, which significantly increases the values of semantic weights and enhance the importance of transfer learning across languages. Formally, the weight of semantic edge of a word pair w_i, w_j is calculated as follows:

$$weight_{semantic} = \exp(\alpha . cosine\_sim(w_i, w_j)) \qquad (4)$$

$$cosine\_sim(w_i, w_j) = \frac{x_i . x_j}{\|x_i\| \|x_j\|} \qquad (5)$$

where α is a parameter for scaling the similarity between a pair of words, and x_i, x_j denote respectively the aligned word embeddings of the words w_i and w_j. To reduce the complexity, we use bilingual dictionaries to generate a set of linkages across languages rather than computing the similarity between the vocabulary of different languages to select the most similar words.

3.2   Slightly Deep Graph Convolutional Network

The vanilla graph convolutional network [12] is a graph neural network which follows a message-passing mechanism. GCN operates directly on a graph to learn nodes representations by aggregating the feature information from their topological neighbors. Formally, consider an attributed graph $\mathcal{G} = (\mathcal{V}, \mathcal{E}, A)$ where $\mathcal{V}$ is the set of $n = |\mathcal{V}|$ nodes, $\mathcal{E}$ is the set of edges, and $A \in \mathbb{R}^{n \times n}$ is the adjacency matrix of $\mathcal{G}$. For each connected pair of words w_i, w_j, the $A(i,j)$ is the weight of the edge; otherwise, $A(i,j) = 0$. In addition, each node is assumed to be connected to itself, i.e., $\forall w_i \in \mathcal{V}; (w_i, w_i) \in \mathcal{E}$ and $A(i,i) = 1$. The normalized symmetric adjacency matrix of A is $\tilde{A} = D^{-\frac{1}{2}} A D^{-\frac{1}{2}}$, where D denotes the degree matrix of A, and $D_{ii} = \Sigma_j A_{ij}$. Let $X \in \mathbb{R}^{n \times d}$ be a feature matrix containing the embedding vectors of all nodes of $\mathcal{G}$, where d denotes the dimension of the embedding vectors and each row x_i of the matrix represents the embedding vector associated to the node i. The vanilla GCN model $f(X, A)$ learns the nodes embeddings based on two-layers as follows:

$$f(X, A) = softmax(\tilde{A} \; ReLU(\tilde{A} X W^{(0)}) W^{(1)}) \tag{6}$$

here $W^{(0)} \in \mathbb{R}^{d \times h}$ and $W^{(1)} \in \mathbb{R}^{h \times c}$ are the weight matrices of the first and second layer, respectively, with h denotes the dimension of the hidden representation, and c is the number of classes. The activation functions are defined as follows:

$$ReLU(x) = max(x, 0) \tag{7}$$

$$softmax(x_i) = \frac{\exp(x_i)}{\Sigma_j \exp(x_j)} \tag{8}$$

Inspired by the success of the vanilla GCN, we propose a SlideGCN which learns deeper and better nodes representations. We reformulate the vanilla GCN by increasing the depth of the architecture, changing the activation function, and adding a single feed forward neural network layer. We adopt $tanh$ instead of ReLU, because it makes the learning easier, as it is zero centered and its values lie between -1 to 1. However, the limitation faced by ReLU is the dying ReLU problem which decreases the learning ability of the model. Formally, our forward SlideGCN model is described as follows:

$$H^{(1)} = tanh(\tilde{A} H^{(0)} W^{(0)}) \tag{9}$$

$$H^{(2)} = tanh(\tilde{A} H^{(1)} W^{(1)}) \tag{10}$$

$$H^{(3)} = softmax(H^{(2)} W^{(2)} + b) \tag{11}$$

where $W^{(0)} \in \mathbb{R}^{d \times h}$ and $W^{(1)} \in \mathbb{R}^{h \times h}$ denote the weights of the first and second layer, respectively, $W^{(2)} \in \mathbb{R}^{h \times c}$ is the weight matrix of the dense layer and $H^{(0)} = X$. $H^{(i)}$ is the learned representations of nodes in the layer i. The $tanh$ activation function, is defined as:

$$tanh(x) = \frac{\exp(x) - \exp(-x)}{\exp(x) + \exp(-x)} \tag{12}$$

To measure the classification performance of the SlideGCN, we adopt the cross-entropy loss across all labeled documents:

$$\mathcal{L} = - \sum_{d \in \mathcal{Y}_D} \sum_{c=1}^{C} Y_{dc} ln H_{dc}^{(3)} \tag{13}$$

where $\mathcal{Y}_D$ is the set of labeled document indices, and C is the number of classes. The overall MSA-GCN approach is illustrated in Fig. 1.

4 Experiments

To evaluate the performance of the proposed MSA-GCN approach, we conduct several experiments, which aim to discover:

- How the MSA-GCN approach achieves accurate results in different datasets with a variety of languages.
- How the SlideGCN model can learns effective graph node representation.

4.1 Datasets

During the experimental analysis, we adopt four datasets to evaluate the MSA-GCN approach namely, multilingual amazon reviews corpus (MARC) [9], Internet Movie Database (IMDB) [15], Allociné [3], and Muchocine [1].

- **MARC** is a large-scale collection of amazon reviews. It contains reviews from 6 languages: English, Japanese, German, French, Chinese, and Spanish. For each language, there are 200000, 5000, and 5000 reviews for training, validation and test sets respectively.
- **IMDB** is a binary sentiment classification dataset in English. It is made up of 25k highly polar movie reviews for training and another 25k for testing.
- **Allociné** is a sentiment analysis dataset in French. It contains 100k positive and 100k negative movie reviews.
- **Muchocine** is sentiment analysis dataset in Spanish. It has 3872 longform movie reviews, each with a 1–5 scale rating.

Due to the syntactic structural disparities across the 6 languages, we consider reviews from 4 languages namely, English, German, French, and Spanish. Furthermore, we reduce the problem to binary classification. So, we assigned 1 and 2 stars to negative class, 4 and 5 stars to positive class, and 3 stars to neutral class. Then, we construct 6 datasets for amazon reviews based on different language combinations (EN-FR, EN-ES, EN-DE, FR-ES, FR-DE, ES-DE), and one for movie reviews by combining the three movie reviews datasets. Table 1 shows the statistics of each dataset utilized in this study.

Table 1. Statistics of both movie reviews dataset and amazon reviews dataset.

Dataset	Training set	Validation set	Test set
Amazon reviews (single combination)	20000	8000	8000
Movie reviews	10646	2138	2138

4.2 Baselines

We compare the MSA-GCN approach against several baselines and state-of-the-art methods. The baselines include:

- **Bi-LSTM Emb-Non-Static (Bi-LSTM-NS)**: a bi-directional LSTM [14] with trainable word embeddings. Bi-LSTM represents the whole text with the last hidden state. The embeddings are initialized randomly and fine tuned during the training process.
- **CNN Emb-Non-Static (CNN-NS)**: a convolutional neural network [10] with trainable word embeddings. The convolution is applied over the randomly initialized word embeddings to learn the word and text representations.
- **Bi-LSTM Emb-Static (Bi-LSTM-S)**: a bi-directional LSTM [14] with randomly initialized word embeddings. The word embeddings are not optimized during the training process.
- **CNN Emb-Static (CNN-S)**: a convolutional neural network [10] with randomly initialized word embeddings. The word embeddings are not updated throughout the training process.
- **Char-CNN**: a character-level convolutional network for text classification [25]. The designed architecture extracts efficient representation based on the character features.
- **fastText**: a simple baseline method for text classification [8]. The method averages the word features to construct efficient text representation, that will be used for learning a linear classifier.
- **Text level GCN (TL-GCN)**: text level graph neural network for text classification [6]. The method use message passing mechanism to learn representation for each input text modeled with a graph.
- **Graph-GCN**: the vanilla GCN [12], applied to the heterogeneous text graph constructed as described in the Subsect. 3.1.

4.3 Experiment Settings

In our experiments, we apply some common language-independent preprocessing techniques, especially, deleting URLs, lowering, eliminating special characters, and removing HTML tags. The text was then tokenized using a single space. We conduct a set of experiments to select the best hyper-parameters for the proposed approach. We remove low frequency terms that appeared less than 5 times. We investigate different sizes of sliding window (see Fig. 2) and we set the size of the sliding window to 25, as the small window size may not provide enough global word co-occurrence information, whereas large window size could introduce extra edges between words that are not closely connected. As observed in [24], our preliminary experiments reveal that small changes in the window size does not affect much the results. The identity matrix is used to initialize the word embedding, so each node is represented by a one-hot vector. In the semantic edges construction, we investigated several values of α, and we judge that 7 is the best value to select (See Fig. 3). Based on the conducted hyper-parameter tuning, in both layers of SlideGCN, we define the dimension of the node embedding as 200. The SlideGCN is trained for a maximum of 200 epochs by adopting the Adam optimizer [11], with a learning rate of 0.002. We stop the training if the validation loss does not decrease for five consecutive epochs.

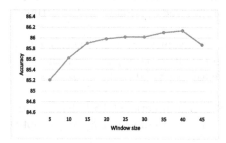

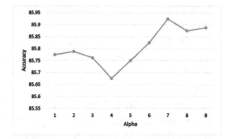

Fig. 2. Test accuracy of MSA-GCN on the EN-FR dataset based on different sliding window sizes

Fig. 3. Test accuracy of MSA-GCN on the EN-FR dataset based on different alpha values

For the baseline models, we use the default hyper-parameter for fastText, TL-GCN, and Char-CNN as described in their original papers. However, we performed hyper-parameter tuning to choose the best hyper-parameter for Bi-LSTM-NS and CNN-NS. So, the results were 100 for the embedding dimension and number of units for Bi-LSTM-NS, 0.5 for dropout, and L_2 for the layer weight regularizer with a weight of 0.01. Also, 100 for the embedding dimension and number of filters for CNN-NS, while the kernel size is 3. We use the same settings for Bi-LSTM-S, CNN-S as Bi-LSTM-NS and CNN-NS respectively. All four models are trained using the Adam optimizer with a learning rate of 0.001. For all baseline models we utilize randomly initialized word embedding rather than pre-trained word embedding.

4.4 Results Analysis

A comprehensive experiment is conducted on several datasets. We evaluate all models 30 times, and we report the mean $\pm$ standard deviation. Tables 2 and 3 report the results of our approach against the baseline models. The achieved results reveals that the proposed MSA-GCN approach significantly outperforms all the baselines (p-value < 0.05 based on student t-test) on 6 datasets over 7, which prove the effectiveness of the proposed approach.

For more detailed performance analysis, we observe that the proposed MSA-GCN achieves best performance in almost all datasets compared with all baselines. Char-CNN achieved the worst results in almost all datasets. This is owing to fact that encoding multilingual texts with just feature characters does not allow the learning of useful information due to the syntactic discrepancy across languages. Bi-LSTM-S and CNN-S models achieved modest results. This is owing to the significant dependency between the learned document representation and the word embeddings. Random word embeddings, on the other hand, do not capture any syntactic or semantic information, making representation learning from them extremely difficult. Also, we observe that FastText achieved modest results in all datasets. This observation demonstrates the limitations of learning insightful document representation for MSA based on simple average of word embeddings. CNN-S performs much better than Bi-LSTM-S, due to its capability to extract local and position-invariant features which are critical for sentiment analysis. Bi-LSTM-NS and CNN-NS obviously beat Bi-LSTM-S and CNN-S,

Table 2. Test accuracy on different multilingual amazon reviews datasets.

Model	EN-FR	EN-ES	EN-DE	FR-ES	FR-DE	ES-DE
Bi-LSTM-NS	84.99 ± 0.72	84.60 ± 1.09	84.97 ± 1.07	86.12 ± 0.97	86.51 ± 0.44	85.76 ± 1.15
CNN-NS	84.72 ± 0.28	84.50 ± 0.18	85.05 ± 0.22	85.87 ± 0.37	86.33 ± 0.25	86.06 ± 0.28
Bi-LSTM-S	64.42 ± 2.41	65.01 ± 1.84	61.83 ± 2.02	67.11 ± 2.40	64.55 ± 2.12	65.24 ± 2.86
CNN-S	73.52 ± 0.69	73.26 ± 0.98	72.89 ± 0.86	75.49 ± 0.51	75.00 ± 0.72	75.07 ± 0.77
Char-CNN	59.67 ± 4.88	55.03 ± 2.26	55.04 ± 1.43	58.55 ± 5.33	56.30 ± 2.48	56.16 ± 2.16
fastText	65.76 ± 4.06	63.82 ± 3.80	64.09 ± 3.81	65.68 ± 3.68	65.96 ± 5.88	66.33 ± 4.18
TL-GCN	83.73 ± 0.14	83.88 ± 0.28	84.52 ± 0.10	84.56 ± 0.22	85.38 ± 0.17	85.58 ± 0.12
Graph-GCN	84.50 ± 0.16	82.69 ± 6.17	84.66 ± 0.16	84.86 ± 0.91	85.62 ± 0.14	85.26 ± 0.13
MSA-GCN	**85.79 ± 0.12**	**85.00 ± 0.15**	**85.68 ± 0.12**	86.29 ± 0.13	**87.03 ± 0.13**	**86.48 ± 0.15**

Table 3. Test accuracy on the multilingual movie review dataset.

Model	Accuracy
Bi-LSTM-NS	78.29 ± 1.57
CNN-NS	79.78 ± 0.92
Bi-LSTM-S	53.24 ± 0.05
CNN-S	63.99 ± 1.40
Char-CNN	54.02 ± 0.57
fastText	67.68 ± 3.66
TL-GCN	82.86 ± 1.08
Graph-GCN	87.55 ± 0.26
MSA-GCN	**88.14 ± 0.26**

respectively, since they can learn valuable words representations in a supervised manner.

The graph-based methods TL-GCN and Graph-GCN perform quite well, and show competitive performances, indicating the efficacy of the graph-based approaches. This suggests the efficiency of modeling the entire multilingual corpus with a single graph and its ability to capture the relations across words and documents. Also, modeling each document with a graph allow to capture the relations between words by learning more expressive edges. The proposed MSA-GCN approach consistently outperforms TL-GCN in all datasets, indicating the effectiveness of the way proposed for constructing the graph from the entire multilingual corpus. We also observed that MSA-GCN outperforms Graph-GCN, which prove the performance of the proposed SlideGCN to learn efficient representation for words and documents better than the vanilla GCN.

4.5   Ablation Study

We conduct ablation studies to further investigate our approach, and the results are shown in Figs. 4, 5.

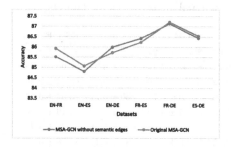 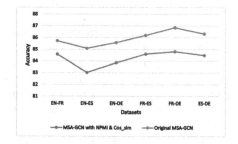

Fig. 4. Test accuracy of the original MSA-GCN and MSA-GCN without semantic edges on 6 datasets.

Fig. 5. Test accuracy of the original MSA-GCN and MSA-GCN with normalized word-word edges.

In the first ablation, we compare the original MSA-GCN constructed as detailed in Sect. 3, with MSA-GCN without building the semantic edges. Figure 4 shows that eliminating the semantic edges makes MSA-GCN perform slightly worse on 4 of the 6 datasets, which highlights the importance of semantic edges. This is supported by the fact that the semantic edges help to bridge the gap between languages.

In the second ablation study, we use the normalized pointwise mutual information (NPMI) to calculate the weights of the sequential edges. Also, we utilize the cosine similarity to compute the weights of the semantic edges. From Fig. 5, we can observe that MSA-GCN with weights calculated with NPMI and cosine similarity achieve worst results compared with the original MSA-GCN. This indicate the strength of our proposed method (see Eq. 4) to compute the weights of semantic edges against the normalized weights.

4.6 Discussion

The experimental results show that the proposed MSA-GCN approach can learn predictive word and document representations, as well as achieve high multilingual sentiment analysis performance. There are two major factors which can explain the supremacy of the MSA-GCN. First, the proposed method for modeling the entire corpus with a single heterogeneous graph based on the semantic, sequential, and statistical information. The constructed graph can capture the global information about words relations in the whole corpus, as well as allows the propagation of the sentimental information between documents based on the word's connections. Also, it bridges the gap between languages through the semantic linkages. Second, the SlideGCN allows to learn a slightly deep predictive embeddings for both words and documents by propagating the feature information from surrounding nodes. This considerably increases the multilingual sentiment analysis performance.

From the achieved results, we can see that Bi-LSTM-NS achieves competitive results to MSA-GCN in all MARC datasets. This is owing to the fact that Bi-LSTM explicitly model consecutive word sequences in both directions, whereas MSA-GCN do not consider word ordering that are extremely important in sentiment analysis.

Even though the MSA-GCN produced efficient results and outperformed the state-of-the-art model, it has limitations. On one hand, MSA-GCN builds a single graph for the entire corpus, meaning that the size of the graph scaled with the corpus size, resulting in excessive memory usage in big corpora. On the other hand, MSA-GCN is intrinsically transductive, therefore it cannot support the online testing, as the structure of the graph and its parameters depend on the initial corpus used for training.

5 Conclusion and Future Works

We have introduced an approach, namely MSA-GCN, for multilingual sentiment analysis. The proposed approach starts with modeling the entire multilingual corpus with a single heterogeneous graph in which words and documents serve as nodes. Then, it learns representations for nodes based on the proposed model, namely SlideGCN. Experiments on different datasets with a variety of languages show that MSA-GCN is capable of learning predictive representations for multilingual sentiment analysis. In addition, the achieved results reveal the robustness of MSA-GCN against language variation. In this setting, this approach significantly outperforms various recently proposed methods. However, a limitation of MSA-GCN is that it is intrinsically transductive, therefore it cannot support the online testing, as the structure of the graph and its parameters depend on the initial data used for training. Also, the size of the graph scaled with the size of data, which results in excessive memory usage in big data.

Future work concerns the generalization of MSA-GCN to inductive settings.

Acknowledgements. The authors would like to thank Dr.Mohammed Amine Koulali for his fruitful discussions and comments on the earlier versions of this paper.

References

1. Spanish movie reviews. http://www.lsi.us.es/fermin/index.php/Datasets
2. Abudawood, T., Alraqibah, H., Alsanie, W.: Towards language-independent sentiment analysis. In: 2018 21st Saudi Computer Society National Computer Conference (NCC), pp. 1–6. IEEE (2018)
3. Blard, T.: French sentiment analysis with bert. GitHub repository. https://github.com/TheophileBlard/french-sentiment-analysis-with-bert
4. Cai, H., Zheng, V.W., Chang, K.C.C.: A comprehensive survey of graph embedding: problems, techniques, and applications. IEEE Trans. Knowl. Data Eng. **30**(9), 1616–1637 (2018)
5. Conneau, A., Lample, G., Ranzato, M., Denoyer, L., Jégou, H.: Word translation without parallel data. arXiv preprint arXiv:1710.04087 (2017)
6. Huang, L., Ma, D., Li, S., Zhang, X., Wang, H.: Text level graph neural network for text classification. In: EMNLP-IJCNLP, pp. 3444–3450 (2019)
7. Jaballi, S., Zrigui, S., Sghaier, M.A., Berchech, D., Zrigui, M.: Sentiment analysis of tunisian users on social networks: overcoming the challenge of multilingual comments in the Tunisian dialect. In: Computational Collective Intelligence: 14th International Conference, ICCCI 2022, Hammamet, Tunisia, September 28–30, 2022, Proceedings, pp. 176–192. Springer (2022). https://doi.org/10.1007/978-3-031-16014-1_15
8. Joulin, A., Grave, E., Bojanowski, P., Mikolov, T.: Bag of tricks for efficient text classification. In: Proceedings of the 15th Conference of the European Chapter of the Association for Computational Linguistics: Volume 2, Short Papers, pp. 427–431, April 2017
9. Keung, P., Lu, Y., Szarvas, G., Smith, N.A.: The multilingual Amazon reviews corpus. In: EMNLP, pp. 4563–4568, November 2020
10. Kim, Y.: Convolutional neural networks for sentence classification. In: EMNLP, pp. 1746–1751 (2014)
11. Kingma, D.P., Ba, J.: Adam: a method for stochastic optimization. In: Bengio, Y., LeCun, Y. (eds.) ICLR (2015)
12. Kipf, T.N., Welling, M.: Semi-supervised classification with graph convolutional networks. arXiv preprint arXiv:1609.02907 (2016)
13. Liao, W., Zeng, B., Liu, J., Wei, P., Cheng, X., Zhang, W.: Multi-level graph neural network for text sentiment analysis. Comput. Electr. Eng. **92**, 107096 (2021)
14. Liu, P., Qiu, X., Huang, X.: Recurrent neural network for text classification with multi-task learning. In: Proceedings of the Twenty-Fifth International Joint Conference on Artificial Intelligence, pp. 2873–2879 (2016)
15. Maas, A.L., Daly, R.E., Pham, P.T., Huang, D., Ng, A.Y., Potts, C.: Learning word vectors for sentiment analysis. In: Proceedings of the 49th Annual Meeting of the Association for Computational Linguistics: Human Language Technologies, pp. 142–150 (2011)
16. Medrouk, L., Pappa, A.: Do deep networks really need complex modules for multilingual sentiment polarity detection and domain classification? In: IJCNN, pp. 1–6 (2018)
17. Mercha, E.M., Benbrahim, H.: Machine learning and deep learning for sentiment analysis across languages: a survey. Neurocomputing **531**, 195–216 (2023)

18. Mikolov, T., Chen, K., Corrado, G., Dean, J.: Efficient estimation of word representations in vector space. In: ICLR 2013, Workshop Track Proceedings
19. Peng, H., et al.: Large-scale hierarchical text classification with recursively regularized deep graph-CNN. In: Proceedings of the 2018 World Wide Web Conference, pp. 1063–1072 (2018)
20. Shakeel, M.H., Faizullah, S., Alghamidi, T., Khan, I.: Language independent sentiment analysis. In: AECT, pp. 1–5 (2020)
21. Veličković, P., Cucurull, G., Casanova, A., Romero, A., Lio, P., Bengio, Y.: Graph attention networks. arXiv preprint arXiv:1710.10903 (2017)
22. Wang, X., Liu, Y., Sun, C.J., Wang, B., Wang, X.: Predicting polarities of tweets by composing word embeddings with long short-term memory. In: Proceedings of the 53rd Annual Meeting of the Association for Computational Linguistics and the 7th International Joint Conference on Natural Language Processing (Volume 1: Long Papers), pp. 1343–1353 (2015)
23. Wehrmann, J., Becker, W., Cagnini, H.E., Barros, R.C.: A character-based convolutional neural network for language-agnostic twitter sentiment analysis. In: IJCNN, pp. 2384–2391 (2017)
24. Yao, L., Mao, C., Luo, Y.: Graph convolutional networks for text classification. In: Proceedings of the AAAI Conference on Artificial Intelligence, vol. 33, pp. 7370–7377 (2019)
25. Zhang, X., Zhao, J., LeCun, Y.: Character-level convolutional networks for text classification. In: Proceedings of the 28th International Conference on Neural Information Processing Systems - Volume 1, pp. 649–657 (2015)

Toward Machine's Artificial Aesthetic Perception: *Could Machines Appreciate the Beauty?*

Mohand Tahar Soualah, Fatemeh Saveh, and Kurosh Madani[✉]

Université Paris-Est Créteil, LISSI Laboratory EA 3956, Senart-FB Institute of Technology, Campus de Senart, 36-37 Rue Charpak – F-77567, Lieusaint, France
{mohand-tahar.soualah,fatemeh.saveh,madani}@u-pec.fr

Abstract. To make cohabit humans and robots in the same living space, robots have not only to develop rational awareness but also acquire emotional awareness regarding their surrounding environment. Investigating *Emotional Machine-Awareness* we tend to provide response-elements to the following question: "likewise to human's emotional awareness, could the machine acquire *Artificial Aesthetic Perception* proffering it kind of emotional esteem of its environment?" In other words, *could the Machine become aware of the cuteness*?

In this paper we present a computational model of Artificial Aesthetic Perception, founded on art-philosophy's and aesthetics psychology's basements. This allows us, on the one hand, to develop a model based on human being's intellectual and visual mechanisms, and on the other hand, makes the investigated model comprehensive (interpretable) with regard to the subjectivity inherent in the notion of "beauty". Experimental results obtained using two datasets of various art-works (Wiki-Art and a home-made human's eye-fixation based with emotions' annotation) show the pertinence of the investigated model firming up cognitive nature of the presented computational approach.

Keywords: *Aesthetic Perception* · Machine-Awareness · Machine-Learning · Philosophy

1 Introduction and Paradigm's Statement

If the main challenge of robotics during the 19th century has consisted of automating repetitive tasks following needs of industrial air for the massive industrialization of the production, and then, sophistication of these machines through digitization (computerization) of robots throughout the 20th century, the challenge of robotics in the current century will be to make cohabit humans and robots in the same living space. In fact, occupying increasingly diverse roles in humans' domestic life, robots are, and will, growingly share the human's "life space", cooperating and evolving with him within his complex environment. Within such context, and in order to equate human's ability in keeping up in complex environment and interacting with other humans, robots have

© The Author(s), under exclusive license to Springer Nature Switzerland AG 2023
I. Rojas et al. (Eds.): IWANN 2023, LNCS 14135, pp. 104–117, 2023.
https://doi.org/10.1007/978-3-031-43078-7_9

to develop advanced and as human-like as possible artificial wisdom (artificial intelligence): thus, not only bring in rational awareness but also acquire emotional awareness regarding their surrounding environment proffering them autonomy alike to the human's self-sufficiency.

By "emotional awareness" we intend the skill of feeling or the talent of emotions-based appreciation of the perceived items. In fact, alongside his lucid alertness human being develops an emotional responsiveness proffering him an affective reception (appreciation) of perceived items: *"The heart has its reasons that reason ignores"* (Blaise Pascal[1]). What is in fact interesting is that, Blaise Pascal doesn't oppose the rational to the emotional but considers the heart (emotional) and the reason (rational) as two levels of a single faculty of the human's intellect (intelligence). The aesthetic esteem (admiration or appreciation) of visual scenery (landscape, paintings, sculptures or other ranges of artworks) is a typical example of such emotional awareness.

Concerning intelligence and autonomy, in his two philosophical essays (*Critique of Pure Reason* [1] and *Critique of Practical Reason* [2]) Immanuel Kant[2] links up through his concept of *Knowledge Theory* the two above-mentioned paradigms. Kant establishes that knowledge (intelligence) requires, on the one hand, sensitivity (emotionality) that he defines as the faculty of shaping intuitive representations and on the other hand, the understanding (rationality) that he describes as the faculty of forming concepts and applying them to the constructed intuitions. According to Kant's philosophical basements, the so-called *Knowledge* kits out clever-being (symbolized by what Kant denote by *Reasonable Being*) self-sufficiency of intellectual resilience (autonomy).

Investigating *Emotional Machine-Awareness*, we tend to contributing in answering the following question: "likewise to human's emotional awareness, could the machine acquire *Artificial Aesthetic Perception* (AAP) proffering it kind of emotional esteem of its environment?" In other words, *could the Machine appreciate the beauty?* Answer to the aforementioned question as well as elucidation of the related paradigm are far from being trivial and remain trifling tasks. Actually, the notion of the beauty deals with subjective aspects coping with philosophical, cultural and psychological concepts. For this reason and aiming in achievement of a computational model of AAP, we found the investigated model on art-philosophy's and psychology's basements: art-philosophy to handle the subjective aspects relating aesthetic (beauty) and behavioral-psychology for emotions' modeling. Such way of doing allows us, on the one hand, to develop our model on the basis of the concepts relating human being's intellectual mechanisms, and on the other hand, makes the investigated model comprehensive (interpretable) with regard to the subjectivity inherent in the notion of "beauty".

The paper is organized in five sections. The next section gives a brief state-of-art relating the fields involved in the present work. Section 3 describes the investigated AAP model giving its detailed operational structure. Section 4 deals with the implementation of the AAP model providing the experimental results. Finally Sect. 5 concludes the paper.

---

[1] Blaise Pascal: French mathematician, physicist, inventor and philosopher (1623 – 1662).

[2] Immanuel Kant: Prussian philosopher (1724 – 1804).

2 Brief State-of-Art Relating the Involved Issues

2.1 Emotions' Models

Emotion' models are theoretical frameworks, mostly issued from behavioral psychology researches, attempting to describe the nature, meaning and function of emotions. There are several models for categorizing emotions, which can be grouped into two main types: categorical models and dimensional models.

Categorical models assume that emotions are discrete and distinct. Ekman's [3] list of six basic emotions based on facial expressions (happiness, anger, fear, sadness, surprise, and disgust) is a typical example of categorical model. Another significant example of such models is the Plutchik's [4] model. Plutchik founds his emotions' model on the basis of eight main emotions: trust, fear, surprise, sadness, disgust, anger, anticipation, and happiness. The aforementioned emotions are arranged in a wheel-like structure in such a way that two opposite (i.e. contradictory) emotions are positioned across from each others. According to this model complex emotions can be obtained (i.e. modeled) as combination of the above-mentioned main emotions. Figure 1 depicts the Plutchik's Wheel of Emotions, summarizing the Plutchik's model.

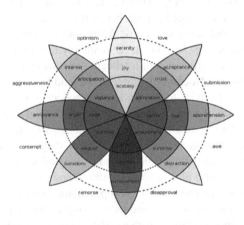

Fig. 1. Plutchik's Wheel of Emotions, summarizing the Plutchik's model.

In contrast with the first category, dimensional models assume that emotions are continuous and dynamic states that can be characterized by N-Dimensional space where each dimension an emotion such as valence, arousal, dominance, or intensity (scale of the involved emotion). Valence/Arousal bi-dimensional model [5], proposed by Russell is a first example of such category of emotions' models. It places specific emotion concepts in a circumflex model of core affect defined by two basic dimensions: Arousal (sized from high to low) and Valence (varying from positive to negative). The same author proposed Pleasure-Arousal-Dominance (PAD) [6], a 3-D Emotional State Model. The model uses Pleasure, Arousal and Dominance (dimensions) to represent other emotions. Finally, Latinjak cube tri-dimensional model [7], proposed by Latinjak in 2012, adds a third dimension (Time Perspective) to the Valence/Arousal bi-dimensional model in order to cope with the overlapping of emotions.

2.2 Visual Aesthetics and Beauty's Esteem

As we mentioned it in the introductory section, we found the investigated model on basis issued from art-philosophy's and psychology's basements. However, aesthetic perception and beauty's appreciation are complex and challenging notions dealing with highly subjective concepts. Thus, in order to establish comprehensive basements for interpretability of our model, we take advantage from aesthetics philosophy dealing with art-philosophy's slant of view concerning those complex notions.

According to G. Grabam [8] the notion of aesthetic appreciation (esteem) of visual artworks (paintings, sculptures etc.) links the pleasure that emerges from emotions activated by them. This is not enough to grasp the challenging concept of the beauty; however, this connects the beauty to the pleasure. Plato[3] emphasizes the beauty as foremost concept in society (*Republic*) [9]: "*If there is anything for which life is worthless, it is the sight of beauty.*" (Plato). Plato introduces the concepts of beauty and art into the philosophical view. Aaccording to him and his philosophical framework, the cultivation of virtue is inseparable from comprehension of the soul and how both beauty and love affect it. In other words, the consciousness (spirit) meets good worth through the feeling (*love* in Plato's terminology) that the beauty procures.

Considering the aforementioned, a first possible slope regarding the notion of the beauty is the frame of Plato's philosophical support (borders). Within this frame and based on the analysis of L. Strauss [10] and W. Tatarkiewicz [11] of the aesthetics of art in ancient Greece, the notion of beauty could be bounded within five associated characteristics: appositeness and proportionality (equilibrium), usefulness of impact (effect), helpfulness in promoting virtue, enjoyable for the eyes and ears and appealing and valuable. The above-mentioned analysis ahead of Plato's philosophy relating art and beauty and Pythagorean concepts of equilibrium and harmony, Tatarkiewicz sends out the possible emergence of the beauty on account of five measurable (or computable) features:

- *Order:* balanced and fine-tuning of the details
- *Harmony in proportions (also known as unity or eurhythmy):* implies a graceful appearance (semblance) details
- *Balance and Symmetry:* joining also the eurhythmy in its symmetrical expression
- *Size*
- *Emphasis (also known as Contrast):* adequate visual appearance regarding intensity and clarity of details and items.

A more sophisticated notion relating aesthetics meets Kant's philosophical framework [12]. Kant argues that beauty is equivalent neither to utility nor perfection, but is still purposive: beauty appears as purposive with respect to our faculty of judgment, but it will have no ascertainable purpose. Kant argues that "pleasure" is a feeling that arises on the achievement of a purpose, or at least the recognition of a purposiveness [13]. In fact, Kant claims that the faculties of the mind are the same: the 'understanding' which is responsible for concepts, and the 'sensibility' which is responsible for intuitions. The difference between ordinary and aesthetic cognition is that in the latter case, there is no

---

[3] Plato: Greek antique philosopher (438 – 348 BC).

one 'determinate' concept that pins down an intuition [14]. In the same way, according to what has been mentioned relating Kant's conceptualization of the aesthetic judgment, a second more sophisticated slant regarding the notion of the beauty is the frame of Kant's philosophical support (borders) leading to more advanced notion of beauty involving high-level features (presented and discussed in Sect. 3).

2.3 Computer Vision Based Emotions' Classification and Aesthetic Ranking

An ever increasing number of research works during the recent decade in field of computer vision have focused emotions' classification from images (and videos) as well as aesthetics ranking (from art-works' images). All of them use either hand-crafted features (i.e. features selected and prepared by human experts) or exploit Deep-Learning based features (i.e. delivered by Deep-Learning based structures).

Hand-Crafted feature based methods encode visual aspects of the image through visual attribute such as colors, textures, etc. The visual features then fed conventional Machine-Learning based approaches (as classifiers or approximators) for performing the aforementioned tasks. Wei-Ning et al. investigated the relationship between line directions and image emotion for emotions' classification [15]. On the other hand, guided by the theories of color psychology, same authors looked proposed three kind of visual features tagging on an orthogonal 3-D emotion space [16]. Yuan et al. [17] propose an image emotion recognition algorithm that includes the detection of facial expressions using eigenface as key feature in determining emotions from image. Eigenfaces are a set of eigenvectors used in the field of computer vision for human face recognition. Also guided by theories of art, Machajdik and Hanbury introduce harmonious composition of features [18]. Finally, Zhao et al. proposed features' extraction based on artistic aspects (such as balance, emphasis, harmony, movement, rhythm, etc.) for image emotion recognition [19].

Deep learning-based methods use robust deep neural networks to learn and encode image from a large number of training images. Cetinic et al. [20] focused on three different levels of image perception: the aesthetic evaluation of the image, the sentiment stirred up by the image and its remembrance. In [21], Achlioptas et al. tackled the prediction of expected emotional distribution of the emotional reaction to an artwork. Recently, Tashu et al. proposed a multi-modal approach based on co-attention for artwork emotion recognition [22]. They use a weighted fusion of information (features) issued from the digitized painting, the title and emotion category. Finally, Bose et al. investigated prediction of emotions from artworks using both textual and visual features [23].

3   Artificial Aesthetic Perception Model

As mentioned in introductory section, we aim achieving a computational model of *Artificial Aesthetic Perception* (AAP), proffering the machine a kind of *emotional awareness* and emotional esteem of its environment. Within this context and taking into account the points highlighted in the previous section dealing with the state-of-art of the fields allied with our objective, we found our model on the basis of three hypotheses:

- Hypothesis-1: decisiveness of the aesthetic degree stems from the emotions activated by attributes characterizing the perceived visual information.
- Hypothesis-2: intensity of the aesthetic degree depends on the feelings emerging from the emotions activated by the perceived visual information.
- Hypothesis-3: patterns activating emotions are those items of perceived visual information that as well likely attract the visual attention as tone with aesthetic traits inherent to the captured information.

Two first hypotheses lead to conceive the target AAP model as a multi-layer computational structure consisting of a layer of emotions and a layer of feelings. As consequence of the third hypothesis, the extracted features representing the perceived visual information have as well characterize aesthetic aspects as typify salient items of the perceived visual information. Thus the AAP model we propose is organized in five computational layers (depicted in Fig. 2):

- the first layer (Perception Layer), constituting the input layer of the model, is in charge of visual information's acquisition (typically the input image).
- the next layer is responsible of extracting the adequate features satisfying the Hypothesis 3. This layer provides several kinds of features: low-level visual features coping with aesthetic aspects, high-level features representing aesthetic aspects, emotional coordinates based on brightness and saturation and saliency features (relating attention).
- the third layer is devoted to computing the emotional intensity of activated emotions within the interval of [0, 1]. The emotional intensity value of "0" stands for inactivated emotion, while an activated emotion may have its intensity comprised in the interval [0, 1].
- the fourth layer is dedicated to working out the feelings' intensity (of activated feelings) with respect to the interval of [0, 1]. The value of "0" stands for inactivated feeling, while an activated feeling may find its intensity comprised in the interval [0, 1].
- Finally, the fifth layer calculates the decision-indicator's value: a value within the interval [-3, + 3] rating the emotional-aesthetic esteem of the perceived visual landscape [24].

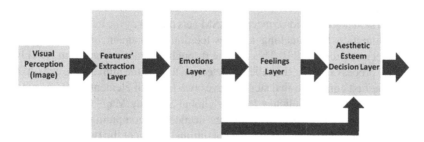

Fig. 2. The operational structure of the proposed Artificial Aesthetic Perception's model.

Wikipedia describes the feeling as a conscious experience created after the physical sensation or emotional experience, whereas emotions are felt through emotional experience. They are manifested in the unconscious mind and can be associated with thoughts, desires, and actions. Feeling is the component of emotion that involves the cognitive functions of the organism, the way of appreciating. The feeling is at the origin of an immediate knowledge or a simple impression. It refers to the perception of the physiological state of the moment.

3.1 Low-Level, High-Level, Emotional and Attention Features

The "Hypothesis-3" asserts that emotional emergence of feelings responsible (together with activated emotions) for the emotional conduct of the target AAP based system requires features as well cope with visual attention as handle aesthetic traits of the perceived landscape.

Accordingly to Plato's and Kant's philosophical frameworks, low-level visual features coping with aesthetic aspects representative of visual aspects may be MPEG-7 color descriptors [23], MPEG-7 contrast features [25], Tamura texture descriptors [26] and Gray-Level co-occurrence matrix descriptors. High-level features representing aesthetic aspects deal with items' representation in the image as: scale, proportion, unity, variety, rhythm, mass, shape, space, balance, volume, perspective, and depth [27]. Emotional features based on brightness and saturation are called emotional coordinates (pleasure, arousal and dominance) defined as following, Where Y are the brightness, and S the saturation:

$$Pleasure = 0.69Y + 0.22S \qquad (1)$$

$$Arousal = -0.31Y + 0.60S \qquad (2)$$

$$Dominance = 0.76Y + 0.3, 2S \qquad (3)$$

Two kinds of features relate aesthetics' attractiveness (attention) of splendor (supposed as inherent to beauty). The first is artificial Visual Attention based Saliency Map (VASM) features [29] and the second is Sun's rate of focused attention (RFA) [30] which measures emphasis.

Based on saliency detection concept, VASM gets underway the integration of human-like visual attention by launching an eyes-fixation mechanism based tuning of the saliency detection process [29]. This is done through a Genetic Algorithm (GA) based evolutionary process shoving the saliency detection toward the human-like eyes-fixation behavior. The first computational step achieves two kinds of elementary saliency maps, representing two different saliency levels: "Global Saliency Map" (GSM) and "Local Saliency Map" (LSM). These two maps are needed for computing the final saliency map. Final saliency map is obtained from conditional fusion of the two abovementioned maps. The final saliency map is then filtered using a Gaussian-Blob Filtering operation stressing the kernel part of human visual field versus the side-line region. The most salient spots are then determined pointing out to Visual Attention Map (VAM).

GSM, the constituents of which are denoted $M_G(x)$, is a result of non-linear fusion of two maps denoted $M_Y(x)$ and $M_{CrCb}(x)$, relating luminance and chromaticity. Equations 4 detail the calculation of each elementary map as well as the resulting GSM. $I_Y(x)$, $I_{Cr}(x)$ and $I_{Cb}(x)$ represent colors values of the image in channels Y, Cr and Cb, respectively in YCrCb color space, where $x \in N^2$ denotes 2D-pixel position. Similarly, $\overline{I_Y}$, $\overline{I_{Cr}}$ and $\overline{I_{Cb}}$ represent average colors' values for each channel throughout the whole image. Finally, $C(x)$ is a coefficient relating the saturation of each pixel in RGB color space.

$$M_Y(x) = \|\overline{I_Y} - I_Y(x)\|$$
$$M_{CbCr}(x) = \sqrt{\left(\overline{I_{Cr}} - I_{Cr}(x)\right)^2 + \left(\overline{I_{Cb}} - I_{Cb}(x)\right)^2} \tag{4}$$
$$M_G(x) = \frac{1}{1+e^{-C(x)}}M_{CrCb}(x) + \left(1 - \frac{1}{1+e^{-C(x)}}M_Y(x)\right)$$

The purpose of local saliency meets the idea of center-surround difference of histograms which the constituents are denoted $M_L(x)$ [31]. $M_L(x)$ involves statistical properties of two centered windows (over each pixel) sliding alongside whole the image [30]. As for GSM, the LSM is also obtained from a nonlinear fusion of two statistical properties-based maps relating luminance and chromaticity of the image ([29] and [30]).

RFA measures the degree (or rate) of attention (focus) headed for going to a gazed image (gazed by humans that watch it). RFA is obtained accordingly to Eq. 5, where W and H denote the width and height of the image, respectively [31]. Saliency(x,y) and Mask(x,y) represent the saliency's and saliency-extraction-mask's value at the pixel located at coordinates (x,y) in the image, respectively. Finally, i = 1,2,... represents different RFA according to different detected salient items in image.

$$RFA(i) = \frac{\sum_{x=1}^{W}\sum_{y=1}^{H} Saliency(x, y)Mask(x, y)}{\sum_{x=1}^{Wid}\sum_{y=1}^{Hei} Saliency(x, y)} \tag{5}$$

4 Implementation and Validation

The implementation of the investigated AAP has been achieved conformably to the operational block diagram depicted by Fig. 3. It implements as well the Plato's-based-AAP as the AAP supported by Kant's philosophical foundations relating aesthetics (Kant's-based-AAP). The so-called the Plato's-based-AAP implements four emotions and two feelings while the Kant's-based-AAP comprises eight emotions and six feelings in its emotional and feelings layers, respectively.

The prediction of as well the activated emotions as emerging feelings are achieved by the use of two MLP artificial neural networks (ANN). Then the final aesthetic judgment (decision on aesthetics' ranking) is obtained as a fusion of two aesthetics' ranking processes each one accomplished through an unsupervised Kohonen-Self-Organizing-Map (SOM) based classifier.

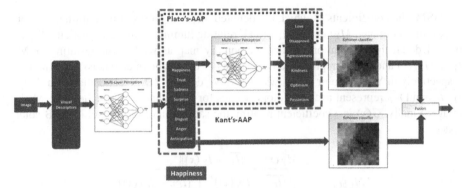

Fig. 3. The implemented Artificial Aesthetic Perception's model.

4.1 Validation

The experimental validation of the investigated concepts and the issued AAP has been done using two databases: Wiki-Art and a home-made human's eye-fixation based database including emotions', feelings' and aesthetic-judgment annotation. The WikiArt Emotions Dataset [24] is a collection of 4105 images of art-works (mostly paintings) that have been annotated for twenty emotions: gratitude, happiness, humility, love, optimism, trust, anger, arrogance, disgust, fear, pessimism, regret, sadness, shame, agreeableness, anticipation, disagreeableness, surprise, shyness, neutral. In addition to emotional labels, the WikiArt Emotions Dataset also includes meta-data such as artist name, title, year of creation, and art style.

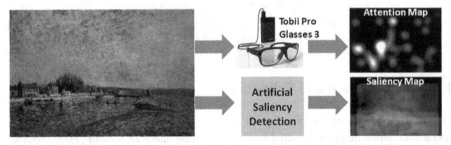

Fig. 4. Example of data provided by Tobii Pro Glasses 3 (left-side) and Artificial Saliency issued attention Map (right-side)

The home-made human's eye-fixation based database is an expansion of the first dataset including a subset of the WikiArt Emotions Dataset enriched with human's eye-fixation data obtained using Tobii Pro Glasses 3 (an eye-fixation maps opto-electronic device devoted to acquisition of human's gaze and human's attention related data). This device captures data related to human visual attention and provides various spatial and temporal measures of eye-fixation. It also allows for the extraction of different representations of eye-fixation and human gazing. Figure 4 shows an example of data provided by Tobii Pro Glasses 3 relative to an image of the WikiArt Emotions Dataset.

4.2   Experimental Results

The data (images) from of the previously-described databases have been divided into two groups of subsets: learning datasets comprising 80% of data and testing datasets containing the resting 20% data. The aforementioned datasets have been exploited for assessing the investigated AAP system within two protocols. In the first one, the AAP model has been calibrated conformably to the so-called Plato's AAP, including four emotions in its emotion-layer and two feelings in its feeling-layer. In the second one the AAP model has been adjusted for matching the so-called Kant's AAP, including eight emotions in its emotion-layer and six feelings in its feeling-layer. Then each model has been trained using the above-described learning datasets and tested using the aforementioned testing-datasets.

Table 1. Emotions, feelings and corresponding colors.

Emotions & corresponding color		Feelings & corresponding color	
Happiness	Red	Love	Red
Trust	Orange	Trust	Orange
Surprise	Yellow	Disapproval	Yellow
Sadness	Green	Aggressiveness	Green
Fear	Blue	Kindness	Blue
Disgust	Brown	Pessimism	Brown
Anger	Beige	Optimism	Purple
Anticipation	Purple		

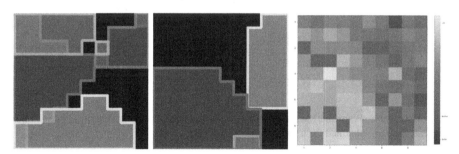

Fig. 5. Clusters of art-works corresponding to each of involved emotion (left-side), to each emerging feeling (middle) and the aesthetic judgment (right-side) for the Plato's AAP.

Figure 5 shows clusters (of images) corresponding to each emotion, to each feeling and the aesthetic judgment for the Plato's AAP. It is pertinent to note that high-scores area corresponding to aesthetic judgment recognized as "good-looking" (beautiful) matches with areas corresponding to "happiness" and "trust" emotions' activation as well as to emergence of the feeling "love". In the same way, low-scores area corresponding to aesthetic judgment recognized as "unattractive" (ugly) matches with areas corresponding to

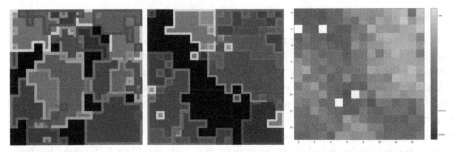

Fig. 6. Clusters of art-works corresponding to each of involved emotion (left-side), to each emerging feeling (middle) and the aesthetic judgment (right-side) for the Kant's AAP.

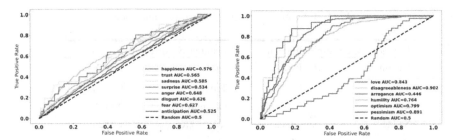

Fig. 7. AUC diagrams for the Kant's AAP relative to emotions (left-side diagram) and feelings (right-side diagram) obtained for training phase.

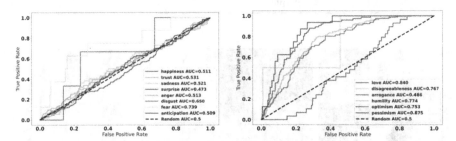

Fig. 8. AUC diagrams for the Kant's AAP relative to emotions (left-side diagram) and feelings (right-side diagram) obtained for testing phase.

"sadness" and "surprise" emotions' activation making emerge the feeling "disapproval". Table 1 gives colors corresponding to emotions and feelings.

Figure 6 depicts clusters (of images) corresponding to each emotion, to each feeling and the aesthetic judgment for the Kant's AAP. Here also, it is relevant to note that high-scores area corresponding to aesthetic judgment recognized as "good-looking" (beautiful) matches with areas of emerging feelings relating "Love", "Kindness" and "Optimism". In the same way, low-scores area corresponding to aesthetic judgment recognized as "unattractive" (ugly) matches with areas linking "Disapproval", "Aggressiveness" and "Pessimism". However, in this case, the impact of activated emotions on aesthetic appreciation is more complex. This behavior of the artificial aesthetic appreciation model is interesting and meets an appeal harmony with Kan's philosophical concept relating what he calls "aesthetic judgment". In fact, if Plato's view on beauty tags along quite basic guidelines (as enjoyableness and preciousness), Kant argues that beauty is equivalent neither to utility nor perfection, but enables the pleasure which according to Kant is defined as a feeling that arises on the achievement of a purpose and thus may involve antagonistic emotions.

Figure 7 and Fig. 8 show Area-Under-Curve (AUC) diagrams for the Kant's AAP relative to emotions (left-side diagrams in Fig. 7 and Fig. 8) and feelings (right-side diagram in Fig. 7 and Fig. 8) obtained for training and testing phases, respectively. One can note that as well for training as for testing the AUC values remain higher than 50% meaning that both emotions' activation and feelings' activation processes are not random and correspond to an emergent activation process issued shaped during the learning phase. In other words, those diagrams confirm that training process proffers the AAP model an emerging behavior following Kant's philosophical basements.

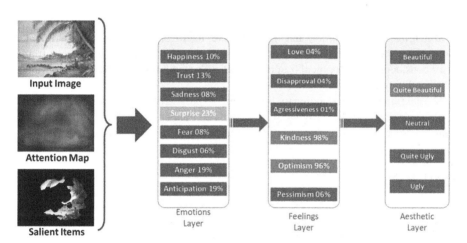

Fig. 9. Example of Kant's AAP's operation showing the aesthetic appreciation ("Quite Beautiful") obtained for an input image (upper picture) and activation degrees of as well emotions as feelings. The middle picture corresponds to the obtained attention map and the lower picture corresponds to salient items.

Figure 9 gives an example of Kant's AAP's operation showing the aesthetic appreciation obtained for an input image (upper picture) of the Wikiart database. The figure depicts the obtained attention map (the middle picture), the salient items obtained from the attention map and the obtained activation degrees of as well emotions as feelings involved in Kant's AAP. As it could be seen in this example the Kant's AAP system concludes that the perceived image is "Quite Beautiful". It is interesting to note that if the dominant emotion (representing an activation degree of 23%) is "Surprise"), however there is no strongly dominant emotion. Actually, "Happiness" and "Trust" are activated slightly with 10% and 13% activation degrees, respectively; "Surprise", "Anger" and "Anticipation" are moderately activated (representing activation degrees of 23%, 19% and 19%, respectively); finally, activation degrees of "Sadness", Fear" and "Disgust" remain lower than 10% meaning that those emotions are weakly activated. On the other hand, concerning the feelings, if activation degrees of off-putting feelings (i.e. "Disapproval", "Aggressiveness" and "Pessimism") remain marginal, there are two strongly dominant feelings (i.e. "Kindness" and "Optimism" with 98% and 96% activation degrees respectively) which emerge from the process while the activation of "Love" (the third upbeat feeling) remains marginal not exceeding 4%.

5 Conclusion and Perspectives

An Artificial Aesthetic Perception model proffering the machine an emotional awareness has been presented. It provides the machine a kind of emotional esteem of its environment. Designed on the basis of aesthetics' philosophy and behavioral psychology basements, the achieved computational model offers advantage of a comprehensive results' interpretability and a human-inspired operability proffering it kind of human-like appreciation of the surrounding environments.

Matching adequate emotional and affective features, the obtained experimental results show the pertinence of the accomplished emotional esteem within the frame of two considered philosophical frameworks: Plato's standpoint of aesthetics and Kant's opinion relating beauty.

The next stage, corresponding to short-term perspective of the presented work, related the implementation of the designed AAP models on Pepper robot. This work opens also perspectives of completing the designed model by using advanced fusion approaches.

References

1. Kant, I.: Critique of Pure Reason. Original ed., Riga, Martin Bodmer Found., Swiss (1781)
2. Kant, I.: Critique of practical reason). Original ed., Riga, Martin Bodmer Found., Swiss (1788)
3. Ekman, P.: An argument for basic emotions. Cogn. Emot. **6**, 169–200 (1992)
4. Kellerman, H., Plutchik, R.: Emotion, Theory, Research, and Experience: Theories of emotion. Academic Press (1980)
5. Russell, J.A.: A circumplex model of affect. J. Personality Soc. Psychol. **39**, I6I-I78 (1980)
6. Mehrabian, A., Russell, J.A.: An approach to environmental psychology. MIT Press (1974)
7. Latinjak, A.T.: The underlying structure of emotions: a tri-dimensional model of core affect and emotion concepts for sports. Revista Iberoamericana de Psicología del Ejercicio y el Deporte **7**, 71–88 (2012)
8. Graham, G.: Philosophy of the Arts: An Introduction to Aesthetics. 3rd edition, published by Routledge, London and New York (2005)
9. Plato: The Republic of Plato. Oxford University Press reprint, London (1970)
10. Strauss, L.: Plato's laws (Two Courses by Leo Strauss). The Department of Political Science, University of Chicago: course offered in the autumn quarter, 1959.© (2016)
11. Tatarkiewicz, W.: Ancient Aesthetics, published by Mouton the Hague. Polish Scientific Publishers (1970)

12. Burnham, D.: Staffordshire University, U.K. https://iep.utm.edu/kantaest/#:~:text=Kant%20argues20that%20beauty% 20is,with%20respect%20to%20determinate%20cognition
13. Critique of Pure Reason. Trans., Werner Pluhar. (Indianapolis: Hackett, 1996)
14. Schaper, E.: Studies in Kant's Aesthetics. Edinburgh University Press, Edinburgh (1979)
15. Wei-ning, W., Ying-lin, Y., Jian-chao, Z.: Image emotional classification: static vs. dynamic. In: IEEE International Conference on Systems, Man and Cybernetics, pp. 6407–6411 (2004)
16. Wei-Ning, W., Ying-Lin, Y., Sheng-Ming, J.: Image retrieval by emotional semantics: a study of emotional space and feature extraction. In: IEEE Internatinal Conference on Systems, Man and Cybernetics, pp. 3534–3539 (2006)
17. Yuan, J., Mcdonough, S., You, Q., Luo, J.: Sentribute: image sentiment analysis from a mid-level perspective. In: Proceedings of the Second International Workshop on Issues of Sentiment Discovery and Opinion Mining, pp. 1–8 (2013)
18. Machajdik, J., Hanbury, A.: Affective image classification using features inspired by psychology and art theory. In: 18th ACM International Conference on Multimedia, pp. 83–92 (2010)
19. Zhao, S., Gao, Y., Jiang, X., Yao, H., Chua, T.S., Sun, X.: Exploring principles-of-art features for image emotion recognition. In: Proceedings of the 22nd ACM International Conference on Multimedia, pp. 47–56 (2014)
20. Cetinic, E., Lipic, T., Grgic, S.: A deep learning perspective on beauty, sentiment, and remembrance of art. IEEE Access 7, 73694–73710 (2019)
21. Achlioptas, P., Ovsjanikov, M., Haydarov, K., Elhoseiny, M., Guibas, L.J.: Artemis: affective language for visual art. In: Proceedings of the IEEE/CVF Conference on Computer Vision and Pattern Recognition, pp. 11569–11579 (2021)
22. Tashu, T.M., Hajiyeva, S., Horvath, T.: Multimodal emotion recognition from art using sequential co-attention. J. Imaging 7(8), 157 (2021)
23. Bose, D., Somandepalli, K., Kundu, S., Lahiri, R., Gratch, J., Narayanan, S.: Understanding of emotion perception from art. arXiv preprint arXiv:2110.06486 (2021)
24. Mohammad, S.M., Kiritchenko, S., n.d. WikiArt Emotions: An Annotated Dataset of Emotions Evoked by Art
25. Manjunath, B., Ohm, J.R., Vasudevan, V., Yamada, A.: Color and texture descriptors. IEEE Trans. Circ. Syst. Video Technol. 11, 703–715 (2001)
26. Tamura, H., Mori, S., Yamawaki, T.: Textural features corresponding to visual perception. IEEE Trans. Syst. Man Cybern. 8, 460–473 (1978)
27. Wang, Z., Ho, S.-B., Cambria, E.: A review of emotion sensing: categorization models and algorithms. Multimed. Tools Appl. 79, 35553–35582 (2020)
28. Valdez, P., Mehrabian, A.: Effects of color on emotions. J. Exp. Psychol. Gen. 123(4), 394 (1994)
29. Madani, K., Kachurka, V., Sabourin, C., Golovko, V.: A soft-computing-based approach to artificial visual attention using human eye-fixation paradigm: toward a human-like skill in robot vision. Soft. Comput. 23, 2369–2389 (2019)
30. Ramík, D.M., Sabourin, C., Moreno, R., Madani, K.: A machine learning based intelligent vision system for autonomous object detection and recognition. J. Appl. Intell. 40(2), 358–375 (2014)
31. Sun, X., Yao, H., Ji, R., Liu, S.: Photo assessment based on computational visual attention model. In: 17th ACM interna. Conf. on Multimedia, pp. 541–544 (2009)

Detection and Visualization of User Facial Expressions

Martyna Wojnar⍟, Tomasz Grzejszczak^(✉)⍟, and Natalia Bartosiak⍟

Silesian University of Technology, Gliwice, Poland
martwoj082@student.polsl.pl, tomasz.grzejszczak@polsl.pl

Abstract. The work covers topics in face detection, prediction of the position of the face landmarks as well as control of the graphical model. The purpose of the work is to create a vision system that detects the user's facial expressions and visualizes them on the created computer model. The scope of work includes detection of the face and face landmarks, creation of a graphic model of a character whose facial expressions can be modified, and control of the created model using data obtained from the camera image. The main objectives of the project are easy accessibility, simplicity of use and low cost of the tools used.

Keywords: facial expression · face landmarks · visualization · computer model · 3D modeling · human-computer interaction

1 Introduction

Facial expressions are closely linked to emotions expressed intentionally or involuntarily. Humans, by their nature, feel better when they communicate with a self-similar creature capable of expressing emotions. This is why social robotics does not use robots similar to manipulators, used on factory floors, or mobile robots, but rather humanoid robots. Human-robot interaction should take place in real time, and the robot itself should respond to stimuli to which a human would react - including changes in facial expressions. According to a study conducted by [5], facial expressions have a strong influence on the empathy felt for a person.

When it comes to constructing a robot capable of communicating on a non-verbal level, including reflecting emotional states, it is worth analyzing how humans express emotions. The expression of emotions consists of facial expressions, but also tone of voice, spoken words, gestures, or biological data [7]. In the literature, the most extensive analysis of the relationship between emotions and the way they are expressed has been carried out for facial expressions, mainly due to the availability of data for analysis (most often a database of photographs) and resistance to cultural differences. This study will also use facial expression analysis, which can be used for further research on the relationship between feeling emotions and their expression. The work uses camera images to detect and then visualize facial expressions.

I. Rojas et al. (Eds.): IWANN 2023, LNCS 14135, pp. 118–129, 2023.
https://doi.org/10.1007/978-3-031-43078-7_10

The problem of creating a character and its animation is not complicated. Such an approach is used in the creation of computer games, or animated films. There are many possibilities for creating a face capable of expressing emotions, including the possibility of using 3D scanners to better reproduce the features of the user's face. However, it is more complicated to create a system capable of reflecting the user's facial expressions on the created model. The proposed solutions [4] are based on extensive models and advanced calculations.

The problem of visualizing real facial expressions itself is mainly used in the computer game and animation film industry, but it can also be useful in the context of studying human behavior from a medical angle.

1.1 Research Aim

The present work consists of two main issues. The first is to create a vision system for detecting facial expressions, while the second is to create a computer model capable of expressing emotions through facial expressions. These issues should be linked together so that the computer model is able to reproduce the detected facial expressions.

2 Materials and Methods

The workflow consist of several steps, each described as a subsection. Each of the subproblems have several applicable approaches. It is important to chose such solutions that can work together and are compatible with each other

2.1 Visualization of Facial Expressions

A human being, when presented with a photo, can distinguish at least six different emotions [6]. This is because each emotion is associated with a different arrangement of facial elements. For example, anger is manifested by pulled-down eyebrows, and fear by wide-open eyes. A person uses facial muscles to express a given expression.

2.2 Vision System

A main issue is the problem of face detection. Three popular existing solutions using are analyzed:

- cascade classifier, using Haar features - a relatively old method (2001) based on detection in the image of Haar features [3]; an existing classifier developed for frontal facing face detection was selected,
- dlib - a library that uses a historam of directed gradients and a linear classifier for face detection, while modifying the processed image [1],
- mediapipe - a library developed by Google that uses artificial intelligence methods extensively, allowing for face detection, motion tracking, or segmentation of objects such as hair.

In the following work, the methods presented are be called Haar method, Dlib method and mediapipe method, respectively.

The detected facial expressions should be mapped in a suitable way. There are many graphics programs that allows to create a model, and among them Blender was chosen, because it allows to control animation using the Python language, it is free, widely available and easy to use.

3 Methods for Detecting Facial Feature Points

Face detection itself plays a fundamental role in facial expression detection. In this work, the proposed methods for face detection and feature point position prediction are compared.

Each of the methods proposed in Sect. 2.2 gives the expected result - a face is found in the input image, as can be seen in Fig. 1, as a purple rectangle, but the marked area differs for each method. The dlib library allows to select the smallest area, while Haar's method does not include the entire beard. The marked area, depending on the method used, also differs in the area of the forehead.

The libraries used allow marking feature points on the face. The dlib method and the Haar method allow 68 feature points to be marked, while the mediapipe method has as many as 468 points. The first two methods do not differ in the position and number of feature points detected. The differences between the methods can be seen in the location of the plotted points, which do not overlap. The mediapipe method allows multiple points to be marked on the cheeks, chin, forehead and nose, which, from the point of view of this work, is not relevant due to data redundancy. A visualization of the applied landmarks is shown in Fig. 1.

Fig. 1. Images with face and feature points detection for methods: dlib, Haar, mediapipe

When approaching the subject of facial expression detection, it is first necessary to implement a face detection solution, which should be positioned as frontally as possible (so that the position of as many feature points as possible can be read). It is also important that a face detected in a given frame of the video stream is also detected in the next frame, preserving the smoothness of the video system. Another important aspect is the misclassification of an area as a face. In such a case, the system will not work, as a non-existent face will be detected, while feature points will not be applied.

3.1 Comparison of Methods

For the tests, 20 approximately ten-second video recordings were used, with different resolution, different lighting, and different number and positioning of faces in the video. Some of the recordings showed slightly obscured faces or faces not positioned frontally. A correctly classified frame was considered one where the number of detected faces matched the actual number of faces visible on the frame, while misclassification occurred when the number of detected faces on a frame did not match the actual number of faces, or when a face was detected where there was actually another object.

The results of the tests are presented in Table 1. The accuracy of the methods was checked, so the table shows results in percentage, where 1 means 100% of effectiveness and 0 means 0% of effectiveness, respectively. The results were measured on $n = 20$ video recordings and the average is presented in last column. Based on the results presented, it can be concluded that the best method of face detection turned out to be the one used by the mediapipe method, but during the tests it was noted that in many cases a face was detected even when it was presented in profile or when it was obscured by a hand or other object. Such situations are undesirable due to the problem considered in this work, since the main purpose of the work is to detect facial expressions, and when part of the face is obscured, facial expressions cannot be correctly detected.

Table 1. Comparison of face detection for $n = 20$ videos for methods: dlib (D), Haar(H) and MediaPipe(M) [%]

n	1	2	3	4	5	6	7	8	9	10	11	12	13	14	15	16	17	18	19	20	$\bar{x}$
M	100	100	96	100	74	53	66	87	100	0	95	100	100	80	92	100	100	91	97	56	**84**
H	100	64	0	59	0	9	23	45	100	0	76	0	67	20	0	3	80	0	83	0	**36**
D	95	91	75	100	71	5	32	55	100	0	3	10	98	98	74	100	96	46	97	0	**62**

4   Implementation

In order to represent the detected facial expressions, a computer model was created, using the program "Makehuman". This program allows to quickly create a character, allowing you to modify each element of the character, such as the shape of the face, eye color, nose length, etc.

4.1   Character Animation

The created model was imported into Blender. Character animation can be done in two ways:

- using bones - each place that should perform movement is equipped with a so-called bone, to which the area to be moved using this bone is assigned;
- using predefined modifiers (blend shape) - multiple independent distortions are created (e.g., a lift of the corner of the mouth, a lift of the eyebrows), and then the degree of use of the selected modifier in the final scene is defined.

Within the scope of the work, the first approach was chosen because it seems to be a logical continuation of feature point detection on the face. The second approach does not seem to be a suitable choice for creating a real-time vision system, but only good for animating predefined positions of individual points. With a view to further work on the created model, bones were applied to the entire body, including the limbs. 68 bones were applied to the face to coincide with the landmarks detected by the method used (Fig. 2). In addition to the bones on the located face, equally important is the bone on the neck, which is responsible for the movement of the entire head.

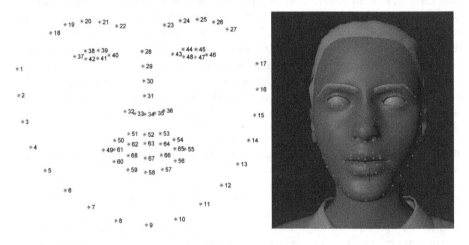

Fig. 2. Feature points used for annotation [2] with placement on the model.

All the bones are connected to each other and form a single skeleton. Within this skeleton, there is a superior bone (parent), which is located at the site of the human spine. The bones subordinate to it, are the limbs and neck, while the neck bone is superior to the 68 facial bones. By using this hierarchy, it is possible to move the facial bones, while moving the neck, while not moving the limbs. It is also worth noting that in modeling a bone is called a rigid element, so the model created contains many bones, even though anatomically the human head contains only one skull bone.

4.2   Assigning Weights to Facial Areas

Defining the area to move with a given bone is done by assigning to each vertex of the object's surface, the appropriate weight, where 0 means no reaction, while 1 means the maximum possible reaction to the bone's movement. Note that when assigning the appropriate weights (in the range $[0;1]$), one needs to take into account every possible component that can move with the movement of a given bone. A good example of a given situation is the movement of the neck bone, when not only the face and head are moved, but also the teeth, tongue, eyebrows, eyes and hair, which are independent components of the character.

Another important issue is the need to assign weights to a given area in such a way that the transition from an area with a weight of 0, to one with a weight of 1, is as smooth as possible, in order to reproduce facial expressions as closely as possible. Although some areas of the face move independently of each other, there is no point that moves without entailing, even subtle, movement of neighboring tissues, for example, moving the jaw, the lips also move.

An example of assigning weights to facial areas can be seen in the Fig. 3. The left side shows the vertices of the created surface, while the right side shows the corresponding weights.

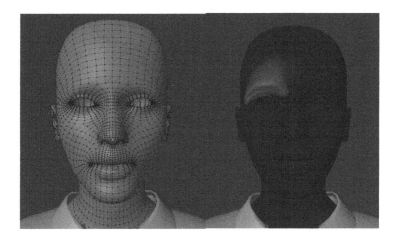

Fig. 3. Head model with edges and assigned weights in case of eyebrow bone.

4.3   Manual Control

The model prepared in the above way, can be animated. It is possible to modify each bone in 3 different ways - displacement, rotation and scaling. The third possibility will not be used, because the human face is not able to change its dimensions. Facial expressions will be represented by changing the position of the bones placed on the face, while rotation of the neck bones will be responsible for the movement of the head. Figure 4 shows an example of modifying facial expressions by manually changing the position of facial bones. The attachment point is located at the character's neck. The transformation of neck to cursor (marked on chin) can be described by the following matrix:

$$T = \begin{bmatrix} -1.0000 & -0.0001 & 0.0009 & -0.0023 \\ 0.0009 & -0.1266 & 0.9919 & 0.0272 \\ -0.0000 & 0.9919 & 0.1266 & 3.7760 \\ 0.0000 & 0.0000 & 0.0000 & 1.0000 \end{bmatrix} \tag{1}$$

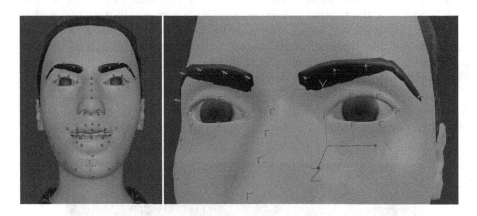

Fig. 4. Manual displacement of bone (left) and local coordinate frames (right)

Each bone is made up of three elements - root, body and tip. The tip of each bone is also the origin of the local coordinate system. Each system created by a face bone is in the same orientation, and they differ in position. Examples of local coordinate systems can be seen in the Fig. 4, and the local coordinate system created by the neck bone can also be seen in the figure.

4.4   Head Movement

The main idea of this work is to visualize facial expressions, but an integral part of expressing emotions, is the movement of the head, for example, when a person is sad - he hangs his head, and when he is happy - the opposite.

The face bones were moved by means of a change in position, while the neck is moved by means of a change in orientation. This approach seems reasonable due to the nature of humans - when expressing facial expressions, the skin is stretched and given points, for example, the eyebrow, change position, while when moving the head, a person is not able to change the length of his neck, and when rotating, all points of the head rotate by the same angle.

4.5   Algorithm for Facial Expression Detection

The mapping of facial expressions is done in several steps as shown in the Algorithm 1. The first and essential step is face detection in the image transmitted from the camera. The selected face detection method returns the position of the vertices of the rectangle bounding the detected face (bounding box). The camera image, as well as the returned position of the vertices, are passed to the face feature point predictor, and these in turn are passed as an argument to the function of the dlib library used, which makes it possible to obtain an image of the aligned face. This alignment involves the appropriate rotation of the image so that the face is in a vertical position and the tip of the nose is exactly in the center of the image. Also used is so-called padding, which means surrounding the face by an additional area of the output image so as to avoid the situation where the detected characteristic points of the face are outside the image of the aligned face. The resulting image should be 400×400 pixels.

After such preprocessing of the image, the feature point predictor can be used again to determine the position of the feature points in the aligned face image.

The point on the face that does not move independently is the point under the nose (point No. 34 in the image 2), so this was considered the base point for further calculations.

Algorithm 1. Visualization of face mimics

1: Get frame and detect faces
2: **if** face is detected **then**
3:      Get feature points
4:      Align face position
5:      Get feature points after alignment
6:      Calculate distance from feature points to base point
7:      Calculate the difference of distances with previous frame
8:      Move appropriate bones in the model
9: **end if**

---

5   Tests

5.1   Distances Between Feature Points

As part of the present study, a personalized solution was proposed for the author of the paper. This personalization consisted of preparing three photos in which the face remained in as neutral a position as possible, so that the input distances between each characteristic point on the face can be calculated. The distance is calculated as an Euclidean distance and is always calculated from the base point, which is the point under the nose. The orientation of the adopted coordinate system is the same as the orientation of the local coordinate systems of each bone of the computer model. The adopted orientation coincides with that of image processing - the upper left pixel is the origin of the coordinate system.

Determination of the distance of feature points from the base point is done for each frame on which a face is detected. The difference between the previous and current distance of each point from the base point is then determined, and the bones of the computer model are moved on this basis. The new position of a given bone is the sum of its previous position and the determined difference, within the vertical z and horizontal x axes. The distances determined in a given frame, become the base for comparison in the next frame.

5.2   Head Orientation

Determining the orientation of the head is a complicated problem because the camera image transmits information in two-dimensional space, while the model is created in three-dimensional space. To solve this problem, the solvePnP function of the OpenCV library was used. This function takes as one of its arguments the position of points in the model, which were defined at the beginning of the program. In order to correctly determine the orientation of an object, at least 6 object points must be passed. It was decided to use for this purpose those points that do not affect the expression of the face. Four points on the nose and the outer corners of the eyes were chosen. Their position was defined as follows:

Having the initial rotation matrix and the rotation matrix for a given frame, the orientation of the neck bones can be determined. For this purpose, Euler angles were used to describe the orientation of a given object, relative to the underlying coordinate system. Euler angles allow the orientation of an object to be described using angles of rotation relative to each axis of the underlying coordinate system. The Euler angles for the neck bones are determined as the difference of the designated angle of rotation about the axis of the global coordinate system and the initial angle of rotation about the same axis.

In order to ensure smooth movement of the animated character, a function has been created to provide this effect. The operation of the function is nothing more than the use of a moving average. The moving average is the average of the last few given values, and it is time-varying. In the case of a change in the orientation of the neck, the calculation of the moving average of the last 5 indications of the determined Euler angles (each angle is independent) was used. This approach avoids sudden jumps in the figure's head position.

6   Visualization of Facial Expressions

When observing the change in facial expressions, it was subjectively found that the points that are most prominent when changing facial expressions are the areas that are colored differently from the skin, i.e. eyebrows and lips. In addition, a very important element is the degree of opening of the mouth, for which the mandible is responsible. Taking these aspects into account, it was decided to control only seven facial bones. The points set consict of: central point of the right and left eyebrow, right and left corner of mouth, the central point of the upper and lower lip and the chin.

Control of the points is implemented as shown in Sect. 4. The bones on the eyebrows, are responsible for the movement of the entire eyebrows and part of the forehead, the corners of the mouth and points on the lips, for the movement of the lips, while the chin bone is responsible for the movement of the jaw. Tongue detection has also not been implemented, which in itself does not seem to be a complicated object detection problem, while moving the tongue can also change its shape, which can effectively prevent detection of its position and orientation. For this reason, the character's tongue moves in the same way as the mandible, remaining at the same distance from the lower teeth at all times. The Algorithm 2 presents the detecting and visualizing the user's facial expressions.

Algorithm 2. Visualization of face mimics

1: **while** $STOP$ **do**
2:     Get frame and detect faces
3:     **if** face is detected **then**
4:         Extract 640x480 face image
5:         Get feature points
6:         Calculate face position and orientation
7:         Rotate neck bone along detected face orientation
8:         Align face position for a frontal view and resize to 400x400.
9:         Get feature points after alignment
10:        Calculate distance between feature points
11:        Move appropriate bones in the model
12:     **end if**
13:     Display image
14: **end while**

---

7   Verification and Validation

Figure 5 shows an example of correct facial expression mapping. Some irregularities can be seen, such as an excessively raised left eyebrow or cheek distortion. However, the orientation of the user's face, as well as the opening of the mouth, has been correctly mapped.

7.1   Speed

One of the assumptions of the project is real-time operation. To test this, it was decided to specify the frequency of changing the character's position. The sampling rate of the camera used is 30 frames per second. Checking the frequency of change in the character's position and orientation involved averaging the 45 values indicated during program execution. The first approach was to manipulate

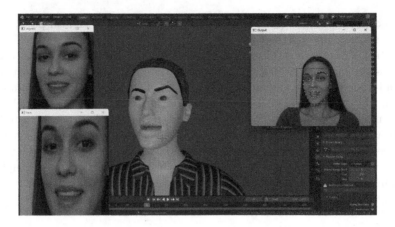

Fig. 5. An example of how the program works.

only the position of the facial bones. In this case, the result of the experiment was 7.12 frames per second. After allowing the neck to rotate, the value decreased to about 6.89 frames per second. The difference between the results is small, and the value itself is more than four times smaller than the capabilities of the camera used. This shows that facial expression detection itself takes the most time in the whole process of facial expression detection and visualization. Presented on the computer model, facial expressions are visibly delayed relative to the user's movement.

7.2   Accuracy

The accuracy of the proposed solution was determined by a subjective, visual evaluation of the results presented on the model. While the visualization of the detected facial expressions itself performs at a satisfactory level, given the number of bones that are controlled, many reservations can be made about the detection of faces and feature points themselves. Although the method is effective in terms of analyzing each frame independently, it does not give satisfactory results when analyzing a sequence of frames. During the operation of the program, the position of the face bounding box and the applied feature points can be well seen.

It was decided to perform a test by placing the phone with the face photo displayed on the screen stationary on a tripod. Such a method was used to test the performance of the algorithm in the absence of user movement. The test showed that even without changing the position and orientation of the face, the dlib method detects the face in slightly different places, and also detects an area of different dimensions. This puts the controlled bones into a slight movement, but it is clearly visible on the model. This different interpretation of the same image introduces interference with the program's operation.

8   Summary and Conclusion

The presented solution, despite its drawbacks, meets all the stated objectives. First of all, it is easy to use, and does not require additional dedicated equipment. The only requirement is a working, publicly available webcam and installed "Blender" program. All the goals of the project have also been achieved - a vision system responsible for detecting facial expressions has been created, as well as a computer model for visualizing the user's facial expressions. The computer model is adapted only to visualize the movement of a few characteristic points, but it should be noted that tensing the facial muscles produces visible facial wrinkles, which were not taken into account when creating the model.

The solution is personalized, which means that it is necessary to initialize the data when the system is used by different users. Each person expresses a given emotion in a different way, which further complicates the assessment of the accuracy of visualization of facial expressions.

The presented solution can be freely improved. First of all, other methods of face and feature point detection can be implemented or a custom model can be created to solve this problem. Another possibility to develop the project is to include the ability to visualize the movement of the human body and posture. The created model has bones responsible for controlling both the head, which was used in this work, but also the rest of the body, so developing the project in the proposed way, it is only necessary to implement a method for detecting body movements. Ultimately, the present work is intended to help solve the problem of visualizing facial expressions on a real robot.

Acknowledgements. This research was funded by the Silesian University of Technology (SUT) through the subsidy for maintaining and developing the research potential.

References

1. http://dlib.net/face_detector.py.html
2. Facial point annotations. https://ibug.doc.ic.ac.uk/resources/facial-point-annotations/
3. Behera, G.S.: Face detection with haar cascade (2020). https://towardsdatascience.com/face-detection-with-haar-cascade-727f68dafd08
4. Ersotelos, N., Dong, F.: Building highly realistic facial modeling and animation: a survey. Vis. Comput. (1), 13–30 (2008)
5. Kovalchuk, Y., Budini, E., Cook, R.M., Walsh, A.: Investigating the relationship between facial mimicry and empathy. Behav. Sci. **12**(8) (2022). https://doi.org/10.3390/bs12080250. https://www.mdpi.com/2076-328X/12/8/250
6. Kraut, R.E., Johnston, R.E.: Social and emotional messages of smiling: an ethological approach. J. Pers. Soc. Psychol. **37**(9), 1539 (1979)
7. Lee, M.S., Lee, Y.K., Pae, D.S., Lim, M.T., Kim, D.W., Kang, T.K.: Fast emotion recognition based on single pulse PPG signal with convolutional neural network. Appl. Sci. **9**(16) (2019). https://doi.org/10.3390/app9163355. https://www.mdpi.com/2076-3417/9/16/3355

Offline Substitution Machine Learning Model for the Prediction of Fitness of GA-ARM

Leila Hamdad[1(✉)], Cylia Laoufi[1,2], Rima Amirat[1,2], Karima Benatchba[2], and Souhila Sadeg[2]

[1] LCSI, Ecole nationale Supérieure en Informatique (ESI ex INI), BP 68M, 16309 Oued Smar, Alger, Algeria
l_hamdad@esi.dz
[2] LMCS, Ecole nationale Supérieure en Informatique (ESI ex INI), BP 68M, 16309 Oued Smar, Alger, Algeria

Abstract. Association rule mining (ARM) is one of the most popular tasks in the field of data mining, very useful for decision-making. It is an NP-hard problem for which Genetic algorithms have been widely used. This is due to the obtained competitive results. However, their main drawback is the fitness computation which is time-consuming, especially when working with huge data. To overcome this problem, we propose an offline approach in which we substitute the GA's fitness computation with a Machine Learning model. The latter will predict the quality of the different generated solutions during the search process. The performed tests on several well-known datasets of different sizes show the effectiveness of our approach.

Keywords: Association rules · Genetic algorithm · Fitness · Substitution model · Off-line

1 Introduction

Industries are, continuously, accumulating valuable amounts of data that are often under-exploited. This data can contain relevant and useful knowledge for prediction and decision-making purposes. Data mining is the heart of the knowledge mining process and Association Rule Mining (ARM) is one of its most popular tasks. It was proposed by [1] for market basket analysis, to find out which items appear simultaneously or in a specific order. Nowadays, it is used in many fields as web mining, image mining, DNA mining, among others.

Extracting AR from a transaction database has been classified as an NP-Hard problem [6]. Several approximate methods have been proposed in the literature to deal with this complexity which increases as the number of transactions and the total number of items increase in the database. However, the computation time required to reach a good solution can quickly become prohibitive. In this

I. Rojas et al. (Eds.): IWANN 2023, LNCS 14135, pp. 130–142, 2023.
https://doi.org/10.1007/978-3-031-43078-7_11

case, the approximate methods constitute an essential alternative for the extraction of ARs. The first and most proposed approximate methods are genetic algorithms (GAs). They evolve populations of solutions organized in generations to optimize a fitness function. This principle was applied for the ARM problem. First, the algorithm starts with a population of rules generated randomly or according to an initialization strategy. Then, gradually, it evolves this population by applying search operators and evaluating the new rules with a well-defined objective function. Among the works that use GA for ARM, we cite **GENAR** [9] and **GAARM** [3]. The main motivation for applying GAs to knowledge mining tasks is that these algorithms are robust and are adaptive search methods that perform a global search. Moreover, GAs allow obtaining feasible solutions without greatly affecting their quality in a reasonable time. However, the computation of GAs' fitness is time-consuming. This slows down greatly the optimization process. To overcome this problem, some recent works in literature propose strategies based on Machine Learning (ML) that exploit the data generated by evolutionary algorithms during the research process to build a model for fitness's prediction. The latter is built using generated solutions and their real evaluation. This prediction speeds up the evaluation process during the search, reducing the overall execution time. These strategies have been applied for many optimization problems using two different approaches: On-line and Off-line. In this work, we are interested in the latter.

We propose to use a surrogate model to predict the fitness of GA-ARM [3], an effective GA for ARM. To our knowledge, no work has used this strategy for this problem before. This model will be built on a training set consisting of generated solutions by GA-ARM and their real evaluation. In GA-ARM, an encoding of variable size is used to represent a solution. This raises a problem when applying ML algorithms for training the model as this step requires structured data in a single, static format. For this, we propose new codifications for a solution.

The organization of this paper is as follows: in Sect. 2, we present related work on evolutionary algorithms assisted by fitness substitution to reduce the computational cost. Section 3 is dedicated to the problem statement. Then, GA-ARM is introduced in Sect. 4 and we describe our proposition in Sect. 5. Finally, in Sect. 6, a summary of performed tests is given. We summarize the work and give a perspective in Sect. 7.

2   Related Work

Evolutionary algorithms are widely used in several real-world problems. This is due to their easiness to adapt to any problem. Moreover, they reduce the optimization time compared to exact methods. However, one of their drawbacks is the time required for the fitness computation which evaluates the quality of obtained solutions. Moreover, a large number of fitness evaluations is generally necessary to obtain a good solution. The computational cost of this function has become a crucial challenge in some problems.

For this, evolutionary algorithms assisted by substitution of the fitness have been developed to reduce their computational cost. Surrogate models are approximate models that can simulate the behavior of the real fitness function. These models are used to predict the fitness values of new solutions instead of computing them. They reduce the computational cost of the optimization process while preserving the quality of the solutions. For this purpose, regression models, such as polynomial regression, regression of the Gaussian process (kriging model), artificial neural networks, support vector machines and random forests are commonly used (see [10]).

Approximate fitness functions determine the approximation of a function using a chosen set of solutions in the search space that are evaluated with the real fitness function. These solutions with their evaluations are learned by an ML approach to approximate the fitness function for newly generated solutions. Approximation methods based on ML are classified into two sub-classes: similarity-based approximation and functional online and offline approximation. In this section, we will focus on the works in literature that have addressed offline approximation.

For some real problems, the determination of fitness during the optimization process is very difficult and computationally expensive. The category of offline algorithms known as Data-driven evolutionary algorithm (DDEA) is more adequate to solve these problems. This offline strategy consists in building a model of fitness before the start of the optimization process. It is based on previous data obtained from optimizations with real computation of fitness. Offline models are not updated and no new data can be sampled during optimization.

As real fitness is not accessible during optimization, offline DDEAs face significant challenges. Research work on offline DDEAs is directed towards improving the quality and quantity of offline data, as it greatly influences the performance of surrogate models [4]. Therefore, many studies focus on additional offline data generation, such as artificial data generation [2], DDEA-PES [7], BDDEA-LDG [8] and TT-DDEA [5].

Furthermore, the quality of artificial data strongly depends on generative models, which can induce errors in the surrogate. Other research work focuses on the use of ensemble learning, which consists of building several models instead of just one. Combining multiple models can improve the performance of the surrogate model. Therefore, many works apply ensemble learning such as DDEA-SE [13], BDDEA-LDG [8], ASMEA [14], CALSAPSO [12] and SRK-DDEA [4].

3   Problem Statement

Given a set of items $I = \{i_1, i_2, , i_n\}$ and a set of N transactions $T = \{t_1, t_2, , t_N\}$, each transaction contains a number of items, and two item-sets A, B of I, an association rule is a causal relation $R : A \to B$, where A is called premise of the rule, B the consequence and $B \cap A = \emptyset$ [1].

The quality of a rule can be evaluated through Support which expresses rules' generality and Confidence which expresses rules' validity:

$$Support(A \cup B) = Supp(A, B) = \frac{|A \cup B|}{N} \tag{1}$$

$$Confidence(R) = Conf(A, B) = \frac{|A \cup B|}{|A|} \tag{2}$$

The interesting rules extracted are those with Support greater than $Minsup$ and Confidence greater than $Minconf$, where $Minsup$ and $Minconf$ are two thresholds given by the user.

4   Genetic Algorithm for ARM (GA-ARM)

Genetic algorithms have proven their efficiency in the extraction of ARs by reducing the number of accesses to the database. [3] studied the impact of all GA operators (types of chromosomes encoding, types of crossover, type of replacement,...) on the quality of returned AR. We chose to use GA-ARM in this work, the algorithm with the operators that have given the best performance. Its pseudo-algorithm is presented below.

Algorithm 1. GA-ARM [3]

Begin
- Generate randomly a population (initial population)
- Compute the fitness of all its chromosomes
    For each generation
        - Select a pair of individuals
        For each selected pair of individuals
            - Crossover with a probability Pc
            - Compute new fitness
            - Replacement
        End for
        - Mutation with a probability Pm.
        - Compute fitness
    End for
End.

To represent a solution (R), the one rule encoding is used. A chromosome represents a rule and its fitness is computed as follows:

$$fitness(R) = \frac{\alpha * Support + \beta * Confiance}{\alpha + \beta} \tag{3}$$

where α and β are two parameters in $[0, 1]$.

In the beginning, a set of rules of different sizes are generated randomly. Then, at each iteration, chromosomes are selected randomly in pairs on which one point crossover is applied with a probability Pc. Two new solutions are generated and the parent replacement is applied. It consists in replacing a parent with its son if the cost of the latter is better. Then each obtained chromosome is mutated with a probability Pm. It consists in changing a random item with another one, selected randomly.

5   Proposed Solution

The purpose of the application of a supervised machine learning algorithm in the GA-ARM is to predict the evaluation of the rules that are generated by the optimization process without the need to access the database each time to compute their support and confidence. This will speed up the optimization process. The training of the model goes through three phases: a collection of data from real optimizations, organization of data according to a well-defined encoding, and finally the training of the model.

5.1   Arranging Surrogate Model Input and Output Data

In Machine Learning, training and test data must have a consolidated form with the same attributes stored in the same table. For this, we propose to add a phase of encoding the learning data and tests before training the model. The one rule encoding is not suitable for the training model. Two issues can be raised firstly, *the Size of the rules*. Indeed, the size of the antecedent part differs from one rule to another as generated rules vary from small sizes to MaxChromosomeSize. For this, rules of different sizes cannot be stored in the same table with the same characteristics. The second issue is the *Order of items*. For association rules, two itemsets are considered: one for the antecedent and another for the conclusion part. The order is not important in both itemsets. However, for the learning data set, the order of the items is important and has an impact on the result.

To overcome these problems, we propose new codifications which are more adapted to the principle of static organization of training and test data. These codifications will result in consolidated input for the model, using the same attributes. A method of static association rules' codification is to make all items of the dataset appear in the representation of a rule. An item can have three states: either it belongs to the set of antecedents or conclusions or it does not appear in the rule. To take into account these categorical characteristics, we consider the items of the dataset as features each having three categories: '*a*' to designate antecedent, '*c*' for conclusion and '*n*' to indicate that it does not appear in the rule. Several works in the literature apply this type of encoding on association rules as [11]. To simplify the categorical representations of the variables in our study, we propose to use three different numerical codifications: LabelEncoding, OneHotEncoding and BinaryEncoding.

LabelEncoding consists in representing each rule as a vector **V** of n elements, such that n represents the total number of items in the database:

- **V**[i]=**0** if item i does not appear in the rule.
- **V**[i]=**1** if item i appears in the antecedents part.
- **V**[i]=**2** if item i appears in the conclusion part.

The use of LabelEncoding allows ML algorithms to assume a natural order between the categories (0, 1 and 2) of the training data.

OneHotEncoding consists in replacing the categorical variable with a number of binary variables equal to the number of the variable's categories. In our case, each item has three categories: **antecedent**, **conclusion** and **not**. So each item is assigned 100 if it is **antecedent**, 010 if it is **conclusion** and 001 if it is **not** in the rule. Therefore, each **item** will be replaced by **three binary variables**. By generalizing on all the items, we will have a data table of size $3 * n + 1$, such that **n** is the total number of items. The last column of the table represents the fitness.

BinaryEncoding. In this type of encoding, we will have $2n + 1$ variables in our dataset, such that n is the total number of items. For each item, it is assigned 10 if it is in **antecedent**, 01 if it is in **conclusion** and 00 if **not part of the rule**. Indeed, each item will be represented by **two variables**.

The major drawback of these codifications lies in the application of genetic operators, namely crossing and mutation to generate new individuals. In order to no complicate the GA-ARM, we decided to use the one rule encoding during the evolution of the population since it eases the application of these operators. And we apply one of the new encoding only when we evaluate individuals with the predicted objective function (by the model).

5.2 Proposed Off-Line Approach for Approximating the Objective Function

GA based on a surrogate model works as follows: first we take a set of points sampled from the search space and evaluated with the real fitness. Then, this data is used by a supervised Ml algorithm to train the fitness prediction model. The model inputs are the association rules generated during the search process and the output represents the associated rule rating or predicted fitness. In our case, we design the offline substitution approach of the fitness function. The fitness approximation model is built before the start of the optimization process, by exploiting the data previously generated by other optimizations with the GA-ARM. Data are the rules and their actual evaluations that are generated throughout the research process. They will be collected and saved in a database.

Then, we split them into two sets: training data set (75%) and test data set (25%). The training data will be used to train the surrogate model offline and the test data to measure the accuracy of the model. After validating the

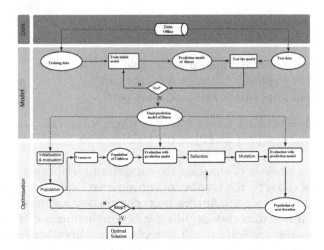

Fig. 1. GA-ARM process with offline prediction of fitness using ML algorithms.

fitness prediction model, it will be ready to be used in a new optimization. And the computation of the fitness of GA-ARM is done by the substitution model. Figure 1 shows the overall operation of this approach.

6   Tests and Results

To conduct our tests, we have used PYTHON as programming language. For our tests, we used the environment of **Google Colab**[1]. Google Colab offers for a simple user a RAM of 12.6 GB, a disk space of 78 GB, single-core hyperthreaded Xeon CPU at 2.3 Ghz, that is to say (1 core, 2 threads) and a cache memory of 56.3 MB. In order to structure the training and test data, we proposed to apply the codifications: label, oneHot and binary. To predict the fitness values of the new association rules, we used the following ML algorithms: Decision Tree Regressor (DTR), Random Forest Regressor (RFR), MLP Regressor (MLPR).

Two types of datasets are used to illustrate our approach and test the performance of the proposed algorithms: real and synthetic. Real datasets are downloaded from the **UCI Machine LearningRepository**[2] directory. Synthetic datasets are data generated artificially by the **IBM Quest Synthetic database**[3] program. Table 1 describes their characteristics. These datasets are classified as small or medium according to the number of transactions and item occurrences.

To evaluate our approach, we conducted a set of quality (rules + model) and performance tests. The quality tests of the resulting rules are carried out using

---

[1] https://colab.research.google.com/.

[2] https://archive.ics.uci.edu/ml/datasets.php.

[3] http://www.almaden.ibm.com/cs/quest/syndata.html.

Table 1. Description of the used datasets

Type	Dataset	Nbr of trans	Nbr of items	Nbr of item occur	Size
Real	Spect-heart	187	86	8 415	Small
Real	Cars	1728	25	12 096	Small
Synthetic	C20D10	2,000	386	40,000	Small
Real	Chess	3,196	75	121,448	Medium
Actual	Mushroom	8,416	128	193,568	Medium
Synthetic	C73D10	10,000	2,178	750,000	Medium
Real	Pumsb-star	49,046	2088	2,524,993	Medium
Actual	Connect	67,557	129	2,972,508	Medium

metrics such as the number of valid rules as well as the average fitness while the predictive accuracy of the ML models is measured with the Root Mean Squared Error (RMSE) and the cross-validation. Performance tests are evaluated through execution time.

6.1   Results of GA Based on Off-Line Substitution Model

The tests conducted are of three types: performance, quality and precision. Their objectives are to study the effect of model type and encoding on the execution time, the number, and extracted rules' average fitness as well as RMSE. For the models, we used the library's default settings: the distribution strategy of each node of the decision tree according to the best characteristic, and 100 trees for random forests. We considered three layers including a hidden one containing 100 neurons and a maximum number of iterations equal to 200 for the convergence of each training data point in the neural network.

Parameters of the GA-ARM are: $\alpha = \beta = 0.5$, number of iterations = 100, population size = 100, probability of crossing = 0.9, probability of mutation = 0.1, maximum size of one rule = 4 items, minimum support = 0.1, minimum confidence = 0.5, Rate of real evaluations initialized to 1. Note that for each test result in the following, we consider the average of 20 executions.

We name **AG-HL-Model-Encoding** all GAs used, where Model can be DTR, RFR and MLPR, and Encoding can be: LE (label), OE (oneHot) and BE (binary).

6.2   Collect Training Data

In order to build the training dataset, we performed several optimizations with the basic GA-ARM to collect enough data. For each optimization, we save the rules generated throughout the optimization process as well as their real computed fitness. At the end of each optimization, the collected data is encoded and organized in the dataset. Duplicates are not taken into account.

6.3   Comparing Encoding According to the Models

In order to compare between encodings, we performed two types of tests: performance and accuracy tests.

Performance Tests. Figures 2, 3, 4 show the performance results of the ML models according to the label, oneHot and binary encodings. Those figures represent comparisons of the training time between the encodings according to the ML models DTR, RFR, and MLPR respectively.

We can deduce that the type of encoding impacts the training time of DTR, RFR, and MLPR models. For DTR and RFR models, the training time increases with the increase of the number of features used. Indeed, if we consider dataset C20d10, it has 386 features (items) when training with label, then we will have $2*386$ features with binary and $3*386$ features with oneHot. Moreover, we notice that the MLPR model gives good performance in terms of training time with encodings of type (0 1): oneHot and binary. Specifically, MLPR model results are better with binary because it manipulates fewer features than oneHot. However, labeling the MLPR dramatically increases training time.

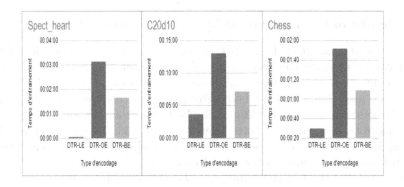

Fig. 2. Training time of AG-HL-DTR

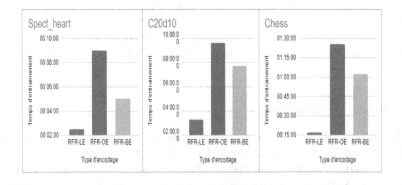

Fig. 3. Training time of AG-HL-RFR

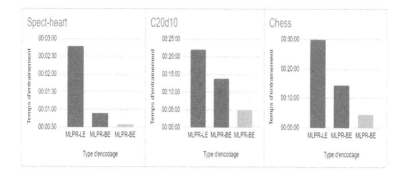

Fig. 4. Training time of AG-HL-MLPR

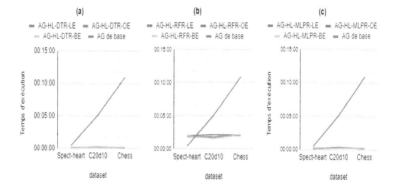

Fig. 5. Time execution of AG-HL according to models: (a) DTR (b) RFR (c) MLPR

Figure 5 shows a comparison of the encodings according to the execution time of the AG-HL-DTR, AG-HL-RFR and AG-HL-MLPR algorithms. We notice that the three algorithms give the same performance for all encodings. Therefore, we can deduce that the encoding time does not influence the prediction time of the ML models.

Accuracy. Table 2 shows the RMSE error of RFR, DTR and MLPR models according to the encodings for the Spect-Heart dataset.

From the results shown in Table 2, we find that for all models the value of the RMSE is very small. This is due to the size of the dataset and the number of collected rules on which the model is trained. Moreover, we find that applying the oneHot and binary encodings reduces the RMSE more than the label encoding. We notice that, for the first tests, we had underfitting, since the model failed to predict the outputs of the new data. With the increase in data, we were able to reduce the error of the models.

Table 2. RMSE errors for the spect-heart dataset according to encoding type

Model	RMSE		
	LabelEncoding	OneHotEncoding	BinaryEncoding
RFR	0.004	0.003	0.003
DTR	0.01	0.005	0.005
MLPR	0.01	0.004	0.003

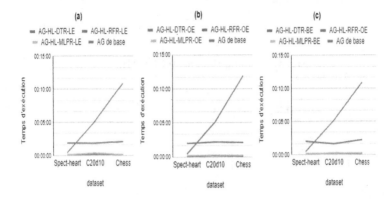

Fig. 6. Time executions of AG-HL-DTR, AG-HL-RFR and AG-HL-MLPR (a) Label encoding (b) OneHot encoding (c) Binary encoding

6.4 Comparative Study of Models According to Encodings

Figure 6 shows comparison in execution times between AG-HL-DTR, AG-HL-RFR and AG-HL-MLPR by varying the type of used encoding.

AG-HL-DTR and AG-HL-MLPR algorithms significantly reduced the execution time for all datasets. Moreover, AG-HL-RFR did not improve the performance of AG-ARM for Spect-heart dataset, and this was for all encodings applied. This is explained by the prediction time of the RFR model which requires 100 predictions made by the trees of the forest.

6.5 Visualization of Average Fitness with AG-HL Model

Figures 7, 8 and 9 show the evolution of the average fitness of the AG-HL-DTR for the spect-heart dataset over the generations according to the Label, oneHot and Binary encodings. The models correctly predict fitness and do not impact rule quality, implying that the models have been sufficiently trained.

For our approach, GA with DTR model and label encoding gave the best results for all test datasets. We have also noticed that the models perform well as the number of transactions increases.

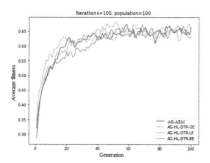

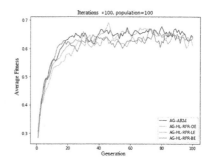

Fig. 7. Average fitness of solutions given by AG-HL-DTR

Fig. 8. Average fitness of solutions given by AG-HL-RFR

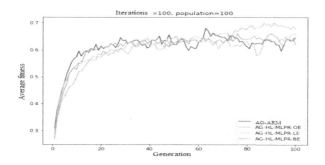

Fig. 9. Average fitness of solutions given by AG-HL-MLPR

7   Conclusion

In this work, we were interesting by speeding up the execution time of GA-ARM to extract ARs in a reasonable time. The computation of the fitness used to evaluate GA-ARM is the most time consuming task. For this, we proposed an offline substitution ML approach to approximate the fitness of GA. This solution allowed us, at first, to make a feasibility study of the integration of substitution model for the extraction of ARs. The obtained results show that the AG-HL-DTR significantly reduced the execution time for all encodings while maintaining the quality. This proves that the model has been sufficiently trained. Our presented work opens several perspectives. Among them, is the design of an on-line approach to approximation of GA.

References

1. Agrawal, R., Srikant, R.: Fast algorithms for mining association rules. In: Proceedings of 20th International Conference on Very Large Data Bases, VLDB, vol. 1215, pp. 487–499 (1994)

2. Guo, D., Chai, T., Ding, J., Jin, Y.: Small data driven evolutionary multi-objective optimization of fused magnesium furnaces. In: IEEE Symposium Series on Computational Intelligence (SSCI), pp. 1–8. IEEE (2016)
3. Hamdad, L., Benatchba, K., Bendjoudi, A., Ournani, Z.: Impact of genetic algorithms operators on association rules extraction. In: Rojas, I., Joya, G., Catala, A. (eds.) IWANN 2019. LNCS, vol. 11507, pp. 747–759. Springer, Cham (2019). https://doi.org/10.1007/978-3-030-20518-8_62
4. Huang, P., Wang, H., Ma, W.: Stochastic ranking for offline data-driven evolutionary optimization using radial basis function networks with multiple kernels. In: IEEE Symposium Series on Computational Intelligence (SSCI). IEEE (2019)
5. Huang, P., Wang, H., Jin, Y.: Offline data-driven evolutionary optimization based on tri-training. Swarm Evol. Comput. **60**, 100800 (2021)
6. Kabir, M.M.J., Xu, S., Kang, B.H., Zhao, Z.: A new evolutionary algorithm for extracting a reduced set of interesting association rules. In: Arik, S., Huang, T., Lai, W.K., Liu, Q. (eds.) ICONIP 2015. LNCS, vol. 9490, pp. 133–142. Springer, Cham (2015). https://doi.org/10.1007/978-3-319-26535-3_16
7. Li, J.-Y., Zhan, Z.-H., Wang, H., Zhang, J.: Data-driven evolutionary algorithm with perturbation-based ensemble surrogates. IEEE Trans. Cybern. **51**(8), 3925–3937 (2020)
8. Li, J.-Y., et al.: Boosting data-driven evolutionary algorithm with localized data generation. IEEE Trans. Evol. Comput. **24**(5), 923–937 (2020)
9. Mata, J., Alvarez, J.L., Riquelme, J.C.: Mining numeric association rules with genetic algorithms. In: Kůrková, V., Neruda, R., Kárný, M., Steele, N.C. (eds.) Artificial Neural Nets and Genetic Algorithms, pp. 264–267. Springer, Vienna (2001). https://doi.org/10.1007/978-3-7091-6230-9_65
10. Talbi, E.G.: Machine learning into metaheuristics: a survey and taxonomy. ACM Comput. Surv. (CSUR) **54**(6), 1–32 (2021)
11. Wakabi-Waiswa, P.P., Baryamureeba, V.: Extraction of interesting association rules using genetic algorithms. Int. J. Comput. ICT Res. **2**(1), 26–33 (2008)
12. Wang, H., Jin, Y., Doherty, J.: Committee-based active learning for surrogate-assisted particle swarm optimization of expensive problems. IEEE Trans. Cybern. **47**(9), 2664–2677 (2017)
13. Wang, H., et al.: Offline data-driven evolutionary optimization using selective surrogate ensembles. IEEE Trans. Evol. Comput. **23**(2), 203–216 (2018)
14. Yu, M., Li, X., Liang, J.: A dynamic surrogate-assisted evolutionary algorithm framework for expensive structural optimization. Struc. Multidis. Optim. **61**, 711–729 (2020)

Deep Learning Applied to Computer Vision and Robotics

Phenotype Discrimination Based on Pressure Signals by Transfer Learning Approaches

Marina Aguilar-Moreno[✉] and Manuel Graña

Computational Intelligence Group, University of the Basque Country (UPV/EHU), San Sebastian, Spain
marina.aguilar@ehu.eus

Abstract. Computational Ethology is the study of the animal behavior using advances in the field of Computer Vision and Artificial Intelligence. This field of research allows scientists to analyse and characterise behaviors and find out the differences that exist between distinct diseases or disorders for pharmacological studies. In this work we will analyse the data recorded with a multisensor system composed of a top video camera and a piezoelectric pressure sensor that records the movements of an animal. Specifically, this work aims to answer the research question of whether it is possible to differentiate phenotype of an animal model using transfer learning over the pressure signal alone. To do this, the piezoelectric signal will be analysed in the frequency domain by computing its spectrogram, and we segment the chunks corresponding to the locomotion events, previously detected. Convolutional neural models previously trained will be used for classification by applying a transfer learning approach. The results show that an accuracy of more than 96% is obtained and the confirmation that it is possible to classify phenotypes with the data obtained with pressure sensors.

Keywords: Animal Ethology · Piezoelectric · Pressure sensor · Locomotion · Animal behavior · Tracking · Signal processing · Binary classification

1 Introduction

Ethology is defined as the discipline that studies the animal behavior in terms of its phenomenological, causal, ontogenetic and evolutionary aspects, in order to provide answers to the causes and development that animal behavior undergoes, as well as to understand how it is performed [1], bearing in mind that behavior is understood as the set of muscular responses of a living being as a consequence of an external stimulus and internal motivation [2]. In this sense, Computational Ethology (CE) incorporates the advances in Computer Vision and Artificial Intelligence (AI) in the study of animal behavior, allowing to automate and

I. Rojas et al. (Eds.): IWANN 2023, LNCS 14135, pp. 145–156, 2023.
https://doi.org/10.1007/978-3-031-43078-7_12

increase the precision in experimentation analysis, and also giving more information to characterize the behaviors. By analysing the behavior is possible to extract characteristics from them that allow us to describe them quantitatively. From a pharmacological point of view, ethological experimentation give another possibility to test new medicines, as it is capable of quantifying behaviors and comparing them between different subjects. In addition, it is possible to generate animal models with genetic modifications that provide study subjects with anatomy, physiology or response to a pathogen that are sufficiently similar to humans to be able to extrapolate the results obtained to them [3]. The most commonly used models in research are rodents, especially mice and rats, zebra fish, amphibians and reptiles, birds and other small animals.

There exist several kinds of sensors to carry out experiments in CE such as depth and RGB video cameras, accelerometers, magnetometers, gyroscopes and pressure sensors, among others. In the study of behavior, it is interesting to use sensors that are as non-invasive as possible. In this sense, pressure sensors can be useful as they allow data to be collected without interfering with the behavior of the animals, such as the Phenotypix platform [4], which offers a high sensitivity for detecting pressure changes, making it possible to detect freezing, breathing and heartbeats in mice.

Apart from sensors, there are numerous techniques developed in the field of Computer Vision and AI that are used to process the experimental data in CE.

DeepLabCut is a markerless motion tracking system based on transfer learning and it is easily tailored to the specific experimental setting. Recently, it has been used to analyse stroke recovery in rodent [5], X-ray video of rodent locomotion [6], cardiac physiology assessment in zebra fish [7], multi animal pose estimation and tracking [8]. Bonsai [9] is another tracking framework that can perform other tasks, such as data acquisition and experiment control. LEAP [10] and SLEAP [11] are both tools based on Deep Neural Network for pose estimation and, being SLEAP specifically designed for social interaction.

JAABA is a well-known application that uses a semi-supervised machine learning algorithm for automatic annotation of animal behavior [12]. There are also neural networks that identify the genetics of a mouse by analyzing grooming behavior [13] and algorithms based on deep learning such as DeepEthogram, which classifies behaviors in a supervised manner from raw pixels [14]. Convolutional neural networks are also very useful for classification and behavior detection using frames and currently also 3D convolutional networks for behavior automatic classification [15] and for grooming detection of mouse [16]. On this matter, transfer learning is a tool that takes advantage of pre-trained networks to develop models instead of training from scratch, decreasing the amount of data and computation time [17,18].

Thus, this paper proposes a research question based on the hypothesis that an strain classifier can be implemented from the pressure signals after segmentation based on the images recorded by a top video camera by using pre-trained models and transfer learning. These results are important in Pharmacology to improve the understanding of the behavior in animals, establishing differences between two phenotypes quantitatively. There are some works in the literature that provide solutions to this behavioral phenotyping problem. One of these classifiers is carried out by using video recordings, where a SVM is fed with a 32-dimensional feature vector that indicates the relative frequency of each of the 8 behaviors of interest [19]. Another example analyses the percentage of time in open arms, total distance traveled and percentage of time in the center with k-Nearest Neighbor from RGB-D images [20].

In the next section, the materials and the recording system is explained, together with the data processing, the classification methods and their evaluation. Next, the results of the validation test are described, showing the accuracy, precision, recall and F1 score for the classification models. Last section makes a discussion about the results of the research question.

2   Materials and Methods

2.1   Animals

The experiments were carried out with 12 mice with two different strains and the dataset are divided as follow:

- 7 wild-type (WT)[1] mice with a total duration of recordings of 4 h and 20 min
- 5 transgenic Fmr1-knockout (Fmr1-KO)[2] mice with a total duration of recordings of 4 h and 20 min

The recordings took place during the light period and it was the first time these models were in contact with the recording system. All animals were bred in the laboratory animal facility in collective cages, and transferred to individual cages for the duration of the experiments. Animals were kept on a 12 h/12 h light/dark cycle, provided with nesting material and food and water *ad libitum*. All experiments were performed during the light period under constant mild luminosity (60–70 Lux). All experimental procedures were performed in accordance with EU directives relating to the protection of animals used for experimental and scientific purposes, in accordance with the reference work [4].

2.2   Behavioral Data Acquisition

The pressure sensor environment used to record the experiments is composed of an opaque-walled cage and a base resting on several piezoelectric sensors at a sampling rate of 20 kHz [4]. This platform is mounted on a table stabilized with pressured air in order to remove environmental motion noise. A Logitech HD Webcam C270 video camera is placed on top of the cage to record the experiment at 25 fps as shown in Fig. 1.

---

[1] Wild-type gene is a term used to describe a gene when it is found in its natural, non-mutated (unchanged) form.

[2] Fmr1-knockout (Fmr1-KO) mice may be useful for studying behavioral and synaptic abnormalities associated with Fragile X Syndrome.

Fig. 1. The experimental platform.

Animals were introduced individually into the platform and the walls and cage were cleaned with 70% ethanol before each recording. To visualise the piezo-electric signal obtained during the recordings, Spike2 software (CED, Cambridge, UK) was connected to a computer where the data were stored for later analysis. At the same time the videos were recorded with the top camera and stored separately from the signals. The data were then processed with Bonsai [9], MATLAB (Mathworks, Natik, MA, USA) and Python [21]. The length of the videos range from 20 min to 1 h.

2.3   Data Processing

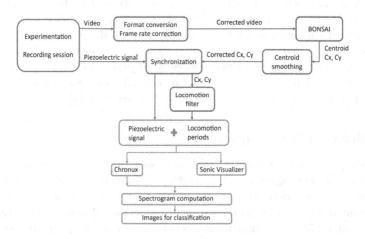

Fig. 2. Data processing pipeline.

Once the data are collected after the acquisition phase, they are processed to obtain the images for the classification algorithms. The pipeline of this process is shown in Fig. 2 and described below:

1. Check whether the video has the correct format and duration. If not the multimedia framework ffmpeg [22] is used to change the video format and correct its duration by changing the frame rate. This framework can benefit from GPU acceleration and its functions can be automated with Python or MATLAB scripts.
2. Compute the x and y position of the animal centroid through videos using Bonsai software carrying out the steps shown in Fig. 3:
   (a) Apply an affine transformation to align the video with the window frame.
   (b) Crop the video to center the region of interest to the platform where the animal will move.
   (c) Apply an binarization to change colors to gray scale. This step facilitates the animal segmentation. The function checks if the color is darker than a predefined threshold to paint the pixel white and otherwise black. The result is an image in black and white where the animal is white and the background is in black. In case there are objects apart from the animal, these object will be seen in white too.
   (d) Find contours based on the color changes.
   (e) Take binary region and select the largest one to avoid objects out of interest.
   (f) Compute the centroid of the largest binary region in pixels.
   (g) Convert centroid from pixels to cm by using a proportional relationship taking into account the size of the recording platform.
3. Smooth the x and y centroid computing the average of 3 consecutive points. This eliminates possible positional errors and filter the trajectory of the animal.
4. Synchronize the time scale from of the centroid and the piezoelectric signal, taking into account the time offset between both time scales and skipping the first seconds of the video before the animal appears on the recording platform.
5. Apply a locomotion filter to detect periods when the animal is walking using a MATLAB script with a minimum velocity of 2.5 cm/s and a minimum distance of 2.5 cm for each period.
6. Obtain the spectrogram using Chronux library and Sonic Visualizer for the whole signal and segment the time periods corresponding to valid locomotion events.
7. Obtain the images from these spectrograms and save them to train and test the classification models.

Table 1. Parameters for spectrogram computation with Chronux library.

Parameters	Value 1	Value 2	Default values
Window size (s)	1	2	–
Windows step (s)	0.1	0.2	–
Tapers	[4, 2]	[3, 5]	[3, 5]: A numeric vector [TW K] where TW is the time-bandwidth product and K is the number of tapers, less than or equal to 2TW-1
Frequency of interest (Hz)	[1.5–40]	[4–112]	[0–Fs/2] (Fs: sampling frequency)

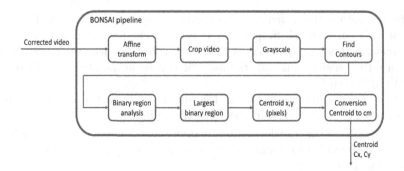

Fig. 3. Bonsai pipeline for video processing.

Table 2. Parameter for spectrogram computation with Sonic Visualizer.

Parameter	Value	Range of values
Colour	Green	[Green, Sunset, ... , Wasp, Ice, ...]
Scale	dB	[Linear, Meter, dB^2, dB, Phase]
Window size	256	[32, 64, 128, 256, 512, ... , 16384, 32768]
Overlap	93.75%	[none, 25%, 50%, 75%, 87.5%, 93.75%]
Show	All bins	[All Bins, Peak Bins, Frequencies]
Scale	Linear	[Linear, Log]

Related to the Chronux library[3], there are some parameters which can be tuned, such as window size, window step, number of tapers or frequency of interest. The window size is how long the number of samples is in every computation and the window step is the size of the displacement to take the next sample. Modifying the number of tapers is a way to control the degree of the smoothing in the spectrogram and the frequency of interest allows to limit the frequency

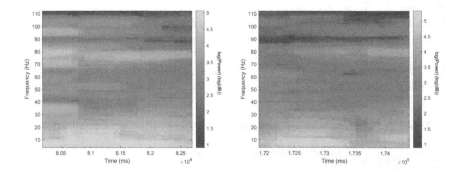

Fig. 4. Samples of spectrogram images computed with Chronux for a segmented signal chunk during locomotion time for a) wild type and b) transgenic Fmr1-KO model.

---

[3] http://chronux.org/.

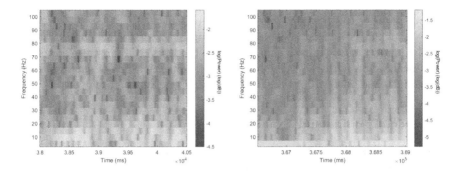

Fig. 5. Samples of spectrogram images computed with Sonic Visualizer for a segmented signal chunk during locomotion time for a) wild type and b) transgenic Fmr1-KO model.

range. Table 1 shows the parameters used in the processing with Chronux library and the Fig. 4 shows two spectrogram chunks obtained from the locomotion periods for a wild type (a) and transgenic Fmr1-KO (b) models with Chronux library. This spectrogram has been calculated for a frequency band of interest of 4–112 Hz, a window size of 1 s, a window step 0.1 s and number of tapers [3, 5].

For the spectrogram computed with Sonic Visualizer, the parameter tuning was carried out manually searching for the configuration that highlights the differences in the spectrogram for different strains. This configuration is shown in the Table 2 and the Fig. 5 shows two spectrogram chunks obtained from the locomotion periods for a wild type (a) and transgenic Fmr1-KO (b) models with Sonic Visualizer software.

2.4   Pre-trained Models

To classify the images between wild type and Fmr1-KO models we have used pre-trained models by applying transfer learning. This process takes advantage of models that have been previously trained with a large amount of data and high hardware resources, so that to adapt it to a new dataset, only the last layer has to be trained instead of starting from scratch. In this work we have used the following convolutional neural networks:

- AlexNet [23] is a well-known convolutional neural network which won the ImageNet Large-Scale Visual Recognition Challenge (ILSVRC) in 2012. ImageNet is a dataset of over 15 million labeled high-resolution images belonging to 22,000 categories. However, a subset of ImageNet with 1000 images for 1000 categories was used for the ILSVRC. The architecture of AlexNet is composed of five convolutional layers and three fully connected layers with a final 1000-way softmax.
- GoogLeNet [24] is a deep convolutional neural network with 22 layers winner of ILSVRC in image classification and detection in 2014. This network is based

on the Inception architecture that consists of making the network wider than deeper in order to take into account different scales.
- ResNet50 [25] is a deep convolutional neural network which uses learning residual functions to train the model. It won the ILSVRC15 in image detection and localization with the ImageNet dataset doing the task of classifying the image into one of the 1000 categories in the ImageNet hierarchy.

2.5 Model Training and Evaluation

The question presented in this contribution is formulated as a binary classification problem that aims to discriminate between wild type (0 class) and Fmr1-KO (1 class) models. The spectrogram images previously generated during the locomotion periods will be used for training the classifiers, described in the Subsect. 2.4, along with the training parameters shown in the Table 3.

Table 3. Training parameters

Parameter	Value
Solver	Adam
Learning rate	0.0001
Mini batch size	52
L2 Regularization	0.0001
Folds for Cross-validation	5

The image dataset is divided into two parts, one for training and one for testing with a proportion of 80% and 20%, respectively. To train the models, a 5-fold cross-validation has been applied over the training dataset, dividing it into 5 folds. For each of the 5 training sessions, a model is trained with 4 folds and the 5th is used as development set, repeating this process alternating the development set and obtaining 5 different models. For validation, the best of the 5 previous models is chosen according to the accuracy and it is applied to the test set.

The image size is variable depending on the model, so the dimensions must be $227 \times 227 \times 3$ for AlexNet and $224 \times 224 \times 3$ for ResNet50 and GoogLeNet.

The models were implemented in MATLAB using an Intel(R) Core(TM) i9-9880H computer with 64 GB of RAM and a NVIDIA Quadro RTX 3000 GPU. The execution times for training and validation, the number of layers and the learnables parameters are showed in the Table 4.

Table 4. Execution time for training and evaluation (cross validation), number of layers and number of learnables parameters.

Algorithm	Execution time	Layers	Total learnables
AlexNet	≈35 min	25 (depth 8)	56 876 418
ResNet50	≈4 h	177 (depth 50)	23 538 690
GoogLeNet	≈2 h	144 (depth 22)	5 975 602

3   Results

After obtaining the images showed in the Sect. 2.3, we have trained the models explained in the Sect. 2.4. In this chapter we show the results of accuracy, precision, recall and F1 score for the validation tests. The following parameters have been used for the experiments:

– Minimum locomotion duration of 2500 ms
– Sampling window size of 2500 ms
– Window step of 100 ms

3.1   Results for Spectrogram Computed with Chronux Library

Table 5 shows the accuracy, precision, recall and F1 score results after using the AlexNet, GoogLeNet and ResNet50 networks for the images calculated using Chronux library with 1 s window size and 0.1 s window step for frequencies from 4 Hz to 112 Hz. An accuracy of value of 99.47% has been reached with GoogLeNet and ResNet50 for a number of tapers of [3, 5] and 100% with AlexNet for a number of tapers of [4, 2]. The precision, recall and F1 score values are very similar and above 0.97 for the three models.

Table 5. Results for spectrogram images computed with Chronux library for 1 s window size and 0.1 s window step.

Tapers	3–5				4–2			
	Accuracy	Precision	Recall	F1 score	Accuracy	Precision	Recall	F1 score
AlexNet	98.40%	1.00	0.97	0.98	**100.00%**	**1.00**	**1.00**	**1.00**
GoogLeNet	**99.47%**	**0.99**	**1.00**	**0.99**	99.47%	1.00	0.99	0.99
ResNet50	**99.47%**	**0.99**	**1.00**	**0.99**	99.47%	0.99	1.00	0.99

Similarly, Table 6 shows the accuracy, precision, recall and F1 score results for 2 s window size and 0.5 s window step. Accuracy of values of 99.47% and 97.86% have been reached with ResNet50 for a number of tapers of [3, 5] and [4, 2], respectively. The precision, recall and F1 score values are very similar and above 0.93 for the three models.

Table 6. Results for spectrogram images computed with Chronux library for 2 s window size and 0.5 s window step.

Tapers	3–5				4–2			
	Accuracy	Precision	Recall	F1 score	Accuracy	Precision	Recall	F1 score
AlexNet	98.93%	0.98	1.00	0.99	96.79%	0.93	1.00	0.97
GoogLeNet	97.86%	0.97	0.99	0.98	97.33%	0.97	0.98	0.97
ResNet50	**99.47%**	**0.99**	**1.00**	**0.99**	**97.86%**	**0.97**	**0.99**	**0.98**

3.2   Results for Spectrogram Computed with Sonic Visualizer

Table 7 shows the accuracy, precision, recall and F1 score results for validation test after training the AlexNet, GoogLeNet and ResNet50 networks with the images from Sonic Visualizer spectrograms. An accuracy value of 100% has been reached with AlexNet and more than 96% with GoogLeNet and ResNet50. In relation to precision, recall and F1 score, all three models return similar values above 0.93.

Table 7. Results for spectrogram images computed with Sonic Visualizer.

Window size - overlap	256 93.75%			
	Accuracy	Precision	Recall	F1 score
AlexNet	**100.00%**	**1.00**	**1.00**	**1.00**
GoogLeNet	96.26%	0.93	0.99	0.96
ResNet50	96.79%	0.95	0.98	0.97

4   Conclusions and Future Work

This paper proposes a research question based on the hypothesis that it is possible to differentiate phenotypes on the basis of pressure signals applying transfer learning. From the data obtained, spectrogram images have been created using the Chronux library and Sonic Visualizer software. These images have fed three different pre-trained models: AlexNet, GoogLeNet and ResNet50.

The classification results for the validation set show that minimum accuracy obtained for all the experiments is 96% reaching 100% for the CNN AlexNet using the images generated with the spectrograms from Sonic Visualizer and some configurations with the Chronux library. This paper also presents the training and evaluation times showing that the computation times for the CNNs range from 35 min with AlexNet to 4 h with ResNet50.

To sum up, it is concluded that it is possible to perform strain classification for animal model, using a recording system composed of a piezoelectric platform measuring pressure and a top camera. This system allows experiments to be carried out in a non-invasive way, offering high sensitivity measurements, making it possible to establish differences between distinct strains and classify them with high accuracy, precision, recall and F1 score values.

As future work, we could obtain spectrogram parameters automatically and integrate this step in the whole algorithm with a grid search to improves the efficiency. Moreover, we plan to apply this approach to an experimental study on healthy ageing in the elderly, where we are collecting data from a walking platform and an electroencephalogram (EEG) to identify gait anomalies.

Acknowledgments. This work has been partially supported by grant PRE2018-085294 funded by MCIN/AEI/10.13039 /501100011033 and by "ESF Investing in your future" through the project TIN2017-85827-P, with the grant PRE2018-085294, PID2020-116346GB-I00, and grant IT1689-22 as university research group of excellence from the Basque Government. We gratefully acknowledge the data shared by Prof. Leinekugel to test our approach.

References

1. Anderson, D., Perona, P.: Toward a science of computational ethology. Neuron **84**, 18–31 (2014)
2. Gomez-Marin, A.: A clash of umwelts: anthropomorphism in behavioral neuroscience. Behav. Brain Sci. **42**, e229 (2019)
3. Datta, S., Anderson, D., Branson, K., Perona, P., Leifer, A.: Computational neuroethology: a call to action. Neuron **104**, 11–24 (2019)
4. Carreño-Muñoz, M., et al.: Detecting fine and elaborate movements with piezo sensors provides non-invasive access to overlooked behavioral components. Neuropsychopharmacology **47**(4), 933–943 (2022)
5. Weber, R.Z., Mulders, G., Kaiser, J., Tackenberg, C., Rust, R.: Deep learning-based behavioral profiling of rodent stroke recovery. BMC Biol. **20**, 232 (2022)
6. Kirkpatrick, N.J., Butera, R.J., Chang, Y.-H.: Deeplabcut increases markerless tracking efficiency in x-ray video analysis of rodent locomotion. J. Exp. Biol. **225** (2022)
7. Suryanto, M.E., et al.: Using deeplabcut as a real-time and markerless tool for cardiac physiology assessment in zebrafish. Biology (Basel) **11** (2022)
8. Lauer, J., et al.: Multi-animal pose estimation, identification and tracking with deeplabcut. Nat. Methods **19**, 496–504 (2022)
9. Lopes, G., Monteiro, P.: New open-source tools: using bonsai for behavioral tracking and closed-loop experiments. Front. Behav. Neurosci. **15** (2021)
10. Pereira, T., et al.: Fast animal pose estimation using deep neural networks. Nat. Methods **16**(1), 117–125 (2019)
11. Pereira, T.D., et al.: Sleap: a deep learning system for multi-animal pose tracking. Nat. Methods **19**, 486–495 (2022)
12. Kabra, M., Robie, A., Rivera-Alba, M., Branson, S., Branson, K.: Jaaba: Interactive machine learning for automatic annotation of animal behavior. Nat. Methods **10**(1), 64–67 (2013)
13. Geuther, B., Peer, A., He, H., Sabnis, G., Philip, V., Kumar, V.: Action detection using a neural network elucidates the genetics of mouse grooming behavior. eLife **10** (2021)
14. Bohnslav, J., et al.: Deepethogram, a machine learning pipeline for supervised behavior classification from raw pixels. eLife **10** (2021)

15. Long, L., et al.: Automatic classification of cichlid behaviors using 3D convolutional residual networks. iScience **23**(10) (2020)
16. Sakamoto, N., Kobayashi, K., Yamamoto, T., Masuko, S., Yamamoto, M., Murata, T.: Automated grooming detection of mouse by three-dimensional convolutional neural network. Front. Behav. Neurosci. **16** (2022)
17. Jin, T., Duan, F.: Rat behavior observation system based on transfer learning. IEEE Access **7**, 62152–62162 (2019)
18. Ruiz, J., Pérez, J., Blázquez, J.: Arrhythmia detection using convolutional neural models. Adv. Intell. Syst. Comput. **800**, 120–127 (2019)
19. Jhuang, H., et al.: Automated home-cage behavioural phenotyping of mice. Nat. Commun. **1**(6) (2010)
20. Gerós, A., Magalhães, A., Aguiar, P.: Improved 3D tracking and automated classification of rodents' behavioral activity using depth-sensing cameras. Behav. Res. Methods **52**, 2156–2167 (2020)
21. Van Rossum, G., Drake Jr., F.L.: Python Reference Manual. Centrum voor Wiskunde en Informatica Amsterdam (1995)
22. Tomar, S.: Converting video formats with FFMPEG. Linux J. **2006**(146), 10 (2006)
23. Krizhevsky, A., Sutskever, I., Hinton, G.E.: Imagenet classification with deep convolutional neural networks. In: Pereira, F., Burges, C., Bottou, L., Weinberger, K. (eds.) Advances in Neural Information Processing Systems, vol. 25. Curran Associates Inc (2012)
24. Szegedy, C., et al.: Going deeper with convolutions, vol. 07–12-June-2015, pp. 1–9 (2015)
25. He, K., Zhang, X., Ren, S., Sun, J.: Deep residual learning for image recognition. CoRR, abs/1512.03385 (2015)

AATiENDe: Automatic ATtention Evaluation on a Non-invasive Device

Felix Escalona, Francisco Gomez-Donoso$^{(\boxtimes)}$, Francisco Morillas-Espejo,
Monica Pina-Navarro, Luis Marquez-Carpintero, and Miguel Cazorla

University Institute for Computing Research,
San Vicente del Raspeig, Alicante, Spain
{felix.escalona,fgomez,francisco.morillas,monica.pina,
luis.marquez,miguel.cazorla}@ua.es

Abstract. The study of student attention is an important topic in education because this type of analysis provides important information to teachers to potentially improve the quality of their classes. In this paper, we present AATiENDe, a system that uses emotion recognition, gaze direction approximation and body posture analysis as features to classify whether students are paying attention to their computer screens. To do this, we use a mixture of deep learning-based techniques and novel machine learning techniques applied to tabular classifiers to produce the final predictions. We also capture and label a customized dataset to train the models. Our approach provides over 90% accuracy using two cameras and over 80% accuracy using only the foreground camera.

Keywords: deep learning · emotion recognition · body pose analysis · attention estimation

1 Introduction

Student feedback is one of the most important pieces of information for a teacher to improve their teaching skills and the quality of their classes. However, the methods that have traditionally been used consist of a battery of questions asked at the end of each semester, whose content does not usually allow us to detect exactly what are the moments or circumstances that cause the lack of attention or motivation of students.

In recent years, tools are being incorporated to obtain real-time information about the general state of attention of the students on the teacher's lesson, allowing the teacher to redirect the class to regain the attention of their students. It also allows them to know exactly which parts of the syllabus are the most boring or complicated to understand, which allow them to rethink those sessions.

This real-time information can be obtained thanks to the enormous recent advances in artificial intelligence, specifically in the field of artificial vision, which makes it possible to process images from color cameras to estimate those aspects of classroom performance that they want to know, such as the posture of the students, their emotions or the direction in which they focus their gaze.

I. Rojas et al. (Eds.): IWANN 2023, LNCS 14135, pp. 157–168, 2023.
https://doi.org/10.1007/978-3-031-43078-7_13

This work proposes AATiENDe (Automatic ATtention Evaluation on a Non-invasive Device) a system capable of evaluating a person's level of attention using color cameras. This system uses information from several cameras including emotion recognition, gaze direction approximation and 2D pose estimation.

Specifically, the main contributions of this work are:

- A real-time system that integrates the outputs of different image processing algorithms and different viewpoints to estimate the attention level of a subject in a non-intrusive manner.
- Ablation study to determine which characteristics most affect the proper recognition of a person's attention.
- A custom dataset and recording methodology to capture the required data to infer a person's attention from regular color cameras.

The rest of the paper is organized as follows: first, we review some significant related works in Sect. 2. Then, we describe our approach in Sect. 3. In Sect. 4, we show the results of the experiments we conducted to validate our proposal. Finally, we draw the conclusions and state the limitations and future work in Sect. 5.

2   Related Works

Recently, assessing student participation in the classroom has become increasingly important and is now considered a crucial aspect for enhancing the learning process and increasing productivity in education.

In their research, [1] separates student engagement into three categories: affective, behavioral, and cognitive. Affective engagement refers to how students feel in their respective class. Behavioral engagement concerns students' participation in class and their attitude towards class tasks and assignments. Lastly, cognitive engagement involves students' thinking during academic tasks.

Student engagement can be quantified through various behaviors or expressions, including eye movements, facial features, or posture.

Previous research has explored the usefulness of analyzing gaze data to determine student engagement during a class or lesson.

The study presented in [2], relates the direction of the student's head to the level of engagement, assuming that, if the student looks directly at the teacher, it indicates that the person is engaged. Abedi and Khan [3] propose a model to track students' faces and their gaze direction, assuming that the more the students' gaze differs from the camera direction, the lower their level of attention. On the other hand, [4] proposes a model to detect people sleeping in class.

Additionally, numerous research studies have concluded that an individual's body posture is a clear and reliable indicator of their level of engagement in learning.

In [5], students' faces are detected using Haar Cascade Classifier [6] and then tracked using TLD (Tracking Learning Detection) face tracking algorithm [7], which locates the students in each frame. Optical flow [8] is determined for each student to assess their engagement by determining their movement from their initial attentive position. Furthermore, students' head movements are analyzed to see where their attention is directed.

[9] proposes a two-stage freehand gesture recognition method for pose estimation and hand gesture recognition.

Facial expressions are one of the most powerful and universal ways for people to communicate their emotional states and intentions. Previous research has emphasized the significance of facial expressions in affective states related to learning.

Selim et al. [10], presents three new end-to-end hybrid models based on EfficientNet B7, which offers more accuracy than the previous ConvNets. In a first model, they integrate EfficientNet B7 [11] together with LSTM [12]. Meanwhile, the second and third proposed models use Bidirectional LSTM [13] and TCN [14] to preserve past and future frame information, allowing the input of frames to be flow in both backward and forward directions.

In this paper, we propose AATiENDe (Automatic ATtention Evaluation on a Non-invasive Device), a system that can evaluate an individual's level of attention through the use of color cameras. This system incorporates data from multiple cameras and features including emotion recognition, gaze direction approximation, and 2D pose estimation.

3 Approach to Evaluate the Attention Level

As mentioned before, we propose AATiENDe, which is a system to evaluate the attention level of a subject that only relies on cameras. Specifically, we use a range of different metrics to evaluate whether they are paying attention to the information displayed in the screen. The metrics we obtain are the facial expression, the tridimensional pose and the gaze direction of the user. These three metrics are fed to a classifier which is in charge of finally stating if the user was paying attention or not. A diagram of the proposed system is displayed in Fig. 1. As shown, two cameras are being used. First, a closeup camera, which is placed on top of the display, is in charge of capturing the facial expressions and the gaze direction. Then, a bird-eye-view (BEV) camera will provide a third person perspective which is more suitable for estimating the 2D pose of the user.

3.1 Emotion Recognition

To recognize the emotion displayed by the user from an RGB image, we use the approach described in [15]. In this work, the authors perform a study of the labeling errors that are present in the most common databases for facial expression recognition (FER2013, AffectNet and NHFI) and propose a methodology to prune misclassified and irrelevant images. With this preprocessing, the authors created a higher quality combined dataset and then trained a CNN with the architecture described in [16]. As this network requires an image of a face with aligned eyes, image preprocessing is required beforehand. The first step

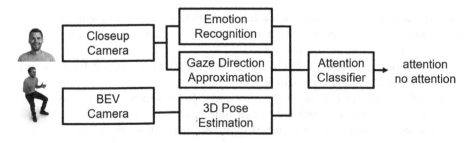

Fig. 1. Diagram of the system. There are two cameras that feed the algorithms to estimate the emotion, the gaze direction and the 2D pose. The outputs are fused and send to an attention classifier that finally states the level of attention shown by the subject.

consists of detecting the face in the image. For this, we have made use of the pre-trained model with the SSD architecture, trained using the Caffe framework, within the OpenCV image processing library [17]. The output of this method provides us with the clipping of each of the faces present in the image along with a probability.

The second step consists in aligning the eyes so that they share the same horizontal line. Although it is not required in the paper [15], this processing is carried out because traditionally many datasets contain images with the faces aligned, so the performance of many face recognition or emotion analysis models can be strongly affected when faces appear skewed in the image. To perform this process, To carry out this process, it has been used the method presented in x [18]. This method provides 68 keypoints relative to different parts of the face such as mouth, eyebrows, eyes, nose and jaw. Once the eyes have been obtained, the rotation matrix is calculated so that both eyes are in the same horizontal position using the nose as a pivot.

Finally, the aligned face is sent to the emotion recognition network described above, which provides the probability for each of the 7 emotions contemplated in the training datasets: Angry, Disgust, Fear, Happy, Neutral, Sad and Surprise.

3.2 2D Pose Estimation

In order to capture the two-dimensional pose, we leveraged the BlazePose GHUM Heavy [19]. This architecture features an encoder-decoder shape that is used to predict heatmaps for all joints, followed by another encoder that regresses directly to the coordinates of all joints. The key insight behind this work is that the heatmap branch can be discarded during inference so it is lightweight and has low computational requirements.

The architecture provides 33 different keypoints corresponding to different joints in the body. The keypoints are provided in the space of the image and, thus, they are two-dimensional. In our case, as the lower part of the body is likely to be under the desk, the corresponding keypoints are discarded. Namely, we only consider the points of the shoulders, the arms, the hips and a single point in the head. The output of this step, thus, is a list of 9 keypoints, corresponding to the joints of the upper body part. Each keypoint is composed of 2 values that are the X and Y position of the joint in the image.

3.3 Gaze Direction Approximation

As mentioned in Sect. 2, most works involves some kind of eye tracking to perform attention estimation. We also adopted it as part of our system, relying in the head pose estimation as an approximation.

MediaPipe [20] has a face mesh solution that estimates 3D locations with a single RGB camera. For that, it leverages two different deep learning models, i.e. a detector that computes face location and a 3D face keypoint detector which predicts the 3D surface via regression. The combination of these models provide 468 different landmarks composed of X, Y and Z coordinates. In addition, it should be mentioned that one of the advantages of using MediaPipe is its high frame rate.

In order to approximate the gaze direction, the Euler angles of the face are calculated. Namely, the face rotation over X (roll), Y (pitch) and Z (yaw) axes. For that, the Perspective-n-Point (PnP) problem must be solved between the landmarks detected in the image and the landmarks of the canonic face the method includes.

After collecting the facial landmarks of the image, the PnP problem is solved and the Euler angles are extracted. Note that the output of this module is a vector with the rotational angles (roll, pitch, yaw).

3.4 Fusion of Features and Classification

The final step of the proposed pipeline is to merge the outputs of the emotion recognition, the gaze direction and the 2D pose estimation systems. To do so, the outputs are linearized together so all the values are arranged in a vector. Thus, for each image we compute a feature vector of 18 (9 joints × 2 components) + 7 (emotions) + 3 (gaze direction). This is, a feature vector of 28 components. The values are normalized so all the values range from 0 to 1 within each of the outputs and not within the whole feature vector to keep the values on the corresponding space while preventing bias in the classifiers due to figures too big.

The normalized feature vector is fed to the classifier. We tested a number of state-of-the-art tabular classifiers, namely XGBoost [21], LightGBM [22], ElasticNet [23], Classic Neural Network [24] and Support Vector Machine [25]. The output of the classifier is a vector of 2 components that states the probability of being paying-attention or not-paying-attention. These classifiers are briefly described next.

The tabular classifier XGBoost is an implementation of gradient boosting, which is based on decision trees as a base. XGBoost uses L1 (Lasso) and L2

(Ridge) regularization to minimize overfitting. Light GBM is another implementation of gradient boosting, it focuses on improving training speed and greater accuracy. ElasticNet is a regularization technique for linear regression models. It allows to combine L1 and L2 techniques, it is especially used to avoid overfitting. Support Vector Machine computes a hyperplane that separates two different classes in the best possible way based on the provided points. Finally, a Classic Neural Network is a machine learning model based on an architecture composed of layers of interconnected nodes.

4   Experiments

In this section, we explain the details relevant to the experiments such as hardware and software setup specifications, dataset collection and the results achieved by our proposal.

4.1   Hardware and Software

All the experiments were run in a desktop workstation that features an Intel i7-8700 processor at 4.6 GHz and 12 cores, 32 Gb DDR4 RAM and a NVIDIA GTX1080Ti for GPU computations mounted on a Z390 Asus Aorus Elite motherboard. The operating system is Ubuntu 18.04. All the algorithms were implemented in Python 3.8, Tensorflow 2.11, OpenCV 4.6 and MediaPipe 0.9. The steps of the pipeline are interconnected using webservices for easy distributed computation if needed.

4.2   Dataset Collection and Description

Despite there are a few public datasets such as [26–28] to train attention classifiers, they are just focused on the gaze direction or in other physical features like electromyographic (EMG) sensors. While they provide varied results, they all share a major drawback: the requirement of a physical, specific and expensive equipment such as eye trackers or high resolution EMG sensors.

Due to this, we chose to build our custom dataset. The hardware setup consists of a computer screen laying on a desk with a Logitech C920 color camera on the top of it. The camera is pointing to the face of the subject, who would be sitting in front of the desk. Another Logitech C920 color camera is placed besides the person providing a BEV of the subject from above and slightly tilted downwards. Thus, the dataset we captured is composed of the two different streams of color frames. Random samples of the dataset are shown in Fig. 2.

The subjects are instructed to be shown a set of exciting and interesting videos and another set of boring videos with no further instructions. The videos are chosen from internet free repositories based on their personal preferences in order to capture attention and no-attention samples. The videos have different lengths, ranging from 2 to 8 min, but we stored the frames corresponding to the center minute of each experience. In total, 7 subjects took part in the dataset,

Fig. 2. Random samples of the dataset depicting both image streams.

obtaining a total of around 20.000 frames distributed among two classes: paying-attention and not-paying-attention. Two different splitting methods are used. First, all the frames are shuffled and randomly divided so the 70% of the frames are used for training and the rest for testing the system. This is the intra-samples version. Then, in the intra-subjects version, we did the same with the participants so 5 subjects were used for training and the rest for testing. This is done to check if the method is able to generalize to different individuals. It is worth noting that we actually used one frame every fifteen instead of using all frames to avoid considering information that is too similar.

4.3   Results for Different Classifiers

Table 1 shows that LightGBM and XGBoost provide similar results and clearly outperform the rest of the methods. This is understandable as both models rely in implementations of classic Gradient Boosting. Their regularization and optimization make them more efficient and accurate compared to the other tested methods. The ElasticNet regularization has been used in a logistic regression model, yielding the worst results in the comparison. The performances of the trained models can be seen in the scores obtained: accuracy, area under the curve (AUC) and F1.

Table 1. Results for different attention classifiers for intra-samples and intra-subjects versions of the dataset.

	Intra-Sample			Intra-Subject		
Classifier	Acc.	AUC	F1	Acc.	AUC	F1
XGBoost	0.948	0.947	0.952	0.877	0.857	0.908
LightGBM	0.954	0.952	0.957	0.877	0.876	0.904
ElasticNet	0.784	0.778	0.809	0.810	0.798	0.852
Classic NN	0.907	0.904	0.916	0.849	0.843	0.882
SVM	0.793	0.789	0.817	0.838	0.827	0.874

In general, the intra-subject dataset, which is trained with a set of users and tested with a different set, yield worse results, due to the variability of behavior between each individual in cases of attention and non-attention. However, in the case of using Support Vector Machines, it seems to be benefited from this. In addition, the Classic Neural Network, composed of a hidden layer and fifty neurons in that layer, resulted in a successful model, placing third, only behind XGBoost and LightGBM.

The confusion matrices (Fig. 3) show a homogeneous distribution of false positives and false negatives, meaning that both models (XGBoost and LigthGBM) do not have a bias towards a particular class.

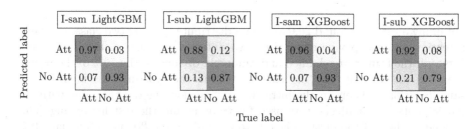

Fig. 3. Confusion matrix for the best performers (LightGBM and XGBoost) for intrasample (I-sam) and intra-subject (I-sub) data setups. The accuracy is shown in each cell as a normalized value.

In the light of the obtained results, we can state that these are very reliable models when categorizing attention based on the position of the face, the position of the body and the emotion, and provide the necessary information to determine with high accuracy whether attention is being paid or not. And although LightGBM seems to be slightly better than XGBoost, the score difference is not significant enough to definitely state the superiority of one model over the other.

4.4 Ablation Study

In this experiment, we will do an ablation study by training the models with different mixes of features in order to provide insight about the effect of each. Note that, in this experiment, only the best two classifiers are considered, that is XGBoost and LightGBM.

As it can be seen in Table 2, although the pose seems to be the most influential element in determining attention, the results would still be good enough to get the subject's attention if the pose is excluded and only emotion and gaze are considered. The results, in this case, are of 83.5% and 82.5% accuracy (for LGBM and XGBoost respectively) in the case of intra-samples, 64.8% and 61.5% for the intra-subjects cases.

Table 2. Results for different mix of features fed to the two best classifiers (X. stands for XGBoost and L. for LightGBM) for intra-samples and intra-subjects versions of the dataset.

Classifier, Features	Intra-Sample			Intra-Subject		
	Acc.	AUC	F1	Acc.	AUC	F1
XGBoost, Emotion	0.773	0.774	0.784	0.559	0.564	0.618
XGBoost, Gaze	0.680	0.675	0.716	0.592	0.608	0.640
XGBoost, Pose	0.918	0.918	0.922	0.894	0.873	0.921
XGBoost, Emotion + Gaze	0.835	0.836	0.842	0.648	0.678	0.683
XGBoost, Emotion + Pose	0.933	0.934	0.937	0.894	0.897	0.927
XGBoost, Gaze + Pose	0.938	0.938	0.942	0.872	0.849	0.904
LightGBM, Emotion	0.794	0.793	0.808	0.570	0.576	0.628
LightGBM, Gaze	0.731	0.731	0.750	0.659	0.663	0.714
LightGBM, Pose	0.943	0.943	0.947	0.899	0.886	0.924
LightGBM, Emotion + Gaze	0.825	0.827	0.830	0.615	0.644	0.650
LightGBM, Emotion + Pose	0.938	0.938	0.942	0.905	0.897	0.927
LightGBM, Gaze + Pose	0.938	0.938	0.942	0.905	0.905	0.926

Excluding the pose from the model would provide additional advantages such as it would eliminate the need for a second camera to detect attention, resulting in a faster and more cost-efficient process.

However, pose seems to be an important factor in determining attention. Although emotion and gaze have a significant impact, pose can provide relevant information about the physical and emotional state of the subject, which can be useful in detecting attention. Additionally, the pose can also be an important indicator in detecting other behaviors such as fatigue, stress, and concentration.

4.5   Effect of Varying the Number of Training Samples

In this experiment, we trained the classifiers with different amounts of data. First, for the intra-sample dataset, we shuffled the data and randomly chose 20%, 40%, 60% and 80% of the training set. For the intra-subject, 1, 3 and 5 users were chosen among those available in the training set. The test sets are always the same, as stated earlier.

In Fig. 4, it can be seen that, overall, the models are prone to provide higher accuracy as it is increased the number of samples. This leads one to think that the results of the models could be further improved by creating and using a dataset with more data and more users. However, in that case, the possibility of overfitting the model would have to be taken into account. The first graph in Fig. 4 shows that no overfitting has been observed and there is a linear trend in the growth of the accuracy as increases the number of samples provided.

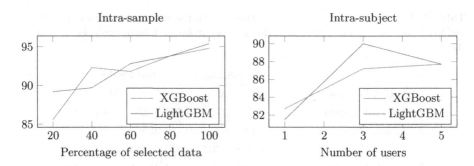

Fig. 4. Accuracy of the XGBoost and LightGBM models varying the number of subjects and the number of samples selected for the study. The accuracy scores are in percentage.

While in the case of the second graph in Fig. 4, it can be observed that as the number of subjects increases, there comes a point where the accuracy stops improving. This is logical because the more users tested, the more robust the model becomes, to the point where it cannot improve between subjects.

5   Conclusions and Future Work

In this work, we introduced AAtiENDe, an Automatic ATtention Evaluation on a Non-invasive Device. It is a methodology and a system to predict whether a person in paying attention to a screen or not from the image stream of two cameras. Our method successfully mixes deep-learning and traditional machine-learning algorithms to perform the task. We achieve around a 95% accuracy by using both cameras, 83% accuracy using only the camera pointing to the face of the subject, and 91% using the BEV camera. Finally, if one camera is to be used, it is preferable to use the closeup camera, as the mixture of gaze and emotion still provides high accuracy and this way the system is independent of the position of the camera.

Regarding the runtime of AATiENDe, the complete pipeline runs in 98ms, namely at 10FPS. However, as our approach only considers one frame every fifteen, it can be considered real-time. Note that the runtimes were measured in the computer itemized in Sect. 4.1

As a future work, we plan to tackle the dependency on the camera position for the pose computation, which is a limitation of the current system. To overcome this issue, we will rely in the 3D pose that we will normalize to abstract it from the position of the sensor. In addition, we plan to improve the dataset by involving more users and more camera setups.

Acknowledgments. This work has been carried out under the framework of the grant CIPROM/2021/17 funded by Prometeo program from Conselleria de Innovación, Universidades, Ciencia y Sociedad Digital of Generalitat Valenciana (Spain). This work has also been funded by a PhD grant under the reference UAFPU21-78 from the University of Alicante (Spain).

References

1. Bosch, N.: Detecting student engagement: human versus machine. In: Proceedings of the 2016 Conference on User Modeling Adaptation and Personalization, UMAP '16, pp. 317–320. Association for Computing Machinery, New York, NY, USA (2016). https://doi.org/10.1145/2930238.2930371

2. Barbadekar, A., et al.: Engagement index for classroom lecture using computer vision. In: Global Conference for Advancement in Technology (GCAT), 2019, pp. 1–5 (2019)

3. Canedo, D., Trifan, A., Neves, A.J.R.: Monitoring students' attention in a classroom through computer vision. In: Bajo, J., et al. (eds.) PAAMS 2018. CCIS, vol. 887, pp. 371–378. Springer, Cham (2018). https://doi.org/10.1007/978-3-319-94779-2_32

4. Li, W., Jiang, F., Shen, R.: Sleep gesture detection in classroom monitor system. In: ICASSP 2019–2019 IEEE International Conference on Acoustics, Speech and Signal Processing (ICASSP), pp. 7640–7644 (2019)

5. Dinesh, A.N.S.D., Bijlani, K.: Student analytics for productive teaching/learning. In: 2016 International Conference on Information Science (ICIS), pp. 97–102 (2016)

6. Viola, P., Jones, M.J.: Robust real-time face detection. Int. J. Comput. Vision **57**(2), 137–154 (2004)

7. Kalal, Z., Mikolajczyk, K., Matas, J.: Tracking-learning-detection. IEEE Trans. Pattern Anal. Mach. Intell. **34**(7), 1409–1422 (2011)

8. Martin, S., Tran, C., Trivedi, M.: Optical flow based head movement and gesture analyzer (ohmega). In: Proceedings of the 21st International Conference on Pattern Recognition (ICPR2012), pp. 605–608. IEEE (2012)

9. Liao, W., Xu, W., Kong, S., Ahmad, F., Liu, W.: A two-stage method for hand-raising gesture recognition in classroom, pp. 38–44 (2019). https://doi.org/10.1145/3318396.3318437

10. Selim, T., Elkabani, I., Abdou, M.A.: Students engagement level detection in online e-learning using hybrid efficientnetb7 together with TCN, LSTM, and Bi-LSTM. IEEE Access **10**, 99:573–99:583 (2022)

11. Tan, M., Le, Q.: Efficientnet: rethinking model scaling for convolutional neural networks. In: International Conference on Machine Learning, pp. 6105–6114. PMLR (2019)

12. Hochreiter, S., Schmidhuber, J.: Long short-term memory. Neural Comput. **9**(8), 1735–1780 (1997)

13. Schuster, M., Paliwal, K.K.: Bidirectional recurrent neural networks. IEEE Trans. Sig. Process. **45**(11), 2673–2681 (1997)

14. Bai, S., Kolter, J.Z., Koltun, V.: An empirical evaluation of generic convolutional and recurrent networks for sequence modeling. arXiv preprint arXiv:1803.01271 (2018)

15. Christian Mejia-Escobar, E.M.-M., Cazorla, M.: Towards a better performance in facial expression recognition: a data-centric approach. In: Computational Intelligence and Neuroscience (2023)

16. Bhallaakshit: Facial expression recognition, September 2020. https://www.kaggle.com/code/bhallaakshit/facial-expression-recognition/notebook

17. Suwarno, S., Kevin, K.: Analysis of face recognition algorithm: Dlib and OpenCV. J. Inform. Telecommun. Eng. **4**(1), 173–184 (2020)
18. Kazemi, V., Sullivan, J.: One millisecond face alignment with an ensemble of regression trees. In: Proceedings of the IEEE conference on computer vision and pattern recognition, pp. 1867–1874 (2014)
19. Valentin Bazarevsky, E.G.B., Grishchenko, I.: Blazepose: on-device real-time body pose tracking. In: CVPR Workshop on Computer Vision for Augmented and Virtual Reality. ACM, August 2020. https://doi.org/10.1145/2F2939672.2939785
20. Lugaresi, C., et al.: Mediapipe: a framework for building perception pipelines (2019). https://arxiv.org/abs/1906.08172
21. Chen, T., Guestrin, C.: XGBoost. In: Proceedings of the 22nd ACM SIGKDD International Conference on Knowledge Discovery and Data Mining. ACM, August 2016. https://doi.org/10.1145/2F2939672.2939785
22. Ke, G., et al.: LightGBM: a highly efficient gradient boosting decision tree. Adv. Neural. Inf. Process. Syst. **30**, 3146–3154 (2017)
23. Zou, H., Hastie, T.: Regularization and variable selection via the elastic net. J. Roy. Stat. Soc. Ser. B (Stat. Methodol.) **67**(2), 301–320 (2005)
24. McCulloch, W.S., Pitts, W.: A logical calculus of the ideas immanent in nervous activity. Bull. Math. Biophys. **5**(4), 115–133 (1943)
25. Cortes, C., Vapnik, V.: Support-vector networks. Mach. Learn. **20**(3), 273–297 (1995)
26. Delvigne, V., Wannous, H., Dutoit, T., Ris, L., Vandeborre, J.-P.: Phydaa: physiological dataset assessing attention. IEEE Trans. Circuits Syst. Video Technol. **32**(5), 2612–2623 (2022)
27. Chong, E., Wang, Y., Ruiz, N., Rehg, J.M.: Detecting attended visual targets in video. In: The IEEE Conference on Computer Vision and Pattern Recognition (CVPR), June 2020
28. Fan, S., et al.: Emotional attention: a study of image sentiment and visual attention. In:. IEEE/CVF Conference on Computer Vision and Pattern Recognition 2018, pp. 7521–7531 (2018)

Effective Black Box Adversarial Attack with Handcrafted Kernels

Petr Dvořáček[(✉)] [ID], Petr Hurtik [ID], and Petra Števuliáková [ID]

Centre of Excellence IT4Innovations,
Institute for Research and Applications of Fuzzy Modeling, University of Ostrava,
30. dubna 22, Ostrava, Czech Republic
{Petr.Dvoracek,Petr.Hurtik,Petra.Stevuliakova}@osu.cz

Abstract. We propose a new, simple framework for crafting adversarial examples for black box attacks. The idea is to simulate the substitution model with a non-trainable model compounded of just one layer of handcrafted convolutional kernels and then train the generator neural network to maximize the distance of the outputs for the original and generated adversarial image. We show that fooling the prediction of the first layer causes the whole network to be fooled and decreases its accuracy on adversarial inputs. Moreover, we do not train the neural network to obtain the first convolutional layer kernels, but we create them using the technique of F-transform. Therefore, our method is very time and resource effective.

Keywords: Black box · Adversarial attack · Handcrafted kernel

1 Introduction

As the utilization of machine learning within the industry continues to rise, concerns emerge regarding its susceptibility to being hacked using adversarial machine learning techniques. The field of adversarial machine learning aims to trick machine learning models by providing misleading input (an adversarial example) and includes both the generation and detection of adversarial examples [4]. The *adversarial example* is an input to machine learning models that an attacker has deliberately designed to cause the model to make a mistake. In particular, it is a corrupted version of a valid input, where the corruption is done by adding a perturbation of a small magnitude to it. The adversarial example aims to appear "normal" to humans but causes misclassification by the targeted machine learning model. An *adversarial attack* is then a method to generate adversarial examples.

We distinguish two kinds of adversarial attacks: a white box attack and a black box attack. In the white box attack, the adversary has complete access to the target model, including the model architecture and parameters, allowing the attacker to use the model's gradient to produce the most effective adversarial examples [20]. On the contrary, in the black box attack scenario, the attackers

I. Rojas et al. (Eds.): IWANN 2023, LNCS 14135, pp. 169–180, 2023.
https://doi.org/10.1007/978-3-031-43078-7_14

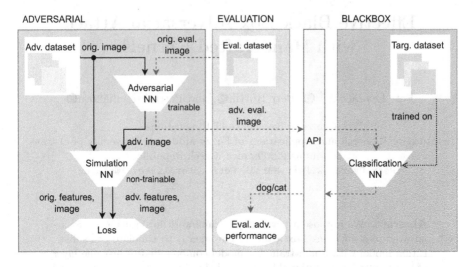

Fig. 1. The proposed scheme. We consider three neural networks (NN): classification, simulation, and adversarial. The classification NN is a black box trained on an unknown 'target' dataset and presented by an unknown architecture. The simulation NN is not trainable and has one layer with 'simulation' kernels and serves for computing loss difference between the original and the adversarial output. The adversarial NN is a simple architecture, highly distinct from the black box one and trained on a different 'adversarial' dataset. Its goal is to create adversarial output that is similar to its input but maximizes the output of the simulation network. The trained adversarial NN is then used to distort images representing an input to the classification NN during the evaluation phase.

do not have full access to the model and can only observe the output of its prediction [3,16].

Current methods to attack a neural network in a black box manner require training a custom substitution network on a dataset labeled with the probabilities from the output of the target network, which was acquired via an API (Application Programmable Interface) [16]. For this purpose, custom images from the same domain as the target model must be acquired by the attacker and then labeled using many API requests to the model's API. Thousands of images, thus API requests are required to train the NN properly. This is not only inefficient but also brings unwanted attention to the attacker. A simple, time- and resource-efficient framework for performing a *first-query* black box attack is presented in this paper. Our framework does not even require images from the same domain as the target model or training a huge classification neural network.

The framework we propose is shown in Fig. 1 and consists of three main components: adversarial, evaluation, and black box. The main idea of this paper is captured in the adversarial component, which utilizes an adversarial dataset with classes different from the target dataset, an adversarial NN, a simulation

NN, and a loss function. The adversarial NN is a trainable CNN generator of adversarial images. Simulation, on the other hand, is a non-trainable CNN with kernels predefined using the F-transform. The loss function consists of three parts: Mean Absolute Error (MAE) to minimize the global intensity difference of the produced image, Structural Similarity Index (SSIM) to force the network to produce an image visually similar to the original, and finally, the variability loss to prevent the trainable adversarial network from changing the color significantly. The framework itself is described in detail in Sect. 4.

2   Related Work

The first effective method of black box adversarial attack is described in [16]. The method involves substituting the targeted model with a custom-trained simulation model that can be fully controlled, allowing the attacker to obtain its gradients. To be able to train a simulation model similar to the target model, the attacker needs to create its own training dataset from the same domain as the model's domain. The labels for this dataset must be obtained as probability scores indicating the degree of association of the input with each class via the model's API.

The first published method to generate an adversarial example is *L-BFGS* (Limited-memory BFGS) [12] which employs the *Broyden-Fletcher-Goldfarb-Shanno* algorithm in the optimizer to create perturbations to the original image under ℓ_2 norm. The BFGS-based optimizer is more resource-consuming but performs better in this task than the Adam or Adadelta optimizer. The creation of an adversarial example itself is slow in comparison with other methods, such as FGSM or BIM.

FGSM (Fast Gradient Sign Method) [4] is much faster than the L-BFGS method mentioned for creating adversarial input. After obtaining the gradients from the neural network by forward and backward pass, the gradient sign in the image is taken, and the image itself is perturbed by a small amount in the direction of the sign. The FGSM method itself is usually applied multiple times to increase its effectiveness. FGSM attack applied multiple times is called the BIM (Basic Iteration Method) attack [10].

The current framework for crafting adversarial inputs comes with several drawbacks:

– It requires a pseudo-labeled dataset to train the simulation neural network. To obtain this dataset, an excessive number of API queries must be sent to the target model, which could alert the victim about the ongoing attack and potentially lead to the discovery of the attacker.
– The training process of the simulation neural network can be time-consuming and computationally demanding, which could limit the feasibility of this attack in specific scenarios.

- Generating adversarial examples using this framework requires multiple forward or backward passes through the network, making execution on embedded devices or systems with limited computational power highly resource intensive and impractical.

We present a new framework for generating adversarial examples, which provides various benefits compared to the current methods. Our framework does not rely on data from the same domain and does not require many API calls to create pseudo-labels. Instead, adversarial examples are generated using a separate neural network that is significantly smaller in size than those used in current methods. This results in faster training times and lower resource requirements to generate adversarial inputs. Our approach offers a more efficient and practical solution for crafting adversarial examples that can be applied in various real-world scenarios.

3   The Proposed Framework

A feed-forward convolutional deep neural network sequentially maps the input to the desired output while starting with low-abstract features and consequently internally using more abstract features. It is known [2] that low-abstract features are produced using kernels that are not task-dependent and serve as feature extractors for deeper layers. The feature representation of the input data should be complete in the sense of a possible reconstruction (backward representation) of any input object. Mathematically, this representation of an object is its approximation.

We aim to design a "simulation" convolutional neural network (CNN) that consists of universal kernels for extracting task-independent features representing input data. In general, most of the CNNs use in the first layer similar kernels that share the following characteristics: Gaussian-like, edge detection with various angle specifications, texture detection, and color spots; for example, see Fig. 2. Therefore, kernels with such characteristics are usually those widely used in image processing, namely, Gaussian, Sobel, Kirsch, etc. Our idea of using universal kernels is based on kernels of possibly different dimensions and rotations. Unfortunately, standard image processing kernels lack such desired properties. Therefore, we propose to establish universal kernels based on the technique of F-transform (and F^m-transform) [17,18] which is a universal approximation technique already used in the field of CNNs [14,15]. More details about the proposed F-transform-based kernels are given in Sect. 3.1.

3.1   F-transform-Based Kernels

F-transform, in general, is a technique that transforms a function (image) into a component representation and back. Initially, the F-transform components were defined as constants representing the average values [17]. Later, the F-transform of higher degree (F^m-transform) was defined [18] where the F^m-transform components are represented by polynomials of degree m. The original F-transform

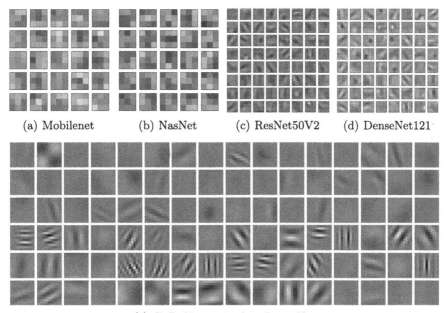

(a) Mobilenet   (b) NasNet   (c) ResNet50V2   (d) DenseNet121

(e) CaffeeNet trained on ImageNet

Fig. 2. Visualization of the first convolutional kernels in different neural network architectures. CaffeeNet model weights visualization was taken from [1]. The others are produced by neural networks used in this paper on CIFAR-10.

is then the F^0-transform. In general, components can be understood as space-localized function features. The components are sufficient to reconstruct the original function with arbitrary precision. The process of generating components is dependent on the so-called basic functions (forming a fuzzy partition), their shape, width, and distance between them (for more details, see [17]). The connection between the F-transform technique and convolution consists of the chosen basic functions; in particular, the basic function support corresponds to the width of the convolution kernel, the basic function shape corresponds to the shape of the convolution kernel, and the distance between the basic function nodes to the convolution stride.

The F^m-transform was already used to create convolutional kernels for CNN [14,15]. The authors there presented that the shapes of the F^m-transform-based kernels are similar to the shapes of the kernels from the most well-known CNNs. In particular, the F^0-transform-based (FT^0) kernels are Gaussian-like, and the F^1-transform-based (FT^1) kernels are (horizontal or vertical) edge detectors that correspond to Sobel. In contrast to Sobel, the advantage is that the F^m-transform-based kernels are functionally defined and therefore can be easily expanded. Moreover, the F^m-transform-based kernels used in the first layer of CNN do not significantly change their shapes during training and are therefore an ideal choice for feature extraction.

In our simulation CNN, we particularly use F^1-transform-based kernels. We follow the theory and construction described in [15] and compute the FT^1 kernels with dimension $d \times d$ (based on the triangular basic function of width d) as follows:

$$FT^1(x, y) = \left(x - \frac{d+1}{2} \right) \left(1 - \frac{|2x - d - 1|}{d + 1} \right) \left(1 - \frac{|2y - d - 1|}{d + 1} \right), \quad (1)$$

$$x, y \in [1, d].$$

We show the visualization of the kernels based on FT^1 with several dimensions in Fig. 3.

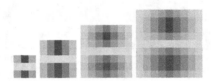

Fig. 3. Visualization of simulation kernels represented by using FT^1. From left: kernel with dimension of 3×3, 5×5, 7×7, and 9×9.

In order to obtain kernels with edge detection features of various angles, we rotate the proposed FT^1 kernels by:

$$K_\alpha = K_x \cos\alpha + K_y \sin\alpha, \quad (2)$$

where K_x, K_y are kernels in the directions x and y, respectively, and K_α is the desired kernel rotated with the angle $\alpha \in [0°, 360°]$. Visualization of an example of the 5×5 FT^1 kernel rotated by $30°$ is shown in Fig. 5.

4   Experimental Verification

An overall visualization of the simulation framework to evaluate the effectiveness of the newly proposed black box adversarial attack is depicted in Fig. 1. It consists of three components: adversarial, black box, and evaluation. The black box component comprises the targeted classification NN model and its corresponding training data. As attackers, we are not granted direct access to the black box, but rather have the ability to invoke its API to obtain a prediction for a given input image. During the evaluation, data belonging to the model's domain (images of cats and dogs) are slightly modified by the adversarial generator to serve as the adversarial input. The targeted model's accuracy is assessed under this evaluation of adversarial attack. The adversarial component incorporates the adversarial image generator, implemented using a convolutional neural network, and a targeted neural network model simulator consisting of the F-transform-based kernels. The adversarial image generator is optimized through backpropagation to generate images that meet two criteria: 1) they should be similar to their original counterpart, and 2) when convolved with kernels from the simulator model, they should produce highly dissimilar vectors.

The CIFAR-10 dataset was used to represent the image classification task due to its low-performance requirements during training and testing. The attack was carried out against four common neural network architectures, namely ResNet-50 V2 [5], DenseNet 121 [7], NASNetMobile [24] and MobileNet [6].

4.1 Data

We utilized CIFAR-10 [9] dataset with standard train/test split. From it, we extracted classes of cat and dog to be the aim of the black box neural network. It means that it is trained on 10,000 images and then tested on 4,000 images where these test images are fed into the network in their original and adversarial version. The remaining part of CIFAR, i.e. 40,000 training images and 6,000 test images, is used to train the adversarial network. The important thing is that the adversarial network does not have access during the training stage to the images that are devoted to the black box neural network. Only after training of the adversarial network is the test set of 4,000 images used to produce adversarial images just for the purpose of the attack performance evaluation. Image samples from the dataset can be seen in Fig. 4.

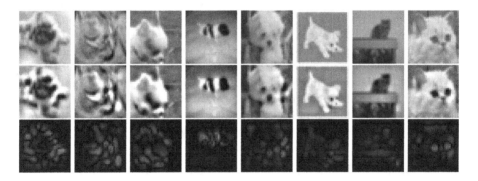

Fig. 4. Samples from the classification's neural network test set. Top row: the original images. Middle row: images modified by the adversarial neural network. Bottom row: the absolute difference between the original and the adversarial image. The differences have a local manner and are connected to the main object in the image despite the adversarial model has been trained on different classes and without the information about the labels.

4.2 Used Neural Networks

Classification Neural Network. We involve a variety of neural networks (see Table 2) in order to evaluate the robustness of the proposed black box attack.

We have selected neural networks that can process small input resolution of the dataset used, which is the reason why we have omitted EfficientNet [21] and ConvNeXt [13] (Table 1). All networks are trained with Adadelta [23] for 40k

Table 1. Overview of the classification architectures used in the benchmark.

Neural Network	Parameters	Depth	First kernels
ResNet-50 V2 [5]	25.6M	103	7×7
DenseNet 121 [7]	8.1M	242	7×7
NASNetMobile [24]	5.3M	389	3×3
MobileNet [6]	4.3M	55	3×3

iterations, where the learning rate is 1.0 for the first 24iterations, 0.5 for the succeeding 10k iterations, and 0.1 for the last 6k iterations. The batch size is 64 and the loss function is binary crossentropy. As augmentation, we have chosen flips, perspective distortion, resize, rotate, and blur.

Adversarial Neural Network. As the objective of the adversarial network is to transform an image input into an image output, an encoder-decoder scheme such as U-Net [19] complemented with a common backbone can be used. Because our dataset has a relatively small resolution, U-Net would map the input to too small feature maps. Therefore, we designed a straightforward architecture consisting of five convolutional layers without maxpooling/stride where each but the last convolutional layer is followed by batch normalization [8] and ReLU activation. The last layer is connected to the sigmoid to obtain the output within the same range as the input image, that is, $[0, 1]$. It is obvious that the architecture used is very distinct from the architectures used for the classification neural network; from this point, it is the real black box adversarial attack.

Simulation Neural Network. the network aims to produce a feature of the processed input image. It consists of n depthwise convolutional layers that are connected into the input image (so that they process the same input on the same scale), and where each of the layers represents one particular FT^1 kernel. In our implementation, we use 5×5 kernels that are mutually rotated by $30°$ so that $n = 12$. For visualization of the kernels, see Fig. 5. The produced feature maps are finally concatenated. The whole network is not trainable to preserve the kernels in the predefined form.

4.3 Loss Function

The loss function must consist of two parts, where the objective of the first is to produce the adversarial image visually similar to the original image, while the

Fig. 5. Visualization of the kernels' weights in the simulation neural network. We used 5×5 FT^1 kernels rotated by $30°$.

objective of the second part is to obtain a dissimilar feature mapping of these two images. Formally, let ψ be the simulation network consisting of FT kernels, ω be the adversarial network, and I be the input image. Then, the compound loss function for training the adversarial network is defined as

$$\ell(I) = \ell_S(I, \omega(I)) + |\psi(I) - \psi(\omega(I))|,$$

where ℓ_S is the similarity loss function given as

$$\ell_S(I_1, I_2) = |I_1 - I_2|\alpha + \text{SSIM}(I_1, I_2)\beta + \text{varc}(I_1, I_2)\gamma.$$

Here, we use α, β, γ as weighting constants, SSIM be a structural similarity [22], and varc a function that computes the variability in the channel dimension. It has been shown [11] that perceptual similarity suits human perception better than similarities derived from MSE, MAE, etc. Unfortunately, the perceptual similarity is based on neural networks, and there is a hypothesis that when our adversarial image confuses a general neural network, it will confuse the network that realizes similarity as well. That is the reason why we combined three partial statistic-based losses for similarity. The first part, MAE, minimizes global intensity difference, SSIM focuses on the texture, and variability of channels preserves colors.

4.4   Numerical Results

In the experiments conducted, all evaluated networks exhibited an accuracy significantly greater than chance (50%) when applied to the unmodified test set. However, a considerable decrease in accuracy was observed for the test set modified by the adversarial network, as shown in Table 2. Notably, the model's accuracy for the 'dog' class did not deviate significantly from chance. Conversely, the accuracy for the 'cat' class increased, which can be attributed to the models' tendency to classify images (comprising both cats and dogs) as 'cat' more frequently.

Visual observation of adversarial inputs visualized in Fig. 4 led to the finding that adversarial inputs have a local increase in contrast. Therefore, we added augmentations of contrast, saturation, hue, and intensity into the training pipeline of the classification networks to observe how robustness to adversarial attacks will be changed. Table 3 shows the results.

Table 2. Overview of the classification architectures used in the benchmark.

| | Accuracy on test set [%] | | | |
| | Original input | | Adversarial input | |
Neural network	cat	dog	cat	dog
ResNet-50 V2 [5]	78.9	83.9	82.2	50.1
DenseNet 121 [7]	79.2	82.7	84.3	57.7
NASNetMobile [24]	74.8	86.6	80.5	46.2
MobileNet [6]	77.1	85.0	80.8	49.6

Table 3. Overview of the classification architectures used in the benchmark. The models were trained with enriched augmentations of contrast, saturation, hue, and intensity.

| | Accuracy on test set [%] | | | |
| | Original input | | Adversarial input | |
Neural network	cat	dog	cat	dog
ResNet-50 V2 [5]	81.7	85.6	77.7	53.5
DenseNet 121 [7]	81.2	85.5	84.5	58.4
NASNetMobile [24]	71.5	83.5	80.5	52.7
MobileNet [6]	79.4	85.6	81.0	51.4

5   Conclusion

We have presented the framework for generating adversarial examples in the black box manner that does not require access to any target model information, its training dataset, or the logit probabilities of its output via an exposed API. This was achieved by simulating the substitution model using the non-trainable model that employed only the single convolutional layer. The convolutional kernels in this layer were handcrafted using the F-transform to simulate the kernels that are usually created by the neural network without the need to train the substitution classification neural network itself. This method of simulating the substitution neural network instead of training seems to be promising, mainly because the need to access the target model logits is eliminated and the creation of the adversarial examples is more time and resource efficient. Verification of attack performance was performed using the CIFAR10 dataset.

Acknowledgements. This work was partially supported by SGS13/PřF-MF/2023.

References

1. Brachmann, A., Redies, C.: Using convolutional neural network filters to measure left-right mirror symmetry in images. Symmetry **8**(12), 144 (2016)

2. Bulat, A., Kossaifi, J., Tzimiropoulos, G., Pantic, M.: Incremental multi-domain learning with network latent tensor factorization. In: Proceedings of the AAAI Conference on Artificial Intelligence, vol. 34, pp. 10470–10477 (2020)

3. Carlini, N., Wagner, D.: Towards evaluating the robustness of neural networks. In: 2017 IEEE Symposium on Security and Privacy (SP), pp. 39–57. IEEE (2017)

4. Goodfellow, I., Shlens, J., Szegedy, C.: Explaining and harnessing adversarial examples. In: International Conference on Learning Representations (2015). http://arxiv.org/abs/1412.6572

5. He, K., Zhang, X., Ren, S., Sun, J.: Identity mappings in deep residual networks. In: Leibe, B., Matas, J., Sebe, N., Welling, M. (eds.) ECCV 2016. LNCS, vol. 9908, pp. 630–645. Springer, Cham (2016). https://doi.org/10.1007/978-3-319-46493-0_38

6. Howard, A.G., et al.: Mobilenets: efficient convolutional neural networks for mobile vision applications. arXiv preprint arXiv:1704.04861 (2017)

7. Huang, G., Liu, Z., Van Der Maaten, L., Weinberger, K.Q.: Densely connected convolutional networks. In: Proceedings of the IEEE Conference on Computer Vision and Pattern Recognition, pp. 4700–4708 (2017)

8. Ioffe, S., Szegedy, C.: Batch normalization: accelerating deep network training by reducing internal covariate shift. In: International Conference on Machine Learning, pp. 448–456. PMLR (2015)

9. Krizhevsky, A., Hinton, G., et al.: Learning multiple layers of features from tiny images (2009)

10. Kurakin, A., Goodfellow, I.J., Bengio, S.: Adversarial examples in the physical world. arXiv abs/1607.02533 (2016)

11. Ledig, C., et al.: Photo-realistic single image super-resolution using a generative adversarial network. In: Proceedings of the IEEE Conference on Computer Vision and Pattern Recognition, pp. 4681–4690 (2017)

12. Liu, D., Nocedal, J.: On the limited memory BFGS method for large scale optimization. Math. Program. **45**(1–3), 503–528 (1989). https://doi.org/10.1007/BF01589116

13. Liu, Z., Mao, H., Wu, C.Y., Feichtenhofer, C., Darrell, T., Xie, S.: A convnet for the 2020s. In: Proceedings of the IEEE/CVF Conference on Computer Vision and Pattern Recognition, pp. 11976–11986 (2022)

14. Molek, V., Perfilieva, I.: Convolutional neural networks with the F-transform kernels. In: Rojas, I., Joya, G., Catala, A. (eds.) IWANN 2017. LNCS, vol. 10305, pp. 396–407. Springer, Cham (2017). https://doi.org/10.1007/978-3-319-59153-7_35

15. Molek, V., Perfilieva, I.: Deep learning and higher degree f-transforms: interpretable kernels before and after learning. Int. J. Comput. Intell. Syst. **13**(1), 1404–1414 (2020)

16. Papernot, N., McDaniel, P., Goodfellow, I.: Transferability in machine learning: from phenomena to black-box attacks using adversarial samples. arXiv preprint arXiv:1605.07277 (2016)

17. Perfilieva, I.: Fuzzy transforms: theory and applications. Fuzzy Sets Syst. **157**(8), 993–1023 (2006). https://www.sciencedirect.com/science/article/pii/S0165011405005804

18. Perfilieva, I., Daňková, M., Bede, B.: Towards a higher degree F-transform. Fuzzy Sets Syst. **180**, 3–19 (2011)

19. Ronneberger, O., Fischer, P., Brox, T.: U-net: convolutional networks for biomedical image segmentation. In: Navab, N., Hornegger, J., Wells, W.M., Frangi, A.F. (eds.) MICCAI 2015. LNCS, vol. 9351, pp. 234–241. Springer, Cham (2015). https://doi.org/10.1007/978-3-319-24574-4_28

20. Szegedy, C., et al.: Intriguing properties of neural networks. arXiv preprint arXiv:1312.6199 (2013)
21. Tan, M., Le, Q.: Efficientnet: rethinking model scaling for convolutional neural networks. In: International Conference on Machine Learning, pp. 6105–6114. PMLR (2019)
22. Wang, Z., Bovik, A.C., Sheikh, H.R., Simoncelli, E.P.: Image quality assessment: from error visibility to structural similarity. IEEE Trans. Image Process. **13**(4), 600–612 (2004)
23. Zeiler, M.D.: Adadelta: an adaptive learning rate method. arXiv preprint arXiv:1212.5701 (2012)
24. Zoph, B., Vasudevan, V., Shlens, J., Le, Q.V.: Learning transferable architectures for scalable image recognition. In: Proceedings of the IEEE Conference on Computer Vision and Pattern Recognition, pp. 8697–8710 (2018)

Implementation of a Neural Network for the Recognition of Emotional States by Social Robots, Using 'OhBot'

Natalia Bartosiak$^{(\boxtimes)}$, Adam Gałuszka , and Martyna Wojnar

Silesian University of Technology, Gliwice, Poland
{natabar462,martwoj082}@student.polsl.pl, adam.galuszka@polsl.pl

Abstract. Convolutional Neural Networks are a popular approach for image classification problem. This article presents an overview of open-source facial expression datasets and performance comparison of CNN models. Evaluated model, trained to detect seven basic emotions, on the combined set of datasets offers 86,7% of accuracy on a validation set and 97,2% on a training set. In this work a system of automated appropriate response to human emotion expression and set of layers that ensure high performance are proposed. The system combines real-time CNN with robotic head OhBot. The information about current emotional state of a person based on its facial expression is the input signal for the subsystem controlling the robotic head, whose task is to react appropriate to the situation.

Keywords: Social robots · Deep Learning · Emotion recognition · OhBot · Human-Robot Interaction · Real-Time Systems · Convolutional Neural Networks

1 Introduction and Related Work

1.1 Social Robots

Due to rapid technology development, human-robot interaction (HRI) becomes increasingly significant. Social robotics and human-robot interaction are currently a popular and rapidly growing topic. Robots are implemented in humans life for years, especially in the industry. Social robot is defined as a robot that interacts with humans [5] and is capable of understanding them. What distinguishes social robots from the standard ones, found in factories is their ability to communicate and support society. Fong [4] defines social robot as capable of expressing and perceiving emotions, and highlights these characteristics as a key aspect of human-robot interaction. Fasola [7] presented in his work robot, that was used to encourage senior participants to perform exercises. They were asked to follow the movements of the robot's arms and mimic them on their own, while the robot vocally navigated and provided guidance. Vocal feedback is beneficial

I. Rojas et al. (Eds.): IWANN 2023, LNCS 14135, pp. 181–193, 2023.
https://doi.org/10.1007/978-3-031-43078-7_15

due to keeping the user engagement, whereas the addition of emotion recognition would allow patients to be monitored during exercise, so that action can be taken when necessary.

Emotion recognition in social robots is particularly important in autism spectrum therapy. Autism spectrum disorder (ASD) is a spectrum of disorders characterized by a lack of social skills, communication difficulties and a lack of imagination. Symptoms are dependent on the individual and last a lifetime, as no cure has yet been discovered. ASD mainly affects children. Popular therapies are toys and pets. The disadvantage of using animals is their unpredictable behaviors that can put a child's health at risk. The solution to these problems is social robots [18]. This type of therapy is referred to as robot- assisted therapy (RAT). OhBot head is the educational version of a social robot that allows to implement basic reactions [1].

1.2 Emotion Recognition

One of the ways of communication between people is facial expression. The ability to recognize emotion is important in the context of social robots, as it makes it possible to adjust the tone of conversation to match the mood of the interlocutor. Human-robot interaction should be smooth and the robot must be able to adjust to the present situation and extract as much information as possible based on user's face analysis. Recognition of human emotions can be accomplished by analyzing facial expression, sound or body language. This paper focuses on emotion recognition based on facial expression. Seven basic emotions are being detected: fear, anger, happy, sad, surprised, disgusted, and neutral. Based on Darwin's theory on the biological origins of facial expression [2] and Ekman's cultural recognition [3] six basic emotions - happy, surprise, disgust, fear, anger and sadness are expressed the same across all cultures. Criteria for classifying emotions include the positioning of the eyebrows and the mouth. Pulled down eyebrows indicate anger or disgust. Raised eyebrows and an open mouth express fear or surprise. It is necessary to remember that the perception of emotions is frequently subjective, and a given facial expression may express more than one emotion. The adoption of some rigid framework of facial expression classification makes it possible to create a consistent set of data.

Emotion classification is a problem that has been developed for a long time, so a number of solutions have already been proposed, starting from Deep Convolutional Neural Networks (DCNN), Recurrent Neural Networks (RNN) [15], to Long Short-Term Memory (LSTM) [16]. In addition to Deep Neural Network solutions, other machine learning methods such as Support Vector Machines (SVMs) are being used [17]. Despite the numerous development of algorithms [8], emotion classification is still is a complicated topic due to the contextual nature and subjectivity of perceived emotions [9] [10]. Facial expression is mostly a product of many factors such as the nature of the person, situation, environment, age and background.

1.3   Contribution

In this work a system of automated appropriate response to human emotion expression is proposed. The system combines real-time CNN with robotic head OhBot. The task of the neural network is to correctly assess the emotional state of a person based on the video signal. Information about the emotion recognized in real time is the input signal for the subsystem controlling the robotic head, whose task will be to react emotionally, through the expression of emotion by the robotic head, and verbally, depending on the implemented reaction scenario. The second section discusses the methodology with Data collection and CNN model, third section presents results of OhBot implementation and classification of facial expressions. Finally, conclusion and future works are indicated.

2   Methodology

2.1   Data Collection

The first step to creating a neural network is the collection of relevant data. It was decided to use open source datasets. In the end, the three most suitable ones were selected: 2013 Facial Expression Recognition (FER-2013), JAFFE and Extended Cohn-Kanade. As a result of further analysis and research, the FER 2013 was rejected. Although 2013 Facial Expression Recognition (FER-2013) is one of the most popular datasets for emotion recognition, it is not balanced, and contains some errors and inaccurately labeled photos [11].

JAFFE dataset contains of 213 images with different face expressions from 10 different Japanese women. Each one of them was asked to make 7 expressions (6 basic and neutral). The images were annotated with average semantic ratings of each facial expression by 60 annotators.

Cohn-Kanade contains 593 video sequences, representing 123 different people. The subjects are of different genders, different backgrounds, and range in age from 18 to 50. Each of the videos shows a change in facial expression from neutral to the target expression. Of the videos, 327 were assigned to one of seven expression categories: anger, contempt, disgust, fear, happiness, sadness, surprise. The CK+ dataset is classified as lab-controlled, resulting in its frequent use in emotion classification problems based on facial expressions.

A Custom Dataset containing 10 video sequences was created, showing 10 people of different sexes, same age and same ethnicity. Each of the video sequences represented a person demonstrating facial expressions in the following order: anger, disgust, happiness, neutrality, sadness, fear, surprise. Sample images were placed in Figure 1. It was decided to create a proprietary dataset due to the scarcity of training data.

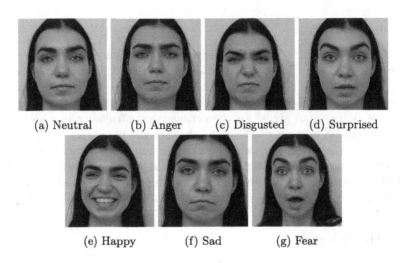

(a) Neutral   (b) Anger   (c) Disgusted   (d) Surprised

(e) Happy   (f) Sad   (g) Fear

Fig. 1. Sample images collected for a custom dataset.

The total number of collections is shown in the Table 1.

Table 1. Volume of data in test and training set.

Label	Test	Train	Sum
Angry	37	153	190
Disgusted	46	186	232
Fear	27	113	266
Happy	59	239	298
Sad	27	109	136
Surprised	58	236	294
Neutral	23	97	120
Total	276	1133	1409

2.2 Convolutional Neural Network Model

One of the aims of the study was to create a Convolutional Neural Network to detect seven emotional states. Four various pre-trained CNN's are used to finally select the most efficient model. The use of previously trained network is a widespread practice of deep learning and is characterized by high performance when operating on small data sets. A pre-trained network is a stored network, that has been learned on a large dataset. As for the models used, it is the ImageNet dataset [12].

The MobileNet architecture enables the creation of very small models that can be applied to designs for mobile applications. In the context of applying neural networks models on social robots, the small size of the model is important, due to the fact that often the hardware at our disposal is not powerful enough. There are various methods to reduce the size of the model - by compressing an already trained network or directly training a small network. Much of the work on small networks focus on size reduction, ignoring speed. MobileNet mainly optimizes latency, which distinguishes this architecture from other solutions.

MobileNetV2 is modeled on and very close to the MobileNet architecture. It differs in the use of residual blocks with bottleneck functions. The other key difference is the number of parameters, which is significantly smaller compared to MobileNet. The MobileNetV2 architecture is also designed for resource-constrained environments.

VGG16 (Visual Geometry Group) is a convolutional neural network that is considered one of the best computer vision models to date. The developers of this model used an architecture with very small (3×3) convolutional filters [13].

ResNet50 (Residual Network) consists of 50 layers - 48 convolutional, one MaxPool and one AvgPool. The depth of the network is crucial [14] [13], especially in the image classification. A problem that arises when learning deep neural networks is the exploding gradient and degradation.

When training the model, problems such as too low accuracy, a rapidly increasing loss function, jumps in the accuracy of the training set, over- or under-fitting appeared. All of the above aspects led to the need for model regularization and fine-tuning of its parameters. Table 2 shows the final set of layers, the selection and order of which were modified at the stage of optimization. Parameter selection was done by analysing loss and accuracy charts. To prevent over-fitting, dropout layers were used. The parameters were initially started with the lowest values and then increased until the best results were obtained. As emotion classification is a multi-class problem, Softmax was adopted as the final layer.

Table 2. Volume of data in test and training set.

Layer	Parameter/Activation	Size of input data
Last layer of base model		(7, 7, 1280)
Dropout	0.5	(7, 7, 1280)
AveragePooling	size (7,7)	(1, 1, 1280)
Flatten		(1280)
Dense	ReLU	(512)
Dropout	0.25	(512)
Dense	ReLU	(256)
Dense	ReLU	(128)
Dense	Softmax	(7)

All images have been resized to 224×224 pixels, due to the input requirements of the pre-trained model. In the case of MobileNet models the inputs pixel values are scaled between -1 and 1, sample-wise. VGG16 and ResNet50 had images converted from RGB to BGR, then each color channel was zero-centered with respect to the ImageNet dataset, without scaling. SGD has been used as an optimizer since Adam gave a worse performance. Further research showed that higher number of epochs did not improve the performance of the model. The configuration presented in Table 2 prevents over-fitting while maintaining model performance.

3   Results

MobileNet achieved a classification accuracy of 95,7% on the training set and 89,1% on the validation set. MobileNetV2 achieved an even higher accuracy on the training set of 97,2%, but its accuracy on the validation set was lower at 86,7%. The classification accuracy of the VGG16 model was 92,9% and 83,1% for the training set and the validation set, respectively. The worst performance was of the ResNet50 model, which achieved an accuracy of 89,9% for the training set and 84,4% for the validation set.

The training process is presented on Figs. 2 and 3. The MobileNet and MobileNetV2 models were selected as the two that performed most effectively. The accuracy for these two models is the highest, on both the training and validation sets. Given the purpose of the models created, the matter of capacity must also be considered. MobileNet models were designed with mobile applications in mind and take up much less storage space. Consequently, model evaluation on devices is likely to be faster.

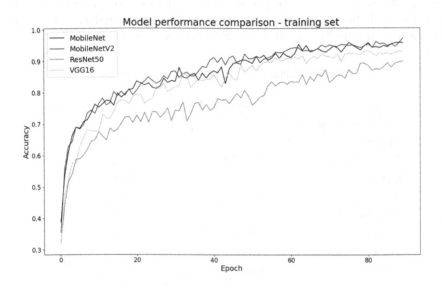

Fig. 2. Training set accuracy.

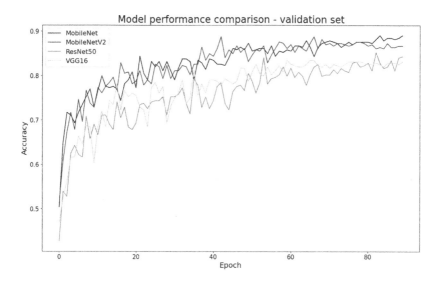

Fig. 3. Validation set accuracy.

Based on further analysis from Fig. 4 of the results from the confusion matrix and measures, a model based on MobileNetV2 was selected, since it made better predictions on the test set. Significant differences can be seen in the Fear category, MobileNet performed predictions with a precision of 79%, while MobileNetV2 achieved a precision of 96% (Table 3).

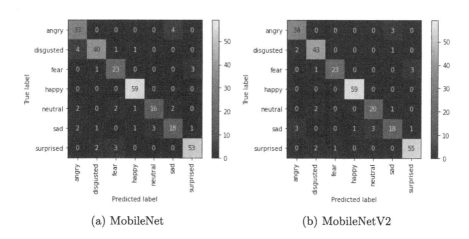

(a) MobileNet   (b) MobileNetV2

Fig. 4. Confusion matrix.

Table 3. Metrics

Label	Model	Precision	Sensitivity	F1 Score	Counts
Angry	MobileNet	87%	89%	85%	37
	MobileNetV2	87%	92%	89%	
Disgusted	MobileNet	91%	87%	89%	46
	MobileNetV2	90%	93%	91%	
Fear	MobileNet	79%	85%	82%	27
	MobileNetV2	96%	85%	90%	
Happy	MobileNet	95%	100%	98%	59
	MobileNetV2	98%	100%	99%	
Neutral	MobileNet	84%	70%	76%	23
	MobileNetV2	87%	87%	87%	
Sad	MobileNet	75%	69%	72%	26
	MobileNetV2	78%	69%	73%	
Surprised	MobileNet	93%	91%	92%	58
	MobileNetV2	93%	95%	94%	
Accuracy	MobileNet			88%	276
	MobileNetV2			91%	
Weighted average	MobileNet	88%	88%	88%	276
	MobileNetV2	91%	91%	91%	

3.1   OhBot Implementation

The purpose of the study was to create and implement a Convolutional Neural Network model using an OhBot-type robotic head. The most important aspect to meet was the real-time performance of the algorithm.

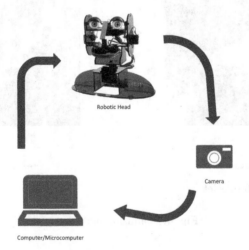

Fig. 5. The process of communication between the program and the robot.

Real-time Emotion Recognition is accomplished in four major steps. They are as follows:

– Collection of dataset,
– Implementation of a CNN architecture,
– Creating an algorithm for a face detection,
– Passing image of detected face as an input to CNN.

The algorithm can be extended to include an agent, in this case an OhBot robot, acting as an intermediary between the program and the human. Figure 5 shows the way the robot interacts with the environment. The robot uses the camera to collect data from the environment, then transmits it to the computer.

Since the problem of detecting the face from an image has already been studied many times, it was decided to use a ready solution to this problem using the Mediapipe library [6]. MediaPipe Face Detection allows for the detection of multiple faces and is equipped with 6 landmarks. Face detection was assumed with a minimum accuracy of 75%. The camera is then initialized, and based on the received image, face detection is performed. The detected face is stored in memory, then scaling of the image to 224×224 pixels is performed, and their values belong to the range $[-1, 1]$. These operations are a necessary and crucial step, as it is important that the model receives the data in the form on which it was trained. In the next step, the model makes a prediction, and the bounding box displays the prediction and the accuracy of the given prediction.

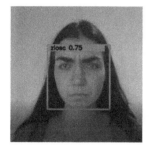 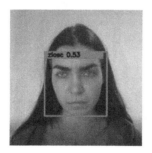 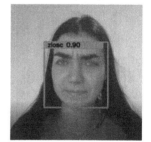

Fig. 6. Anger prediction.

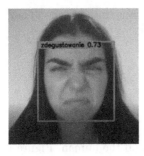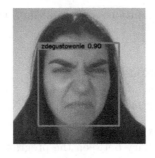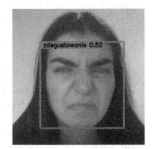

Fig. 7. Disgust prediction.

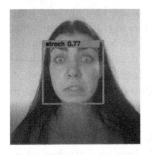

Fig. 8. Fear prediction.

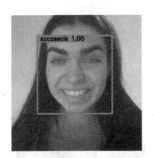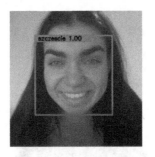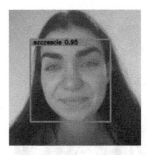

Fig. 9. Happiness prediction.

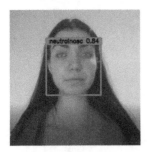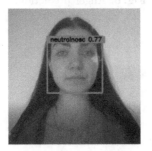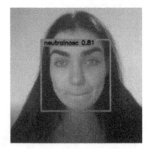

Fig. 10. Neutral prediction.

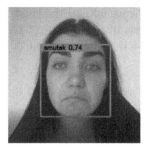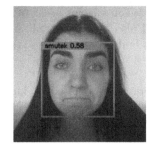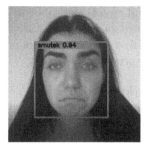

Fig. 11. Sadness prediction.

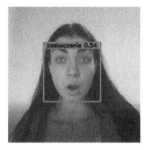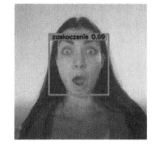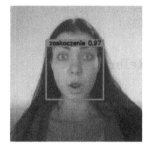

Fig. 12. Surprise prediction.

Figures 6, 7, 8, 9, 10, 11, 12 represent frames from a video demonstrating the prediction of emotional states. The expression of emotion is a dynamic process and the accuracy of the model's predictions varies according to the intensity of the emotion.

The robot carries on a conversation by adjusting its tone to the current mood of the subject. If sadness is detected, the robot is designed to show support and express sympathy. If two extreme emotions are detected, such as joy and sadness or joy and anger, for example, the robot brings the tone of the conversation to neutral. Another five predictions are then tested to rule out misreading. If the labels are consistent, the predicted emotion is considered correct and the robot accepts it as the user's current emotional state.

4   Conclusion and Future Work

In this work, a comparison of pre-trained models in the problem of facial emotion recognition was presented. In addition a set of layers was proposed that ensure high performance. Selected method achieves 97,2% accuracy on training set of combined dataset and 86,7% on validation. To improve performance of CNN and extend this work, it is planned to create large dataset with the assistance of a psychologist. Their support will also be used to improve the robot's response to the identified emotions. It will ensure high-quality human-robot interaction and enable the robot to be used in various areas of life, especially in the treatment of the ASD. ture - will also be considered. Analysis of emotions will be linked to body positioning, as body language is as strong a determinant as facial expression.

Acknowledgements. The work was supported by Mentoring Program realised by the Silesian University of Technology (SUT) (Program Mentorski - "Rozwin skrzydla") and paid from the reserve of the Vice-Rector for Student Affairs and Education: MPK: 60/001 GŹF: SUBD. This work was supported by Upper Silesian Centre for Computational Science and Engineering (GeCONiI) through The National Centre for Research and Development (NCBiR) under Grant POIG.02.03.01-24-099/13. This work was supported by Rector's funds within 4th competition for financing projects of student research clubs under the Initiative of Excellence - Research University in the year 2023.

References

1. Eryka, P., et al.: Application of tiny-ML methods for face recognition in social robotics using OhBot robots, pp. 146–151 (2022). https://doi.org/10.1109/MMAR55195.2022.9874278
2. Darwin, C.: The Expression of the Emotions in Man and Animals, 3rd edn. Fontana Press, London (1999/1872)
3. Ekman, P., Sorenson, E.R., Friesen, W.V.: Pan-cultural elements in facial displays of emotion. Science **164**, 86–88 (1969)
4. Fong, T., Nourbakhsh, I., Dautenhahn, K.: A survey of socially interactive robots. Robot. Auton. Syst. **42**, 143–166 (2003). https://doi.org/10.1016/S0921-8890(02)00372-X
5. Kirby, R., Forlizzi, J., Simmons, R.: Affective social robots. Robot. Auton. Syst. **58**(3), 322–332 (2010)
6. Lugaresi, C., et al.: MediaPipe: A Framework for Building Perception Pipelines (2019). https://google.github.io/mediapipe/ (term. wiz. 03 Jan 2023)
7. Fasola, J., Matarić, M.J.: A socially assistive robot exercise coach for the elderly. **2**(2), 3–32 (2013). https://doi.org/10.5898/JHRI.2.2.Fasola
8. IMotions Facial Expression Analysis. https://imotions.com/facial-expressions. Accessed 12 Dec 2018
9. Abramson, L., Marom, I., Petranker, R., Aviezer, H.: Is fear in your head? A comparison of instructed and real-life expressions of emotion in the face and body. Emotion **17**, 557–565 (2017)
10. Magdin, M., Benko, L., Koprda, Š.: A case study of facial emotion classification using affdex. Sensors **19**, 2140 (2019). https://doi.org/10.3390/s19092140
11. Yaermaimaiti, Y., Kari, T., Zhuang, G.: Research on facial expression recognition based on an improved fusion algorithm. Nonlinear Eng. **11**(1), 112–122 (2022). https://doi.org/10.1515/nleng-2022-0015
12. Deng, J., Dong, W., Socher, R., Li, L.-J., Li, K., Fei-Fei, L.: Imagenet: a large-scale hierarchical image database. In: 2009 IEEE Conference on Computer Vision and Pattern Recognition, pp. 248–255. IEEE (2009)
13. Simonyan, K., Zisserman, A.: Very Deep Convolutional Networks for Large-Scale Image Recognition (2014). https://doi.org/10.48550/ARXIV.1409.1556

14. He, K., Zhang, X., Ren, S., Sun, J.: Delving Deep into Rectifiers: Surpassing Human-Level Performance on ImageNet Classification (2015). https://doi.org/10.48550/ARXIV.1502.01852

15. Saste, S.T., Jagdale, S.M.: Emotion recognition from speech using MFCC and DWT for security system. In: 2017 International Conference of Electronics, Communication and Aerospace Technology (ICECA), vol. 1, pp. 701–704 (2017)

16. Sun, B., Wei, Q., Li, L., Xu, Q., He, J., Yu, L.: LSTM for dynamic emotion and group emotion recognition in the wild. In: Proceedings of the 18th ACM International Conference on Multimodal Interaction, pp. 451–457. Association for Computing Machinery, New York, NY, USA (2016). ISBN 9781450345569

17. Kim, J.-C., Kim, M.-H., Suh, H.-E., Naseem, M., Lee, C.-S.: Hybrid approach for facial expression recognition using convolutional neural networks and SVM. Appl. Sci. **12**, 5493 (2022). https://doi.org/10.3390/app12115493

18. Elissa, K.: Toward socially assistive robotics for augmenting interventions for children with autism spectrum disorders. In: Khatib, O., Kumar, V., Pappas, G.J. (eds.) Experimental Robotics. Springer Tracts in Advanced Robotics, vol. 54, pp. 201–210. Springer, Heidelberg (2009). https://doi.org/10.1007/978-3-642-00196-3_24

Towards a Voxelized Semantic Representation of the Workspace of Mobile Robots

Antonio-Jesus Perez-Bazuelo, Jose-Raul Ruiz-Sarmiento$^{(\boxtimes)}$, Gregorio Ambrosio-Cestero, and Javier Gonzalez-Jimenez

Machine Perception and Intelligent Robotics group (MAPIR), Department of System Engineering and Automation, Malaga Institute for Mechatronics Engineering and Cyber-Physical Systems (IMECH.UMA), University of Malaga, Málaga, Spain
{antonioperez,jotaraul,gambrosio,javiergonzalez}@uma.es

Abstract. The primitives used to model objects in semantic maps heavily influence their suitability for certain robot tasks, as well as the computational load required to process them. This paper contributes a semantic mapping framework that incrementally and efficiently builds a voxelized representation of the robot workspace, providing a balanced trade-off between model expressiveness and computational load. Our proposal detects objects in intensity images coming from an RGB-D camera, and uses depth information to retrieve their point cloud representations. These point clouds are then voxelized and enhanced with their probability of belonging to certain object categories. Finally, voxels are fused with the semantic map in a Bayesian probabilistic framework. Efficiency comes from its client-server design, which allows multiple mobile robots to participate as clients and leaves computationally intensive processes to the server. The proposed framework has been evaluated in both simulated and real environments, yielding accurate voxelized representations.

Keywords: Mobile robots · Semantic maps · Object detection · CNN · Unity · ROS 2 · Voxelization · Client-server · RGB-D Camera

1 Introduction

Mobile robots are progressively landing in diverse fields such as education, healthcare, security, industry, etc., all of them having in common the need for the robot to have some level of understanding of the workspace in which it will operate. That is, the robot has to build and maintain a representation of the elements in its environment, commonly referred as *map*, whose complexity depends on its operation requirements.

Perhaps the most common representation on mobile platforms are geometric maps [20], as they provide the basic capabilities for self-localization and navigation. When combined with information about the topology of the workspace (e.g. how spaces are connected), these *hybrid maps* also permit the planning for

I. Rojas et al. (Eds.): IWANN 2023, LNCS 14135, pp. 194–205, 2023.
https://doi.org/10.1007/978-3-031-43078-7_16

paths to navigate from an arbitrary point A to another B [1]. Nevertheless, the occupancy and connectivity information is not sufficient for a high-level operation beyond localization and navigation. For instance, when commanding a robot to get an item out of a refrigerator, these maps lack the expressiveness to define what a refrigerator is, where it can be found, and how to open it. It is at this point where *semantic maps* come in handy by enhancing geometry and topology with semantics, i.e. with meta-information about the spatial elements and their relationships with the environment [7, 24].

In the field of mobile robotics, *object-oriented* semantic mapping involves creating and maintaining a reliable, high-level representation of the objects within the robot's workspace [19]. This process links geometric information, such as the object's pose, size, and shape, with its semantic information, which includes object types, functionalities, and relations [2, 13]. Examples of the latter could be that *ovens* are objects typically found in kitchens, close to other appliances, and can be used to cook or heat food. Semantic maps, thanks to the level of understanding that they provide, also allow the robot to interact with humans in a more natural way [21] and to perform tasks with greater ease and efficiency [26].

A cornerstone for the population of said maps is object detection, which deals with the identification of objects and their integration into the map. Modern object detection techniques are mainly Deep Learning-based approaches, with trendy backbones including *Convolutional Neural Networks* (CNNs, like Mask-RCNN [10] or YOLO [11]), and *transformers* (Swin Transformer [16]), also existing works combining both (e.g. D-DETR [32]). This way, the typical pipeline in object-oriented semantic mapping involves using one of these networks to detect objects in images, then retrieving the localization of each detected object with respect to the map, and finally integrating this information also taking into account previously inserted detections. The objects' geometric information usually considered in this process are their bounding boxes, a lightweight representation that permits the encoding of objects' extension, position and orientation, also enabling the utilization of simple methods for the integration of detections over time. Although valid for performing a variety of high-level tasks, this coarse representation of objects' geometry shows some limitations: it poorly codifies objects' surfaces needed for tasks like pick and place; bounding boxes' vertices flicker due to partially observed objects, occlusions, and the fact that a 3D representation is built from 2.5D information that can change from different viewpoints [19], compromising their use for tasks such as semantic SLAM; it is not enough to support HRI applications, or it does so at a very basic level.

More fine-grained representations of objects exist that partially overcome these limitations, like point clouds or meshes-based ones. However these approaches suffer from a poor scaling with space size, and present more complex information integration techniques. A promising alternative that is currently being explored is such of voxelized representations, which provide a good trade-off between expressiveness and efficiency [30]. First works exploring them build 3D occupancy grid maps for obstacle avoidance purposes [17]. Others consider semantic segmentation but are limited to one-shot operation [28], not integrating information over time and space, and those doing so are computationally demanding and not suitable for running on resource-constrained platforms like mobile robots [9,22,23,25], so there is room to explore.

This paper presents a semantic mapping framework that incrementally and efficiently builds a voxelized representation of the robot workspace. In a nutshell, our proposal employs a CNN to detect objects in intensity images coming from an RGB-D sensor. The registration between RGB and depth images is used to extract the depth information of detected objects, which is converted into point clouds that are later voxelized. These voxels also incorporate the probability that the region of the space they represent belongs to a set of object categories, information that is propagated from the network output. This results in semantically annotated voxels, which are later fused with the information already present in the map. This is accomplished within a Bayesian probabilistic framework that integrates information over time and space. The efficiency comes from its client-server design, which permits the instantiation of an arbitrary number of mobile robots (clients) sharing workspace and whose collected information is processed and fused in the server. Our proposal sets mobile robots as mere sensors/actuators carrying out basic tasks like localization and navigation, while more heavy duties are left to the server side, which in our implementation runs at the edge. This way the constrained resources of these platforms are not compromised with demanding tasks like object detection, voxels integration, etc.

The proposed framework relies on the Robot Operating System 2 (ROS 2) [18] to orchestrate the operation of robots. For object detection and voxelization we resort to state-of-the-art techniques from Detectron2 [29] and Open3D [31], respectively, although we are not fixed to them. The Bayesian probabilistic framework is implemented in Unity [12], which also enables Human-Robot Interaction (HRI) and Virtual Reality applications, among others. These developments have been integrated into the ViMantic robotic architecture [4].

To perform an initial validation of the presented framework we have carried out experiments in both simulated and real environments. On the one hand, for the simulated test we have relied on the Robot@VirtualHome [5], a collection of 30 virtual houses replicating the layout and objects occurrence of real ones hosted on Idealista, a popular real estate website in Spain. On the other hand, the real experiments were carried out with a Giraff robot [8,27] in an office setting. The experiments yielded promising qualitative results with voxelized representations close to the modeled scenarios.

2   Framework Description

This section delves into the proposed framework for building voxel-based semantic maps. Figure 1 summarizes its workflow from raw sensor data up to the final semantic map. The presented pipeline starts on the robot side (client) with the capture of RGB-D images and 2D laser scans from the sensors mounted on it.

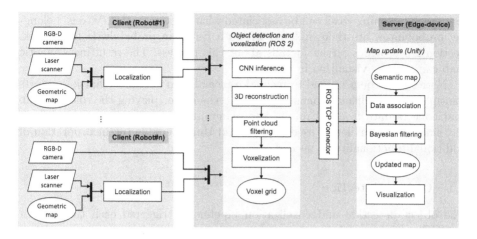

Fig. 1. Overview of the proposed framework illustrating its client-server design and main components. Ovals represent data while rectangles stand for processes.

Laser scans are used to localize the robot within a geometric map, and both RGB-D and localization data are broadcast to an edge device (server), which is responsible for performing the remaining processes. This includes object detection, voxelization, semantic map building and visualization. With this design, the robot is offloaded from computationally intensive tasks, leaving its resources for critical tasks (e.g. , safe navigation or localization).

Image and localization data then go through an object detection and voxelization process. First, a CNN is used to detect objects in the intensity image, providing for each detection: a mask that indicates the image pixels that belong to the object, and a discrete probability distribution over the possible object belonging categories (e.g. , *Table, Chair, TV, Microwave*, etc.). The registration between intensity and depth images is then used to retrieve the depth information associated with each mask, which is converted to a 3D point cloud. Point clouds are filtered to remove outliers, projected into the same reference frame as the geometric map using the robot localization, and finally voxelized. Objects' belonging probabilities are also propagated to voxels. It is worth mentioning that pixels not belonging to any detection are assigned to a special class (*Clutter*) and also go through this process. Section 2.1 and Sect. 2.2 give more information about object detection and voxelization, respectively.

All the processes discussed so far take place in the ROS 2 ecosystem, while the semantic map itself is built and maintained using Unity. The ROS TCP Connector package[1] is employed to exchange information between both ecosystems, concretely the voxelized scene. The Unity reference system is set to match that of the geometric map so, in order to integrate new information into the semantic

---

[1] https://github.com/Unity-Technologies/ROS-TCP-Connector.

map, each incoming voxel can be associated with its counterpart. Voxels' belonging probabilities are then updated within a Bayesian probabilistic framework. Section 2.3 provides further details about this process. The resulting semantic map can then be visualized in Unity.

Notice that the proposed framework is scalable in the sense that it supports multiple robots (clients) concurrently, with the server playing the role of a hub that collects data from all robots and maintains the semantic map, and flexible, since the technologies used (ROS 2 and Unity) enable the incorporation of additional functionality as needed.

2.1   Object Detection

The object detection and voxelization pipeline is triggered each time a pair $< RGB - Dimage, robot localization >$ is available. The first step, object detection, is carried out by Detectron2 [29], a DL framework equipped with an implementation of the popular Mask-R-CNN network pre-trained on the Microsoft COCO dataset [15]. Specifically, this dataset includes 80 object categories, to which we have added an additional one, *Clutter*. The purpose of this category is twofold. On the one hand, it adds uncertainty to the detection results in order to model false detections, i.e. a region of the image that does not contain a detectable object but is identified as such. On the other hand, it permits the image regions not detected as objects to be part of the environment voxelization, hence being incorporated to the semantic map and also appearing in its visualization. The network output is a set of object detections $\mathcal{D} = \{d_0, \ldots, d_n\}$ where each detection d_i provides a pixel mask $\mathcal{M}_i$ covering the region in the image containing the object, and a discrete probability distribution $P(C_{d_i}|o)$ over the considered object categories, being o the observed image.

The pixel masks produced by CNNs tend to be inaccurate and include pixels that belong to adjacent elements in the image. This can pose a significant challenge when creating a 3D representation of objects. To overcome this issue, we pre-process each pixel mask $\mathcal{M}_i$ by applying a thinning morphological operation [14], resulting in $\mathcal{M}'_i$. This process removes potential outliers at the object' boundaries while preserving their topological skeleton.

Since the intensity and depth images from RGB-D cameras are registered, we can directly get the depth of the object projected into a pixel in the intensity image by accessing to the pixel with the same coordinates in the depth one, that is, a mapping $(\Omega \in \mathbb{N}) \rightarrow \mathbb{R}$ exists on the image domain Ω. Having masks $\mathcal{M}'_i$ and depth information, a 3D representation of the object in the form of a point cloud $\mathcal{P}^C_i = [X^C, Y^C, Z^C]^T$ expressed in the camera reference frame (hence the superscript C) can be retrieved as:

$$
\begin{bmatrix} X^C \\ Y^C \\ Z^C \end{bmatrix} = Z^C K^{-1} \begin{bmatrix} x' \\ y' \\ 1 \end{bmatrix} = Z^C \begin{bmatrix} \frac{1}{f} & 0 & \frac{-x_0}{f} \\ 0 & \frac{1}{f} & \frac{-y_0}{f} \\ 0 & 0 & 1 \end{bmatrix} \begin{bmatrix} x' \\ y' \\ 1 \end{bmatrix} = Z^C \begin{bmatrix} \frac{x'-x_0}{f} \\ \frac{y'-y_0}{f} \\ 1 \end{bmatrix} \tag{1}
$$

being K the matrix containing the camera intrinsic parameters (focal length f and camera center (x_0, y_0)), (x', y') the coordinates of the pixels in the mask, and Z^C their depth.

Despite the application of the thinning morphological operator to the mask, the point cloud produced in the previous step can still show outliers. This is specially frequent in objects with *holes*, like chairs with bars in the back. To filter these outliers we rely on a spatial density-based clustering (DBSCAN [3]), resulting in the point cloud $\mathcal{P}'_i$. Once the point cloud for each object detection d_i is obtained and filtered, they form the set $\mathcal{P}' = \{\mathcal{P}'_0, \ldots, \mathcal{P}'_n\}$.

As mentioned before, in order to consider possible wrong detections and build a comprehensive representation of the robot workspace also including non-detected regions, a new object category called *Clutter* is introduced. To include it, a point cloud is built by applying the concatenation operator to the point clouds in $\mathcal{P}'$, and extracting the result from the point cloud of the whole image $\mathcal{P}^{\mathcal{G}}$, that is $\mathcal{P}_{clutter} = \mathcal{P}^{\mathcal{G}} \ominus \{\mathcal{P}'_0 \oplus \mathcal{P}'_1 \oplus \cdots \oplus \mathcal{P}'_{n-1} \oplus \mathcal{P}'_n\}$, so now $\mathcal{P}' = \{\mathcal{P}' \cup \mathcal{P}_{clutter}\}$. The probability distribution associated with this *artificially* generated detection is defined as:

$$P(C_{d_{clutter}}|o) = \begin{cases} C_c & \text{for the } Clutter \text{ label} \\ (1 - C_c)/(N_c - 1) & \text{otherwise} \end{cases} \tag{2}$$

being N_c the total amount of object categories considered, and C_c a configuration parameter called *Clutter confidence* used to specify the confidence level of the *Clutter* class. The higher this value, the higher the confidence in the (non)detection output, and vice versa. In addition to this, the *Clutter* category is also introduced into the probability distribution of the other detections by assigning to it a configurable probability value and normalizing accordingly.

2.2 Voxelization

After generating the set of point clouds $\mathcal{P}'$, the Open3D library [31] is used to voxelize it in the form of a voxel grid $\mathcal{V} = \{v_0, \ldots, v_m\}$. In this representation a voxel is occupied if at least one point from the set of point clouds falls within it, that is $\mathcal{V} = \{v_i \mid \exists \mathcal{P}'_j(k) \in v_i\}$, with $\mathcal{P}'_j \in \mathcal{P}'$ and $k \in [0, |\mathcal{P}'_j| - 1]$ being an index to retrieve the coordinates of point k. The voxel color is calculated as the average color of all the pixels within the voxel. Finally, the probability distribution $P(C_{d_i}|o)$ is propagated to $\mathcal{V}$ by assigning to each voxel the probability of the detection with higher occurrence within it, hence $P(C_{v_i}|o) = P(C_{d_j}|o) \implies (|P'_j \in v_i| > |P'_k \in v_i|) \forall j \neq k$. In summary, voxels are characterized by: i) their spatial coordinates, ii) a color, and iii) a probability distribution over the possible object categories that the region they represent can belong to.

The voxel grid can be generated specifying the resolution of each voxel (voxel size). This parameter establishes a trade-off between the level of detail of the representation (small sizes result in more fine-grained representations, while large ones produce coarse models) and computational complexity (memory requirements and query/update times).

In order to fuse new voxel grids into a global semantic map, the coordinates of both must be expressed in the same reference frame. We employ the same reference frame for the semantic map as for the geometric map, commonly called *world frame*, so voxel grids must be referenced to it. For that it is needed to know the rigid transformation between the RGB-D camera frame and the world frame T_C^W when the images were taken. Since the transformation between camera and robot frames T_C^R is typically known (or can be retrieved by an extrinsic calibration procedure [33]), and the transformation between camera and world frames can be computed by $T_C^W = T_R^W T_C^R$, the problem is reduced to obtain the robot-world relative transformation T_R^W. This is retrieved by means of the Adaptive Monte Carlo Localization (AMCL) method [6], a widely-used localization technique available in ROS2. This method provides an accurate localization of the robot in the environment, allowing us to determine the robot's pose in the world frame. Once T_C^W is calculated, the voxels previously obtained are referred w.r.t. the map by applying this transformation. Globally located voxels are then sent to Unity through the *ROS TCP Connector* node.

2.3   Semantic Map Building and Maintenance

As commented, the semantic map is created and maintained in the Unity ecosystem. Different approaches exist for updating the semantic map with incoming voxel grids. The trivial one would be that the new voxels directly replace the previous ones sharing the same position, including their probability distributions $P(C_{v_i}|o)$, but this could result in inconsistent semantic labeling between successive frames. Here we propose a Bayesian probabilistic framework that recursively updates these probability distributions over multiple consecutive voxel grids. For better understanding, let's introduce the following definitions:

- Let $L = \{l_1, \ldots, l_{N_c}\}$ be the set of object categories that a voxel can belong to (e.g. ., clutter, sofa, chair, table, bottle, etc.).
- Define $O = [o_1, ..., o_{(t-1)}, o_t]$ as a vector containing the observations where a certain voxel v appears, being o_t the most recent one.
- Let C_v be a discrete random variable modeling the belonging category of voxel v and taking values from the set L.

Thus, the probability that a certain voxel v belongs to an object category given its observations is calculated as:

$$P(C_v|o_1, \ldots, o_t) = \frac{P(o_t|C_v, o_1, \ldots, o_{t-1})P(C_v|o_1, \ldots, o_{t-1})}{P(o_t|o_1, \ldots, o_{t-1})} \quad (3)$$

Assuming first order Markov properties, i.e. independence between observations given the object category, and applying the Bayes' Theorem to the resulting likelihood term, it can be rewritten as:

$$P(C_v|o_1,\ldots,o_t) = \frac{P(o_t|C_v)P(C_v|o_1,\ldots,o_{t-1})}{P(o_t|o_1,\ldots,o_{t-1})}$$

$$P(C_v|o_1,\ldots,o_t) = \frac{P(C_v|o_t)P(o_t)P(C_v|o_1,\ldots,o_{t-1})}{P(C_v)P(o_t|o_1,\ldots,o_{t-1})} \tag{4}$$

Based on the assumption that the probability of finding any type of object is the same for each object, we can consider the expression $P(C_v)$ (prior probability) as constant, while $P(o_t)$ and $P(o_t|o_1,\ldots,o_{t-1})$ are shared for all the possible object categories. Therefore, we can obtain the final expression:

$$P(C_v|o_1,\ldots,o_t) \propto \mu P(C_v|o_t)P(C_v|o_1,\ldots,o_{t-1}) \tag{5}$$

Analyzing it we can see how $P(C_v|o_1,\ldots,o_{t-1})$ is a term computed recursively, while $P(C_v|o_t)$ is the probability distribution propagated from the network output for voxel v (recall Sect. 2.2). μ stands for a normalization factor.

3 Results

In order to validate the proposed semantic mapping framework, we have designed and conducted two different sets of experiments. The first one considers a robot operating in virtual environments from the Robot@VirtualHome dataset [5] (see Sect. 3.1). In this experiment, since the robot operation is simulated, all software components including ROS 2 nodes run in the same machine. The second one takes place in a real office environment with a robot (see Sect. 3.2). In this case, most of the framework runs on an edge device, except for critical nodes for the robot operation (e.g. , those managing sensors, localization and navigation) that remain on the mobile robot.

Regarding implementation details, we rely on Detectron2 for object detection (model `mask_rcnn_R_50_FPN_3x` with a ResNet-50 + FPN backbone and a 3x schedule [~37 COCO epochs], having 80 detectable categories plus *Clutter*), ROS 2 for software components intercommunication (linked through a local network when needed), and ViMantic [4] as an integration platform. ViMantic is a robotic software architecture tailored to semantic mapping that provides: a distributed execution via a client-server design, a formal model for defining and managing semantic information, user interfaces for visualization and interaction with maps, and public availability, among other features. For the experiments, a robot (simulated or real) was instantiated as a client under this architecture, and its collected data were used to build voxelized semantic maps (recall Fig. 1).

3.1 Operating in Virtual Environments

The first set of experiments was conducted using Robot@VirtualHome [5] as a testbed. This is a publicly available ecosystem composed by 30 virtual environments recreated from real houses containing objects belonging to 60 different categories, most of them present in the COCO dataset. The ecosystem also

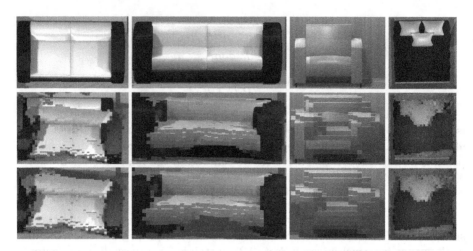

Fig. 2. Voxelized reconstruction of some objects from Robot@VirtualHome. The first row shows the original object, the second row the object with the voxelized reconstruction overlapped, and the third one the voxelized object alone. The first two columns show a couch from two different point of view, whereas the third and fourth show an arm couch and a bed.

includes a virtualized version of the Giraff robot [8,27] which is equipped with a 2D laser scan and an RGB-D camera (resolution: 640 × 480 pixels, working frequency: 30 Hz). For validation purposes, we have employed the proposed framework to build voxelized representations of a number of selected houses. This way, by taken a closer look at the resulting representations we can get a feel about its correctness. Figure 2 yields some examples of such voxelized representations for objects found in House 30. These were built using a voxel size of 5cm, which in our experiments reported a balanced trade-off between representation accuracy and computational load. Also, they are only shown the voxels whose highest belonging probability was *Couch* in the first three columns, and *Bed* and *Pillow* in the last one.

These images allow us to draw some initial conclusions. First, voxelized representations provide more accurate information about the size and shape of objects than the bounding box-based counterpart. Analyzing each individual case, the voxelized version of the couch largely retains the original shape of the object, with only the upper part not well reconstructed due to the limited field of view of the camera, as seen in the top view. Regarding the arm couch, the reconstruction is generally of a high quality. However, certain parts, such as the arms of the couch, may be less accurate than others. Currently, our method creates a voxel when it is first detected and remains in the scene permanently, even if it is not observed in subsequent frames. This approach does not deal with the uncertainties in sensor measurements and robot poses that may be causing these inaccuracies. We recognize that this is an area for improvement and we plan to address it in future work. Finally, the reconstruction of the bed is also accurate, except for the pillows, which are harder to model because of their rounded shape.

Fig. 3. Left image, real photo from the offices. Right image, reconstructed environment.

3.2 Real Experiments

The second set of experiments relies on a real Giraff robot to navigate through and build a semantic map of an entire office. Figure 3-left shows part of the inspected environment, while Fig. 3-right reports its voxelized version. In this case, they are also visualized voxels with a high probability of belonging to the *Clutter* category (recall Sect. 2.1). Again a voxel size of $5cm$ is used.

We can see how the voxelized map fairly represents objects' shapes, although spurious points can be observed upon closer examination. These phenomena can be due to the additive building approach aforementioned. To reduce the negative effect of the uncertainty inherent in sensor measurements, depth images are smoothed with a Gaussian kernel to remove noise, and far measurements are discarded since they are prone to exhibit large errors. This leads to a reduction in the observable area of the scene. As a consequence, certain objects, such as the blue flag, do not appear in the reconstruction. Nevertheless, other objects, including the chairs, the computer, and the cupboard on the right, have been properly reconstructed. We consider that incorporating occupancy information, extending the concept of 2D occupancy grid maps to voxelized representations, could help to mitigate these issues.

4   Conclusions

This paper has introduced an initial framework for the building of voxelized semantic representations of the workspace of mobile robots from sensor data. Concretely, RGB-D images coming from a camera mounted on the robot go through an object detection and voxelization pipeline, which ends up with coloured voxels enhanced with semantic annotations and localized w.r.t. a geometric map. For said localization both the camera extrinsic calibration and the robot pose are used. Finally, this information is fused into a voxelized semantic map using a Bayesian probabilistic framework, which allows us to manage the uncertainty inherent in the object detection and categorization processes. The proposed framework has been designed on a client-server architecture, ViMantic, which enables the instantiation of multiple robots as well as the execution of highly demanding tasks in external devices, off-loading the robots' resources. ROS 2 has been used for the creation and integration of nodes in charge of managing sensors, carrying out localization and navigation, and performing the object detection and voxelization pipeline. The semantic map building and visualization is carried out using Unity. The suitability of our proposal has been evaluated in both real and simulated environments, showing accurate reconstructions.

For future work we plan to continue evolving the proposed framework by also considering occupancy information in addition to semantic one. Moreover, the voxel representation could be optimized to reduce the computational cost of rendering larger scenes. This may involve exploring techniques such as level of detail (LOD) algorithms, occlusion culling, or other optimization strategies to improve the performance of our system.

Acknowledgements. Work partially supported by the research projects ARPEGGIO ([PID2020-117057GB-I00]) and HOUNDBOT ([P20-01302]), funded by the Spanish Government and the Regional Government of Andalusia with support from the ERDF (European Regional Development Funds), respectively.

References

1. Blanco, J.L., González, J., Fernández-Madrigal, J.A.: Subjective local maps for hybrid metric-topological slam. Robot. Auton. Syst. **57**(1), 64–74 (2009)
2. Chatila, R., Laumond, J.: Position referencing and consistent world modeling for mobile robots. In: Proceedings. 1985 IEEE International Conference on Robotics and Automation, vol. 2, pp. 138–145. IEEE (1985)
3. Ester, M., Kriegel, H.P., Sander, J., Xu, X., et al.: A density-based algorithm for discovering clusters in large spatial databases with noise. In: KDD, vol. 96, pp. 226–231 (1996)
4. Fernandez-Chaves, D., Ruiz-Sarmiento, J.R., Petkov, N., Gonzalez-Jimenez, J.: Vimantic, a distributed robotic architecture for semantic mapping in indoor environments. Int. J. Knowl.-Based Syst. **232**, 107440 (2021)
5. Fernandez-Chaves, D., Ruiz-Sarmiento, J.R., Jaenal, A., Petkov, N., Gonzalez-Jimenez, J.: Robot@VirtualHome, an ecosystem of virtual environments and tools for realistic indoor robotic simulation. Expert Syst. Appl. **208**, 117970 (2022)
6. Fox, D., Burgard, W., Dellaert, F., Thrun, S.: Monte carlo localization: efficient position estimation for mobile robots. AAAI/IAAI **1999**(343–349), 2–2 (1999)
7. Galindo, C., Fernández-Madrigal, J.A., González, J., Saffiotti, A.: Robot task planning using semantic maps. Robot. Auton. Syst. **56**(11), 955–966 (2008)
8. González-Jiménez, J., Galindo, C., Ruiz-Sarmiento, J.: Technical improvements of the giraff telepresence robot based on users' evaluation. In: 2012 IEEE RO-MAN, pp. 827–832 (2012)
9. Grinvald, M., et al.: Volumetric instance-aware semantic mapping and 3D object discovery. IEEE Robot. Autom. Lett. **4**(3), 3037–3044 (2019)
10. He, K., Gkioxari, G., Dollár, P., Girshick, R.: Mask R-CNN. In: Proceedings of the IEEE International Conference on Computer Vision, pp. 2961–2969 (2017)
11. Jocher, G., Chaurasia, A., Qiu, J.: YOLO by Ultralytics (2023). https://github.com/ultralytics/ultralytics

12. Juliani, A., et al.: Unity: a general platform for intelligent agents. arXiv preprint arXiv:1809.02627 (2018)
13. Kuipers, B.: Modeling spatial knowledge. Cogn. Sci. **2**(2), 129–153 (1978)
14. Lam, L., Lee, S.W., Suen, C.Y.: Thinning methodologies-a comprehensive survey. IEEE Trans. Pattern Anal. Mach. Intell. **14**(09), 869–885 (1992)
15. Lin, T.-Y., et al.: Microsoft COCO: common objects in context. In: Fleet, D., Pajdla, T., Schiele, B., Tuytelaars, T. (eds.) ECCV 2014. LNCS, vol. 8693, pp. 740–755. Springer, Cham (2014). https://doi.org/10.1007/978-3-319-10602-1_48
16. Liu, Z., et al.: Swin transformer: hierarchical vision transformer using shifted windows. In: ICCV, pp. 10012–10022 (2021)
17. Macenski, S., Tsai, D., Feinberg, M.: Spatio-temporal voxel layer: a view on robot perception for the dynamic world. Int. J. Adv. Robot. Syst. **17**(2) (2020)
18. Macenski, S., Foote, T., Gerkey, B., Lalancette, C., Woodall, W.: Robot operating system 2: design, architecture, and uses in the wild. Sci. Robot. **7**(66), eabm6074 (2022)
19. Matez-Bandera, J.L., Fernandez-Chaves, D., Ruiz-Sarmiento, J.R., Monroy, J., Petkov, N., Gonzalez-Jimenez, J.: LTC-Mapping, enhancing long-term consistency of object-oriented semantic maps in robotics. Sensors **22**(14), 5308 (2022)
20. Milstein, A.: Occupancy grid maps for localization and mapping. Motion Plann. 381–408 (2008)
21. Mutlu, B., Roy, N., Šabanović, S.: Cognitive human–robot interaction. In: Siciliano, B., Khatib, O. (eds.) Springer Handbook of Robotics, pp. 1907–1934. Springer, Cham (2016). https://doi.org/10.1007/978-3-319-32552-1_71
22. Nakajima, Y., Saito, H.: Efficient object-oriented semantic mapping with object detector. IEEE Access **7**, 3206–3213 (2018)
23. Narita, G., Seno, T., Ishikawa, T., Kaji, Y.: Panopticfusion: online volumetric semantic mapping at the level of stuff and things. In: 2019 IEEE/RSJ International Conference on Intelligent Robots and Systems (IROS), pp. 4205–4212. IEEE (2019)
24. Nüchter, A., Hertzberg, J.: Towards semantic maps for mobile robots. Robot. Auton. Syst. **56**(11), 915–926 (2008)
25. Rosinol, A., Abate, M., Chang, Y., Carlone, L.: Kimera: an open-source library for real-time metric-semantic localization and mapping. In: ICRA (2020)
26. Ruiz-Sarmiento, J.R., Galindo, C., Gonzalez-Jimenez, J.: Building multiversal semantic maps for mobile robot operation. Knowl.-Based Syst. **119**, 257–272 (2017)
27. Ruiz-Sarmiento, J.R., Galindo, C., González-Jiménez, J.: Robot@ home, a robotic dataset for semantic mapping of home environments. Int. J. Robot. Res. **36**(2), 131–141 (2017)
28. Song, S., Yu, F., Zeng, A., Chang, A.X., Savva, M., Funkhouser, T.: Semantic scene completion from a single depth image. In: CVPR, pp. 1746–1754 (2017)
29. Wu, Y., Kirillov, A., Massa, F., Lo, W., Girshick, R.: Detectron2 repository (2023). https://github.com/facebookresearch/detectron2/. Accessed March 30 2023
30. Xiang, Y., Choi, W., Lin, Y., Savarese, S.: Data-driven 3D voxel patterns for object category recognition. In: ICVPR
31. Zhou, Q.Y., Park, J., Koltun, V.: Open3D: a modern library for 3D data processing. arXiv:1801.09847 (2018)
32. Zhu, X., Su, W., Lu, L., Li, B., Wang, X., Dai, J.: Deformable DETR: deformable transformers for end-to-end object detection. In: ICLR 2021 (2021)
33. Zuñiga-Noël, D., Ruiz-Sarmiento, J.R., Gomez-Ojeda, R., Gonzalez-Jimenez, J.: Automatic multi-sensor extrinsic calibration for mobile robots. IEEE Robot. Autom. Lett. **4**(3), 2862–2869 (2019)

Intersection over Union with Smoothing for Bounding Box Regression

Petra Števuliáková[ID] and Petr Hurtik[✉][ID]

Centre of Excellence IT4Innovations, Institute for Research and Applications
of Fuzzy Modeling, University of Ostrava, 30. dubna 22, Ostrava, Czech Republic
{Petra.Stevuliakova,Petr.Hurtik}@osu.cz

Abstract. We focus on the construction of a loss function for the
bounding box regression. The Intersection over Union (IoU) metric is
improved to converge faster, to make the surface of the loss function
smooth and continuous over the whole searched space, and to reach a
more precise approximation of the labels. The main principle is adding a
smoothing part to the original IoU, where the smoothing part is given by
a linear space with values that increases from the ground truth bounding box to the border of the input image, and thus covers the whole
spatial search space. We show the motivation and formalism behind this
loss function and experimentally prove that it outperforms IoU, DIoU,
CIoU, and SIoU by a large margin. We experimentally show that the
proposed loss function is robust with respect to the noise in the dimension of ground truth bounding boxes. The reference implementation is
available at https://gitlab.com/irafm-ai/smoothing-iou.

Keywords: Bounding box regression · Intersection over Union ·
Object detection · Noisy labels

1 Problem Formulation

Object detection is an essential part of computer vision and is presented in areas
of (identity) object tracking [10], optical quality evaluation [6], or autonomous
driving [7] to name a few. The current state of the art is exclusively given by
data-driven approaches, i.e., deep neural networks (convolutional or transformer-
based types are used the most) that replaced older model-driven methods that
suffered for precision and robustness. The detection itself consists of three parts:
confidence regression, object classification, and bounding box regression; regardless, we speak about one-step [13] or two-step [14] deep learning approaches. The
three mentioned parts appear in a compound loss function and thus are critical
for the training of a neural network.

In this paper, we focus on the construction of a new loss function for the
bounding box regression. The current way of research improves the well-known
Intersection over Union (IoU) metric defined as the similarity between two arbitrary shapes to converge faster, to make the surface of the loss function smooth

I. Rojas et al. (Eds.): IWANN 2023, LNCS 14135, pp. 206–216, 2023.
https://doi.org/10.1007/978-3-031-43078-7_17

IoU   DIoU   IoU+smooth

Fig. 1. An illustration of the proposed loss given by a child game of throwing rings to a stick. The ring and the stick represent the predicted box and the loss function, respectively. The goal is to place the ring in order to centre it with respect to the stick. The standard IoU loss represents the simple stick, the DIoU loss uses a rubber between the ring and the stick to move (converge) it faster. The proposed IoU with smoothing is represented by a cone covering the loss space and continuously navigating the ring.

and continuous, and to reach a more precise approximation of the labels. Similarly, our approach aims at these aspects but is motivated by the data-centric [5] approach: we assume that the training data set is small and that some labels are not perfect, so *the loss must converge efficiently and be noise-robust*. Note that these are the requirements that naturally appear during cooperation with industry sector.

Here, we propose to enrich the standard IoU loss function with a smoothing part motivated by label smoothing [9], whose purpose is twofold. Firstly, it guides the positioning through the whole domain (image) and, secondly, it weakens the effect of noisy labels. See Fig. 1 that shows the intuition behind the proposed approach. We show the motivation and formalism behind it in Sect. 3 and demonstrate in Sect. 4 that it outperforms the other IoU variants by a large margin. Noise robustness is benchmarked up to noise of 60% of the side size of the bounding box and shows that the decrease in test accuracy is minor. This is a valuable property because the integration of the loss is simple and does not require changes in the architecture compared to the teacher/student scheme that is commonly used when such robustness is required.

2 Related Work and Preliminaries

Currently, most of the loss functions for the bounding box regression fall into two categories: ℓ_n-norm losses and IoU-based losses. Here, we recall them.

2.1 Overview of ℓ_n-Norm Losses and IoU-Based Losses

The category of ℓ_n-norm losses is mainly based on the ℓ_1-norm and the ℓ_2-norm which have some drawbacks. ℓ_1 is less sensitive to outliers in the data, but is not differentiable at zero. Whereas ℓ_2 is differentiable everywhere, but is highly sensitive to outliers. Therefore, Fast R-CNN [2] and Faster R-CNN [14] default use a *Smooth ℓ_1* loss (originally defined as Huber loss [4]), which is differentiable everywhere and less sensitive to outliers than the ℓ_2 loss used in the precedent object detection network, R-CNN [3]. A disadvantage of the *Smooth* ℓ_1 loss [2] is that it depends on a positive real parameter (controlling the transition from ℓ_1 to ℓ_2) that must be selected. Based on the *Smooth* ℓ_1 loss, there are other modifications, for example, *Dynamic smooth* ℓ_1 loss [20] or *Balanced* ℓ_1 loss [11]. However, the main disadvantages of using the ℓ_n-norm in general are ignoring the correlations between the four variables of the bounding boxes $(x; y; w; h)$, which is inconsistent with reality, and bias against large bounding boxes, which basically obtain large penalties in the calculation of the localization errors.

The second category, IoU-based losses, jointly regresses all the bounding box variables as a whole unit; they are normalized and insensitive to the scales of the problem. The original IoU loss function [18] was directly derived from the IoU metric. The main issue is that it does not respond to difference when the bounding boxes do not overlap; for such cases, the maximal loss value is produced. Meanwhile, a Generalized IoU (GIoU) loss [15] resolves the regression issue for the non-overlapping cases. However, both the IoU and the GIoU losses have a slow convergence. Distance IoU (DIoU) loss [22] considers a normalized distance between the central points of the boxes. In addition, Complete IoU (CIoU) loss [22] assumes three geometric components: the overlap area, the distance between the central points, and the aspect ratio. CIoU significantly improved the localization accuracy and convergence speed. However, the aspect ratio is not yet well defined. Based on CIoU, there are other modifications, for example, Improved CIoU (ICIoU) loss [17] that utilizes the ratios of the corresponding widths and heights of the bounding boxes; Efficient IoU (EIoU) loss [21] redefines the ratios of widths and heights between the boxes; A Focal EIoU loss [21] was designed to improve the performance of the EIoU loss. Other IoU-based loss functions and improvements came, for example, with a SCYLLA IoU (SIoU) loss [1] where four cost functions are considered: the IoU cost, the angle cost, the distance cost and the shape cost; or a Balanced IoU (BIoU) loss [12] where the parameterized distance between the centers and the minimum and maximum edges of the bounding boxes is addressed to solve the localization problem.

2.2   Closer Look on the Selected IoU-Based Losses

Intersection over Union (IoU) [18] is a measure of comparison of the similarity between two arbitrary shapes $A, A' \subseteq \mathbb{S} \in \mathbb{R}^n$

$$IoU = \frac{|A \cap A'|}{|A \cup A'|}. \tag{1}$$

IoU as a similarity measure is independent of the space scale of $\mathbb{S}$ and can be transformed to the distance $(1 - IoU)$ that satisfies all the standard metric properties. Therefore, it is popular for evaluating many 2D/3D computer vision tasks, mainly for image segmentation. However, IoU also has weaknesses, so it does not reflect different alignments of A and A' as long as their intersection is equal. Moreover, if there is no intersection between A and A', IoU is always zero and does not reflect any additional information, for example, the distance between A and A', their different sizes, areas, etc.

In general, there is no simple and fast analytic solution to calculate the intersection between two arbitrary non-convex shapes. Fortunately, for the 2D object detection task, where the aim is to compare two axis-aligned bounding boxes, the solution is straightforward. Furthermore, IoU can be used directly as a loss function [18], i.e., $\mathcal{L}_{IOU} = 1 - IoU$ to optimize deep neural network-based object detectors. However, $\mathcal{L}_{IOU}$ still suffers from the weaknesses described above and its convergence speed is slow. Therefore, there are many modifications to the original IoU loss that aim to overcome its drawbacks and improve the accuracy of localization and convergence speed. Generally, IoU-based loss functions can be commonly defined as follows:

$$\mathcal{L}(B^g, B^p) = \mathcal{L}_{IOU}(B^g, B^p) + \mathcal{R}(B^g, B^p), \tag{2}$$

where $\mathcal{L}_{IOU}(B^g, B^p) = 1 - IoU$ is the standard IoU loss between the ground truth B^g and the predicted bounding box B^p. Then $\mathcal{R}(B^g, B^p)$ denotes the penalty term that specifies the particular modification. Our loss function proposed in Sect. 3 also follows Formula (2). In the following, we briefly describe the most commonly used IoU-based loss functions, which are further used for comparison.

The Generalized IoU (GIoU) loss [15] considers a minimum convex area C that contains both boxes B^g and B^p. The penalty term $\mathcal{R}(B^g, B^p)$ then provides a ratio of the area difference between the area C and the union of the boxes. Therefore, the GIoU loss significantly enlarges the impact area by considering the non-overlapping cases of B^g and B^p. Similarly to IoU, the GIou loss is invariant to the scale of the regression problem, and as a distance it is again a metric. But there are also drawbacks. In the cases at horizontal and vertical orientations it still carries large errors. The penalty term aims to minimize the area difference between C and the union of B^g and B^p, but this area is often small or zero (when two boxes have inclusion relationships), and then the GIoU loss almost degrades to the IoU loss. This yields a very slow convergence.

The Distance IoU (DIoU) loss [22] minimizes a distance between B^g and B^p. In particular, the penalty term $\mathcal{R}(B^g, B^p)$ provides a ratio of the Euclidean distance between the central points of the two boxes B^g and B^p and the diagonal length of a minimum convex area C containing both boxes. It is again scale-invariant to the regression problem. In contrast to GIoU, for cases with inclusion of boxes B^g and B^p, or in horizontal and vertical orientations, the DIoU loss can cause very fast convergence. However, in the case of inclusion and the central points of the boxes aligned with each other, the DIoU loss again degrades to the IoU loss.

The Complete IoU (CIoU) loss [22] considers three important geometric factors, i.e., the overlap area, the central point distance, and the aspect ratio. In this case, the penalty term $\mathcal{R}(B^g, B^p)$ is composed of the penalty term for DIoU (central point distance ratio) and the aspect ratio measured by a relationship of a width-to-height ratio of B^g and a width-to-height ratio of B^p. The convergence speed and bounding box regression accuracy of the CIoU loss are significantly improved compared to the previous loss functions. However, the aspect ratio is not yet well defined. It just reflects the discrepancy of the width-to-height ratios between the two boxes, rather than the real relations between the corresponding widths and heights of the boxes. This can lead to cases where the width-to-height ratios of both boxes are the same, even when the predicted box is smaller or larger.

The SCYLLA IoU (SIoU) loss [1] considers four cost functions: IoU cost, angle cost, distance cost, and shape cost. In particular, the penalty term $\mathcal{R}(B^g, B^p)$ consists of a distance-based term and a shape-based term. The distance-based term works with the distance between the central points of the two boxes B^g, B^p and the angles between the vector of central points and the axes x, y. The idea is first to bring the prediction B^p to the closest axis x or y and then to continue the approach along the relevant axis. The shape-based term works with ratios of the corresponding widths and heights of the boxes B^g, B^p and the degree of attention that should be paid to the shape cost. Since SIoU loss introduces the vector angle between the required regressions, it can accelerate the convergence speed and improve the accuracy of the regression.

3   Adding Smoothing Part to IoU Loss

We assume $B = \{x_1, y_1, x_2, y_2\}$ to be a bounding box, where we use standard notation and consider x_1, y_1 to be the top-left point and x_2, y_2 the bottom-right point, together determining the rectangular area. In particular, we assume a couple of bounding boxes B^g, B^p representing the ground truth bounding box and the predicted bounding box, respectively. The objective is to refine B^p to match B^g by minimizing the value of the corresponding loss function. We propose

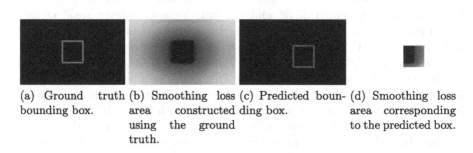

(a) Ground truth bounding box.

(b) Smoothing loss area constructed using the ground truth.

(c) Predicted bounding box.

(d) Smoothing loss area corresponding to the predicted box.

Fig. 2. 2D visualization of how the smoothing part is constructed. The smoothing loss value is computed as the mean from the smoothing loss area (d). The formalism given in Sect. 3 describes how such a process is carried out analytically. The collapsing of prediction into a small area inside the ground truth is prevented, primary by the fact that the loss inside the ground-truth part is zero, so the decreasing size of the predicted box does not decrease the loss, and secondary by adding the standard IoU part.

using a novel *smoothing* version of the original IoU loss defined by the general formula (2) and specified as follows:

$$\mathcal{L}_S(B^g, B^p) = \mathcal{L}_{IOU}(B^g, B^p) + \mathcal{R}_S(B^g, B^p), \tag{3}$$

where $\mathcal{L}_{IOU}(B^g, B^p)$ is the standard IoU loss [18] as mentioned above and the penalty term $\mathcal{R}_S(B^g, B^p)$ denotes a *smoothing* part of the loss that is the proposal of this study. The purpose of the smoothing part is to obtain resistance to noisy labels (similar to label smoothing [9]) and navigate the gradient to converge faster, so the motivation is similar to the other variants of losses based on IoU.

The smoothing part $\mathcal{R}_S(B^g, B^p)$ is assumed to be a linear space with values that increase from the ground truth bounding box to the border of the input image, and thus cover the entire spatial search space; see Fig. 2. In particular, we consider the smoothing part to be specified by the offsets between the boxes B^p and B^g. In addition, the offsets are loaded with weights designated by the distance between B^p and B^g. The offsets and their weights are defined below.

Right offset:

$$\underline{\mathbf{d}}^R = (\underline{x}^R, \underline{y}^R) = (\max(x_1^p - x_2^g, 0), \max(y_2^p - y_1^g, 0)), \tag{4}$$

$$\overline{\mathbf{d}}^R = (\overline{x}^R, \overline{y}^R) = (\max(x_2^p - x_2^g, 0), \max(y_1^p - y_1^g, 0)), \tag{5}$$

$$\mathbf{d}^R = (d_x^R, d_y^R) = \overline{\mathbf{d}}^R - \underline{\mathbf{d}}^R. \tag{6}$$

Left offset:

$$\underline{\mathbf{d}}^L = (\underline{x}^L, \underline{y}^L) = (\max(x_1^g - x_1^p, 0), \max(y_2^g - y_2^p, 0)), \tag{7}$$

$$\overline{\mathbf{d}}^L = (\overline{x}^L, \overline{y}^L) = (\max(x_1^g - x_2^p, 0), \max(y_2^g - y_1^p, 0)), \tag{8}$$

$$\mathbf{d}^L = (d_x^L, d_y^L) = \underline{\mathbf{d}}^L - \overline{\mathbf{d}}^L. \tag{9}$$

Let $b_c^g = [x_c^g, y_c^g]$ be the central point of B^g and

$$\mathbf{d}^c = (d_x^c, d_y^c) = (\max(x_c^g, 1 - x_c^g), \max(y_c^g, 1 - y_c^g)).$$

We define the weights of the right and left offsets, respectively, as follows:

$$\omega^R = (\omega_x^R, \omega_y^R) = \left(1 - \frac{x^R + \overline{x}^R}{2d_x^c}, 1 - \frac{y^R + \overline{y}^R}{2d_y^c}\right), \tag{10}$$

$$\omega^L = (\omega_x^L, \omega_y^L) = \left(1 - \frac{x^L + \overline{x}^L}{2d_x^c}, 1 - \frac{y^L + \overline{y}^L}{2d_y^c}\right). \tag{11}$$

Finally, $\mathcal{R}_S(B^g, B^p)$ is given by

$$\mathcal{R}_S(B^g, B^p) = 1 - \frac{\ell_x \ell_y}{4}, \tag{12}$$

where

$$\ell_x = (1 - d_x^R)\omega_x^R + (1 - d_x^L)\omega_x^L, \tag{13}$$

$$\ell_y = (1 - d_y^R)\omega_y^R + (1 - d_y^L)\omega_y^L. \tag{14}$$

The proposed loss function mimics the standard properties of a distance, namely:

1. The proposed loss function $\mathcal{L}_S(B^g, B^p)$ is invariant to the scale of the regression problem.
2. When the bounding boxes perfectly match, then

$$\mathcal{L}_S(B^g, B^p) = \mathcal{L}_{IOU}(B^g, B^p) = 0.$$

   When the boxes are far away, then

$$\mathcal{L}_S(B^g, B^p) \to 2.$$

3. The smoothing part $\mathcal{R}_S(B^g, B^p)$ is always non-negative and therefore,

$$\mathcal{L}_S(B^g, B^p) \geq \mathcal{L}_{IOU}(B^g, B^p).$$

4 Experimental Evaluation

The current benchmarks are heavily dependent on the famous COCO dataset [8], which is huge and evaluates the entire object detector, including the classification and confidence parts. To omit the influence of these parts, we propose our own lightweight dataset consisting of 304 images, where each image includes exactly one object to be detected. The task is then solved by training a backbone that produces four values that are necessary to perform the regression. The object is the red box; see Figs. 3 and 4. From a practical point of view, the red box is projected by a red laser and marks an area that has to be further inspected.

The detailed setting of the experiment is as follows: 50% fixed train/test split without data leak, fixed resolution of 512×384px, backbone Efficient-NetB2V2 [16], batch size 16, 6000 iterations, Adadelta optimizer [19] with default LR (1.0) and the decrease to 0.4 and 0.1 after 3000th and 4500th iterations. Each of the tested losses was used in three separate runs, i.e., trained from scratch.

The detailed results are shown in Table 1 for the original dataset and in Tables 2, 3 and 4 for synthetically added noise of 20–60% of the side size of the bounding box. For an illustration of the images, see Figs. 3 and 4 for the clean and noisy version. The notation± used in the tables expresses the difference between the best and the worst run. Noise means that each coordinate is added to $n \sim \mathbb{U}(-\mu s, \mu s)$ where $\mu \in [0, 1]$ is the noise level and s corresponding side size; width for the x coordinates and height for the y coordinates.

The interpretation of the results and highlights is as follows:

- Smoothing IoU yields the best results on both clean and noisy dataset.
- Smoothing IoU achieves the lowest overfit on clean dataset.
- Smoothing IoU achieves the largest underfit on noisy dataset, i.e., it is least sensitive to noisy labels among the benchmarked IoU variants.
- With an increase in the noise level, the regression accuracy on testset remains stable for smoothed IoU while strongly decreasing for other IoU variants.
- The proposed loss function is so superior that, when trained on 40% noisy dataset, it still obtains a better test accuracy than the other losses trained on a clean dataset.

Fig. 3. Images from the training dataset where the goal is to detect the red box. The ground truth label is visualized by the green color. (Color figure online)

Fig. 4. Images from the 40% noisy training dataset where the goal is to detect the red box. The *noisy* ground truth label is visualized by the green color. (Color figure online)

Table 1. Clean dataset

Loss type	Measured IOU similarity				
	Train avg	Test avg	Test best	Test ±	Overfit
IOU	0.635	0.536	0.559	0.051	0.099
SIOU [1]	0.668	0.378	0.495	0.270	0.290
DIOU [22]	0.668	0.504	0.615	0.228	0.164
CIOU [22]	0.656	0.526	0.579	0.082	0.130
IOU + smooth	0.713	0.684	0.729	0.073	0.029

Table 2. Noisy dataset, 20%. Because the dataset is noisy, train avg and overfit are informative only.

Loss type	Measured IOU similarity			Test ±	Overfit
	Train avg	Test avg	Test best		
IOU	0.509	0.471	0.555	0.138	0.038
SIOU [1]	0.551	0.550	0.591	0.064	0.001
DIOU [22]	0.521	0.442	0.482	0.079	0.113
CIOU [22]	0.534	0.494	0.532	0.087	0.040
IOU + smooth	0.535	0.588	0.628	0.090	-0.053

Table 3. Noisy dataset, 40%. Because the dataset is noisy, train avg and overfit are informative only.

Loss type	Measured IOU similarity			Test ±	Overfit
	Train avg	Test avg	Test best		
IOU	0.340	0.386	0.453	0.112	-0.046
SIOU [1]	0.399	0.444	0.562	0.232	-0.045
DIOU [22]	0.363	0.442	0.530	0.163	-0.079
CIOU [22]	0.383	0.423	0.507	0.166	-0.040
IOU + smooth	0.387	0.546	0.574	0.054	-0.159

Table 4. Noisy dataset, 60%. Because the dataset is noisy, train avg and overfit are informative only.

Loss type	Measured IOU similarity			Test ±	Overfit
	Train avg	Test avg	Test best		
IOU	0.202	0.294	0.362	0.156	-0.092
SIOU [1]	0.271	0.345	0.378	0.086	-0.074
DIOU [22]	0.250	0.328	0.347	0.040	-0.078
CIOU [22]	0.258	0.349	0.417	0.149	-0.091
IOU + smooth	0.241	0.509	0.581	0.107	-0.268

5   Summary

In this contribution, we have designed the smoothing modification of the standard IoU loss function for the bounding box regression. The proposed smoothing part navigates the gradient through the whole image to converge faster. The exact analytical formalism of the smoothing part is also described in order to be simply integrated to the standard architectures. Based on experimental evolution, we show that the smoothing version of IoU outperforms the benchmarks IoU, SIoU, DIoU, and CIoU. Moreover, the proposed smoothing IoU loss function is resistant to noisy labels. In comparison with the mentioned benchmarks where the regression accuracy strongly decreases for increasing noisy labels, the smoothing IoU remains stable. In industrial applications where noisy labels often appear and therefore robustness is required, the simple implementation of the smoothing IoU loss is the great advantage. The reference implementation is available at https://gitlab.com/irafm-ai/smoothing-iou.

References

1. Gevorgyan, Z.: Siou loss: more powerful learning for bounding box regression. arXiv preprint arXiv:2205.12740 (2022)
2. Girshick, R.: Fast R-CNN. In: Proceedings of the 2015 IEEE International Conference on Computer Vision (ICCV). ICCV 2015, USA, pp. 1440–1448, IEEE Computer Society (2015). https://doi.org/10.1109/ICCV.2015.169
3. Girshick, R., Donahue, J., Darrell, T., Malik, J.: Rich feature hierarchies for accurate object detection and semantic segmentation. In: Proceedings of the IEEE Computer Society Conference on Computer Vision and Pattern Recognition (2013). https://doi.org/10.1109/CVPR.2014.81
4. Huber, P.J.: Robust estimation of a location parameter. Ann. Math. Stat. **35**, 492–518 (1964)
5. Jarrahi, M.H., Memariani, A., Guha, S.: The principles of data-centric AI (DCAI). arXiv preprint arXiv:2211.14611 (2022)
6. Lei, R., Yan, D., Wu, H., Peng, Y.: A precise convolutional neural network-based classification and pose prediction method for PCB component quality control. In: 2022 14th International Conference on Electronics, Computers and Artificial Intelligence (ECAI), pp. 1–6. IEEE (2022)
7. Li, G., Ji, Z., Qu, X., Zhou, R., Cao, D.: Cross-domain object detection for autonomous driving: a stepwise domain adaptative yolo approach. IEEE Trans. Intell. Veh. **7**(3), 603–615 (2022)
8. Lin, T.-Y., et al.: Microsoft COCO: common objects in context. In: Fleet, D., Pajdla, T., Schiele, B., Tuytelaars, T. (eds.) ECCV 2014. LNCS, vol. 8693, pp. 740–755. Springer, Cham (2014). https://doi.org/10.1007/978-3-319-10602-1_48
9. Müller, R., Kornblith, S., Hinton, G.E.: When does label smoothing help? In: Advances in Neural Information Processing Systems, vol. 32 (2019)
10. Pal, S.K., Pramanik, A., Maiti, J., Mitra, P.: Deep learning in multi-object detection and tracking: state of the art. Appl. Intell. **51**(9), 6400–6429 (2021). https://doi.org/10.1007/s10489-021-02293-7
11. Pang, J., Chen, K., Shi, J., Feng, H., Ouyang, W., Lin, D.: Libra r-cnn: Towards balanced learning for object detection. In: 2019 IEEE/CVF Conference on Computer Vision and Pattern Recognition (CVPR), pp. 821–830. IEEE Computer Society (2019). https://doi.org/10.1109/CVPR.2019.00091, https://doi.ieeecomputersociety.org/10.1109/CVPR.2019.00091
12. Ravi, N., Naqvi, S., El-Sharkawy, M.: Biou: an improved bounding box regression for object detection. J. Low Power Electron. Appl. **12**, 51 (2022). https://doi.org/10.3390/jlpea12040051
13. Redmon, J., Farhadi, A.: Yolov3: an incremental improvement. arXiv preprint arXiv:1804.02767 (2018)
14. Ren, S., He, K., Girshick, R.B., Sun, J.: Faster R-CNN: towards real-time object detection with region proposal networks. In: Cortes, C., Lawrence, N.D., Lee, D.D., Sugiyama, M., Garnett, R. (eds.) NIPS, pp. 91–99 (2015)
15. Rezatofighi, S.H., Tsoi, N., Gwak, J., Sadeghian, A., Reid, I.D., Savarese, S.: Generalized intersection over union: a metric and a loss for bounding box regression. CoRR abs/1902.09630 (2019). http://arxiv.org/abs/1902.09630
16. Tan, M., Le, Q.: Efficientnetv2: smaller models and faster training. In: International conference on machine learning, pp. 10096–10106. PMLR (2021)
17. Xufei, W., Song, J.: ICIOU: improved loss based on complete intersection over union for bounding box regression. IEEE Access PP, 1–1 (2021). https://doi.org/10.1109/ACCESS.2021.3100414

18. Yu, J., Jiang, Y., Wang, Z., Cao, Z., Huang, T.: Unitbox: an advanced object detection network. In: Proceedings of the 24th ACM International Conference on Multimedia. MM 2016, New York, NY, USA, pp. 516–520. Association for Computing Machinery (2016). https://doi.org/10.1145/2964284.2967274, https://doi.org/10.1145/2964284.2967274

19. Zeiler, M.D.: Adadelta: an adaptive learning rate method. arXiv preprint arXiv:1212.5701 (2012)

20. Zhang, H., Chang, H., Ma, B., Wang, N., Chen, X.: Dynamic R-CNN: towards high quality object detection via dynamic training. In: Vedaldi, A., Bischof, H., Brox, T., Frahm, J.-M. (eds.) ECCV 2020. LNCS, vol. 12360, pp. 260–275. Springer, Cham (2020). https://doi.org/10.1007/978-3-030-58555-6_16

21. Zhang, Y.F., Ren, W., Zhang, Z., Jia, Z., Wang, L., Tan, T.: Focal and efficient IOU loss for accurate bounding box regression. Neurocomputing **506**, 146–157 (2022). https://doi.org/10.1016/j.neucom.2022.07.042, https://www.sciencedirect.com/science/article/pii/S0925231222009018

22. Zheng, Z., Wang, P., Liu, W., Li, J., Ye, R., Ren, D.: Distance-IOU loss: faster and better learning for bounding box regression. Proceedings of the AAAI Conference on Artificial Intelligence, vol. 34, no. 07, pp. 12993–13000 (2020). https://doi.org/10.1609/aaai.v34i07.6999, https://ojs.aaai.org/index.php/AAAI/article/view/6999

Artificial Vision Technique to Detect and Classify Cocoa Beans

Luis Zhinin-Vera[1,4,5(✉)]📵, Jonathan Zhiminaicela-Cabrera[3,4]📵,
Elena Pretel[5,6]📵, Pamela Suárez[2]📵, Oscar Chang[1,4]📵,
Francesc Antón Castro[1,4]📵, and Francisco López de la Rosa[6,7]📵

[1] School of Mathematical and Computational Sciences, Yachay Tech University,
100650 Urcuqui, Ecuador
[2] School of Biological Sciences and Engineering, Yachay Tech University, 100119
Urcuqui, Ecuador
[3] Faculty of Agricultural Sciences, Universidad Técnica de Machala, 070151 Machala,
Ecuador
[4] MIND Research Group - Model Intelligent Networks Development, Urcuqui,
Ecuador
[5] LoUISE Research Group, I3A, University of Castilla-La Mancha, 02071 Albacete,
Spain
luis.zhinin@uclm.es
[6] Instituto de Investigación en Informática de Albacete (I3A), Universidad de
Castilla-La Mancha, Avenida de España s/n, 02071 Albacete, Spain
[7] Department of Electrical, Electronic, Automatic and Communications Engineering,
Universidad de Castilla-La Mancha, Avenida de España s/n, 02071 Albacete, Spain

Abstract. This article discusses the use of Artificial Intelligence (AI)
to classify cocoa beans as healthy or diseased based on established classification criteria, given the challenges faced by the cocoa industry due
to the impact of diseased beans on quality and grading. The proposed
method uses YOLOv5 and achieved an 94.5% accuracy rate. The article
also outlines the development of an affordable and easy-to-implement
prototype system that cocoa farmers can use to grade and assure bean
quality. The results suggest that the proposed system is successful, and
increasing the amount of data improves its reliability, which could help
farmers improve their competitiveness in the market.

Keywords: cocoa beans classification · computer vision · smart
farming

1 Introduction

Ecuador is a major cacao exporter with a long history of production since 1600,
ranking fourth worldwide in 2019 [4,8]. Despite this success, price fluctuations
related to social, political, and biological crises have been a challenge. In addition,
insect pests and diseases present a significant issue for the country's current cocoa
production [1].

I. Rojas et al. (Eds.): IWANN 2023, LNCS 14135, pp. 217–228, 2023.
https://doi.org/10.1007/978-3-031-43078-7_18

Cocoa bean quality depends on various factors such as soil type, climate, genetics, and management practices [26]. To ensure quality, cocoa beans are evaluated using INEN standards 175, 176, and 177 [10], which cover cutting tests, classification and quality requirements, and sampling procedures. A detailed quality protocol for beans and their derivatives includes physical, chemical, sensory, and spectrometric analyses to ensure excellent quality [18].

Cocoa beans have varying shapes depending on their genotypes: *forasteros* are flattened and smaller, *criollo* beans are large and ovoid, while *trinitarios* show a mixture of both [5]. The beans also vary in color: *criollos* are white, *forasteros* are violet, and *trinitarios* have a medium violet color. In Ecuador, the national genotype cocoa beans tend to have pinkish tones [22]. Grading cocoa beans requires experienced labor, making it a time-consuming process. Ecuadorian company SIRCA has developed size-based alternatives for grading cocoa beans, but these methods do not address issues such as foreign genotypes and insect pests, which could be identified using mathematical models or AI techniques, thereby enhancing grading efficiency.

To classify cocoa beans for optimal use in by-product production, it is necessary to assess their quality. The Honduran Foundation for Agricultural Research proposes characteristics for this classification in their manual for the Evaluation of Cocoa Bean Quality [3]. These characteristics distinguish between fresh cocoa beans (with mucilage) and dry beans (ready for export) and serve as a useful tool for classification.

2   Related Work

Many studies have focused on classifying cocoa beans, as summarized in Table 1. The degree of fermentation of the beans is a critical factor in determining their quality and aroma. Hence, artificial intelligence methods can aid in accurately classifying cocoa beans by leveraging this variable. In [25], an affordable and fast method based on a machine learning electronic nose system was developed to classify the fermentation degree of cocoa beans. Six machine learning techniques (bootstrap forest, boosted tree, decision tree, artificial neural network (ANN), naive Bayes, and k-nearest neighbors) were used for this purpose.

The article [19] proposes an AI-based approach to classifying cocoa fermentation degrees using the "Random decision forests" technique, which achieved high accuracy. The study highlights the potential of artificial vision as an analytical method for the cocoa and chocolate industry. In [2], a methodology for textural feature analysis of digital images of cocoa beans is presented, comparing the performance of Convolutional Neural Network (CNN) and gray level co-occurrence matrix (GLCM) features for feature extraction. The results suggest that GLCM feature extraction is more reliable than CNN feature extraction for classification.

On other hand, in [6], a machine vision and machine learning model is proposed for the classification of fermented and unfermented cocoa beans. The model uses ANN models for segmentation, calculation, and classification based on color features, specifically the average values of RGB and $L*a*b$. Results show that the MLP of the ANN outperforms other models, achieving a training and validation accuracy of 94%. In [17], morphological parameters were used to classify beans based on their physical characteristics, such as area, perimeter, and aspect ratio. The Multiclass Ensemble Least Squares Support Vector Machine (MELS-SVM) was used as the classification model, achieving 99.705% accuracy for classifying grains based on their morphological features.

Table 1. Summary of the results of the main related works that propose to develop the classification of cocoa beans using Artificial Intelligence techniques.

Algorithm	Main Results	Reported Accuracy	Ref.
Bootstrap forest, Boosted and decision tree, ANN and Naïve Bayes, k-nearest neighbors	Application of ML based electronic nose system indetermining fermentation degree of cocoa bean.	Bootstrap forest 90.6% ANN 87.2% Boosted tree 86.4%	[25]
CNN	GLCM texture feature extraction is more reliable results than CNN	SVM 59.14% CNN XGBoost 56.99% CNN SVM 61.04% GLCM XGBoost 65.08% GLCM	[2]
Random Forest	Results greater than 91%	Imbalanced: 93% Balanced: 92%	[19]
ANN and MLP	Classification into 2 classes: fermented, unfermented	Accuracy: 94%	[6]
Resnet18, Resnet50 SVM and Random Forest	Comparison of traditional Computer Vision System and Deep Computer Vision	ResNet 18: 96.82%. SVM 85.71%	[12]
Multiclass Ensemble Least-Squares SVM	Classification into 4 classes: Normal, broken, fractured and skin damaged beans	Accuracy: 99.7%	[17]

3   Materials and Methods

3.1   Bean Classification Criteria *Theobroma Cacao* L.

Two stages with three classification criteria were considered for the classification of the beans in the dataset:

1. **Initial stage morphology and plant health:** The FHIA manual for evaluating cocoa bean quality, outlined in [3], classifies fermented and dried cocoa beans into quality groups based on the presence of defective or damaged beans. The categories include good, small, broken, twin, flat, germinated, sultanas, and trash beans, and take into account impurities in the jute sacks used for export. Additionally, beans affected by pests or disease were also included in this classification stage, as per [24].

2. **Evaluation stage by cocoa beans cutting test:** To evaluate the degree of fermentation of the beans, they were cross-cut after the initial stage of grouping. The National Institute of Agricultural Research (INIAP) information booklet for cocoa beans was used to guide the evaluation [14], which considers fermentation as a key factor in determining cocoa liquor quality. The different fermentation indices, as per [3], were represented in Fig. (1b) with an identifier for each index: **A.** Brown beans with crumbs, brittle texture and their testa is easily separated, in the image. **B.** Brown bean with violet, less cracks, less brittle and somewhat compact aspect. **C.** Intense violet beans, compact or semi compact aspect. **D.** Blackish gray beans, opaque, very compact, unpleasant and prolonged flavor. **E.** Whitish color and sometimes greenish or yellowish. **F.** Presence of insects, eggs and insect excrement, deteriorated up to 90%.

Limitations of the cross-cutting test used to evaluate cocoa beans can result in variations in determining bean quality. Visual observation alone may not suffice, as factors such as damage, pathogens, and sensory characteristics may also impact classification. Proposed models utilizing artificial intelligence for cocoa bean classification must consider these criteria, as noted in [19].

Physical, chemical, and biological conditions can influence cocoa fermentation rates, which in turn impact the organoleptic quality of chocolate [13,20]. Several factors, including bean maturity [9], fermentation method [21], bean health [10], and drying temperature [7], among others, can affect the fermentation of cocoa beans.

The classification of cocoa beans into good and poor quality groups was based on morphological characteristics and fermentation indices. Good quality beans were characterized by the FHIA morphological criteria and fermentation indices of good, medium, and intense violet. Conversely, the poor quality group included small beans unrelated to the trinitario genotype, and beans affected by deficiencies in physiological fruit development, impurities, and unfavorable fermentation indices like slaty, moldy, and infested beans. Table 2 summarizes the classification criteria and groups. The inclusion of both morphological development and fermentation rate criteria is crucial for a more comprehensive classification of cocoa beans, as they directly impact the final quality of cocoa liquor.

Table 2. Selection criteria used for the classification of beans of *Theobroma cacao* L. based on the main characteristics that determine that a bean is of good quality.

Classification Criteria	Healthy beans	Diseased beans
FHIA [3]	Beans of good morphology independent of genotype	Small, broken, twins, flat, sprouts, sultanas and trash
INIAP [14]	Good, medium and violet fermentation	Fermentation index deep violet, Slate, Mouldy and Infested
Plant health charac. [24]	Healthy beans without incidence of pathogens	Diseased bean with incidence of pathogens

3.2   Dataset

This study analyzed a dataset of 3268 images of cocoa beans, which included whole beans and their cross-sectional views (Fig. 1c), resulting in 1634 unique beans. Based on morphological development and fermentation rate, 1261 beans were classified as good quality, while 373 were classified as poor quality. These beans were collected from the La Maná region of Ecuador, which is known for cocoa cultivation. The dataset was obtained from a realistic production setting, where mixing of healthy and diseased beans and variations in post-harvest processing are common, emphasizing the need for on-site analysis to obtain accurate data. The original and processed dataset is available for research purposes at www.kaggle.com/cacao-beans-for-classification.

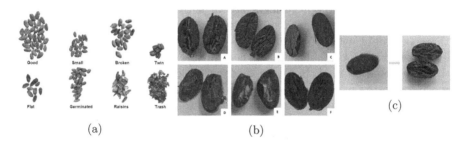

(a)   (b)

Fig. 1. (a) Characteristics of cocoa beans with defects and damage. (b) Classification of dried cocoa beans by degree of fermentation. Source: [3]. (c) The degree of fermentation is determined by cutting each bean in half.

3.3   Proposed Algorithm: YOLOv5

In this study, computer vision algorithms are proposed to classify cocoa beans in real time, using a subset of the dataset for training and the rest for evaluation. This study employs YOLO <You Only Look Once> [23]. The YOLOv5 algorithm is used for object detection and classification, dividing images into grids and detecting objects within each grid cell. The algorithm is tested using a camera and evaluated using conventional metrics, with its performance explored on beans in varying conditions. YOLOv5 is written in PyTorch and is much lighter and easier to use than previous versions of the algorithm.

4   Proposed Classifier System

The 4th industrial revolution has brought about advancements in the integration of IoT and AI in industrial food and agriculture [16]. Image-based monitoring systems have been developed for plantations, including complex systems such as Vision-Based detection for Strawberry diseases [15]. We designed a prototype system that learns to differentiate between healthy and diseased beans for research and testing purposes. The system is built on a robotic prototype previously designed for reinforcement learning in Tic-Tac-Toe [11]. The adapted prototype enables real-time bean detection and execution of actions in the physical world.

4.1 Prototype Robotic Arm

A prototype robot is designed to carry out the physical actions resulting from the training process of healthy and diseased bean recognition. This prototype presents a 3-axis mechanical scheme: shoulder, elbow, finger and servomotors, power source, Arduino board and connections from computer to servomotor.

The system uses a convolutional neural network to detect the location of diseased beans, and this information is sent to an Arduino board that controls a three-axis robot with servomotors. The Arduino board has software that converts the network coordinates to local coordinates for the robot to position itself over the infected element. In future versions, a pickup element will be added to remove the rotten piece. A visual explanation of the process is shown in Fig. 2a. Figure 2b shows the final robot. The prototype robotic arm designed for healthy and diseased bean recognition was found to be accurate, and its cost-effective implementation makes it a valuable tool for didactic purposes.

4.2 Prototype Classifier System

The study proposes a more advanced system for automatic classification of cocoa beans. This system requires a bit more investment but uses the same basic principle as the previous prototype, which is to use a robotic arm to physically remove the infected elements. The proposed system is shown in Fig. 2c and is considered an efficient solution to the problem of classifying cocoa beans. In the same illustration, the steps of this system are shown. The beans are deposited into a funnel (A), then gravity-fed through a narrowing tube (B) until they are aligned for processing by the camera as they pass through a specific point (C). The camera detects healthy and diseased beans, triggering a signal in a servomotor that can be implemented in Arduino to block or allows the path of beans according to their class (D). Ultimately, the trained algorithm successfully classifies each bean (E).

5 Implementation and Results

5.1 Labeling, Preprocessing and Training

Preprocessing and labeling methods are critical in computer vision tasks, and are used to enhance the quality and structure of images before they are fed into

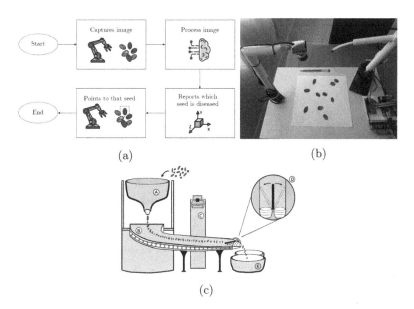

(a)   (b)

(c)

Fig. 2. (a) General scheme of the operation of the robotic prototype of the cocoa classification system. (b) Robotic prototype of the cocoa classification system adapted from a previous article. (c) Illustration of the prototype classifier system.

a machine learning algorithm. Theses techniques play a critical role in achieving good results in computer vision tasks, and careful consideration of these techniques can make a significant impact on the performance of the model.

The bean images were previously classified using the above criteria. Then, using an annotation tool, we proceed to mark with bounding box annotations around the objects we want to detect. Once the entire dataset has been annotated, it must be exported in YOLOv5 format. Figure 3a shows an example of the annotated dataset. Then with the labeled dataset (divided into training (85%), validation (10%) and testing set (5%)) we use the following techniques: resizing, cropping, normalization, and augmentation. Resizing is used to adjust the dimensions of an image to a specific size, while cropping removes unwanted areas of the image (Fig. 3b). Normalization is used to adjust the color and brightness of an image, which helps to reduce the effects of lighting variations. The dataset is imbalanced, for this reason we use Augmentation techniques, such as flipping, rotating, and adding noise, are used to create additional training images and to improve the robustness of the model. The size of the dataset generated after applying these techniques is 3432 images for training, 327 images for validation and 153 for testing phase.

This training phase requires the previously annotated and pre-processed database. The export format is YOLOv5 PyTorch. The algorithm is implemented in Google Colaboratory more commonly referred to as "Google Colab". The algorithm is executed using Python>=3.7.0 environment, including PyTorch>=1.7.

The algorithm was optimized using Stochastic gradient descent (SGD) and trained with 100 epochs and a batch size of 16, which were determined based on multiple experiments and expert recommendations for this dataset.

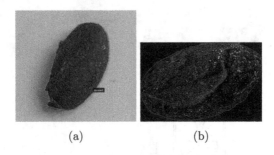

(a)   (b)

Fig. 3. (a) Annotation of classes: disease and healthy, in each image of the dataset. (b) Beans processed using different techniques.

5.2   Results

The performance of the computer vision algorithm for detecting healthy and diseased seeds is evaluated using various metrics such as confusion matrix, F1 curve, Precision-Recall curve, mAP@0.5, and error metrics for training and validation sets. The confusion matrix is shown in Fig. 4a and the results of the evaluation metrics are presented below. Images with several seeds were loaded to test the system's effectiveness in classification (Fig. 4b).

The class of healthy seeds shows a correct prediction of 95% while the diseased seeds show a correct prediction of 94%. The F1 curve obtained shows a value for all classes of 0.76 at 0.472. This suggests that the algorithm performs well in terms of precision and recall for all classes. On the other hand, the Precision-Recall curve shows a high value of 0.943 for healthy seeds and 0.81 for diseased seeds. This indicates that the algorithm has a high precision for healthy seeds, but a relatively lower precision for diseased seeds (Fig. 5).

The mAP@0.5 for all classes shows a value of 0.821, which is a good measure of overall performance (Fig. 6). In addition to the above evaluation metrics, multiple error metrics such as mAP (mean Average Precision) when IOU are at 0.5 (50%) and 0.95 (95%) are calculated. The algorithm has achieved an mAP of 0.821 when IOU is at 0.5 and 0.473 when 0.95. These metrics are useful in detect the areas that require improvement in the algorithm. Overall, the computer vision algorithm performs well in detecting healthy and diseased seeds with a high precision rate. However, there is room for improvement in the recall rate for all classes.

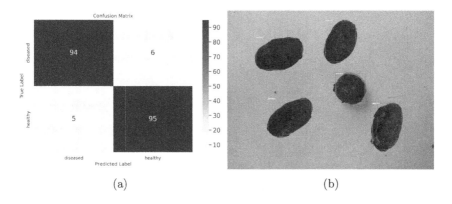

Fig. 4. Confusion Matrix and a sample of classified cacao beans.

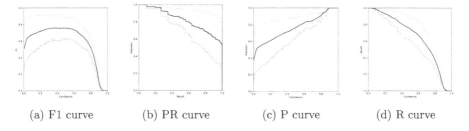

(a) F1 curve   (b) PR curve   (c) P curve   (d) R curve

Fig. 5. Performance metrics of the proposed algorithm. *Blue*: all clases, *orange*: healthy and *green* diseased. (Color figure online)

6 Discussion

The process of evaluating the quality of beans is known as **bean quality assessment**. Physical characteristics, such as grain size, shell percentage, fat content, butter hardness, and humidity, influence the selection of cocoa quality by chocolate manufacturers. These parameters help control undesirable flavors caused by aspects such as astringency, mold, and acidity. The color of the beans was considered in this study as it is a physical characteristic that can provide quality information. The goal of this research was to create a prototype of a robotic arm capable of identifying and differentiating good quality beans from poor quality beans using computer vision techniques.

Regarding the **imbalanced dataset**, grading is a crucial aspect of any approach, and in this article, the main focus is on healthy seeds as they are more prevalent in the final product. The data collection process involved a classification process based on established criteria, which extended the analysis period. While our resulting database is a preliminary version intended to validate the system's objectives, it is part of a larger database aimed at addressing imbalanced data issues, reducing dependence on data augmentation and preprocessing techniques. However, the study findings show the need for more data to infer

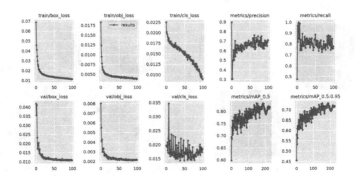

Fig. 6. Training losses and multi-metric results

visual characteristics than traditional algorithms. Therefore, the classification is not entirely accurate as our labels are not solely based on external information.

Bean classification system based solely on cutting tests may not be sufficient as it does not take into account important morphological and health characteristics that affect bean quality. For instance, a bean that passes the cutting test may still be of poor quality and produce low-quality cocoa paste. Sorting out poor quality beans based on sensory tests can significantly enhance the overall quality of chocolate. A **mixed classification criteria** is needed.

7   Conclusions

Object detection and imbalanced classification pose significant challenges in computer vision due to limited information from public databases. However, by utilizing data augmentation and preprocessing techniques, algorithmic approaches can achieve improved results. In this study, we utilized a database of 1634 seeds, classified based on external appearance, odor, and degree of internal fermentation, to ensure accurate classification. Our YOLOv5 system can detect relevant characteristics in real-time that may go unnoticed by cocoa farmers. Our experiments resulted in an overall system accuracy of 94.5%, with the potential for significant improvements through class balance correction and additional image data. We evaluated our algorithm on a prototype robotic hand and propose a classifier system for larger-scale use. Our study highlights the importance of computer vision systems in identifying diseased cocoa beans that could impact the final product and provides valuable information for future studies on cocoa bean sorting. Despite these advancements, cocoa bean classification remains a complex challenge, particularly in exporting countries like Ecuador.

Finally, regarding future work, since smell and taste are determinants for classifying a cocoa bean, the algorithm could also be improved to identify beans with traceability problems that affect the aroma and taste of the final product. In addition, the robotic arm can identify the factor causing the poor quality of the beans, such as fungi, bacteria or insects.

References

1. Abad, A., Acuña, C., Naranjo, E.: El cacao en la costa ecuatoriana: estudio de su dimensión cultural y económica. Estudios de la Gestión: revista internacional de administración (7), 59–83 (2020)
2. Adhitya, Y., Prakosa, S.W., Köppen, M., Leu, J.S.: Feature extraction for cocoa bean digital image classification prediction for smart farming application. Agronomy **10**(11), 1642 (2020)
3. Aguilar, H.: Manual para la evaluación de la calidad del grano de cacao. Editorial FHIA, La Lima, Honduras (2016)
4. Alcívar-Córdova, K., Quezada-Campoverde, J., Berrezueta-Unda, S., Garzón-Montealegre, V., Héctor, C.R.: Análisis económico de la exportación del cacao en el ecuador durante el periodo 2014–2019. Polo del Conocimiento **6**(3), 2430–2444 (2021)
5. Álvarez, C., et al.: Evaluación de la calidad comercial del grano de cacao (theobroma cacao l.) usando dos tipos de fermentadores. Revista científica UDO agrícola **10**(1), 76–87 (2010)
6. Anggraini, C.D., Putranto, A.W., Iqbal, Z., Firmanto, H., Riza, D.F.A.: Preliminary study on development of cocoa beans fermentation level measurement based on computer vision and artificial intelligence. In: IOP Conference Series: Earth and Environmental Science, vol. 924, no. 1, p. 012019 (2021)
7. Ortiz de Bertorelli, L., Rovedas, G., Graziani de Fariñas, L.: Influencia de varios factores sobre índices físicos del grano de cacao en fermentación. Agronomía tropical **59**(1), 81–88 (2009)
8. Blare, T., Useche, P.: Competing objectives of smallholder producers in developing countries: examining cacao production in northern Ecuador. Environ. Econ. **4**, 71–79 (2013)
9. Cabrera, J.B.Z., Encalada, C.M., Guerrero, J.Q., Reyes, S.H., Castillo, A.M., Toro, J.L.: Influencia de la madurez de las mazorcas de cacao: Calidad nutricional y sensorial del cultivar CCN-51. Revista Bases de la Ciencia. **6**(2), 27–40 (2021). e-ISSN 2588-0764
10. Chang, J.F.V., Torres, C.V., Morán, D.E.P., Véliz, J.M., Remache, R.R., Rodríguez, W.M.: Atributos físicos-químicos y sensoriales de las almendras de quince clones de cacao nacional (theobroma cacao l.) en el ecuador. Ciencia y Tecnología **7**(2), 21–34 (2014)
11. Chang, O., Zhinin-Vera, L.: A wise up visual robot driven by a self-taught neural agent. In: Arai, K., Kapoor, S., Bhatia, R. (eds.) FTC 2020. AISC, vol. 1288, pp. 606–617. Springer, Cham (2021). https://doi.org/10.1007/978-3-030-63128-4_47
12. Fernandes, J., Turrisi da Costa, V., Barbin, D., Cruz-Tirado, J., Baeten, V., Barbon Junior, S.: Deep computer vision system for cocoa classification. Multimedia Tools Appl. **81**, 41059–41077 (2022)
13. Gutiérrez-Correa, M.: Efecto de la frecuencia de remoción y tiempo de fermentación en cajón cuadrado sobre la temperatura y el índice de fermentación del cacao (theobroma cacao l.). Revista Científica UDO Agrícola **12**(4), 914–918 (2012)
14. Jiménez Barragán, J.C., Amores Puyutaxi, F.M.: Clasificación de almendras de cacao por el grado de fermentación (2008)
15. Kim, B., Han, Y.K., Park, J.H., Lee, J.: Improved vision-based detection of strawberry diseases using a deep neural network. Front. Plant Sci. **11**, 559172 (2021)
16. Kim, S.S., Kim, S.: Impact and prospect of the fourth industrial revolution in food safety: mini-review. Food Sci. Biotechnol., 1–8 (2022)

17. Lawi, A., Adhitya, Y.: Classifying physical morphology of cocoa beans digital images using multiclass ensemble least-squares support vector machine. J. Phys: Conf. Ser. **979**, 012029 (2018)

18. Loor Solórzano, R.G., Casanova Mendoza, T.d.J., Plaza Avellán, L.F.: Mejoramiento y homologación de los procesos y protocolos de investigación, validación y producción de servicios en cacao y café (2016)

19. Oliveira, M.M., Cerqueira, B.V., Barbon, S., Jr., Barbin, D.F.: Classification of fermented cocoa beans (cut test) using computer vision. J. Food Compos. Anal. **97**, 103771 (2021)

20. Portillo, E., Graziani de Fariñas, L., Betancourt, E.: Efecto de los tratamientos post-cosecha sobre la temperatura y el índice de fermentación en la calidad del cacao criollo porcelana (theobroma cacao l.) en el sur del lago de maracaibo. Revista de la Facultad de Agronomía **22**(4), 394–406 (2005)

21. Portillo, E., et al.: Influencia de las condiciones del tratamiento poscosecha sobre la temperatura y acidez en granos de cacao criollo (theobroma cacao l.). Rev. Fac. Agron **28**, 646–660 (2011)

22. Quevedo Guerrero, J.N., Ramírez Villalobos, M., Zhiminaicela Cabrera, J., Noles León, M.J., Quezada Hidalgo, C., Aguilar Flores, S.: Diversidad morfoagronómica: caracterización de 650 árboles de theobroma cacao L. Revista Universidad y Sociedad **12**(6), 14–21 (2020)

23. Redmon, J., Farhadi, A.: Yolo9000: better, faster, stronger (2016)

24. Solís Hidalgo, Z.K., Peñaherrera Villafuerte, S.L., Vera Coello, D.I.: Las enfermedades del cacao y las buenas prácticas agronómicas para su manejo (2021)

25. Tan, J., Balasubramanian, B., Sukha, D., Ramkissoon, S., Umaharan, P.: Sensing fermentation degree of cocoa (theobroma cacao l.) beans by machine learning classification models based electronic nose system. J. Food Process Eng. **42** (2019)

26. Vera-Chang, J.F., Torres-Navarrete, Y.G., Vallejo-Torres, C.A.: Bolletín técnico n° 1. In: Murillo-Campuzano, G., et al., (eds.) Guía para el mejoramiento del cacao nacional, pp. 1–23. UTEQ (2017)

Efficient Blind Image Super-Resolution

Olga Vais[1] and Ilya Makarov[2,3(✉)]

[1] HSE University, Moscow, Russia
oevays@edu.hse.ru
[2] Artificial Intelligence Research Institute (AIRI), Moscow, Russia
[3] AI Center, NUST MISiS, Moscow, Russia
iamakarov@misis.ru

Abstract. A hybrid method to Single Image Super-resolution is proposed. We used zero-shot super-resolution method to reconstruct high-resolution image from low-resolution one based on the degradation trained on unpaired high-resolution and low-resolution samples. This approach gives the benefits of internal networks, such as extracting features from a particular picture, as well as external methods working with high-resolution and low-resolution image distributions. The proposed scheme would be of high-interest for super-resolution of single images from a specific devices with the same degradations.

Keywords: Image Super-Resolution · Blind Upscaling · ZSSR

1 Introduction

Single Image Super-resolution (SISR) is an important class of ill-posed problems in computer vision and image processing [13,18,21–23,26,33,44]. This problem is aimed to recover high-resolution (HR) image from low-resolution (LR) one that is of high interest for increasing the resolution of medical images, Earth-observation remote sensing images, images of astronomical observations, for the purposes of biometric information identification [31,44] and to improve the quality of synthetic images [8,19,20,24,25,29].

HR recovering can be realized by different approaches, which can be divided to the following categories: interpolation-based, reconstruction-based and learning-based methods. Interpolation based methods, such as bicubic interpolation [11] and Lanczos resampling [7], are speedy and straightforward, but have a low accuracy. Reconstruction-based methods use a prior knowledge to generate flexible and sharp details [5,36]. However, the quality of these methods rapidly decreases during increasing the scale factor. Learning-based methods analyze statistical relationships between a LR image and its corresponding HR

I. Makarov—The work of Ilya Makarov was made in the framework of the strategic project "Digital Business" within the Strategic Academic Leadership Program "Priority 2030" at NUST MISiS.

I. Rojas et al. (Eds.): IWANN 2023, LNCS 14135, pp. 229–240, 2023.
https://doi.org/10.1007/978-3-031-43078-7_19

counterpart from training samples. It can be realized through neighbor embedding [4], super-resolution forests [34], naive bayes super-resolution forest [27], which shows high speed and comparable high quality. Some methods use image partitioning on atoms, such as image super-resolution via sparse representation [42] or anchored neighborhood regression [37]. Moreover image super-resolution methods can be based on combination of reconstruction-based and learning based ones.

At the same time, deep learning SISR algorithms have already demonstrated great superiority to reconstruction-based and other learning-based methods. This work is devoted to such methods of single-image super-resolution. Related works are discussed in Sect. 2. Section 3 is dedicated to the method proposed, while we show the results of our experiments in Sect. 4.

2   Related Works

There are different deep learning methods solving SISR task that have already shown results exceeding another approaches discussed in Introduction. The first huge class of DL approaches refers to supervised methods [30], when neural network is trained with both low-resolution and corresponding high-resolution images. One of the main part of them is upsampling layer, which can be interpolation-based, such as NN-interpolation, bilinear interpolation, bicubic interpolation, and learning-based, which are using transposed convolution layer, sub-pixel layer or meta upscale module. The DL methods can be classified based on upsampling layer position relatively CNNs (Convolutional Neural Networks) that significantly impacts on the quality of image reconstruction and training strategy [41]. The application of a similar approach using attention convolution neural network for medical image super-resolution reconstruction is discussed in [39] where the fuzzy set theory [32] is applied to characterize the uncertainty of pixels.

Pre-upsampling approach, which is used in SRCNN [6], is the most straightforward, when firstly LR is upscaled, after that the result is refined with using deep neural networks. However predefined upsampling can introduced additional distortions. Moreover pre-upsampling leads to most computations in high-dimensional space that is contrast to post-upsampling approach (such as EDSR, ESRGAN [17,40]), where the computational efficiency is significantly improved and the feature extraction occurs already in low-dimensional space. Although such methods are mainstream and have simple learning strategies, they have difficulties in transformations with large scaling factors. Then progressive upsampling NN, such as Laplacian pyramid SR network (LapSRN) [14], can be used. At the same time, they require more complicated learning strategies. To capture the mutual dependency of LR-HR pairs, iterative up-and-down sampling neural networks (DBPN [10], SRFBN [15]) can be used. They consist of sequential upscaling and downscaling layers, and results are based on all intermediately reconstructions.

Considered methods require paired LR-HR samples, while in real SISR tasks such pairs do not exist. If there are unpaired samples of LR and HR images, weakly-supervised Super-resolution methods can be used. In such algorithms, firstly, by using unpaired HR-LR samples, HR-to-LR GAN (Generative Adversarial Network) is trained to learn degradation. After that LR-to-HR is trained based on LR-HR pairs, where corresponding LR images were generated by HR-to-LR GAN [3]. Moreover, this strategy allows simultaneously learning LR-to-HR and HR-to-LR mappings as an alternative to the two-stage learning degradation process. However such approach requires some advanced strategies for reducing the training difficulty and instability (CycleGAN [46]).

The models considered above can be also called as externally trained networks, i.e. trained externally supplied examples relatively to a single image. However if degradation kernel of LR images is significantly differ from one of the training sample, HR-result can be closer to one received by Bicubic interpolation [35]. For these purposes, internal networks can be used [35]. Internal networks are fully unsupervised and use just LR image, which should be transformed. Since they take into account entropy contained in the image (not in all dataset), these model have lower numbers of features. Although such models should be trained again for every new image, the total time of both training and predicting phases is significantly less than one for externally trained networks. One more approach is called Deep Image Prior [38], where randomly initialized CNN tries to generate the target HR image from a random vector. This untrained network captures the low-level statistics of the image. Although this method is outperformed by learning-baser SR methods, it is better than the non-trained approaches.

3   Method

All methods discussed above are based on either statistics of paired/unpaired HR-LR samples or internal statistics of reconstructed picture. However for low-resolution images from a specific device (camera, tomograph or another image detector) with the same degradation kernel, a hybrid approach can be of high interest. We propose a model based on HR-to-LR GAN, which was the part of neural network used in [3], to extract degradation kernel from unpaired high-resolution and low-resolution samples. The HR data can be received from devices with higher resolution, i.e. target quality. After learning HR-to-LR GAN, its generator is used for downsampling images in ZSSR-model to reconstruct LR images also taking into account their own internal statistics.

3.1   ZSSR

Zero-shot Super-resolution model (ZSSR) refers to internal trained methods, which uses inner statistics of transformed picture to convert it in HR image [35]. It is trained on LR-HR pairs extracted from the test image itself. Each pair is represented by "HR parent" and "LR child", which was given by downscaling by the desired SR scaled-factor the first one, while every "HR fathers" are the result of downscaling the initial LR image. The further augmentation is done by image rotations and their mirror reflections in the vertical and horizontal directions.

After training, Image-Specific CNN is applied to the initial LR image to produce its HR output. Figure 1 shows architecture of this model. Downscaling can be realized by some kind of analytical kernels or kernel predicted by another neural network [2, 16].

ZSSR can be improved with downscaling based on an image-specific Internal-GAN, which is called "KernelGAN" [2]. This model is also trained on the LD image and learns its internal patch distributions. Its Generator, which is represented by Deep Linear network (sequence of linear layers with no activations), is trained downscaling LR image to cheat its Discriminator, which should distinguish the patch distribution of downscaled image from the patch distribution of original one. In this approach after training, the generator implicitly contains the correct image-specific SR-kernel, which can be also extracted in explicit form. After that, the generator (downscaling operator with the correct kernel) is used for downscaling image in the other algorithms. In the paper [2], the authors used this downscaling SR-Kernel to generate "HR fathers" and "LR sons" in ZSSR. The degradation kernel also can be reconstructed by a normalizing flow-based kernel prior (FKP), was recently suggested for kernel modeling including to replace Deep Linear network in KernelGAN (KernelGAN-FKP, [16]).

Image-Specific CNN has a simple architecture based on convolution layers with ReLU activation on each layer. For training, L_1 pixel loss is used with ADAM optimizer [12]. The result of the image transformation highly depends on downsampling method used in training, that is why ZSSR must be strengthened by a suitable approach for degradation-kernel prediction. In our model we use HR-to-LR GAN for this purposes.

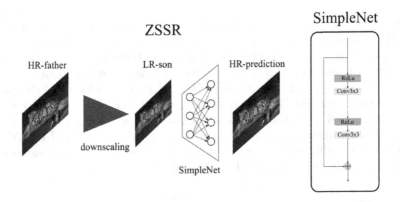

Fig. 1. ZSSR model: an image-specific CNN (SimpleNet) is trained on examples extracted from an initial image by downsampling.

3.2   HR-to-LR GAN

Unpaired LR-HR samples allow using weakly-supervised methods to reconstruct degradation kernel. Our approach is based on HR-to-LR GAN, which was a part of neural networks with LR-to-HR GAN for image super-resolution. In that model HR-to-LR GAN was trained to learn degradation, which were used for generating LR part for existing HR samples. After that, LR-to-HR is trained based on LR-HR pairs [3]. In such approach, final HR images satisfy the distribution of real HR as well as LR images are also based on real LR distributions (not only degradation kernel). We in turn separated HR-to-LR GAN from it for reconstruction of the degradation based on LR-image statistics. Its generator and discriminator are schematically shown in Fig. 2.

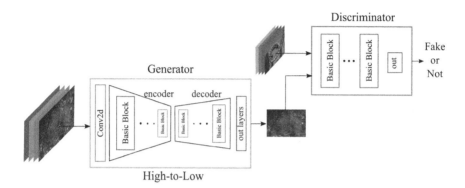

Fig. 2. HR-to-LR GAN: the generator with encoder-decoder architecture and the discriminator, both of them consist of Basic Blocks.

Both the generator and the discriminator were implemented on the basic block with pre-activation. The generator has an encoder-decoder structure and consists of convolution layer, Basic blocks (two convolution layers with batch normalization and ReLU pre-activation) and out part of the neural network (sequence of two convolution layers with ReLU and Tanh activations respectively). Architectures of the generator and Basic Blocks are shown in Fig. 3. HR-to-LR discriminator is consists of 6 Basic blocks without batch normalization, which are followed by a fully connected layer.

Similarly to the paper [3], these networks were trained with a total loss as a combination of GAN loss and an L_2 pixel loss:

$$l = \alpha L_2 + \beta L_{GAN}, \tag{1}$$

where $\alpha = 1, \beta = 0.05$ are the weights. GAN loss were calculated as:

$$L_{GAN} = \mathbb{E}[\min(0, -1 + D(x))] + \mathbb{E}[\min(0, -1 - D(\hat{x}))], \tag{2}$$

where x is the sample of LR-distribution, $\hat{x}$ is the output of HR-to-LR generator. Optimization were realized by ADAM method.

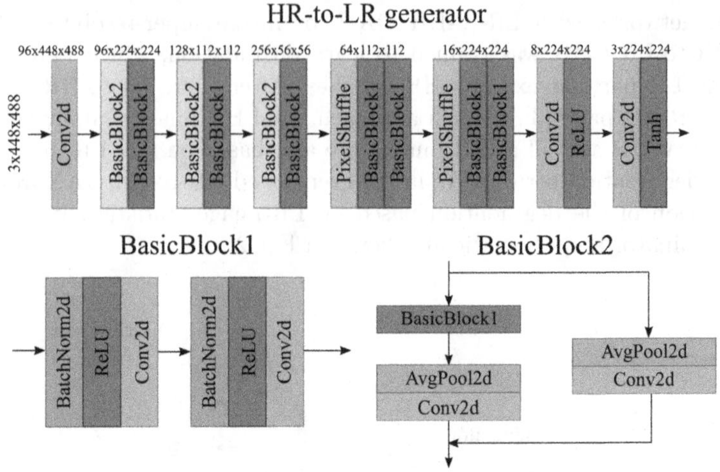

HR-to-LR generator

Fig. 3. The encoder-decoder architecture of HR-to-LR generator which in/out-layers and architecture of Basic Block.

4 Experiments

In this work, we use Berkeley Segmentation Dataset 500 (BSDS500, [1]), which consists of 500 natural images showing animals, buildings, food, landscapes, people or plants in JPEG format with the average resolution of (432, 370). Some examples of HR images are shown in Fig. 4. We compare the results of our approach with ones received by bilinear interpolation implemented in OpenCV and ZSSR with different degradation kernel: bilinear, cubic, sinc and reconstructed by KernelGAN [2].

4.1 Training Data

For training HR-to-LR GAN, 200 images were randomly chosen as LR examples, which were downscaled by means of given kernel (the left image in Fig. 5). Another 200 images from the dataset were used as high-resolution samples. High-resolution images were transformed to square images of 448×448 size, while low-resolution ones were resized to 224×224. Both sizes exceed an image size along the largest edge. All images had BGR (OpenCV standard) color channels, which were normalized with 0.5 mean and 0.5 std.

4.2 Metrics

To measure reconstruction quality and compare results with ones received by another approaches, we use the peak signal-to-noise ratio (PSNR) and the structural similarity index (SSIM), which are defined in the following way [43]:

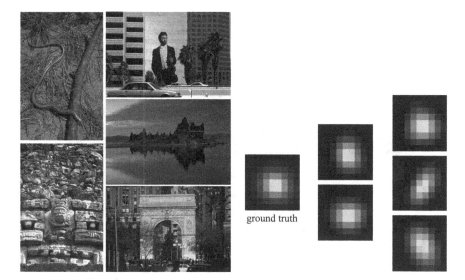

Fig. 4. Examples of high-resolution images from BSDS500 dataset [1]. The red squares show positions of the samples from Fig. 6

Fig. 5. The ground-truth degradation kernel, which was used for preparing low-resolution images. Degradation kernels predicted by KernelGAN [2] for low-resolution images corresponding to HR examples from Fig. 4

$$PNSR = 10\log_{10}(L^2/MSE), \quad SSIM(I,\hat{I}) = \frac{2\mu_I\mu_{\hat{I}} + k_1}{\mu_I^2 + \mu_{\hat{I}}^2 + k_1} \cdot \frac{\sigma_{I\hat{I}} + k_2}{\sigma_I^2 + \sigma_{\hat{I}}^2 + k_2}, \tag{3}$$

where I and $\hat{I}$ are ground truth and reconstructed images, MSE is the mean squared error, L is the dynamic range of the images (1 or 255, dependently on the data representation). μ and σ^2 are the mean and the variance of an image, $\sigma_{I\hat{I}}$ is the covariance between I and $\hat{I}$, k_1 and k_2 are the constants.

4.3 Benchmarks

The leftover 100 images from BSDS500 dataset were used for checking the quality of the model and its comparison with another approaches. These images were as high-resolution targets, while low-resolution test images were received by downscaling them by means of the given kernel, which was used to prepare training dataset.

4.4 Results

We have compared the results received by the proposed model with images reconstructed by ZSSR method with another degradation kernels (bilinear, cubic, sinc or predicted by KernelGAN) or by resizing function with bilinear interpolation (the left pictures in Fig. 6).

Our calculations show that, for given degradation kernel (the left image in Fig. 5), bilinear degradation gives the best result as compared with the other analytical kernels (see Table 1), while using sinc-function reconstructs the images with the same quality as a simple bilinear interpolation (Fig. 6).

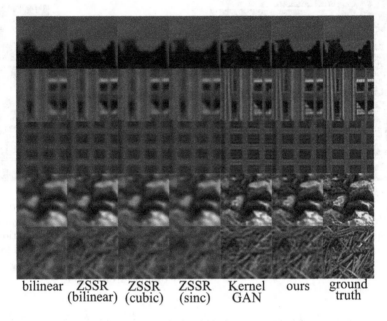

bilinear   ZSSR   ZSSR   ZSSR   Kernel   ours   ground
(bilinear) (cubic) (sinc) GAN   truth

Fig. 6. Visual comparison achieved on Berkeley Segmentation Dataset 500 [1] for ×2 SR. The chosen squares correspond to ones shown in Fig. 4 by red squares.

Reconstructed images, which were received by ZSSR model using the degradation kernel predicted by KernelGAN, have sharper edges than previous ones. However the PSNR results have the lower values than in the case of the bilinear kernel. It can be explained by significant differences between predicted kernel and one, which were used for degradation, that is shown in Fig. 5. Our approach is trained to downscale images based on some LR examples. It allows reconstructing images with sharper edges as well as with better PSNR/SSIM results, which are closer to ones received for the bilinear kernel.

Table 1. PSNR and SSIM results achieved on 100 test LR and HR samples from BSDS500 dataset.

degrad.	ZSSR (bilinear)	ZSSR (cubic)	ZSSR (sinc)	Kernel GAN	ours
PSNR	**25.43**	25.06	24.94	24.94	25.31
SSIM	0.71	0.68	0.68	0.71	**0.73**

5   Conclusion

This work is devoted to the problem of Single Image Super-Resolution. Here we proposed a hybrid model based on two neural networks: weakly-supervised HR-to-LR GAN and unsupervised ZSSR. Our approach improves the reconstruction quality using LR distributions and HR images with the target resolution that would be of high interest for images from a specific device.

Both neural networks were trained on BSD500 dataset, which contains images showing animals, buildings, food, landscapes, people or plant, that refers to a wide range of possible applications. Experimental results show a good visual quality as well as high values of PSNR/SSIM metrics, which corresponds to more accurate degradation kernel prediction than in the methods using one image for this purpose.

However, using neural network for image degradation significantly slows down ZSSR training, i.e. image transformation to its high-resolution representation. Thus, further development of proposed method consists of HR-to-LR GAN optimization by the variation of the number of layers and features and speeding-up the inference in a neural network [28]. It is also necessary to study the applicability of the data-free compression techniques [9] for the SISR models [45].

References

1. Arbeláez, P., Maire, M., Fowlkes, C., Malik, J.: Contour detection and hierarchical image segmentation. IEEE Trans. Pattern Anal. Mach. Intell. **33**(5), 898–916 (2011)
2. Bell-Kligler, S., Shocher, A., Irani, M.: Blind super-resolution kernel estimation using an internal-GAN (2019)
3. Bulat, A., Yang, J., Tzimiropoulos, G.: To learn image super-resolution, use a GAN to learn how to do image degradation first. In: Ferrari, V., Hebert, M., Sminchisescu, C., Weiss, Y. (eds.) ECCV 2018. LNCS, vol. 11210, pp. 187–202. Springer, Cham (2018). https://doi.org/10.1007/978-3-030-01231-1_12
4. Chang, H., Yeung, D.Y., Xiong, Y.: Super-resolution through neighbor embedding. In: Proceedings of the 2004 IEEE Computer Society Conference on Computer Vision and Pattern Recognition, 2004, CVPR. IEEE (2004)
5. Dai, S., Han, M., Xu, W., Wu, Y., Gong, Y., Katsaggelos, A.: SoftCuts: A soft edge smoothness prior for color image super-resolution. IEEE Trans. Image Process. **18**(5), 969–981 (2009)
6. Dong, C., Loy, C.C., He, K., Tang, X.: Learning a deep convolutional network for image super-resolution. In: Fleet, D., Pajdla, T., Schiele, B., Tuytelaars, T. (eds.) ECCV 2014. LNCS, vol. 8692, pp. 184–199. Springer, Cham (2014). https://doi.org/10.1007/978-3-319-10593-2_13
7. Duchon, C.E.: Lanczos filtering in one and two dimensions. J. Appl. Meteorol. **18**(8), 1016–1022 (1979)
8. Feygina, A., Ignatov, D.I., Makarov, I.: Realistic post-processing of rendered 3D scenes. In: Proceedings of the 45th ACM International Conference SIGGRAPH (SIGGRAPH'18), pp. 1–2. ACM, New York (2018)

9. Grachev, A.M., Ignatov, D.I., Savchenko, A.V.: Neural networks compression for language modeling. In: Shankar, B.U., Ghosh, K., Mandal, D.P., Ray, S.S., Zhang, D., Pal, S.K. (eds.) PReMI 2017. LNCS, vol. 10597, pp. 351–357. Springer, Cham (2017). https://doi.org/10.1007/978-3-319-69900-4_44

10. Haris, M., Shakhnarovich, G., Ukita, N.: Deep back-projection networks for super-resolution. In: Proceedings of the IEEE Conference on Computer Vision and Pattern Recognition (CVPR), pp. 1664–1673 (2018)

11. Keys, R.: Cubic convolution interpolation for digital image processing. IEEE Trans. Acoust. Speech Signal Process. **29**(6), 1153–1160 (1981)

12. Kingma, D.P., Ba, J.: Adam: a method for stochastic optimization (2014)

13. Korinevskaya, A., Makarov, I.: Fast depth map super-resolution using deep neural network. In: Proceedings of the 17th IEEE International Symposium on Mixed and Augmented Reality (ISMAR'18), TU Munich, pp. 117–122. IEEE, New York (2018)

14. Lai, W.S., Huang, J.B., Ahuja, N., Yang, M.H.: Fast and accurate image super-resolution with deep Laplacian pyramid networks (2017)

15. Li, Z., Yang, J., Liu, Z., Yang, X., Jeon, G., Wu, W.: Feedback network for image super-resolution. In: Proceedings of the IEEE Conference on Computer Vision and Pattern Recognition (CVPR), pp. 3867–3876 (2019)

16. Liang, J., Zhang, K., Gu, S., Gool, L.V., Timofte, R.: Flow-based kernel prior with application to blind super-resolution (2021)

17. Lim, B., Son, S., Kim, H., Nah, S., Mu Lee, K.: Enhanced deep residual networks for single image super-resolution. In: Proceedings of the IEEE Conference on Computer Vision and Pattern Recognition Workshops (CVPRW), pp. 136–144 (2017)

18. Makarov, I., Aliev, V., Gerasimova, O., Polyakov, P.: Depth map interpolation using perceptual loss. In: Proceedings of the IEEE International Symposium on Mixed and Augmented Reality (ISMAR'17), Ecole Centrale de Nantes, France, pp. 93–94. IEEE, New York (2017)

19. Makarov, I., Bakhanova, M., Nikolenko, S., Gerasimova, O.: Self-supervised recurrent depth estimation with attention mechanisms. PeerJ Comput. Sci. **8**(e865), 1–25 (2022)

20. Makarov, I., Borisenko, G.: Depth inpainting via vision transformer. In: Proceedings of the 19th IEEE International Symposium on Mixed and Augmented Reality (ISMAR'21), INSA/IRISA, pp. 286–291. IEEE, New York (2021)

21. Makarov, I., Korinevskaya, A., Aliev, V.: Fast semi-dense depth map estimation. In: Proceedings of the ACM Workshop on Multimedia for Real Estate Tech (RETech'18), University of Tokyo, pp. 18–21. ACM, New York (2018)

22. Makarov, I., Korinevskaya, A., Aliev, V.: Sparse depth map interpolation using deep convolutional neural networks. In: Proceedings of the 41st IEEE International Conference on Telecommunications and Signal Processing (TSP'18), Brno University of Technology, pp. 1–5. IEEE, New York (2018)

23. Makarov, I., Korinevskaya, A., Aliev, V.: Super-resolution of interpolated downsampled semi-dense depth map. In: Proceedings of the 23rd ACM International Conference on 3D Web Technology (Web3D'18), University of Economics and Business, pp. 1–2. ACM (2018)

24. Makarov, I., et al.: On reproducing semi-dense depth map reconstruction using deep convolutional neural networks with perceptual loss. In: Proceedings of the 27th ACM International Conference on Multimedia (MM'19), CNRS-IRISA, pp. 1080–1084. ACM, New York (2019). https://doi.org/10.1145/3343031.3351167

25. Makarov, I., Polonskaya, D., Feygina, A.: Improving picture quality with photo-realistic style transfer. In: Campilho, A., Karray, F., ter Haar Romeny, B. (eds.) ICIAR 2018. LNCS, vol. 10882, pp. 47–55. Springer, Cham (2018). https://doi.org/10.1007/978-3-319-93000-8_6
26. Maslov, D., Makarov, I.: Fast depth reconstruction using deep convolutional neural networks. In: Rojas, I., Joya, G., Català, A. (eds.) IWANN 2021. LNCS, vol. 12861, pp. 456–467. Springer, Cham (2021). https://doi.org/10.1007/978-3-030-85030-2_38
27. Salvador, J., Perez-Pellitero, E.: Naive Bayes super-resolution forest. In: 2015 IEEE International Conference on Computer Vision (ICCV). IEEE (2015)
28. Savchenko, A.V.: Fast inference in convolutional neural networks based on sequential three-way decisions. Inf. Sci. **560**, 370–385 (2021)
29. Savchenko, A.V.: MT-EmotiEffNet for multi-task human affective behavior analysis and learning from synthetic Data. In: Karlinsky, L., Michaeli, T., Nishino, K. (eds.) ECCV 2022. Lecture Notes in Computer Science, vol. 13806, pp. 45–59. Springer, Cham (2023)
30. Savchenko, A.V., Belova, N.S.: Statistical testing of segment homogeneity in classification of piecewise-regular objects. Int. J. Appl. Math. Comput. Sci. **25**(4), 915–925 (2015)
31. Savchenko, A.V., Belova, N.S.: Unconstrained face identification using maximum likelihood of distances between deep off-the-shelf features. Expert Syst. Appl. **108**, 170–182 (2018)
32. Savchenko, A.V., Savchenko, L.V.: Towards the creation of reliable voice control system based on a fuzzy approach. Pattern Recogn. Lett. **65**, 145–151 (2015)
33. Savchenko, A.: Maximum-likelihood dissimilarities in image recognition with deep neural networks. Comput. Opt. **41**(3), 422–430 (2017)
34. Schulter, S., Leistner, C., Bischof, H.: Fast and accurate image upscaling with super-resolution forests. In: 2015 IEEE Conference on Computer Vision and Pattern Recognition (CVPR). IEEE (2015)
35. Shocher, A., Cohen, N., Irani, M.: "zero-shot" super-resolution using deep internal learning (2017)
36. Sun, J., Xu, Z., Shum, H.Y.: Image super-resolution using gradient profile prior. In: 2008 IEEE Conference on Computer Vision and Pattern Recognition. IEEE (2008)
37. Timofte, R., De, V., Gool, L.V.: Anchored neighborhood regression for fast example-based super-resolution. In: 2013 IEEE International Conference on Computer Vision. IEEE (2013)
38. Ulyanov, D., Vedaldi, A., Lempitsky, V.: Deep image prior. Int. J. Comput. Vision **128**(7), 1867–1888 (2020)
39. Wang, C., Lv, X., Shao, M., Qian, Y., Zhang, Y.: A novel fuzzy hierarchical fusion attention convolution neural network for medical image super-resolution reconstruction. Inf. Sci. **622**, 424–436 (2023)
40. Wang, X., et al.: ESRGAN: enhanced super-resolution generative adversarial networks. In: Leal-Taixé, L., Roth, S. (eds.) ECCV 2018. LNCS, vol. 11133, pp. 63–79. Springer, Cham (2019). https://doi.org/10.1007/978-3-030-11021-5_5
41. Wang, Z., Chen, J., Hoi, S.C.H.: Deep learning for image super-resolution: a survey (2019)
42. Yang, J., Wright, J., Huang, T.S., Ma, Y.: Image super-resolution via sparse representation. IEEE Trans. Image Process. **19**(11), 2861–2873 (2010)
43. Yang, W., Zhang, X., Tian, Y., Wang, W., Xue, J.H.: Deep learning for single image super-resolution: a brief review (2018)

44. Yue, L., Shen, H., Li, J., Yuan, Q., Zhang, H., Zhang, L.: Image super-resolution: the techniques, applications, and future. Signal Process. **128**, 389–408 (2016)
45. Zhang, Y., Chen, H., Chen, X., Deng, Y., Xu, C., Wang, Y.: Data-free knowledge distillation for image super-resolution. In: Proceedings of the IEEE/CVF Conference on Computer Vision and Pattern Recognition, pp. 7852–7861 (2021)
46. Zhu, J.Y., Park, T., Isola, P., Efros, A.A.: Unpaired image-to-image translation using cycle-consistent adversarial networks. In: Proceedings of the IEEE International Conference on Computer Vision (ICCV), pp. 2223–2232 (2017)

LAPUSKA: Fast Image Super-Resolution via LAPlacian UpScale Knowledge Alignment

Aleksei Pokoev[1] and Ilya Makarov[2,3(✉)]

[1] HSE University, Moscow, Russia
[2] AI Center, NUST MISiS, Moscow, Russia
iamakarov@misis.ru
[3] Artificial Intelligence Research Institute (AIRI), Moscow, Russia

Abstract. Single image super-resolution is important part of computer vision open problems. Recently, deep neural networks have demonstrated excellent performance in this problem. In this work, several cutting edge methods for super-resolution problem using deep neural networks will be considered. Comparison of their effectiveness and evaluation of neural networks architectures with respect to different metrics is one of our main goals for this research. Modern deep learning methods often require large computational cost and load a lot of computer memory, which affects the ease of use of neural networks and the time of generation super-resolution results. In addition to the existing models, we propose a new architecture of neural networks based on best properties of considered architectures and designed to eliminate their shortcomings. Furthermore, we compare the quality of all considered deep learning methods with baseline method of bicubic interpolation.

Keywords: Image Super-resolution · Computer Vision · Deep CNN · GAN

1 Introduction

Recently, tasks related to image processing (such as object recognition, style transfer, mask overlay, etc.) have become especially popular [34]. The use of neural networks has helped to achieve significant progress and high quality results in many areas of computer vision [41].

One such task is a single image super-resolution (SR). Its aim is to reconstruct a high-resolution (HR) image from a single low-resolution (LR) input. This topic is often found in both scientific literature and in works of art. One of the examples is the science fiction film Blade Runner in which the main character, Detective Rick Deckard, enlarges the photo several times and thereby finds evidence. There

The work of Ilya Makarov was made in the framework of the strategic project "Digital Business" within the Strategic Academic Leadership Program "Priority 2030" at NUST MISiS.

I. Rojas et al. (Eds.): IWANN 2023, LNCS 14135, pp. 241–253, 2023.
https://doi.org/10.1007/978-3-031-43078-7_20

exist also many applications in for biometric information identification [36,44] and to improve the quality of synthetic images [35].

Various images consist of many details, each of which contains certain information, so the larger the image resolution, the more details can be detected. The quality of the resulting images in many photo and video devices is limited [38]. Enhancing image resolution can help to overcome these limitations. Therefore, this task is very important and particularly useful in many practical applications, in which digital image processing is used, such as face recognition, image compression, digital zoom in camera, and many others.

The **object** of the study are SR methods that use deep learning techniques, namely various neural networks models, and the **subject** of research is their performance. The goal of this work is to consider some of the state-of-the-art SR models and compare the quality of their work. Conducting various experiments with networks architecture is also main objective of the research.

Since the quality of SR methods that use convolutional neural networks (CNN) have been significantly boosted, given work does not aim to outperform others. Nevertheless, a lot of state-of-the-art models of neural networks have disadvantages, including image blurring and large load of the computer memory. In this work, there will be an attempt to remove these drawbacks.

At the beginning of the work, a brief overview of the articles offering different SR methods will be made. Then, the structures of some models will be considered, and the architecture of the neural network proposed in this research will be described. Finally, the results of the experiments will be described and the quality of the various SR methods will be compared.

1.1   Related Work

There are many articles presenting different SR methods. Until recently, methods that did not use CNN were prevalent. Examples may be the articles of [8,10], in which methods of HR image restoration based on statistics [37] are described, and articles of [9,20], in which example-based SR techniques were used.

Additionally, there are several methods associated with a similar depth SR topic. Among them there are filtering methods, which include a bilateral filtering [21] and guided image filtering [12]. In addition, there are also optimization methods used to solve this problem. These include models based on Markov random fields [5] and dictionary-based models [25].

Recently, the methods using neural networks have become very popular achieving state-of-art performance. One of the first method that use CNN was proposed in the paper of [6], where 3-layer deep network at first upscale image using bicubic interpolation and then extracts new features and combines them to reconstruct HR image.

The next trend in the works was developing architectures with more layers (that is, deeper CNN). So, a deeper VDSR network [18] has achieved better results compared to SRCNN [6] by increasing the number of layers from 3 to 20. [19] proposed a network (DRCN) with several 16-recursion layers.

A deeper layer architecture leads to increased computational complexity making training difficult and increases learning time. Using residual blocks [13] and skip-connections [19] makes it easier to learn and shows good performance. The idea of the second concept is that the output of one layer was added to the output of one or multiple following layers. Residual blocks are based on skip-connections. They can help in cases where increasing the depth of networks leads to degradation of quality.

The methods mentioned above apply bicubic interpolation to the LR image as preprocessing step before taking it as an input to the neural network, which increases the computational cost. There are several alternative ways to upscale LR image. [7,23] use transposed convolutional layers. Another good option is using sub-pixel convolution [39,42]. Both methods allow to achieve real-time speed, but they have difficulties with large scale factors.

One of the best performance was shown in work [24], where the popular generative adversarial network (GAN) [11] was used. GAN consists of two networks: one is a Generator that produce candidate and the other is Discriminator that evaluate the candidate. This technique has shown excellent results in the task of creating photo-realistic images, such as elements of industrial design, clothing, scenes.

Another model achieving real-time speed and high performance was described in the paper [23]. The authors proposed a neural network architecture based on the Laplacian pyramid. It consists of several levels and allow to progressively predict at each level twice bigger HR image than at previous one. Trained neural network can simultaneously output several images with factors equals to different degrees of 2.

Interesting direction of research may lie in applying depth estimation algorithms [22,26–33] for blind super-resolution [4], but these ideas lie out of the scope of the current paper.

In many papers, pixel-wise mean squared error (MSE) as a loss function is used in order to optimize networks. It provides good solutions for optimization problem, but the obtained HR images are often oversmoothed and poorly correlate with the human perception [16,23,24]. In [24] the authors propose a perceptual loss function, which is the sum of content loss and adversarial loss. The first summand differs from the MSE in that instead of calculating difference between ground truth and reconstructed images, it calculates difference between their feature mapping within the VGG network [40]. Compared to the second term, perceptual loss helps to get more convincing images. Another option is to use Charbonnier loss function, which helps to improve SR performance [23].

2   Method

In order to check the quality of different methods, it is necessary to implement the models of neural networks proposed in them. In this work, the programming language Python 3 was chosen because it is one of the most popular and convenient language for developing. Besides, it has several popular frameworks, such as TensorFlow or Keras, in which many models of neural networks have already been implemented.

In this paper, cutting-edge neural network models described in articles [23, 24] will be considered. Their short names are SRGAN and LapSRN, respectively. These SR methods were chosen because they are among the most advanced at the time of this study and have shown excellent results. Furthermore, the authors of the above-mentioned neural networks did not compare their performance in their articles. In addition to these models, our own neural network architecture will also be described.

2.1   SRGAN

The SRGAN network based on a previously mentioned GAN model. It consists of two networks: the first network (Generator) is trying to create such HR images that the second network (Discriminator) evaluates whether they are genuine.

At the input of the Generator, an LR image is sent, after which it outputs an HR image. The resulting image is also fed to the input of the Discriminator, which attempts to determine whether the image is real. In addition to the obtained generated images, the Discriminator is separately trained on real images to distinguish between false and real images.

An important part of the Generator network are R residual blocks, each of which consists of 2 convolutional layers together with the subsequent batch normalization layers and parametric ReLU activation after first convolutional layer. It is worth noting that a sub-pixel convolution layers are used to upscale the image. The Discriminator network consists of 8 convolutions, each second of which has 2 times more filters than the previous layer (from 64 to 512). Increasing a stride length of the convolution is used in order to decrease the image resolution after each enhancement number of filters.

This architecture has shown great performance in the SR problem and allows to get a fairly good photo-realistic images. One of the reasons for success is the perceptual loss function, which is used to train the network. It is defined as:

$$\ell_{\text{perc}} = \ell_{\text{cont}} + 10^{-3}\ell_{\text{adv}}, \tag{1}$$

where ℓ_{cont} is a content loss, ℓ_{adv} is an adversarial loss and are defined as:

$$\ell_{\text{cont}/i,j} =$$
$$\frac{1}{W_{i,j}H_{i,j}} \sum_{x=1}^{W_{i,j}} \sum_{y=1}^{H_{i,j}} \left(\varphi_{i,j}(I^{HR})_{x,y} - \varphi_{i,j}(G(I^{LR})_{x,y}) \right)^2 \tag{2}$$

$$\ell_{\text{adv}} = \sum_{n=1}^{N} -\log D(G(I^{LR})) \tag{3}$$

Here, I^{LR} and I^{HR} are LR and HR images, respectively; D and G are outputs of Discriminator and Generator networks, respectively; $\varphi_{i,j}$ is the feature map in the VGG19 [40] network (i indicates that map is before i-th maxpooling and j mean that it is after j-th convolution layer); $W_{i,j}$ and $H_{i,j}$ equals to the dimensions of the feature maps and N is a number of samples.

However, SRGAN also has shortcomings, namely, using a large enough VGG network to calculate content loss significantly increases the memory consumption.

2.2 LapSRN

The LapSRN model uses a different approach and its architecture is not similar to SRGAN. LapSRN is a pyramidal structure with $\log_2 S$ levels (where S is an upscale factor), that allows to progressively predict 2 times larger HR image at each level. In this work, architecture with 2 levels for obtaining x4 HR image was used.

At each level, there is feature embedding network, upsample blocks and convolutional layer for obtaining residual image (that is, the difference between the SR image and the HR image). Feature embedding network consist of R recursive blocks, each of which contains stack of D convolutional layers preceded by ReLU activation layer. At the first level, there is additional convolutional layer to convert LR input to feature maps that feature embedding network takes as an input. At others levels, it takes upscaled at previous level features. In this model, transposed convolutional layers are used to upsample images and feature maps. At the output of each level, 2 times bigger image is obtained by element-wise summing residual image and upscaled image from previous level. In addition, in order to reduce the number of parameters, the weights of each component (feature embedding network, upscale blocks and convolutional layer) are shared between each level.

Unlike the previous model, LapSRN is optimized with the Charbonnier loss function. Suppose y_l and $\hat{y}_l$ are ground truth HR and SR images at level l (y_l is obtained by downscaling the HR images y using bicubic kernel with the corresponding downscaling factor). The loss function is defined as:

$$\ell_{\mathrm{C}} = \frac{1}{N} \sum_{i=1}^{N} \sum_{l=1}^{L} \rho(y_l^{(i)} - \hat{y}_l^{(i)}), \qquad (4)$$

where $\rho(x) = \sqrt{x^2 + \epsilon^2}$, $\epsilon = 10^{-3}$, N is the number of samples, $L = \log_2 S$ is the number of levels, S is the upscale factor.

Using MSE as a loss function often leads to blurry results. In an effort to address this shortcoming, the authors of this model decided to use Charbonnier loss function, which they claim leads to better performance over the MSE loss and is better at handling outliers [23].

The pyramidal structure of LapSRN network together with Charbonnier loss function allows to achieve qualitative result, but the architecture of the neural network also has drawbacks: each level l of the network requires ground truth HR y_l images, which leads to additional computational costs (for obtaining y_l images by bicubic downsampling).

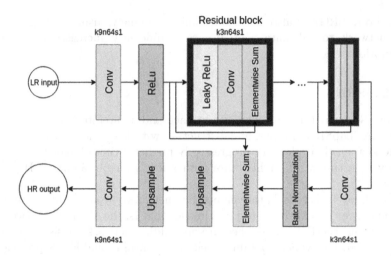

Fig. 1. Architecture of proposed model. Each convolutional layer has detailed description of the structure: the kernel size (k), number of filters (n), and strides (s).

2.3 Proposed Model

SRGAN and LapSRN have achieved excellent performance, but also have short-comings. In an attempt to preserve part of the structure of the networks described above and minimize their disadvantages, the neural network model proposed in this work was developed. The detailed architecture of this model is shown on the Fig. 1.

At the beginning of the network, convolutional layer with ReLU activation is used in order to convert LR input into feature maps. An important part in the network architecture as in the previous models is the R residual blocks. The structure of the each residual block is similar to that of the feature embedding block in LapSRN, but consists of only one convolutional layer and Leaky ReLU activation ($\alpha = 0.2$) is used. In addition, skip-connection is added to the residual block, because, as shown in [24], its presence improves the metric value and does not affect the learning time.

The next part of the network containing the convolutional layer followed by batch normalization layer with element-wise addition is similar to the one in SRGAN. Using batch normalization for training deep networks reduces training time and helps to achieve higher accuracy [15].

After element-wise summing outputs of batch normalization layer and ReLU activation of the first convolutional layer (obtained by skip-connection), there are two upsample blocks. The structure of each one is the same as in SRGAN. It consists of convolution layer, sub-pixel convolution that proposed model uses to increase image resolution and parametric ReLU activation. At the end of the network, there is one additional convolutional layer to obtain HR output.

There are several convolutional layers with different structure in the network. The first and the last layers have 64 filters with kernel size 9×9, layer in the upscale block have 256 filters with small kernel size 3×3, and the rest layers have 64 filters with kernel size 3×3.

The structure of the network is commonly based on that of SRGAN, which showed the best results during the training, and use post-upsampling strategy, where features are directly extracted from LR input by set of residual blocks and the image is upscaled in the end of the propagation.

It was decided to use Charbonnier loss function, since perceptual loss loads a lot of computer memory. In LapSRN model each level has its own ground truth image y_l and loss function, but in this network only the x4 SR image is present in the loss function. This reduces computing costs. So, the overall loss function is as follows:

$$\ell_C = \frac{1}{N} \sum_{i=1}^{N} \rho(y^{(i)} - \hat{y}^{(i)}) \tag{5}$$

3   Experiments

3.1   Training Data

An important point in the process of learning implemented models is the training data. In this paper, it was decided to use **DIV2K** [1] dataset. It contains 800 training color RGB HR images with corresponding downscaled LR images with different factors (2x, 3x, 4x). The resolution of all images in the dataset have 2000 pixels on at least one of the axes.

For training, pairs of LR and corresponding 4x HR images were selected (that is, 16x difference in total resolution of images). All the images were preprocessed: center-cropped, after which, the LR picture size became 96×96 and HR picture size became 384×384.

3.2   Metrics

After implementing multiple architectures, they need to be trained and tested, which one performs best. This can be done by calculating value of peak signal-to-noise ratio (PSNR) and structural similarity (SSIM) index. Both metrics compare ground truth and obtained HR images by calculating numerical criteria. The value of PSNR is inversely proportional to MSE, and is computed in decibels (dB). Its small value provides high differences between images.

The SSIM index is another quality metric that evaluates the similarity of pictures. Although MSE (and PSNR, respectively) is simple in computation and one of the most obvious metrics for image comparison, it does not match with the human perception of image quality [43]. SSIM applies a different strategy to calculate similarity by taking into account the texture of the images. SSIM index was calculated locally in a window that slides pixel-by-pixel across picture. The SSIM value of whole picture then can be obtained by averaging local SSIM scores.

Table 1. Comparison of metrics values obtained by SR methods on benchmarks. The best values of the metrics are in bold.

Algorithm	Set5		Set14		Urban100		BSD100	
	PSNR	SSIM	PSNR	SSIM	PSNR	SSIM	PSNR	SSIM
Bicubic interpolation	26.697	0.8650	24.252	0.7872	21.702	0.7387	24.655	0.7545
SRGAN	26.336	0.8736	24.058	0.7793	22.097	0.7654	23.214	0.7410
LapSRN	**26.990**	0.8770	**24.513**	0.8034	**22.529**	**0.7811**	23.369	0.7760
Lapuska (ours)	26.855	**0.8868**	24.317	**0.8055**	22.317	0.7755	**24.801**	**0.7862**

Table 2. Comparison of values of loss functions obtained by SR methods on benchmarks. The pixel values were normalized from 0 to 1. The best values are in bold.

Algorithm	Set5		Set14		Urban100		BSD100	
	MSE	MAE	MSE	MAE	MSE	MAE	MSE	MAE
Bicubic interpolation	0.0031	0.0328	0.0049	0.0445	0.0087	0.0535	0.0071	0.0578
SRGAN	0.0025	0.0326	0.0047	0.0435	0.0081	0.0526	0.0059	0.0498
LapSRN	**0.0024**	**0.0306**	**0.0041**	**0.0405**	**0.0072**	**0.0486**	0.0058	0.0517
Lapuska (ours)	**0.0024**	0.0320	0.0043	0.0428	0.0075	0.0513	**0.0041**	**0.0405**

3.3 Benchmarks

Usually, the quality of trained models is checked on individual sets of images that are not included in the training data. **Set5** [3], **Set14** [45], **BSD100** [2] and **Urban100** [14] are the most common benchmark datasets encountered in almost all papers associated with the SR problem. All images were preprocessed in the same way as described in "Training data" section similar to other references.

3.4 Results

All the models described above were implemented in Python 3 and are trained on the data described in Sect. 3.1. In SRGAN, it was decided to use $R = 16$ residual blocks (as suggested in the article), the same number of residual blocks in the proposed model. In LapSRN, the number of blocks is $R = 10$ with $D = 1$

Table 3. Comparison of runtime of SR methods. The second column shows the average of all benchmark datasets. The runtime is obtained on images with an average resolution of 664×664. The best values are in bold.

Algorithm	Average time, sec
SRGAN	0.8818
LapSRN	0.7458
Lapuska (ours)	**0.3693**

convolutional layer, because the increase in the number of convolutional layers did not lead to better results in this work.

Comparison of implemented SR methods (as well as bicubic interpolation) is shown in Table 1. It consists of the values of metrics obtained by these methods on the benchmark datasets. For comparison, the values of different loss functions (MSE and MAE) are given in Table 2. MAE is similar to MSE and is defined as:

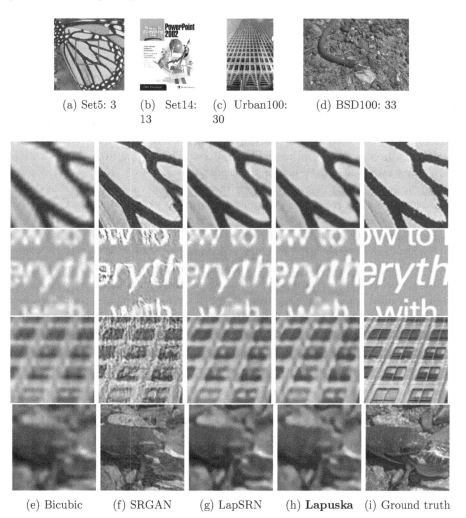

(a) Set5: 3   (b) Set14:   (c) Urban100:   (d) BSD100: 33
              13   30

(e) Bicubic   (f) SRGAN   (g) LapSRN   (h) **Lapuska**   (i) Ground truth

Fig. 2. Visual comparison of SR methods. The first row contains images from the benchmarks. The caption indicates the name of the benchmark and the number of the image in it. The following rows show the parts of the corresponding image (which are highlighted by a red square in the first row) obtained by different methods. SR methods from left to right: bicubic interpolation (e), SRGAN (f), LapSRN (g), Lapuska (ours) (h) and ground truth HR image (i).

$$\text{MAE} = \frac{1}{CMN} \sum_{i=1}^{M} \sum_{j=1}^{N} \sum_{k=1}^{C} |G(i,j,k) - T(i,j,k)| \tag{6}$$

The best values for all datasets from LapSRN and proposed model. But the best performance does not mean the best overall quality. It has already been mentioned that images with better MSE and PSNR values may be blurred.

Visual examples of some reconstructed images from benchmarks obtained by different SR methods and corresponding ground truth HR image are shown in Fig. 2. It can be seen that bicubic interpolation has the worst quality. The image obtained by this method is the most unsharp. Despite the best metrics values, LapSRN and the proposed model reconstructed images turned out to be more smooth than SRGAN ones. Because of this, some details of the images are lost (for example, the line in the lower left corner of the picture of a butterfly wing). On the other hand, some of the images produced by SRGAN are noisy (among them the second and the third images). The quality of the recovered image obtained by proposed model is similar to that of LapSRN, being a little blurry, but more clear than bicubic interpolation.

The average time taken by different models to generate one image is shown in Table 3. It shows that the best runtime with a NVIDIA GeForce GTX 1080 GPU shows the proposed model, which is more than 2 times faster than the nearest competitor.

So, the closest to the ground truth HR image was obtained by SRGAN, but proposed model reconstruct images without using additional VGG network for training, as SRGAN model uses it. Also, it does not require the presence of intermediate ground truth y_l images, unlike LapSRN, which affects the running time. Hence, the experiment with the new architecture can be considered successful.

4   Conclusion

In this work, the state-of-the-art SR methods based on deep neural networks were considered and compared. Selected models, including SRGAN and LapSRN, were trained on DIV2K dataset and tested by calculating average metrics value on different benchmarks. Despite the best metrics in LapSRN, the visual comparison showed that the quality of the restored image is better for SRGAN, which was quite anticipated.

In addition to the existing methods, in this paper we proposed a new model based on SRGAN and LapSRN. In an attempt to eliminate such drawbacks of these models, as large computational costs and occupied computer memory, a new architecture was developed. The overall quality of reconstructed images is slightly worse than that of SRGAN, but comparable to LapSRN. Thus, despite not the best values of metrics, SRGAN demonstrated the best performance. Nevertheless, the developed model showed a quality comparable with both state-of-art methods. Also, it should be noted that all considered methods showed much better results than non-neural network method of bicubic interpolation used as simple baseline.

The further work may include training neural network for obtaining images with higher upscaling factor and the use of new additional data, as well as adding multilateral filtering for decoder parts of the network. Also, one of the extensions of the research is conducting experiments with the depth of the models considered by using more residual blocks. Another directions would be to integrate visual transformers into hybrid neural network architecture [17] and specify images' domain on which training SR may provide much better results.

References

1. Agustsson, E., Timofte, R.: NTIRE 2017 challenge on single image super-resolution: dataset and study. In: The IEEE Conference on Computer Vision and Pattern Recognition (CVPR) Workshops (2017)
2. Arbelaez, P., Maire, M., Fowlkes, C., Malik, J.: Contour detection and hierarchical image segmentation. IEEE Trans. Pattern Anal. Mach. Intell. **33**(5), 898–916 (2011)
3. Bevilacqua, M., Roumy, A., Guillemot, C., Alberi-Morel, M.L.: Low-complexity single-image super-resolution based on nonnegative neighbor embedding. In: Proceedings of BMVC 2012 (2012)
4. Cheng, X., Fu, Z., Yang, J.: Zero-shot image super-resolution with depth guided internal degradation learning. In: Vedaldi, A., Bischof, H., Brox, T., Frahm, J.-M. (eds.) ECCV 2020. LNCS, vol. 12362, pp. 265–280. Springer, Cham (2020). https://doi.org/10.1007/978-3-030-58520-4_16
5. Diebel, J., Thrun, S.: An application of Markov random fields to range sensing. In: Advances in Neural Information Processing Systems, pp. 291–298 (2006)
6. Dong, C., Loy, C.C., He, K., Tang, X.: Learning a deep convolutional network for image super-resolution. In: Fleet, D., Pajdla, T., Schiele, B., Tuytelaars, T. (eds.) ECCV 2014. LNCS, vol. 8692, pp. 184–199. Springer, Cham (2014). https://doi.org/10.1007/978-3-319-10593-2_13
7. Dong, C., Loy, C.C., Tang, X.: Accelerating the super-resolution convolutional neural network. In: Leibe, B., Matas, J., Sebe, N., Welling, M. (eds.) ECCV 2016. LNCS, vol. 9906, pp. 391–407. Springer, Cham (2016). https://doi.org/10.1007/978-3-319-46475-6_25
8. Fattal, R.: Image upsampling via imposed edge statistics. ACM Trans. Graphics (TOG) **26**(3), 95 (2007)
9. Freeman, W.T., Jones, T.R., Pasztor, E.C.: Example-based super-resolution. IEEE Comput. Graphics Appl. **22**(2), 56–65 (2002)
10. Glasner, D., Bagon, S., Irani, M.: Super-resolution from a single image. In: ICCV, pp. 349–356. IEEE (2009)
11. Goodfellow, I., Pouget-Abadie, J., Mirza, M., Xu, B., Warde-Farley, D., Ozair, S., Courville, A., Bengio, Y.: Generative adversarial nets. In: Advances in Neural Information Processing Systems, pp. 2672–2680 (2014)
12. He, K., Sun, J., Tang, X.: Guided image filtering. In: Daniilidis, K., Maragos, P., Paragios, N. (eds.) ECCV 2010. LNCS, vol. 6311, pp. 1–14. Springer, Heidelberg (2010). https://doi.org/10.1007/978-3-642-15549-9_1
13. He, K., Zhang, X., Ren, S., Sun, J.: Deep residual learning for image recognition. In: CVPR, pp. 770–778 (2016)
14. Huang, J.B., Singh, A., Ahuja, N.: Single image super-resolution from transformed self-exemplars. In: CVPR, pp. 5197–5206 (2015)

15. Ioffe, S., Szegedy, C.: Batch normalization: accelerating deep network training by reducing internal covariate shift. arXiv preprint arXiv:1502.03167 (2015)
16. Johnson, J., Alahi, A., Fei-Fei, L.: Perceptual losses for real-time style transfer and super-resolution. In: Leibe, B., Matas, J., Sebe, N., Welling, M. (eds.) ECCV 2016. LNCS, vol. 9906, pp. 694–711. Springer, Cham (2016). https://doi.org/10.1007/978-3-319-46475-6_43
17. Karpov, A., Makarov, I.: Exploring efficiency of vision transformers for self-supervised monocular depth estimation. In: 2022 IEEE International Symposium on Mixed and Augmented Reality (ISMAR), pp. 711–719. IEEE (2022)
18. Kim, J., Kwon Lee, J., Mu Lee, K.: Accurate image super-resolution using very deep convolutional networks. In: Proceedings of the IEEE Conference on Computer Vision and Pattern Recognition, pp. 1646–1654 (2016)
19. Kim, J., Kwon Lee, J., Mu Lee, K.: Deeply-recursive convolutional network for image super-resolution. In: CVPR, pp. 1637–1645 (2016)
20. Kim, K.I., Kwon, Y.: Example-based learning for single-image super-resolution. In: Rigoll, G. (ed.) DAGM 2008. LNCS, vol. 5096, pp. 456–465. Springer, Heidelberg (2008). https://doi.org/10.1007/978-3-540-69321-5_46
21. Kopf, J., Cohen, M.F., Lischinski, D., Uyttendaele, M.: Joint bilateral upsampling. ACM Trans. Graph. (ToG) **26**(3), 96 (2007)
22. Korinevskaya, A., Makarov, I.: Fast depth map super-resolution using deep neural network. In: ISMAR2018, pp. 117–122. IEEE, New York, USA (2018)
23. Lai, W.S., Huang, J.B., Ahuja, N., Yang, M.H.: Fast and accurate image super-resolution with deep Laplacian pyramid networks. arXiv:1710.01992 (2017)
24. Ledig, C., et al.: Photo-realistic single image super-resolution using a generative adversarial network. arXiv preprint arXiv:1609.04802 (2016)
25. Li, Y., Xue, T., Sun, L., Liu, J.: Joint example-based depth map super-resolution. In: ICME, pp. 152–157. IEEE (2012)
26. Makarov, I., Aliev, V., Gerasimova, O.: Semi-dense depth interpolation using deep convolutional neural networks. In: ACM Multimedia (MM2017), pp. 1407–1415. ACM, New York, USA (2017)
27. Makarov, I., Bakhanova, M., Nikolenko, S., Gerasimova, O.: Self-supervised recurrent depth estimation with attention mechanisms. PeerJ Comput. Sci. **8**(e865), 1–25 (2022)
28. Makarov, I., Borisenko, G.: Depth inpainting via vision transformer. In: ISMAR2021, pp. 286–291. IEEE, New York, USA (2021)
29. Makarov, I., Korinevskaya, A., Aliev, V.: Fast semi-dense depth map estimation. In: Proceedings of the ACM Workshop on Multimedia for Real Estate Tech (RETech2018), pp. 18–21. ACM, New York, USA (2018)
30. Makarov, I., Korinevskaya, A., Aliev, V.: Sparse depth map interpolation using deep convolutional neural networks. In: TSP2018, pp. 1–5. IEEE (2018)
31. Makarov, I., et al.: On reproducing semi-dense depth map reconstruction using deep convolutional neural networks with perceptual loss. In: ACM Multimedia (MM2019), pp. 1080–1084. ACM, New York, USA (2019)
32. Maslov, D., Makarov, I.: Online supervised attention-based recurrent depth estimation from monocular video. PeerJ Comput. Sci. **6**(e317), 1–22 (2020)
33. Maslov, D., Makarov, I.: Fast depth reconstruction using deep convolutional neural networks. In: Rojas, I., Joya, G., Català, A. (eds.) IWANN 2021. LNCS, vol. 12861, pp. 456–467. Springer, Cham (2021). https://doi.org/10.1007/978-3-030-85030-2_38
34. Savchenko, A.V.: Fast inference in convolutional neural networks based on sequential three-way decisions. Inf. Sci. **560**, 370–385 (2021)

35. Savchenko, A.V.: MT-EmotiEffNet for multi-task human affective behavior analysis and learning from synthetic data. In: Karlinsky, L., Michaeli, T., Nishino, K. (eds.) Computer Vision – ECCV 2022 Workshops. ECCV 2022. LNCS, vol. 13806. Springer, Cham (2023). https://doi.org/10.1007/978-3-031-25075-0_4

36. Savchenko, A.V., Belova, N.S.: Statistical testing of segment homogeneity in classification of piecewise-regular objects. Int. J. Appl. Math. Comput. Sci. **25**(4), 915–925 (2015)

37. Savchenko, A.V., Savchenko, L.V.: Towards the creation of reliable voice control system based on a fuzzy approach. Pattern Recogn. Lett. **65**, 145–151 (2015)

38. Savchenko, A., Khokhlova, Y.I.: About neural-network algorithms application in viseme classification problem with face video in audiovisual speech recognition systems. Optical Mem. Neural Netw. **23**(1), 34–42 (2014)

39. Shi, W., et al.: Real-time single image and video super-resolution using an efficient sub-pixel convolutional neural network. In: CVPR, pp. 1874–1883 (2016)

40. Simonyan, K., Zisserman, A.: Very deep convolutional networks for large-scale image recognition. arXiv preprint arXiv:1409.1556 (2014)

41. Sokolova, A.D., Kharchevnikova, A.S., Savchenko, A.V.: Organizing multimedia data in video surveillance systems based on face verification with convolutional neural networks. In: van der Aalst, W.M.P., et al. (eds.) AIST 2017. LNCS, vol. 10716, pp. 223–230. Springer, Cham (2018). https://doi.org/10.1007/978-3-319-73013-4_20

42. Wang, Y., Wang, L., Wang, H., Li, P.: End-to-end image super-resolution via deep and shallow convolutional networks. arXiv preprint arXiv:1607.07680 (2016)

43. Wang, Z., Bovik, A.C.: Mean squared error: love it or leave it? A new look at signal fidelity measures. IEEE Signal Process. Mag. **26**(1), 98–117 (2009)

44. Yue, L., Shen, H., Li, J., Yuan, Q., Zhang, H., Zhang, L.: Image super-resolution: the techniques, applications, and future. Signal Process. **128**, 389–408 (2016). https://doi.org/10.1016/j.sigpro.2016.05.002

45. Zeyde, R., Elad, M., Protter, M.: On single image scale-up using sparse-representations. In: Boissonnat, J.-D., et al. (eds.) Curves and Surfaces 2010. LNCS, vol. 6920, pp. 711–730. Springer, Heidelberg (2012). https://doi.org/10.1007/978-3-642-27413-8_47

Deep Learning Approaches Applied to MRI and PET Image Classification of Kidney Tumours: A Systematic Review

Sandra Amador[1], José Perona[1], Claudia Villalonga[2], Jorge Azorin[1], Oresti Banos[2], and David Gil[1(✉)]

[1] Department of Computer Science Technology and Computation, University of Alicante, Alicante, Spain
dgil@dtic.ua.es

[2] Department of Computer Architecture and Computer Technology, Research Center for Information and Communication Technologies, University of Granada, Granada, Spain

Abstract. Are deep learning techniques being applied to image classification of kidney tumours? This systematic review aims to explore papers using this novel scientific method of deep learning (DL) for the detection of kidney tumours in patients. We can start by arguing that this set of techniques is not widespread as only 8 papers have been found in the last 5 years that meet these requirements. This paper analyses the data source, the datasets used, the complexity and limitations of these datasets that drive the use of novel techniques such as data augmentation. We analyse the most advanced deep learning architectures used such as ResNet as well as the possibilities conferred by transfer learning to use the basis of these architectures and then adjust them to the needs of each case.

Keywords: Deep Learning · kidney Tumor · Image Classification · Machine Learning · Artificial Intelligence

1 Introduction

Cancer incidence has been steadily increasing worldwide due to a combination of factors, including population growth, aging and unhealthy changes that have affected diet, lifestyle, obesity, the environment. Efforts to prevent, detect, and treat cancer are critical to addressing this growing health concern and reducing the global burden of cancer.

Kidney cancer, also known as renal cancer, is one of the ten most common cancers in adults. The are a significant and growing health concern worldwide, with increasing incidence and mortality rates in recent years. According to the American Cancer Society[1], an estimated 81,800 new cases of kidney cancer will

---

[1] https://www.cancer.org/cancer/kidney-cancer/about/key-statistics.html.

© The Author(s), under exclusive license to Springer Nature Switzerland AG 2023
I. Rojas et al. (Eds.): IWANN 2023, LNCS 14135, pp. 254–265, 2023.
https://doi.org/10.1007/978-3-031-43078-7_21

be diagnosed in the United States in 2023, and approximately 14,890 people will die from the disease.

The prevalence of kidney cancer varies by age, sex, and geographic region. Most people with kidney cancer are older. The average age of people when they are diagnosed is 64 with most people being diagnosed between ages 65 and 74. Kidney cancer is very uncommon in people younger than age 45. The incidence of kidney cancer is much higher in men than in women, about twice as common in men than in women. Moreover, the incidence of kidney cancer varies widely across the world. According to the Global Cancer Observatory[2], the highest age-standardized rate (ASR) in 2020 was reported in North America (12.2), Europe (9.5), and Oceania (8.8), and lower rates in Africa (1.8), Asia (2.8) and Latin America and the Caribbean (4.7). Also, it is important to highlight the estimate number of prevalent cases (5-year) in 2020 according to the income levels being the high and upper middle income population about the 90%.

The rapid advances in medical imaging technology from the 20th Century has revolutionized medicine and has supposed a major advancement in the diagnosis and treatment of tumors. The identification and classification of tumors in images rely on human experts, being commonly the radiologist as the person who usually interprets the images obtained during the study. The radiologist writes a report on the results and sends the report to the physician. The accurate image classification can be challenging due to the many factors, including complex and heterogeneous nature of tumors.

The use of artificial intelligence (AI) techniques may represent a new major progress in the accurate detection of tumors. AI, and specifically computer vision allows the recognition of complex patterns in images and thus offers the opportunity to transform image interpretation from a purely qualitative and subjective task to a quantifiable and effortlessly reproducible one. Moreover, computer vision quantifies information from images that is not detectable by humans and thereby complement clinical decision making (Bi et al. 2019).

Nowadays, the computer vision solutions are attached to Deep learning (DL) because they have shown its superiority among different approaches. DL uses deep artificial neural networks to learn, classify or regress from complex data having several advantages over traditional machine learning methods. The most important could be the ability to automatically learn features from image data, being able to detect, characterize, and monitor tumors in medical images. However, despite these promising results, several challenges and limitations remain in the use of deep learning for image classification of kidney tumors.

In this paper, we present a systematic review of the literature on deep learning approaches for image classification of kidney tumors. Our review aims to provide a comprehensive and critical evaluation of the existing research on this topic. Several databases have been used for studies that met our inclusion criteria and assessed the quality of the studies using predefined criteria. The results show that the understanding of predictive models based on DL in kidney tumours have been covered so far by few studies.

---

[2] https://gco.iarc.fr.

The remainder of the paper is organized as follows. In the Methods section, we describe our search strategy and methods for evaluating the studies. In the Results section, we summarize the key findings from the studies we included in our review. In the Discussion section, we interpret and compare the findings from different studies and highlight the implications of our review for clinical practice and future research. Finally, in the Conclusion section, we summarize our main findings and provide recommendations for future research.

2   Methods

In this study where a systematic review is carried out for studies that apply deep learning in kidney cancer diseases with MRI and PET images, the instructions of Preferred Reporting Items for Systematic Reviews and Meta-Analyses (PRISMA) 2020 have been followed. The following sections describe the specific methodology followed.

2.1   Eligibility Criteria

This review has focused on the different studies related to the application of deep learning techniques for kidney cancer diseases in patients from the analysis of MRI and PET images.

The inclusion criteria applied to the selected studies are as follows: 1) they must be related to kidney cancer images; 2) the images must be of MRI or PET type; 3) deep learning techniques must be applied to those images.

In addition to meeting the three aforementioned criteria, the studies included in this review had to meet a series of eligibility requirements regarding the characteristics of the report: 1) the report had to be published in the last 5 years; 2) it had to be a scientific article and not a review.

2.2   Information Sources

A systematic electronic search was performed to identify eligible studies in the Scopus, PubMed, and Web of Science reference databases. The search was conducted in March 2023.

2.3   Search Strategy

The search keywords were chosen based on the review framework. The primary concepts for the search were "cancer" or "tumor", and "kidney" along with "deep learning". Additionally, we included "Magnetic Resonance Imaging" or "Positron Emission Tomography", and their acronyms "MRI" or "PET". We also considered words that start with "diagno" as we were searching for diagnoses or diagnostics. The exclusion criteria applied to the search were for the last

5 years. The resulting queries were then executed on Scopus, PubMed, and Web of Science, which are the three main databases for bibliographic references and periodical citations globally.

Scopus: *TITLE-ABS-KEY(("cancer" OR "tumor*") AND "diagnos*" AND "deep learning" AND "kidney") AND (LIMIT-TO(PUBYEAR, 2019) OR LIMIT-TO(PUBYEAR, 2020) OR LIMIT-TO(PUBYEAR, 2021) OR LIMIT-TO (PUBYEAR, 2022) OR LIMIT-TO(PUBYEAR, 2023)) AND (LIMIT-TO (EXACTKEYWORD, "Magnetic Resonance Imaging") OR LIMIT-TO (EXACTKEYWORD, "Positron Emission Tomography") OR LIMIT-TO(EXACTKEYWORD, "MRI") OR LIMIT-TO (EXACTKEYWORD, "PET"))*

Web of Science: *(TS=(("cancer" OR "tumor*") AND "diagnos*" AND "deep learning" AND "kidney" AND ("Magnetic Resonance Imaging" OR "Positron Emission Tomography" OR "MRI" OR "PET"))) AND (PY==("2023" OR "2022" OR "2021" OR "2020" OR "2019"))*

PubMed: *(("cancer"[Title/Abstract] OR "tumor*"[Title/Abstract]) AND "diagnos*"[Title/Abstract] AND "deep learning"[Title/Abstract] AND ("kidney" [Title/Abstract]) AND (y_5[Filter])) AND (("Magnetic Resonance Imaging" [Title/Abstract] OR "Positron Emission Tomography"[Title/Abstract] OR "MRI" [Title/Abstract] OR "PET" [Title/Abstract])) AND (y_5[Filter])*

2.4 Selection Process

After retrieving records from the databases and conducting a manual search, the papers were imported into a SharePoint library. The identification of duplicate records was done manually by comparing the authors and titles of the papers. The retrieved papers were additionally analyzed in detail, their eligibility was evaluated against the criteria described in the "Eligibility Criteria" section, and finally, a set of papers were selected for the review based on a majority consensus.

2.5 Data Collection Process

The reviewers evaluated the studies that were included in the analysis. To conduct the systematic review of the included studies, a Microsoft Excel file was utilized to gather the necessary data. The resulting document is a matrix where each article is represented by a row, and the data elements to be analyzed are indicated by the columns.

2.6   Data Items

The columns defined in the collaborative file corresponded to the data extracted from the analyzed paper. The specific columns defined were: title, authors, type of article, purpose of the paper, deep learning architecture, origin of the dataset used, whether the dataset is open access or not, number of patients, number of images analyzed, if data augmentation technique used or not, transfer learning technique used or not and year of publication.

3   Results

In this section we present the results derived from the systematic review of the application of deep learning in MRI and PET medical images with patients diagnosed with kidney cancer. The remainder of the section provides a detailed description of the research objectives, purpose of the paper, deep learning technique used, the origin of datasets, whether the dataset used is public or private, the number of patients and images used in the studies, year of publication, and the number of articles utilizing data augmentation and transfer learning techniques.

3.1   Study Selection

A bibliographic search was conducted, yielding a total of 59 records. Scopus search returned 25 records, Web of Science 31, and PubMed only 3. After removing 14 duplicates, 45 records remained for consideration. Of these, 18 were excluded as systematic reviews, 7 for not being related to deep learning or cancer, and 11 for not being exclusive to kidney cancer The workflow with the detailed process is shown in Fig. 1.

3.2   Year of Publication

In the systematic review, the years of publication of the selected studies were analysed (Fig. 2). One paper published in 2019 (Müller et al. 2019), two papers in 2020 (Zhao et al. 2020, Xi et al. 2020), three in 2021 (Schulz et al. 2021, Zheng et al. 2021 Nikpanah et al. 2021), and two in 2022 (Xu et al. 2022, Parvathi and Jonnadula, 2022).

3.3   Research Goal

This section discusses the objectives of the selected papers (Fig. 3). 75% (6/8) (Zhao et al. 2020, Xi et al. 2020, Schulz et al. 2021, Zheng et al. 2021, Nikpanah et al. 2021, Xu et al. 2022) of the papers aimed at the classification of kidney tumours while the remaining 25% (2/8) (Müller et al. 2019, Parvathi and Jonnadula 2022) aimed at image segmentation.

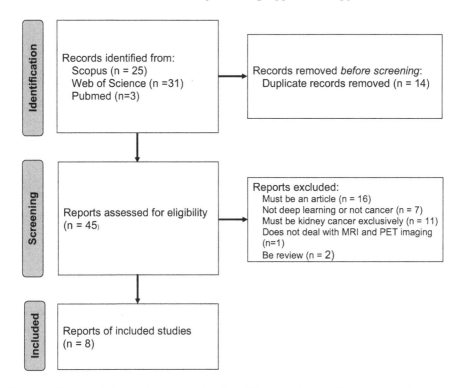

Fig. 1. Flow of information through the different phases of a systematic review (PRISMA methodology).

3.4 Number of Patients

The number of patients used in the selected articles is shown in Fig. 4. One paper (12.50%) (Müller et al. 2019) used less than 50 patients; one paper (12.50%) (Nikpanah et al. 2021) had between 50 and 99 patients; three papers (37.50%) (Zheng et al. 2021, Xu et al. 2022, Schulz et al. 2021) had a number of patients between 100 and 299; one paper (12.50%) (Zhao et al. 2020) had a number of patients between 300 and 499; and finally, one paper (12.50%) (Xi et al. 2020) had more than 500 patients in the study. Only one paper (12.50%) (Parvathi and Jonnadula, 2022), did not specify the number of patients used in their study.

3.5 Number of Images

The number of images used in the selected papers is shown in Fig. 4. Two papers (25%) (Parvathi and Jonnadula, 2022, Müller et al. 2019) used a number of images less than 100; two papers (25%) (Nikpanah et al. 2021, Schulz et al. 2021) used between 100 and 299 images; one paper (12.50%) (Zhao et al. 2020)

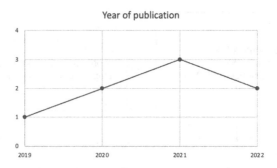

Fig. 2. Year of publication

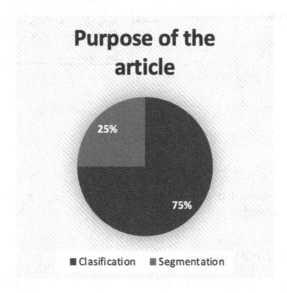

Fig. 3. Research goal

used between 300 and 499 images; and finally, two papers (25%) (Xi et al. 2020, Zheng et al. 2021) used a number of images greater than 500. One of the selected papers (12.5%) (Xu et al. 2022) did not specify the number of images used in their study.

3.6   Data Origin

The selected studies used different databases (Fig. 5). Six times they used databases from hospitals or centres (Müller et al. 2019, Zhao et al. 2020, Xi et al. 2020, Zheng et al. 2021, Xu et al. 2022, Schulz et al. 2021); on three occasions they used the TCGA-KIRC database (Parvathi and Jonnadula, 2022,

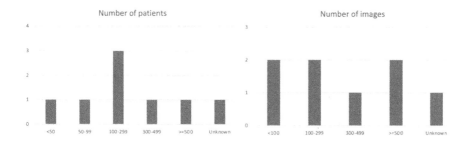

Fig. 4. Number of patients. Number of images

Schulz et al. 2021, Zhao et al. 2020); two cases they used the TCIA database Zhao et al. (2020Xi et al. 2020); and lastly, one paper used the UOB-NCI database (Nikpanah et al. 2021).

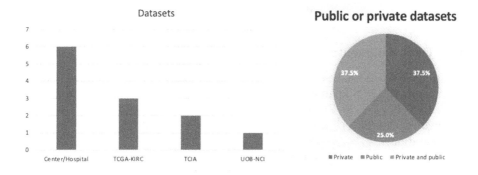

Fig. 5. Data Origin. Open Datase

3.7 Open Dataset

The selected studies used public datasets, private datasets or both (Fig. 5). 37.5% (3/8) (Zheng et al. 2021, Xu et al. 2022, Müller et al. 2019) of the studies used exclusively private datasets; 25% (2/8) (Parvathi and Jonnadula, 2022, Nikpanah et al. 2021) used only public datasets and 37.5% (3/8) (Schulz et al. 2021, Zhao et al. 2020, Xi et al. 2020) used both public and private datasets.

3.8 Deep Learning Architecture

The papers analysed employed different deep learning architectures (Fig. 6). Three papers (37.5%) (Zheng et al. 2021 Xu et al. 2022, Schulz et al. 2021) used the ResNet 18 architecture; Two papers (25%) (Zhao et al. 2020, Xi et al.

2020) used the ResNet 50 architecture; two papers (25%) (Parvathi and Jonnadula, 2022, Müller et al. 2019) used the U-Net architecture; and finally, one paper (12.5%) (Nikpanah et al. 2021) made use of the AlexNet architecture.

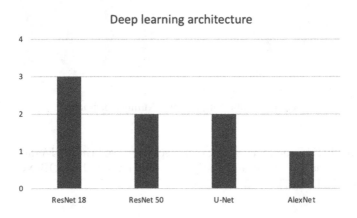

Fig. 6. Deep Learning Architecture

3.9   Data Augmentation and Transfer Learning

There were papers analysed that used the data augmentation technique (Fig. 7). Six papers (75%) (Nikpanah et al. 2021, Parvathi and Jonnadula, 2022, Zhao et al. 2020, Xi et al. 2020, Zheng et al. 2021, Xu et al. 2022) use data augmentation and two papers (25%) (Müller et al. 2019, Schulz et al. 2021) did not use data augmentation.

Regarding transfer learning (Fig. 7), half of the papers did use the technique (Nikpanah et al. 2021, Xi et al. 2020, Zheng et al. 2021, Xu et al. 2022) and half did not (Parvathi and Jonnadula, 2022, Zhao et al. 2020, Müller et al. 2019, Schulz et al. 2021).

Fig. 7. Data Augmentation. Transfer Learning

4   Discussion

Nowadays, it is no surprise that the latest machine learning techniques, especially deep learning, are being applied in our daily situations. In the area of healthcare, we have gradually become accustomed to computer science and technology being fully integrated, controlling, monitoring and supervising all actions. However, the use of models based on deep learning to aid medical diagnosis is at its moment of greatest emergence and development. The main problem in the field of medical imaging, and in particular in the study of MRI and PET is the lack of publicly available datasets. While it is true that there are some datasets available on the Internet, the number of images specific and necessary for our problem is very limited. Therefore, in the most recent literature, data augmentation technique which of using extensive data enhancement by augmenting the data with various parameters and different techniques to fill the lack of data and make the generated system and models invariant to noise and transformations is widely used (Sajjad et al. 2019, Chlap et al. 2021). It is relatively common, and it has been in several of the papers reviewed in this work, to see data augmentation and transfer learning techniques combined. Transfer learning is one of the most recently used machine learning methods that learns previous knowledge applied to solve a problem and reuses it in other similar problems. The transfer learning process could be divided into two main steps: pre-trained model selection, problem size and similarity (Khan et al. 2019, Rai and Sisodia, 2021).

The ResNet architecture is the most used, both in 18 and 50 layers, because residual convolutional networks are very effective in image recognition. The datasets used are usually specific to the center/hospital where the study is being conducted, being this type the most common, although it is often complemented with an external and public dataset. This is usually done in order to obtain a greater amount of data for training the CNN or for validation purposes. Another way to obtain a greater number of images, as has already been discussed in the respective section, is to apply the data augmentation technique to achieve better results, which is why it is a very popular technique.

4.1   Limitations

A broad search was carried out looking for papers on kidney tumours, but this search was restricted to studies where deep learning was applied to very specific images, MRI or PET. Although this search is very specific, it is quite possible that some interesting studies were left out of the search.

As already mentioned in the sections on the dataset and its provenance, a large number of databases are private. The increase in public databases and open data in general will benefit future research, although this is progressing slowly. The technique of data augmentation, as seen in the section itself, has made it possible to train the models with a larger number of images. Finally, it should be noted that although the number of papers studied gives an overview of the current situation, it is short and should be extended in the future both in terms of types of images and other types of cancer.

5 Conclusions

Deep learning techniques have made enormous progress in image classification but still represent a major challenge when it comes to the accuracy of medical image classification due to their complex and heterogeneous nature. The aim of this systematic review is to explore papers using deep learning for classification of kidney tumour images.

The first thing that emerges is the novelty of these techniques since only 8 papers have been found in the last 5 years that meet these requirements. Nevertheless, this work has been deepened, analysing the data source, the datasets used, the complexity and limitations of these datasets that drive the use of novel techniques such as data augmentation. Furthermore, the most advanced deep learning architectures used such as ResNet are analysed as well as the possibilities conferred by transfer learning to use the basis of these architectures and adjust them to the needs of each case. As discussed in the previous section, this search should be extended as future lines for both imaging and cancer types. Furthermore, to increase the set of papers to be included in the study in future lines, it is intended to include other types of tumour images such as CT images.

Acknowledgement. This article is based upon work from COST Action HARMON-ISATION (CA20122). This research has been partially funded by the Spanish Government by the project PID2021-127275OB-I00, FEDER "Una manera de hacer Europa". This work was partially supported by the Spanish State Research Agency (AEI) under grant PID2020-119144RB-I00 funded by MCIN/AEI/10.13039/501100011033. This research was partially funded by the Andalusian Ministry of Economic Transformation, Industry, Knowledge and Universities under grant P20_00163.

References

Bi, W.L., et al.: Artificial intelligence in cancer imaging: clinical challenges and applications. CA Cancer J. Clin. **69**(2), 127–157 (2019)

Chlap, P., Min, H., Vandenberg, N., Dowling, J., Holloway, L., Haworth, A.: A review of medical image data augmentation techniques for deep learning applications. J. Med. Imaging Radiat. Oncol. **65**(5), 545–563 (2021)

Khan, S., Islam, N., Jan, Z., Din, I.U., Rodrigues, J.J.C.: A novel deep learning based framework for the detection and classification of breast cancer using transfer learning. Pattern Recogn. Lett. **125**, 1–6 (2019)

Müller, S., et al.: Benchmarking Wilms' tumor in multisequence MRI data: why does current clinical practice fail? which popular segmentation algorithms perform well? J. Med. Imag. **6**(3), 034001 (2019)

Nikpanah, M., et al.: A deep-learning based artificial intelligence (ai) approach for differentiation of clear cell renal cell carcinoma from oncocytoma on multi-phasic MRI. Clin. Imaging **77**, 291–298 (2021)

Parvathi, S.S., Jonnadula, H.: An efficient and optimal deep learning architecture using custom U-net and mask R-CNN models for kidney tumor semantic segmentation. Int. J. Adv. Comput. Sci. Appl. **13**(6) (2022)

Rai, R., Sisodia, D.S.: Real-time data augmentation based transfer learning model for breast cancer diagnosis using histopathological images. In: Rizvanov, A.A., Singh, B.K., Ganasala, P. (eds.) Advances in Biomedical Engineering and Technology. LNB, pp. 473–488. Springer, Singapore (2021). https://doi.org/10.1007/978-981-15-6329-4_39

Sajjad, M., Khan, S., Muhammad, K., Wu, W., Ullah, A., Baik, S.W.: Multi-grade brain tumor classification using deep CNN with extensive data augmentation. J. Comput. Sci. **30**, 174–182 (2019)

Schulz, S., et al.: Multimodal deep learning for prognosis prediction in renal cancer. Front. Oncol. **11**, 788740 (2021)

Xi, I.L., et al.: Deep learning to distinguish benign from malignant renal lesions based on routine MR imaging deeplearning for characterization of renal lesions. Clin. Cancer Res. **26**(8), 1944–1952 (2020)

Xu, Q., et al.: Differentiating benign from malignant renal tumors using t2-and diffusion-weighted images: a comparison of deep learning and radiomics models versus assessment from radiologists. J. Magn. Reson. Imaging **55**(4), 1251–1259 (2022)

Zhao, Y., et al.: Deep learning based on MRI for differentiation of low-and high-grade in low-stage renal cell carcinoma. J. Magn. Reson. Imaging **52**(5), 1542–1549 (2020)

Zheng, Y., Wang, S., Chen, Y., Du, H.-Q.: Deep learning with a convolutional neural network model to differentiate renal parenchymal tumors: a preliminary study. Abdominal Radiol. **46**, 3260–3268 (2021)

General Applications of Artificial Intelligence

Comparison of ANN and SVR for State of Charge Regression Evaluating EIS Spectra

Andre Loechte[✉][iD], Jan-Ole Thranow[iD], Felix Winters[iD], Andreas Heller[iD], and Peter Gloesekoetter[iD]

FH Muenster University of Applied Sciences, 48329 Steinfurt, Germany
a.loechte@fh-muenster.de

Abstract. The demand for energy storage is increasing massively due to the electrification of transport and the expansion of renewable energies. Current battery technologies cannot satisfy this growing demand because they are difficult to recycle, because the necessary raw materials are mined under precarious conditions, and because the energy density is insufficient. Metal-air batteries offer a high energy density because there is only one active mass inside the cell and the cathodic reaction uses the ambient air. Various metals can be used, but zinc is very promising because of its disposability, non-toxic behavior, and because operation as a secondary cell is possible. Typical characteristics of zinc-air batteries are flat charge and discharge curves. On the one hand, this is an advantage for the subsequent power electronics, which can be optimized for smaller and constant voltage ranges. On the other hand, the state determination of the system becomes more complex, since the voltage level is not sufficient to determine the state of the battery. In this context, electrochemical impedance spectroscopy is a promising candidate since the resulting impedance spectra depend on the state of charge, working point, state of aging, and temperature. Therefore, in this publication, electrochemical impedance spectroscopy is combined with multiple machine learning techniques to also determine successfully the state of charge during charging of the cell at non-fixed charging currents.

Keywords: electrochemical impedance spectroscopy · artificial neural networks · support vector regression · zinc-air battery · state estimation · state of charge

1 Introduction

The growth and importance of battery technology has risen significantly in recent years. Due to topics such as the growth of renewable energies or the Internet of Things, a further increase in battery technology is also expected for the future [1]. To meet the increasing demand, battery technologies with particularly high energy density are necessary [4,14]. Existing battery technologies are already close to their theoretical energy density, so new alternatives have to be found [6].

Funded by organization EFRE-0801585.

Metal-air cells are a promising recent candidate. Since the oxygen in the ambient air is also used as an active component, these cells offer a much higher theoretical energy density than lithium-ion technology [7]. Due to the good availability and non-toxicity of the cell materials used, zinc-air technology is of particular interest [12,13]. Figure 1 shows a demonstrator for a photovoltaic storage system based on rechargeable zinc-air cells, which was developed at FH Münster University of Applied Sciences. The storage system uses a server rack as housing and provides a capacity of 7.2 kW h.

Fig. 1. Demonstrator of a photovoltaic storage system developed at the FH Münster University of Applied Sciences which provides a capacity of 7.2 kW h.

1.1 Problem

Adapted battery management systems must be developed for the new cell technologies, as existing methods for determining the state of charge, i.e. voltage measurement, no longer work [5,11]. The big challenge is that the cell voltage is almost constant over the entire range of the state of charge and only shows very small changes, if any. Electrochemical impedance spectroscopy is a promising method for determining the state of charge of battery cells despite very flat voltage curves [2]. The conventional method for evaluating impedance spectra is very complex because, for instance, an accurate electrochemical cell model has to be known [3,8]. The principle here is that an electrochemical model of the cell is created and the parameters of the model are fitted on the basis of the measured impedance spectra. However, practical tests of this method on zinc-air batteries with particularly flat voltage curves have shown that this method can only detect the charge termination with sufficient accuracy, but not the state of charge during the charging process. As an alternative, the measured impedance spectra can be used as input data of a machine learning model. One difficulty here is that the measured impedance spectra most significantly depend on the DC current applied in parallel during the measurement for charging or discharging the cell while the state of charge has rather a small influence. Previous methods and publications have therefore the disadvantage that a defined charging or discharging current has to be applied during the impedance spectroscopy, because the model was also trained with impedance spectra at the very same dc cell current [9,10].

This paper therefore analyzes whether machine learning methods can be used to determine the state of charge based on impedance spectra measured at previously untrained cell currents. Rechargeable zinc-air cells, as shown in the demonstrator in Fig. 1, are used as an example to address this question.

2   Feature Extraction

In impedance spectroscopy, a small sinusoidal current of a certain frequency is applied in addition to any direct current that may be present (for charging or discharging the battery). As a result, the cell voltage will also follow a sinusoidal course. Both signals are measured and related in the form of impedance. This means that the impedance indicates the ratio of the amplitudes and the phase position between the signals. This type of measurement is repeated for different frequencies, resulting in a whole spectrum. As shown in Fig. 2, the measurement points at different frequencies of the impedance spectra can be connected to form curves and neighboring frequencies have similar measurement values. Therefore, there is a strong correlation of the individual measured values as well. For this reason, a principle component analysis (PCA) is performed to reduce the dimension of the input vector without significantly reducing the variance of the measurement data. When applying PCA, one question is how many principal components should be used. The eigenvalues can be used to determine the explained variance, which is shown in Table 1 depending on the number of principal components. It can be seen well that the individual measurement frequencies are indeed strongly correlated. Thus, 3 principal components are sufficient to explain 90% of the variance.

3   Data Generation

The measured impedance spectra depend not only on the state of charge of the cells, but also on other parameters such as temperature, state of health and electrolyte concentration. In fact, the most significant dependence is on the DC current applied in parallel during the measurement for charging or discharging the cell. For example, larger currents lead to smaller semicircles in the spectra. A

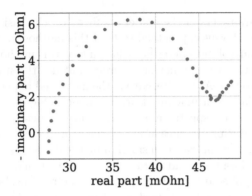

Fig. 2. Sample data of an electrochemical impedance spectroscopy measurement.

Table 1. Explained variance depending on number of principle components.

No. princ. comp	Explained variance
1	46%
2	76%
3	90%
4	95%
5	97%
7	98%
9	99%

robust model is therefore characterized by the fact that it can also determine the state of charge for DC current components that are not part of the trianings data set (generalization). Therefore, the procedure in Fig. 3 was used to generate data, in which the cell is regularly charged or discharged with different currents. As can be seen in the blue curve, the DC portion of the current is alternately set to 1 A, 2 A and 3 A. After each adjustment of the charging current, a measurement of the impedance spectrum is performed. However, the DC component of the cell voltage is also affected by the change in current. Thus, during the charging process, the voltage is slightly higher at a charging current of 3 A than at a charging current of 1 A. This voltage change cannot happen instantaneously. Therefore, the EIS measurement starts with a slight delay after a current change, so that the voltage changes can decay and do not influence the measurement.

Since no breaks with unknown self-discharge occur in this scenario, it is possible to integrate the applied current and define the state of charge. These values can later be used as target values for supervised learning. Once the cell is full, the discharge process starts using the identical technique.

In contrast to common practice, the measured data here are not randomly split into training and test data. Instead, the measurements at 1 A and 3 A form

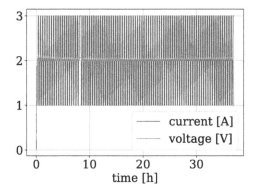

Fig. 3. Charging cycle at varying charge currents.

the training data set and the measurements at 2 A represent the test data set. This provides an actual generalization of the working point during impedance spectroscopy. Since the measurements have been performed alternating, the ratio of training to test data is 2/3. The hyperparameters of each method are optimized by cross-validation of the training data.

The final feature vector is defined by

$$x_i = [I_{\text{bat,DC}}, \Re\{Z_{i,1}\}, \ldots, \Re\{Z_{i,m}\}, \Im\{Z_{i,1}\}, \ldots, \Im\{Z_{i,m}\},$$
$$|Z_{i,1}|, \ldots, |Z_{i,m}|, \phi(Z_{i,1}), \cdots, \phi(Z_{i,m})]. \quad (1)$$

So real and imaginary part as well as magnitude and phase of all measured frequencies are used. In addition, the working point is also part of the feature vector.

4 Results

4.1 Regression Using Artificial Neural Networks

Initially, multilayer perceptron networks with a hidden layer in addition to the input and output layer are utilized for regression. To ensure optimal performance, each feature is scaled to remove its mean and scaled to a variance of $\sigma^2 = 1$ as neural networks are sensitive to feature scaling. Since impedance values at similar frequencies have strong correlations, a principal component transformation is investigated to reduce the feature vector size and improve generalization. However, this transformation is only applied to the measurement data of the spectrum, as the DC operating point is crucial for generalization, but has minimal variance in the training dataset due to the two different values.

The quality of the resulting model is significantly dependent on hyperparameters used, including the number of neurons and L2 regularization term α. The optimal values of these hyperparameters vary with the number of principal components used. Therefore, a Bayesian optimization procedure in combination with cross-validation is employed to find optimized hyperparameter values for each number of principal components. Here, a model is trained with different parameter combinations on four out of five groups of the training data and testing it on the remaining group. This process is repeated five times, and the average of the coefficient of determination R^2 over these five runs is used as the performance of the hyperparameter combination.

The optimized hyperparameter values for each number of principal components are presented in Table 2, along with the performance values for the best model in each case. The performance when using a single principal component is poor, with an R^2 for cross-validation with training data of 0.313. However, using two or more principal components leads to much better performance, with the optimal number of neurons in the hidden layer being below 10 in all cases except for three principal components to prevent overfitting. Figure 4 illustrates the cross-validation performance for the training data graphically, with error bars representing the standard deviation of each run during cross-validation.

Table 2. R^2 on unseen testdata and optimized hyperparameters of current generalization using artificial neural networks depending on number of principal components.

No. princ. comp.	R^2	hidden neurons	α
No PCA	0.995	4	$6.2 \cdot 10^{-4}$
1	0.313	8	$2.7 \cdot 10^{-1}$
2	0.980	7	$5.2 \cdot 10^{-6}$
3	0.989	10	$1.2 \cdot 10^{-5}$
4	0.995	7	$4.3 \cdot 10^{-5}$
5	0.997	7	$1.7 \cdot 10^{-2}$
7	0.998	8	$1.1 \cdot 10^{-2}$
9	0.999	9	$2.5 \cdot 10^{-3}$

Reducing the input dimensionality can also lead to an improvement in model performance, with a partial reduction of the dimensionality resulting in an R^2 value higher than the value when PCA is not applied, starting with five principal components. The best result is obtained with nine principal components.

Next, models are trained with the complete set of training data using the optimized hyperparameters. The resulting performance is sufficiently accurate for the most part, as shown in Table 3, which summarizes the determination coefficient as a function of the number of principal components. In most cases, the performance on the unseen working point is at a similar level or even better than when cross-validating the training data. The networks with three to five principal components performs the best, with an R^2 value of 1.0, which is even significantly better than for the training data. However, the network that uses nine principal components overfits to the training data, resulting in worse performance.

Figures 5 and 6 show the predicted data and the difference from the target value with three components, respectively. It can be seen that the generalization of the DC working point for the model with three components works very well.

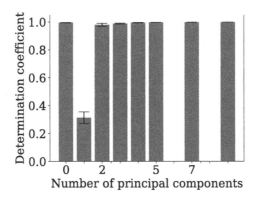

Fig. 4. R^2 and standard deviation of crossvalidated training data depending on number of principal components.

Table 3. Determination coefficient on unseen validation data during current generalization using artificial neural networks depending on number of principal components.

No. princ. comp.	R^2
No PCA	0.979
1	0.153
2	0.991
3	1.00
4	1.00
5	1.00
7	0.997
9	0.996

4.2 Regression Using Support Vector Regression

The process of building a model using support vector regression for determining the state of charge of zinc-air batteries is similar to building a model using artificial neural networks as described in Sect. 4.1. Figure 7 displays the R^2 values of the cross-validation with training data, with the error bars indicating the standard deviation of the determination coefficient within the five combinations. The performance increases with increasing number of principal components, with a principal component number of nine resulting in the best performance.

After the optimal hyperparameters are found, the complete set of training data is used to fit the model and the performance on the test data is determined using the determination coefficient as the performance metric. Table 4 shows the resulting metrics. The test data results are generally better than the cross-validation with the training data, possibly due to the larger number of training data used without cross-validation. Additionally, the results appear to be better when the value of the DC current falls within the range of the training data.

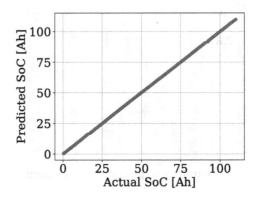

Fig. 5. Predicted data of best ANN as function of target values.

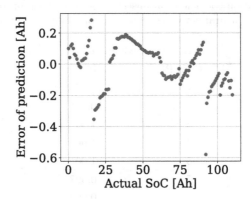

Fig. 6. Difference of predicted data and target values of best ANN.

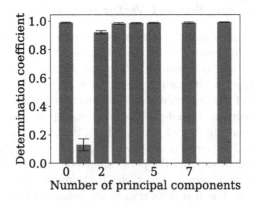

Fig. 7. Determination coefficient and standard deviation of crossvalidated training data depending on number of principal components.

Table 4. Determination coefficient on unseen testdata and optimized hyperparameters of current generalization using Support Vector Regression depending on number of principal components.

No. princ. comp.	R^2	ϵ	σ	C
No PCA	1.00	0.009	8.34E–4	23.7
1	0.999	0.197	0.961	30.4
2	0.996	0.191	0.956	31.6
3	0.996	0.032	0.118	2.96
4	1.00	0.018	0.026	14.7
5	1.00	0.006	0.016	30.7
7	1.00	0.021	0.015	25.0
9	0.998	5.81E-4	0.011	31.0

Figures 8 and 9 show the predicted data and the difference from the target value of the best model using four principal components. The test data are still approximated well. However, a closer look at the differences shows that the performance is worse compared to the best ANN model.

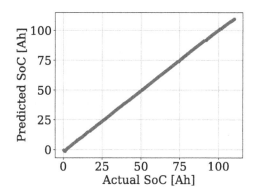

Fig. 8. Predicted data of best SVR model as function of target values.

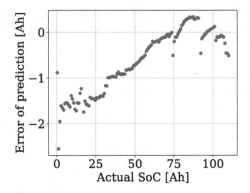

Fig. 9. Difference between predicted state of charge values and the actual values of polynomial and radial basis model.

5 Conclusion

In the coming years, there is expected to be a substantial increase in battery technology, primarily due to the growing need for electric vehicles. However, it is evident that the expansion of battery technology cannot rely solely on current battery technologies. Therefore, new and promising alternatives need to be explored. One such option is zinc-air batteries. Not only do they possess a greater theoretical energy density, but they are also more recyclable than existing battery technologies, making them an ideal solution for future battery expansion.

Nevertheless, the operation of zinc-air cells can be more intricate than traditional batteries due to small voltage differences during charging and discharging, which cannot be handled by existing battery management systems. To address this, a new approach has been developed using electrochemical impedance spectroscopy (EIS) to detect the state of charge of zinc-air cells accurately.

This method employs artificial neural networks and support vector regression to determine the state of charge based on the measured EIS spectra. Additionally, it can also be applied to unknown DC currents during impedance measurement. The accuracy of both approaches was tested by optimizing hyperparameters using the Bayesian optimization based on the training dataset to avoid any data leakage. The combination of EIS and machine learning techniques shows great potential for accurately detecting the state of charge of zinc-air batteries, providing a promising alternative for future battery technology expansion.

Acknowledgement. Funded by EFRE-0801585.

References

1. ADAC: Zufahrtsbeschränkungen in Europa (2021). https://www.adac.de/verkehr/abgas-diesel-fahrverbote/fahrverbote/fahrv/

2. Barsoukov, E., Macdonald, J.R. (eds.): Impedance spectroscopy: theory, experiment, and applications. Wiley, Hoboken, NJ, third edition edn. (2018)
3. Boukamp, B.A.: Impedance Spectroscopy, Strength and Limitations (Impedanzspektroskopie, Stärken und Grenzen). tm - Technisches Messen **71**, 454–459 (2004). https://doi.org/10.1524/teme.71.9.454.42758
4. Bundesministerium fuer Wirtschaft and Bundesministerium fuer Bildung und Forschung: Plattform Industrie 4.0 - Was ist Industrie 4.0 (2021). https://www.plattform-i40.de/IP/Navigation/DE/Industrie40/WasIndustrie40/was-ist-industrie-40.html
5. Chakkaravarthy, C., Waheed, A., Udupa, H.: Zinc-air alkaline batteries - a review. J. Power Sources **6**(3), 203–228 (1981)
6. Dufo-López, R., Cortés-Arcos, T., Artal-Sevil, J.S., Bernal-Agustn, J.L.: Comparison of lead-acid and li-ion batteries lifetime prediction models in stand-alone photovoltaic systems. Appl. Sci. **11**(3), 1099 (2021). https://doi.org/10.3390/app11031099. https://www.mdpi.com/2076-3417/11/3/1099
7. of Energy, U.D.: Energy storage database (2021). https://www.sandia.gov/ess-ssl/global-energy-storage-database/
8. Huang, J.: Diffusion impedance of electroactive materials, electrolytic solutions and porous electrodes: Warburg impedance and beyond. Electrochimica Acta **281**, 170–188 (2018)
9. Loechte, A., Gebert, O., Gloesekoetter, P.: End of charge detection of batteries with high production tolerances, p. 6. Granada (2019)
10. Loechte, A., Gebert, O., Gloesekoetter, P.: End of charge detection by processing impedance spectra of batteries. In: Valenzuela, O., Rojas, F., Herrera, L.J., Pomares, H., Rojas, I. (eds.) ITISE 2019. CS, pp. 163–176. Springer, Cham (2020). https://doi.org/10.1007/978-3-030-56219-9_11
11. Mainar, A.R., et al.: An overview of progress in electrolytes for secondary zinc-air batteries and other storage systems based on Zinc. J. Energy Storage **15**, 304–328 (2018)
12. Melzer, A.: Materialien für Zink und Zink-Luft Batterien (2010). https://docplayer.org/5093744-Materialien-fuer-zink-und-zink-luft-batterien.html
13. Sun, W., et al.: A rechargeable zinc-air battery based on zinc peroxide chemistry. Science **371**(6524), 46–51 (2021). https://doi.org/10.1126/science.abb9554
14. für Sonnenenergie-und Wasserstoff-Forschung Baden-Württemberg (ZSW): Bestand an Elektro-Pkw weltweit (2021). https://www.zsw-bw.de/mediathek/datenservice.html

An Approach to Predicting Social Events via Dailies Tracking

Renata Avros[(⊠)], Dan Lemberg, Elena V. Ravve[iD], and Zeev Volkovich[iD]

Braude College, Karmiel, Israel
{ravros,lemberg,cselena,vlvolkov}@braude.ac.il

Abstract. The study is devoted to a novel method for a real-time prediction of significant discontinuities in social states using the automatic analyses of Arabic dailies' overall logical structures. The paper introduces the novel, named the super-frequent *N-gram* approach and the Regression Mean Rank Dependency characteristic and presents their arrangement with a new model. It makes it possible to reliably forecast social changes based on the dailies' semantic content's high-level repercussions. An evaluation of the approach in a prominent Arabic daily demonstrates its ability to reflect changes in the social state and expose significant events of the "Arab Spring". The study shows that the resulting *N-gram* models corresponding to different sizes of *N* can provide a more vital prediction tool so that the methodology can consistently predict substantial changes in the social state in an online simulation fashion.

Keywords: Text Mining · Forecast Social Changes · Arabic dailies

1 Introduction

The so-called "mediated culture" phenomenon is an outstanding aspect of the modern age. Many kinds of mass media mirror the people's intentions and comprehend specific themes via the massive unstructured documents amounts, distributed in the virtual space. Monitoring such activity is crucial for each multimedia system. It turns it into a proper forecasting instrument in various fields like election results, cf. [1, 2], crime activity, cf. [3, 4], social events, cf. [5–7] asf.

As usual, intelligent summarizations provided for this purpose are created, lacking any assay of the linguistic matter of the documents since the handled corpus is typically associated with a collectively created resource created by informal language with genuine content. For example, methodologies intended to predict stock market prices often employ text mining methods cf. [8, 9] and use data from many sources. For this reason, the linguistic information in a studied text is typically disregarded.

In this paper, we suggest a new approach for predicting significant changes in the social state through alterations in the linguistic content with a case study of an Arabic daily generating the attitude suggested in [10] and [11].

Each text issue is presented as a histogram of the suitably chosen N − grams occurrences inside the famous Vector Space Model [12]. The critical notation pioneered here is the Mean Rank Dependency, designed as the mean Spearman's correlation between the current histogram issue and several previously published ones. This measure reveals a time series plot of the publishing process. Many methods have been recommended to deal with this problem. Social Media data prediction methods can be mentioned in this connection, cf. [13].

The Principal Component Regression methodology (a regression model based on the Principal Component Analysis) is applied in this paper. The dailies under consideration are accumulated. A time series plot is composed using introduced here the super-frequent N-grams model and the Regression Mean Rank Dependency. The correlation p − values are treated as a dissimilarity measure. A value of the Regression Mean Rank Dependency close to one would indicate a similarity between the style of the current issue and its precursors. On the other hand, a poor connection that is close to zero can specify a change in linguistic content. Small, i.e., close to zero p − values, signify a tight connection. The current p − value is examined to be an outlier (anomaly) employing the modified Thompson Tau test, cf. [14], within the time series of the p − values constructed through the previously published issues.

Arabic, is one of the Semitic languages, which about 300 million people speak (according to the Egyptian Demographic Center, 2000). The Arabic lettering "ligature." is one of the many causes of very complex Arabic morphology, especially in comparison with European languages like English or Russian (see, e.g. [15]). In order to treat this problem, we propose a new version of the character N-grams named the super-frequent N-grams model for Arabic. The common N-gram methodology is standard in text retrieval, particularly in Arabic text mining, cf. [16–18].

A straightforward construction of an N-grams model could lead to challenging classification tasks because of many stylistic patterns and characters. One of the ways to overcome this difficulty is a subsequent text normalization, cf. [2]. Here it is done by considering only N − grams formed by the most frequent letters. No relevant training data are available in the proposed method's framework. For this reason, all calculations are provided for numerous N-gram sizes, accompanied by configurations of other model parameters. Using simulated real-time tracking, our research is evaluated on editorial texts published in the most widely circulating Egyptian daily newspaper, "Al-Ahraam" in the Arab Spring period.

The rest of the paper is structured in the following way. Section 2 presents the novel time-series patterning of newspapers based on a dynamic replica of the human writing process. Section 3 introduces the proposed approach. Section 4 provides the partial experimental study results. Section 5 is devoted to discussion and conclusions.

2 Time Series Newspapers Patterning

This section describes a novel time-series representation of newspapers. First, we introduce the notion of the Regression Mean Rank Dependency following the notions stated in [10] and [11] constructed to indicate the significant global changes in the appropriate social state via linguistic behavior.

2.1 Regression Mean Rank Dependency

Let us consider a series of m sequential issues of a newspaper

$$D = \{D_1, D_2, \ldots, D_m\}$$

and take a vocabulary of terms $\mathbf{m\_D} = \{t_1, t_2, \ldots, t_n\}$ to represent an issue, say D_i, within the Vector Space Model, as a histogram $h(D_i)$ of the inner terms frequencies Introduce the group of its T "precursors." $\Delta_{i,T} = \{D_j, j = i - T, \ldots, j = i - 1\}$ Our model is constructed to detect and forecast changes in the content of a newspaper. Thus, we predict the current issue (more precisely, its histogram), based on the histograms of its T precursors and compare the result with the issue's histogram. Agreement between these values indicates the stability of the style at the current point. Otherwise, it can be concluded that the content significantly changes as follows:

- Construct the response variable as $\mathbf{Y} = h(D_i)$ as a histogram of the current issue D_i for $i > T$.
- Compose the predictors set $\mathbf{X}_{i,T} = \{\mathbf{X}_1, \ldots, \mathbf{X}_T\}$ from the histograms

$$\mathbf{X}_{j+T+1-i} = \{h(D_j), j = i - T, \ldots, i - 1, D_j \in \Delta_{i,T}\}$$

Choose the leading components $\mathbf{W_q}(\mathbf{X}_{i,T}) = \{\mathbf{W}_1, \ldots, \mathbf{W_q}\}$ of $\mathbf{X}_{i,T}$.

- Construct a regression $\mathbf{Y} = \beta_1 \mathbf{W}_1 + \cdots + \beta_p \mathbf{W_q}$
- Calculate the regression prediction $h(\hat{D}_i)$ of $h(D_i)$.
- The Regression Mean Rank Dependence $RZV_T(D_i, \Delta_{i,T}, \eta)$ is defined as $RZV_T(D_i, \Delta_{i,T}, \eta) = \eta(h(D_i), h(\hat{D}_i))$,

where η is a similarity measure expressing the histograms' resemblance.

The selection of a similarity measure η is essential in the proposed approach. Consistent with our perception, the writing style of each issue under consideration is outlined by the appropriate histogram of terms. For this reason, a similarity measure has to be appropriately chosen to reflect the relationship between styles. Formally speaking, let us take two sets of random variables, X, and Y, and test the hypothesis about association significance. A test is usually specified by a test statistic S, whose value is evaluated using the drawn samples. The decision is made by associating the test's threshold value, which can be interpreted as the probability of the current samples occurring if the null hypothesis is correct. This $p-$ value can be comprehended as a random variable attained from the drawn samples. A rise in the $p-$ value points to a weakening relationship. In this paper, we consider the following known association measures:

$$F_1 = \left\{ f_i^{(1)}, i = 1, \ldots, m \right\}, F_2 = \left\{ f_i^{(2)}, i = 1, \ldots, m \right\}$$

The Spearman's Rank Correlation Coefficient (the Spearman's ρ) being a variant of the usual Pearson's correlation coefficient calculated for the data transformed to rankings. A function R maps the sets F_j, $j = 1, 2$ onto $(1, \dots, n)$ such that $R(f_i^{(j)})$ is the rank (position) of $f_i^{(j)}$ in the arranged array F_j. The significance of this measure is also evaluated using the test statistic

$$S = r\sqrt{\frac{m-2}{1-\rho^2}},$$

2.2 Modified Thompson Tau Test

The Modified Thompson Tau test, cf. [14], verifies if outliers exist in a sample of scalar quantities. One potential outlier is tested at a time using a version of the Student $t - test$. Let X be a vector of size n. Denote by $\overline{X}$ the average of X and by $\sigma(X)$ the standard deviation of X. The rejection threshold is given as

$$\text{rej} = \frac{t_{\alpha/2,n-2}(n-1)}{\sqrt{n\left(n-2+t_{\alpha/2,n-2}^2\right)}},$$

where $t_{\alpha/2,n-2}$ is the critical value of the Student distribution corresponding to the significance level α and degree of freedom $n - 2$. For the following value

$$\delta(x) = \frac{|x - \overline{X}|}{\sigma(X)}$$

settles if a point x is an outlier by the comparison if $\delta(x) > \text{rej}$. If it is true, then x is recognized as an outlier; else (if $\delta(x) \le \text{rej}$), x is not considered an outlier

3 Methodology

We assume that a vocabulary of the terms **Term_D** is given. The suggested dictionary construction method is discussed later. When each issue D_i, $i > T$ (the delay parameter) is represented as a histogram of **Term_D**, it becomes possible to calculate the appropriate values of RZVT according to early explained by $RS_i = RZV_T (D_i, \Delta_{i,T}, \eta)$ together with their $p - values$ p_i. The suggested dictionary construction method is discussed later.

Algorithm 1 Detection phase

Input:

- T - Delay parameter.
- η - Similarity measure.
- L - Sliding window size.
- $\{p_j, j = i - L, \ldots, i - 1\}$ - Precalculated set of the previous p −values.
- $\Delta_{i,T,0} = \{D_j, \, j = i - T, \ldots, j = i\}$ - A collection including the current issue D_i together with its T predecessors.
- **Term_D** - Current dictionary.
- α_{Th} - Significance level of the Modified Thompson Tau test.
- α_0 - Significance level of p −value.

Output:

- p −value p_i.
- Flag $= 1$ if D_i is recognized as a change point, and Flag=0 otherwise.

Procedure:

1. Calculate histograms of $\Delta_{i,T,0}$ using the current dictionary **Term_D**.
2. Calculate the attitude $RS_i = RZV_T(D_i, \Delta_{i,T}, \eta)$ and its p −value p_i.
3. Perform the Modified Thompson Tau test for the set $\{p_{i-L+1}, \ldots, p_i\}$ at the significance level α_{Th}.
4. If p_i is identified as a potential outlier, then check if $p_i > \alpha_0$
5. If this is true, then D_i is recognized as a change point, and p_i and Flag $= 1$ are returned, otherwise p_i and Flag $= 0$ are returned.

The central assumption is that if the style of a newspaper does not significantly change, then the association between each current issue and its regression prediction is sufficiently high, and the corresponding p − value is sufficiently small. The values of the constructed feature chronologically, considered across the time axis of "i," form the desired time series. In turn, an outlier, appearing at the end of the evaluated p − values sequence, can indicate a considerable modification of the style since RZVT decreases at this point are caused by possible forthcoming transformations in the social state. We use the modified Thompson Tau test to recognize the desired anomalies.

In order to illustrate these matters, let us consider an example of an Egyptian newspaper, "Al-Ahraam". Figure 1 represents a a graphs of RZV_T (in the top panel) and the corresponding p − value (in the bottom panel), constructed based on the newspaper's issues, published during a crucial period of the Arab Spring in Egypt.

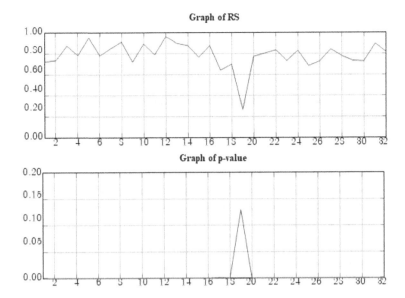

Fig. 1. Example of RZV_T and $p - value$ graphs.

This figure exhibits that the indicator RZV_T falls before an expected change in the social situation. Sequential issues located within a sliding window with the length L are mutually tested, aiming to check if the style of the last item is considerably different (i.e., an outlier) from most of the others in the investigated period. The style might stabilize after this point at another level of RZV_T. This fact is also reflected more explicitly by the behavior of the corresponding $p - value$, sharply jumping at the appropriate position. Such $p - value$ outliers are recognized in the detection step of our method. When a point is recognized as a potential outlier, the corresponding $p - value$ is compared to a given threshold, say *0.01*, aiming to be convinced that the $p - value$ is sufficiently significant, and only after the point is accepted as an outlier.

This work proposes a new method for enriching the model based on the so-called super-frequent $N - grams$. The suggestion is that the style of the media is stable within an initial period containing $L > T$ opening issues. Afterward, successive issues are considered in a sliding window with size L.

Algorithm 2 Dictionary construction
   Input:
   - L - Sliding window size.
   - N - N −grams size.
   - f - Fraction of the occurrences of the chosen super-frequent N −grams in the total occurrences.
   - N_{min} - Minimal size of the constructed dictionary.

   Procedure:
   1. Concatenate all issues in the first sliding window $\{D_j, j = 1, ..., j = L\}$ and obtain a text D.
   2. Omit all superfluous characters in D.
   3. Count occurrences of all characters in D.
   4. Construct the list D_N of the top N most frequently occurring characters.
   5. Construct the set $\mathbb{D}_N$ of N −grams including at least $N - 1$ characters belonging to D_N.
   6. Sort the elements of $\mathbb{D}_N$ in descending order according to their occurrences.
   7. Choose the subset $\mathbb{D}_{N,f} \subset \mathbb{D}_N$ of the top most frequently occurring elements covering the f fraction of the total occurrences of $\mathbb{D}_N$.
   8. If the size of $\mathbb{D}_{N,f}$ is less than N_{min} then take $\mathbb{D}_{N,f}$ as the set of the N_{min} most frequently occurring elements of $\mathbb{D}_N$.
   9. If the size of $\mathbb{D}_N$ is less then N_{min} print ("A dictionary cannot be constructed") and Stop.
   10. Return **Term_D** $= \mathbb{D}_{N,f}$.

At the preprocessing step, all characters that are not the language letters are omitted. All single characters are sorted in descending order according to their occurrence in the corpus. A dictionary **Term_D** is constructed using a certain fraction of the most frequent N − grams, including at least the $N - 1$ most popular characters, resting on L preliminary issues (see, Algorithm 2).

An example of the super-frequent N − grams portions among the most frequent regular N − grams once their numbers grow from three to fifty is given in Fig. 2 for $N = 2, 3$, and 4. As it can be seen, the super-frequent N − grams form a compact and dense subset within the regular *N-grams*. Selecting an appropriate fraction of the most common super-frequent N − grams strengthens these properties, stabilizes the classification process, and makes it possible to estimate the number of the N − grams, involved in the model design more accurately. It is especially evident in the case of longer N − grams leading to sparser representations. Two graphs of $RZV_T (D_i, \Delta_{i,T}, \eta)$, which are calculated for the example mentioned earlier for $N = 4$, are given in Fig. 3. The one, marked in blue, is constructed employing the super-frequent N − grams approach, while the second graph, marked in red, is built via the same technique but using the regular N − grams. The related descent of $RZV_T (D_i, \Delta_{i,T}, \eta)$ is higher in the first case. Namely, the first method more explicitly distinguishes the style changes. The p − values are 0.38 and 0.002 correspondingly, with dictionaries having sizes of 30 (the minimal possible amount in this example) and 169.

Nevertheless, once single words or characters are merged into N − gram words or characters, the frequency of tokens generally fits Zipf's Law approximately with the slope close to (-1). The distribution of the most common $3 - grams$ in Arabic texts,

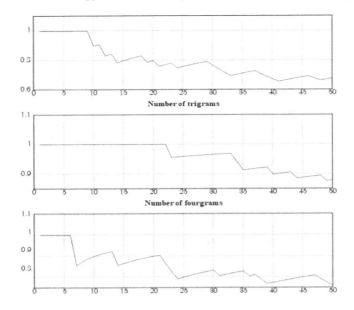

Fig. 2. Fractions of the super-frequent N-grams.

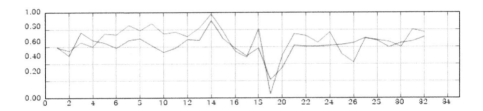

Fig. 3. Graphs of $RZV_T(D_i, \Delta_{i,T}, \eta)$

belonging to the same category also follows this law, cf. [15]. Zipf's Law entails that a $N-$ gram, based classification procedure, should not be very responsive concerning distribution truncation at a particular rank. In other words, the conclusion can be made using a relatively limited number, say .min of the most super-frequent $N-$ grams. All procedures are performed for various sizes of N, conveyed by specific arrangements of other factors. We use the ensemble methodology and combine several parallel predictors, as mentioned earlier.

4 Findings

In our experiments, we study the prediction process by a simulation procedure, where sequential issues of a newspaper are considered within a sliding window to forecast the following issue histogram.

4.1 Parameters Selection

All calculations are provided for the *N-gram* sizes of $N = 2, 3,$ and 4, accompanied by the values *0.2* and *0.3* of the fraction of the most common super-frequent *N-grams* and with the minimal size of the dictionary *Nmin* = *20*. We set Significance level of the Modified Thompson Tau test $\alpha = 0.01$ and Significance level of outlier p − value $\alpha_0 = 0.01$.

For large values of the delay parameter *T*, the resulting *RZVT* is expected to be smoother. However, the necessary information could be lost. It should also be noted that, on the other hand, an increase in parameter *T* would improve the model-based forecast. A balance point between these contrary factors could provide a reasonable estimation of *T*. We choose $T = 10$ as such a poise point, with the appropriate size of the sliding window as $L = 2T$. The experiments are delivered the Spearman's ρ. The total percent of the explained variance in the used *PCR* model is 80%. The studied dataset consists of 909 issues of the "Al-Ahraam," published in: 1.1.2010 - 31.12.2011 and 1.1.2014 - 30.6.2014.

Figure 4 demonstrates results presenting the locations of the detected change-points with the corresponding clarifications given in Table 1. Figure 5 demonstrates the principal components numbers found for different sizes of *N*. The figures' top, middle, and bottom panels exhibit results for each size of the considered *N*-grams, summarized across the fraction values.

Table 1. Change-points detected for "Al-Ahraam" newspaper using the Spearman's similarity.

Num	45	69	173	199	312	358	372	424	522	537	545	549	639	784
Year	2010	2010	2010	2010	2011	2011	2011	2011	2011	2011	2011	2011	2011	2014
Month	2	3	6	7	11	12	1	2	6	6	6	7	10	2
Day	14	10	22	18	8	24	7	28	6	21	29	3	1	23
Prob.	0.02	0.01	0.01	0.01	0.01	0.1	0.03	0.01	0.02	0.01	0.06	0.02	0.02	0.05

Let us consider the alteration in the social states connected to the presented positions:

- 14.2.2010. This change point is close to 24.2.2010, when Mohamed El-Baradei, with several other opposition leaders, founded a new apolitical movement called the "National Association for Change".[1]
- 18.6.2010 and 18.7.2010. On 25.6.2010, Mohamed El-Baradei headed a multitudinous demonstration in Alexandria over Saeed's death.[2]
- 7.1.2011. This point indicates 25.1.2011 ("Day of Revolt") when protests exploded throughout the country, demanding the resignation of President Hosni Mubarak. This event is correctly predicted by our method. [3]

---

[1] http://news.bbc.co.uk/2/hi/middle_east/8534365.stm.

[2] http://edition.cnn.com/2010/WORLD/africa/06/25/egypt.police.beating/index.html.

[3] https://www.bbc.com/news/av/world-middle-east-12282585/three-reported-dead-after-egypt-s-day-of-revolt.

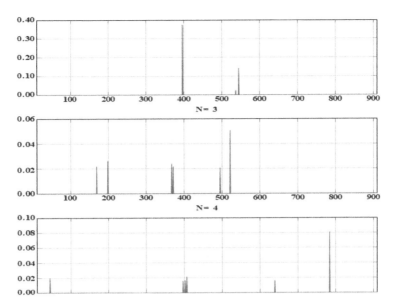

Fig. 4. Change-points locations found for different sizes of the *N-grams* using the Spearman's similarity

- 28.2.2011. This date is close to 2.3.2011 when the constitutional referendum was set on 19.3.2011.[4].
- (6, 21, and 29) in 6. 2011. These change points are strictly associated with a social explosion that appeared in the summer of 2011.[5]
- 23.2.2014. Many commentators agree that after the July overthrow (there is no data about this period in the considered newspaper collection) of the elected President-Islamist Mohammed Mursi and the following repressions against Islamists and liberals, the military authorities decided to return to the era of Hosni Mubarak
- On 24.3.2011, an Egyptian court condemned 529 supporters of the Muslim Brotherhood to death. [6]

It is curious to track the number of leading components in predicting. If $N = 2$, then the component number always equals 1; i.e., the predecessors' histograms, and the issues themselves, are permanently highly associated. This relationship is blurred with increases in N, so the component number becomes 3 in almost half of the cases (see, Table 2). Figure 6 demonstrates the very diverse behavior of this quantity for different sizes of N, calculated for the fraction $f = 0.3$. Thus, the method is more sensitive with greater values of N, where slighter style fluctuations can be detected. This fact can also be understood from Fig. 4, where only the most essential two changes were found for $N = 2$, while for $N = 4$, more minor effects are likewise discovered.

---

[4] https://www.thestar.com/news/world/2011/03/20/constitutional_amendments_approved_in_egypt_referendum.html.

[5] https://www.theatlantic.com/photo/2013/07/millions-march-in-egyptian-protests/100543/

[6] http://uk.reuters.com/article/uk-egypt-politics-idUKBREA1N0KM20140224.

Table 2. Cumulative distribution of the leading components found for $N = 4$.

Value	1	2	3	4	5
Frequency	0.34	23.60	50.79	24.38	0.90

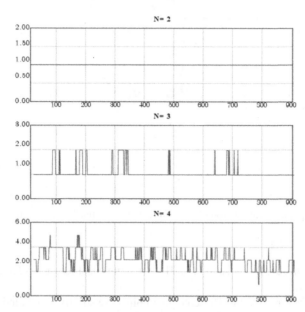

Fig. 5. The number of the leading components found for different sizes of the *N-grams* using the Spearman's similarity for "Al-Ahraam".

5 Discussions and Conclusions

This paper proposes a novel method for online real-time prediction of significant discontinuities in social states using the linguistic development of Arabic. The method describes its combination with a new super-frequent *N-grams* model, making it possible to forecast social changes reliably using the daily stylistic content and demonstrates its ability to provide highly beneficial linguistic models for linguistically complex languages. The study shows that the resulting $N - grams$ models correspond, in general, to different sizes of N. Apparently, newspapers are affected by varieties in vocabulary (in the domain of the language of mass media) due to lexical borrowing and the policies of editorial boards depending on the state authorities, as well as different types of target readers.

References

1. Gerber, M.: Predicting crime using Twitter and kernel density estimation. Decis. Supp. Syst. (Elsevier) **61**, 115–125 (2014). https://doi.org/10.1016/j.dss.2014.02.003
2. Sallam, R.M., Mousa, H.M., Hussein, M.: Article: improving Arabic text categorization using normalization and stemming techniques. Int. J. Comput. Appl. **135**(2), 38–43 (2016)
3. Harrag, F., Al-Qawasmah, E.: Improving Arabic text categorization using neural network with SVD. JDIM **8**(2), 125–135 (2010)
4. Yu, S., Kak, S.: A survey of prediction using social media. CoRR abs/1203.1647 (2012)
5. Kendall, M., Gibbons, J.: Rank correlation methods. Edward Arnold (1990)
6. Khreisat, L.: A machine learning approach for Arabic text classification using N-gram frequency statistics. J. Informet. **3**(1), 72–77 (2009)
7. Leiter, D., Murr, A., Ramrez, E.R., Stegmaier, M.: Social networks and citizen election forecasting: the more friends the better. Int. J. Forecast. **34**(2), 235–248 (2018)
8. Kalyanam, J., Quezada, M., Poblete, B., Lanckriet, G.: Prediction and characterization of high-activity events in social media triggered by real-world news. PLoS ONE **11**(12), 1–13 (2016)
9. Lin, W.-C., Tsai, C.-F., Chen, H.: Factors affecting text mining based stock prediction: text feature representations, machine learning models, and news platforms. Appl. Soft Comput. **130**, 109673 (2022), ISSN 1568–4946
10. Wang, X., Brown, D.E., Gerber, M.S.: Spatio-temporal modeling of criminal incidents using geographic, demographic, and twitter-derived information. In: Zeng, D., Zhou, L., Cukic, B., Wang, G.A., Yang, C.C (eds.) ISI, pp. 36–41. IEEE (2012)
11. Amelin, K.S., Granichin, O.N., Kizhaeva, N., Volkovich, Z.: Patterning of writing style evolution by means of dynamic similarity. Pattern Recogn. **77**, 45–64 (2018)
12. Salton, G., Wong, A., Yang, C.S.: A vector space model for automatic indexing. Commun. ACM **18**(11), 613–620 (1975)
13. Kalampokis, E., Tambouris, E., Tarabanis, K.: Understanding the predictive power of social media. Internet Res. **23**(5), 544–559 (2013)
14. Thompson, R.: A note on restricted maximum likelihood estimation with an alternative outlier model. J. Royal Stat. Soc. Ser. B (Methodol.) **47**(1), 53–55 (1985)
15. Volkovich, Z., Granichin, O., Redkin, O., Bernikova, O.: Modeling and visualization of media in Arabic. J. Informet. **10**(2), 439–453 (2016)
16. Franch, F.: (wisdom of the crowds)2: 2010 UK election prediction with social media. J. Inform. Tech. Polit. **10**(1), 57–71 (2013)
17. Al-Thubaity, A., Alhoshan, M., Hazzaa, I.: Using word N-Grams as Features in Arabic Text Classification. In: Lee, R. (ed.) Software Engineering, Artificial Intelligence, Networking and Parallel/Distributed Computing, pp. 35–43. Springer International Publishing, Cham (2015)
18. Korolov, R., et al.: On predicting social unrest using social media, pp. 89–95. Institute of Electrical and Electronics Engineers Inc., United States (2016)

Data Fusion for Prediction of Variations in Students Grades

Renata Teixeira[1], Francisco S. Marcondes[1], Henrique Lima[2],
Dalila Durães[1(✉)], and Paulo Novais[1]

[1] LASI/ALGORITMI Centre, University of Minho, Guimarães, Portugal
pg47603@alunos.uminho.pt, francisco.marcondes@algoritmi.uminho.pt,
{dad,pjon}@di.uminho.pt
[2] Codevision, S.A., Braga, Portugal
henrique.lima@e-schooling.com

Abstract. Considering the undeniable relevance of education in today's society, it is of great interest to be able to predict the academic performance of students in order to change teaching methods and create new strategies taking into account the situation of the students and their needs. This study aims to apply data fusion to merge information about several students and predict variations in their Portuguese Language or Math grades from one trimester to another, that is, whether the students improve, worsen or maintain their grade. The possibility to predict changes in a student's grades brings great opportunities for teachers, because they can get an idea, from the predictions, of possible drops in grades, and can adapt their teaching and try to prevent such drops from happening. After the creation of the models, it is possible to suggest that they are not overfitting, and the metrics indicate that the models are performing well and appear to have high level of performance. For the Portuguese Language prediction, we were able to reach an accuracy of 97.3%, and for the Mathematics prediction we reached 95.8% of accuracy.

Keywords: Data fusion · Academic performance · Education · Computer science · Machine Learning

1 Introduction

The relevance of education in our lives is remarkable. From a very young age we enter school and start learning not only about writing and reading, but also about history and science. Education is an essential aspect that plays a huge role in the modern and industrially driven society. People need a good education to be able to survive in this competitive world [1]. Educated people are better able to form opinions about various aspects of life, and they also have better job opportunities. Education helps us grow and develop.

Despite the importance of education, it is undeniable that there are still students who fail, and there aren't there many ways to predict whether or not a

I. Rojas et al. (Eds.): IWANN 2023, LNCS 14135, pp. 292–303, 2023.
https://doi.org/10.1007/978-3-031-43078-7_24

student is at risk of failing, one can only conclude this when they get negative grades or even fail the year. One way to try to prevent this problem may come from early prediction of school failure, since the accurate detection of students at risk of failing is of vital importance for educational institutions, it can provide feedback to support educators in making decisions about how to improve student's learning and enable them to apply intervention measures and learning strategies aimed at improving the academic performance of students [2].

The present world is marked by the abundance of diverse sources of information, which makes it difficult to ignore the presence of multiple possibly related datasets, since they may contain valuable information that will be lost if these relationships are ignored [3]. Data Fusion can take advantage of the large amounts of data to help create more complete and consistent datasets. This method is also used in the area of education, namely the area of student performance prediction, due to the amount of different data that can affect the performance of students, like academic grades, parents' education level, interest in school, prospects for the future, *etc.*

In this paper we propose to predict variations in student's Portuguese language and Mathematics' classes' grades from one trimester to another, in order to be able to know beforehand the possibility of a student's grade getting worse, which, in more worrisome cases, can be very useful.

The rest of this paper is organized as follows. In Sect. 2, related work is reviewed using the PRISMA statement and checklist. The proposal is presented in Sect. 3, providing in-depth details about how the dataset was created and processed, about the building of the predictive models, and an analysis of the experimental results through relevant metrics. Finally, the conclusions and further considerations are summarized in Sect. 4.

2   Related Work

2.1   Methodology

This review of articles on Data Fusion in the field of education, namely those that use Data Fusion to predict student's academic performance, was based on the PRISMA (Preferred Reporting Items for Systematic Reviews and Meta-Analyses) [4]. This choice is justified by the fact that PRISMA is widely accepted by the scientific community in engineering and computing.

The literature search was conducted on March 2023 in the popular database for computer science: SCOPUS. Considering the field of the study was based on school failure and student performance, the keywords *Student Performance, Academic Performance, School Failure, School Dropout* and *Academic Failure* were added to the query to specify the fields of this work. And since this work focus mainly on data fusion, the keyword *Data Fusion* was added using a conjunction, while all the other keywords were aggregated using disjunctions.

The query shown below was used for the search in the SCOPUS repository applied to the title, keywords and abstract of the documents.

> TITLE–ABS–KEY (''Data Fusion''_AND_(_''Student_Performance''
> OR ''Academic Performance''_OR_''School_Failure''
> OR ''School Dropout''_OR_''Academic_Failure''))

In order to eliminate unwanted articles among the articles found, some exclusion criteria were defined. Thus, the documents are excluded if they fall in one of these:

EC1 Do not come from the field of computer science or engineering;
EC2 Not freely accessible;
EC3 Do not focus on the variables studied or is out of context.
EC4 Were not written in English or Portuguese, as these are the languages the authors understand.
EC5 Does not show results.

The final query resulted from adding some authors and article titles:

AC1 Articles already studied and considered important for this analysis.

2.2   Data Search Results

The search in the SCOPUS repository identified 15 articles to which the inclusion and exclusion criteria were subsequently applied.

The inclusion criteria **AC1** was introduced because these articles were already studied and considered essential for this analysis. As a result, two articles were introduced, which led us to 17 articles in total.

On the search page of the database, the documents that met the first exclusion criteria **EC1** were filtered out, leaving us with 16 articles that came from the field of computer science or engineering. Of these documents received, the title and abstract were read and it was found that even though all were written in English (**EC4**), 2 documents were not freely available(**EC2**). The remaining 14 articles were read in full, and the third exclusion criteria **EC3** was responsible for eliminating 5 more articles that were considered to be out of the context of this study.

In Fig. 1 is the PRISMA flow diagram related to this study, which helps in understanding the whole process described above.

Below, an overview of the remaining 9 articles is given.

The article [5] merged collected and manually aligned student behavior data with textual data of the course comments to predict student performance, using a designed multimodal data fusion approach. The empirical research indicated that the proposed method could fuse two different types of data and achieves the best classification performance compared to the base methods. The study outcomes show that the classification method can achieve better classification results in terms of RECALL, F1-measure (F1) and the area under the receiver operating characteristic curve (AUC).

The study conducted in [6], intended to scientifically evaluate the effect of video teaching mode, find out its advantages and find common rules. For this

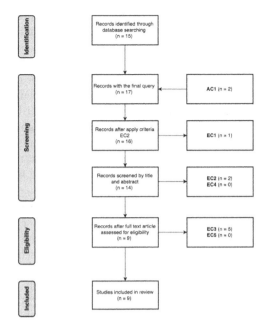

Fig. 1. PRISMA Flow Diagram

purpose, two classes with 30 students were selected, and different teaching methods were adopted for the two classes: one class adopted the traditional teaching mode, and the other class used the video teaching method. It was concluded that in video image teaching, the students could acquire knowledge faster, and their understanding had been greatly improved. According to the article, adding data fusion also helps teachers and students to improve teaching methods, to provide more targeted assistance, so that teachers' teaching efficiency is continuously improved, and students' learning outcomes are constantly increasing.

The article [7] focuses on engineering students using real-life data and focuses on creating a predictive model, based solely on academic data. The authors of this paper believe that applying Feature Engineering techniques has a great impact on learning predictive systems and discuss numerous processes on data, such as dealing with redundancy, correlation, missing values, feature creation or deletion, and data fusion, for example. For part of the study, they worked with the three datasets in parallel, applying different data processing and feature engineering techniques, however, at one point they considered it essential to merge the datasets.

The paper [8] was also accepted in this review as it does not fall in any exclusion criteria but it is written by the same authors as [9], and it is considered an initial approaching to solve the proposed problem, so, a review of the article [9] is presented below.

Data Fusion is also an important role in [9], as this paper proposes to use a data fusion and mining methodology for predicting students' final performance starting from multi- source and multimodal data. It gathers data from several sources: theory classes, practical sessions, online sessions with Moodle, and the course final exam. It also applies some pre-process tasks for generating datasets in two formats: numerical and categorical. Secondly, it uses different data fusion approaches like merging all attributes, selecting the best attributes, using ensembles, and using ensembles and selecting the best attributes, and several white-box classification algorithms with the datasets. Finally, after comparing the predictions produced by the models, the authors conclude that the best result was produced using the fourth approach of using ensembles and selection of the best attributes.

The study [13] aimed to develop, train, and test classification models to predict whether students would persist into the second semester beyond traditional measures of performance. According to the authors of this article, data that has been aggregated over time can provide insight into which students are most likely to fail and may need closer attention, while individual student-level performance data can be used to flag students who may be diverging from success to failure. Multiple characteristics such as each student's academic performance, engagement, and demographic background were made available for this project from various sources, which were then compared at the student level and merged into a single dataset.

The article [14] does not aim to predict student performance or failure, instead it proposes a detection framework for detecting students' mental health. It was not eliminated from this review because it was still considered relevant, since it uses data fusion as a main and important part of the development of the work, it is within the context of education and because we believe that mental health and student performance are very closely related. The first step of this work was using representation learning for the fusion of students' multimodal information, like social life, academic performance, and physical appearance.

Similarly to the case above, the paper [15] does not predict school failure, it actually builds and tests prediction models to track middle-school student distress states during educational gameplay. It was considered relevant due to the fact that it is within the context of education and uses data fusion as a main part of the prediction. In this study multiple types of data from 31 students was collected during a gameplay session. With the collected multimodal data, a multimodal data fusion was implemented in order to predict changes in the outcome variable, which was the state of distress. As a result of this experiment, it was concluded that the classifier with multimodal data fusion outperformed the prediction by each of those classifiers with unimodal data sources. These results led the authors to affirm that the improved performance of the fused classifier in this study corroborates the usefulness of multimodal data fusion when building a learning analytics system.

Finally, the article [16] aimed to predict university students' learning performance using different sources of performance and multimodal data from an

Intelligent Tutoring System. In this paper multimodal data was collected and preprocessed, and in total three different data fusion approaches and six white-box classification algorithms were used.

3   Proposal

The importance of being able to predict students' performance in school is well established and, as it was possible to confirm in the review above, data fusion is used in such predictions and allows to reach good results and conclusions.

Figure 2 provides visual representation of the entire system for predict student failure. It includes fusion data, the pre-processing steps and the model used for prediction. The system architecture comprises three main components, namely fusion, pre-processing, and ML model.

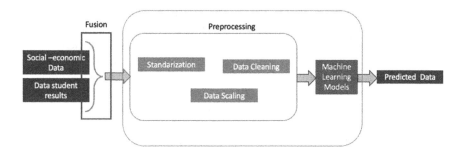

Fig. 2. Visual representation for predicted failure.

3.1   Data Fusion

We used information from 845 middle school students from a school of Northern Portugal. We collected student-related data provided by the school, with data from a questionnaire to which the students responded, that had questions about school motivation, long and short-term self-regulation, mastery and performance motivation, school value, family, friends and teachers' support and about school engagement.

The resulting dataset is a fused dataset consisted of 845 instances and 123 attributes. These attributes included: sex, age, school year, parent's education level, parent's marriage status, Portuguese and Mathematics classes' grades from the first and second trimester, failure history, satisfaction with grades, tutoring, number of times of tutoring per week, familiar support, friends' support, teachers' support and more.

3.2 Preprocessing of Data

We took advantage of the pandas library to handle data as when performing Data Engineering operations.

The first step of the pre-processing was the standarization. So, after fusing the data related to each student into the same dataset, we started treating missing values. We noticed the attributes *"Q9-tipo de explicação"*, type of tutoring, and *"Q10-quantos dias semana"*, how many days of tutoring per week the student has, had 568 and 567 null values each, which meant that 568 students did not get tutoring. For that matter we decided to replace the null values with the value 0. We also noticed that some grades from the first trimester were missing, in this case those rows were dropped. The attributes *"bullying vítima"*, victim of bullying, and *"bullying bully"* also had missing values, we decided to replace those values with the value -1.

Following the process, the second step was data cleaning of handling missing values. We began dropping columns that we didn't consider to be relevant to the context and purpose of this prediction. We noticed the attribute *"Autorização questionário"*, authorization to participate in the questionnaire, had the same value for every instance, so it was dropped. The same happened with the attribute *"Autorização notas"*, authorization to provide grades. The attribute *"Código participação"*, code of participation, was different for each student but considered irrelevant to this study. The attribute *"Escola"*, school that the student attends, was also dropped, as it was considered insignificant.

With the previous task finished, and since most machine learning algorithms only work with numeric values, the next step the data scaling. It was convert categorical features into numerical. The features called *"Q24 Sexo"*, gender of the student, *"Q29-existência de reprovação"*, failure occurrence, *"Q31-escolaridade mãe"*, mom's education level and *"Q35-escolaridade pai"*, dad's education level, were converted to numeric values.

As the resulting dataset had information regarding grades of two different subjects, and the purpose of the study was to attempt to predict the variation of the grades of each subject from the first to the second trimester, the team decided it would be wise to create two identical datasets which would be similar to the original dataset but each would only have information about either Math or Portuguese Language second trimester grades, let's call them, *Port_dataset* and *Math_dataset*.

Since the dataset didn't have a target feature, we created one. For each dataset we subtracted the first trimester grade of one subject from the second trimester grade of that subject and stored the resulting value in a variable. Then we checked whether the variable was greater than, less than, or equal to 0, which would mean the student had improved, worsened or maintained the subject's grade from one trimester to the other, and the target value would be 0,1 or 2 for each possibility described.

Then, in each dataset, we dropped the column *"Q20-nota port. 2.ºperiodo"*, second trimester grades of the Portuguese Language subject, and the column

"*Q26-nota mat 2.ºperiodo*", second trimester grades of the Mathematics subject, respectively.

After studying the amount of each target value, we noted that the dataset *Port_dataset* was made of 671 instances with target value 1, 134 with target value 2 and just 39 with the value 0, meanwhile the dataset *Math_dataset* had 635 instances with target value 1, 134 with target value 2 and 75 with the value 0. Having balanced data for model training is very important, it gives us the same amount of information to help predict each class and therefore gives a better idea of how to respond to test data, therefore, our data needed to be balanced. We decided to apply data augmentation to solve the problem using random oversampling, resulting in two datasets with 2013 and 1905 instances each.

Finally, as part of the data pre-processing, we performed feature selection on both datasets, in order to compare the results given by the resulting datasets with the results with the bigger datasets. The best features of each dataset were selected and two new datasets were created with them. For this, a function with the target name and the number of features desired as input was created, where the correlation matrix was calculated, and the correlation values of the target variable with all other features extracted. Then, the top n features with the highest absolute correlation coefficients were selected and the resulting dataframe returned. We decided to select the best 20 features for both datasets.

The features resulting from the feature selection for predicting Portuguese grades were: the risk of failing, first trimester Portuguese grade, first and second trimester math grade, satisfaction with grades, the three different total values attributed to the support given by friends, the answer to the first, second, fifth, sixth and seventh questions about friend's support, the answer to the first and fourth questions about short term self-regulation, the total value attributed to the short-term self-regulation, the answer to the sixth question about family support, sex, and the answer to the second question about the student's relationship with peers.

In turn, the features resulting from the feature selection for predicting Mathematics grades were: first trimester Math grade, the answer to the seventh question about long-term self-regulation, personal perspective about grades, the answer to the fifth question about short-term self-regulation, the risk of failure, the answer to the fourth question about motivation for mastery, satisfaction with grades, age, how many days a week they have tutoring, school year, the two different total values assigned to school valuation, second trimester Portuguese grade, total value attributed to the mastery motivation, the answer to the first and third questions about school valuation, the total value attributed to the long-term self-regulation, the answer to the seventh question about teacher's support and the answer to the fifth question about school engagement.

3.3 Building of Predictive Models

Two different experiments were conducted in this study, using several classification algorithms. These being using the classification models with hyper-parameter optimization, Gridsearch , on the *Port_dataset* and *Math_dataset*, or using the same classification algorithms for the datasets resulting from feature selection. After those experiments, the results were compared.

Regarding the large number of available classification algorithms and the enormous possible values to assign to the algorithm hyperparameters, considerations that clearly affect the performance of the models, we decided on some of the algorithms which have been proved successful in related work, which were *Random Forest* [10], *XGBoost*, [11] and *Decision Tree* [12].

First we started by creating the test and training datasets for each case of study. This process consisted of splitting the datasets into 2 parts: the training dataset containing 80% of the instances of the total dataset, and the test dataset making up the remaining 20% of the instances.

As already mentioned above, a feature selection function was created, which had as output the resulting dataframe with the number of features that were given as input. After several experiments and results comparisons, eventually the team decided to choose the best 20 features of both datasets, and for that matter, we finally ended up with a total of 4 datasets, the *Port_dataset* and *Math_dataset* and the two dataframes made up of only the 20 selected features, were split into training and test datasets, as already stated.

For each classification algorithm, in order to find the best hyperparameters, we decided to apply grid search instead of random search, since grid search looks at every possible combination of hyperparameters to find the best model and random search only selects and tests a random combination of hyperparameters, instead of conducting an exhaustive search. Therefore, for every model, we defined the parameters to be tuned, created an instance of GridSearchCV and fit the model, and with the best parameters found, we then made the predictions for test data and evaluated the results.

In short, we applied 3 different classification algorithms to the four datasets. The results are shown in the tables below.

In Table 1 we show the results for the *Port_dataset* and for the one with the selected 20 features of this dataset, which we well call *PFS_dataset*.

Table 1. Results for the Portuguese Language Prediction

Datasets	Models	Accuracy	F1 Score	Recall	Precision
Port_Dataset	**XGBoost**	**0.888**	0.867	0.888	0.895
	Decision Tree	0.834	0.834	0.834	0.841
	Random Forest	0.817	0.740	0.817	0.809
PFS_Dataset	**XGBoost**	**0.973**	0.973	0.973	0.974
	Decision Tree	0.948	0.947	0.948	0.952
	Random Forest	0.970	0.970	0.970	0.971

In Table 2 the results of the Mathematics predictions are shown, for both datasets that contain information about this specific subject. The dataset resulting from the feature selection we call *MFS_dataset*.

Table 2. Results for the Mathematics Prediction

Datasets	Models	Accuracy	F1 Score	Recall	Precision
Math_Dataset	**XGBoost**	**0.775**	0.712	0.775	0.699
	Decision Tree	0.763	0.671	0.763	0.599
	Random Forest	0.775	0.677	0.775	0.601
MFS_Dataset	**XGBoost**	**0.958**	0.957	0.958	0.960
	Decision Tree	0.908	0.904	0.908	0.916
	Random Forest	0.942	0.941	0.942	0.943

For each algorithm that obtained the best results for each dataset we then applied cross-validation, in order to evaluate the performance of the models, in terms of overfitting. This way we can assure that the models would also perform well on new, unseen data. The results are presented in Table 3.

Table 3. Cross-validation

Datasets	Mean Accuracy	Standart Deviation
Port_Dataset	0.849	0.034
PFS_Dataset	0.966	0.009
Math_Dataset	0.745	0.030
MFS_Dataset	0.940	0.006

Considering the mean accuracy in cross-validation for all four models, the scores indicate that the models are performing well across multiple folds, with high values of mean accuracy and relatively low standard deviations. This can suggest that the models are generalizing well to new data, and are not overfitting to the training data.

Considering that the models are not overfitting, looking at the values in the Tables 1 and 2, it is possible to state that the models appear to have a high level of performance in terms of accuracy, F1 score, recall, and precision. It is also possible to notice that the models performed better with the datasets that consisted only of the 20 selected features instead of the original 117.

4   Conclusions and Future Work

It is of high importance for educational institutions to predict students' grades in order to create innovative strategies taking into account the students and their specific cases, since education is key for the formation of opinions and for the world of employment itself.

In this paper we use data fusion to fuse information about students, their relatives and a questionnaire to which they participated and responded, and predict variations in middle school student's grades in Mathematics and Portuguese Language. If it becomes possible to predict these variations, teachers and educators gain an advantage that allows them to shape their teaching according to the predictions, and instead of being surprised by decreases in grades, they can avoid them.

After the creation of the models, we can suggest that they are not overfitting, and the metrics indicate that the models are performing well and appear to have high level of performance. For the Portuguese Language prediction, we were able to reach an accuracy of 97.3%, and for the Mathematics prediction we reached 95.8% of accuracy. Even though the best results were achieved using the smaller datasets, with the 20 selected features, these features were not only answers given to the questionnaire, but were also grades and characteristics of the students, information that was initially merged, which proves the importance and relevance of data fusion in this study.

As a prospect for future work, we believe it would be interesting to take this to a real-life experience to actually verify whether the model's predictions would really help teachers to accurately predict school failure of their students and whether they would be successful in trying to avoid this failure when shaping their teaching according to individual student needs. One drawback of this study is that the predictions were made manually. However, the plan for future research is to develop a decision support system that can use the prediction results automatically to guide or alert teachers, parents, and other school personnel. Also it is possible to address some issues like discrimination issues.

Acknowledgement. This work is supported by: FCT - Fundação para a Ciência e Tecnologia within the RD Units Project Scope: UIDB/00319/2020 and the Northern Regional Operational Programme (NORTE 2020), under Portugal 2020 within the scope of the project "Hello: Plataforma inteligente para o combate ao insucesso escolar", Ref. NORTE-01-0247-FEDER-047004.

References

1. Prasad, C., Gupta, P.: Educational impact on the society. Int. J. Novel Res. Educ. Learn. **7**, 1–7 (2020). ISSN 2394–9686
2. Romero, C., Ventura, S.: Data Mining in Education. Data Mining and Knowledge Discovery, Wiley Interdisciplinary Reviews (2013)
3. Lahat, D., Adali, T., Jutten, C.: Multimodal data fusion: an overview of methods, challenges, and prospects. Proc. IEEE **103**(9), 1449–1477 (2015)
4. Asar, S., Jalalpour, S., Ayoubi, F., Rahmani, M., Rezaeian, M.: PRISMA. Preferred Reporting Items for Systematic Reviews and Meta-Analyses. JRUMS (2016)
5. Qu, Y., Li, F., Li, L., Dou, X., Wang, H.: Can we predict student performance based on tabular and textual data? IEEE Access **10**, 86008–86019 (2022)

6. Zou, W., Li, Y., Shan, X., Wu, X.: Application of data fusion and image video teaching mode in physical education course teaching and evaluation of teaching effect. Secur. Commun. Netw. **2022**, 8584350 (2022)
7. Fernández-García, A.J., Preciado, J.C., Melchor, F., Rodriguez-Echeverria, R., Conejero, J.M., Sánchez-Figueroa, F.: A real-life machine learning experience for predicting university dropout at different stages using academic data. IEEE Access **9**, 133076–133090 (2021)
8. Chango, G., Cerezo, R., Romero, C.: Predicting academic performance of university students from multi-sources data in blended learning. In: DATA 2019: Proceedings of the Second International Conference on Data Science, E-Learning and Information Systems, pp. 1–5 (2019). https://doi.org/10.1145/3368691.3368694
9. Chango, W., Cerezo, R., Romero, C.: Multi-source and multimodal data fusion for predicting academic performance in blended learning university courses. Comput. Electr. Eng. **89**, 106908 (2021). ISSN 0045–7906
10. Diaz, P., Salas, J.C., Cipriano, A., Nunez, F.: Random forest model predictive control for paste thickening. Miner. Eng. **163**, 106760 (2021)
11. Ogunleye, A., Wang, Q.G.: XGBoost model for chronic kidney disease diagnosis. IEEE/ACM Trans. Comput. Biol. Bioinf. **17**(6), 2131–2140 (2019)
12. Tong, W., Hong, H., Fang, H., Xie, Q., Perkins, R.: Decision forest: combining the predictions of multiple independent decision tree models. J. Chem. Inf. Comput. Sci. **43**(2), 525–531 (2003)
13. Aguiar, E., Ambrose, G.A.A., Chawla, N.V., Goodrich, V., Brockman, J.: Engagement vs performance: using electronic portfolios to predict first semester engineering student persistence. J. Learn. Analyt. **1**, 7–33 (2014)
14. Guo, T., Zhao, W., Alrashoud, M., Tolba, A., Firmin, S., Xia, F.: Multimodal educational data fusion for students' mental health detection. IEEE Access **10**, 70370–70382 (2022)
15. Moon, J., Ke, F., Sokolikj, Z., Dahlstrom-Hakki, I.: Multimodal data fusion to track students' distress during educational gameplay. J. Learn. Analyt. **9**(3), 75–87 (2022)
16. Chango, W., Cerezo, R., Sanchez-Santillan, M., et al.: Improving prediction of students' performance in intelligent tutoring systems using attribute selection and ensembles of different multimodal data sources. J. Comput. High. Educ. **33**, 614–634 (2021)

Designing a Fault Detection System for Wind Turbine Control Monitoring Using CEP

Enrique Brazález, Gregorio Díaz$^{(\boxtimes)}$, Hermenegilda Macià, and Valentín Valero

School of Computer Science, Universidad de Castilla-La Mancha, Albacete, Spain
{Enrique.Brazalez,Gregorio.Diaz,Hermenegilda.Macia,
Valentin.Valero}@uclm.es

Abstract. Renewal energies are key to face the challenges of climate change. The power generation using Wind Turbines (WT) is among the technologies with higher growth during the last year. The Operation and Maintenance (OM) of WT using condition-monitoring (CM) to minimize failures determine the cost of the produced energy and therefore its efficiency. In this work, we present the design of a Fault Detection System (FDS) for WTCM using Complex Event Processing (CEP) technology to analyze the data streams of a wind farm in real-time. Data streams are provided by the sensors and the Supervisory Control and Data Acquisition (SCADA) system installed in the WT farms. This information is analyzed to determine failures using the stability in the produced power. Changes in this stability are detected by CEP patterns deployed in a CEP engine. A real case scenario is used to illustrate this design. It consists of 30 WTs operated by a private company and shows how this approach can help to plan operation and maintenance actions.

Keywords: Complex Event Processing · Event Processing Language · Wind Energy · Wind Turbine Maintenance · Condition Monitoring · Fault Detection System

1 Introduction

A. Betz discovered the theory behind wind energy and published it in his book 'Wind-Energie' in 1926 [3], although he could not imagine how this technology would affect our current society. Nowadays, deep emissions of gas pollutants to the atmosphere must be reduced globally, as established by the Intergovernmental Panel on Climate Change (IPCC), and this recommendation includes all sectors and regions. Otherwise, the objective of maintaining the warming below 1.5 °C will be impossible to achieve. The development of renewable energies and

This work was supported in part by the Spanish Ministry of Science and Innovation and the European Union FEDER Funds under grants PID2021-122215NB-C32, and the UCLM group research grant cofinanced with European Union FEDER funds with reference 2022-GRIN-34113.

I. Rojas et al. (Eds.): IWANN 2023, LNCS 14135, pp. 304–314, 2023.
https://doi.org/10.1007/978-3-031-43078-7_25

other associated technologies, such as power batteries are a promising solution to alleviate this problem. Figure 1 shows the electricity generated in the world from 2000 classified by its source. Wind energy is classified among the renewable technologies with a higher grow rate.

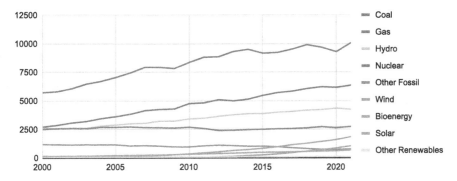

Fig. 1. World electricity generation by source (data source: EMBER, CC BY-SA 4.0)

In this context of high demand, the cost reduction at Operation and Maintenance (OM) level plays a significant role in Wind Turbine (WT) maintenance strategies. The OM reaches 3% of WT lifetime and a share between 10% to 20% of the total cost to produce the electricity [17]. Two strategies are followed in this scenario: the development of better WT designs and the application of condition-monitoring (CM) techniques. We focus on the Wind Turbine Condition Monitoring (WTCM) in this paper. The designed strategies for WTCM can use the streams of data generated by the Supervisory Control and Data Acquisition (SCADA) systems integrated in WTs [12,16]. In recent years, several approaches have been envisioned to analyze these streams using Artificial Intelligence (AI) [20], for instance using Big Data (BD) or Machine Learning (ML) technologies [15]. BD has become a main factor for the evolution of information and communication technology, and Complex Event Processing (CEP) [7,11] is a key technology in this field. CEP allows the correlation of data using patterns to detect relationships between data streams. It is especially useful for analyzing large volumes of data detecting situations of interest in real-time. It is considered as a *Fast Data* technology. CEP has been already integrated in industrial environments to detect errors and anomalies conforming a Fault Detection System (FDS) [2,10,14].

In this work, we present an early FDS in the context of WTCM using CEP technologies to analyze the data streams of a wind farm. WTs are equipped with sensors for monitoring purposes providing the data streams to the SCADA system. In most cases, the information gathered by these sensors is only used in case of an incidence several months afterward to perform a postmortem analysis. This situation can be reversed using CEP technologies to provide real-time analysis and detect anomalies before they can develop into a failure. For instance, we can detect exhaustion problems using the information regarding the production. A FDS capable of detecting this anomaly can assist in the decision-making process and provide the maintenance team with a tool to anticipate failures. Our proposal uses a CEP engine with real-time capabilities to produce alerts. These alerts are detected using patterns specified in an Event Processing Language (EPL) [6]. A set of event patterns[1] are introduced to detect sudden changes in WT throughput, which is a clear sign of WT exhaustion.

The contribution of this work is an early FDS with CEP capabilities to analyze the information streams of a wind turbine using the patterns proposed. This system will provide significant information to maintenance teams to perform a preventive maintenance, which can be scheduled in advance during periods of low wind gusts. Next, we summarize these contributions:

- the design of an early FDS for WTCM,
- the inclusion of a CEP engine in the FDS,
- a set of complex event patterns to detect anomalies in the WT throughput,
- the analysis of a real case scenario.

The structure of this work is as follows. Related works are presented in Sect. 2. A brief introduction to WTs and CEP is presented in Sect. 3. The methodology we propose and the event patterns defined are presented in Sect. 4. A use case is then shown in Sect. 5, and finally, the conclusions and the plans of future work are presented in Sect. 6.

2   Related Work

A broad state of the art of WTCM was made by Tchakoua et al. [17], and more recently by Ma and Juan [12]. CM has also been studied using the information gathered by the SCADA system. For instance, Yang et al. in [22] analyzed the correlation between the different parameters gathered by this system using a laboratory test bed consisting of a WT simulator of 30 kW.

Regarding studies of signal processing over time, Marti-Puig et al. [13] analyzed a set of WTs deployed in the same geographical area. They found that it is possible to detect failures by crossing the information obtained from neighbor WTs. This finding supports the patterns we have considered in this paper, because they use a similar neighbor-based strategy. Wen et al. [21] followed a similar approach, but they analyzed the wind speed data in correlation with adjacent WTs using a dynamic power curve fitting to detect deviations. They performed a statistical analysis extracted from a real wind farm to show the effectiveness of the technique proposed to detect failures.

The use and integration of CEP in FDS is not new, as a valuable tool to detect errors and anomalies in the industry. Ait-Alla et al. [2] used CEP to detect machine failures, Khodabakhsh et al. [10] applied CEP to Oil Refineries, and Mehdi et al. [14] applied it in the scope of industrial turbines through KPIs (Kep Performance Indicators). However, to our knowledge, there is no work applying CEP to design a FDS for WTCM.

---

[1] In this paper we use ESPER EPL to define and process the patterns.

3   Background

This section shows the main concepts regarding the WT maintenance and CEP.

3.1   Wind Turbine Maintenance

The maintenance strategy for WTs is the so-called Reliability-Centered Maintenance (RCM), which is based on preventive maintenance guided by parameters,
such as performance. In this context, the balance between corrective and scheduled maintenance is known as Condition-Monitoring (CM) [9]. Typically, the life
of a WT is 20 years, and the failure rate during that time shows a bathtub curve,
with failures concentrated in the early and late periods. Statistical studies shows
that reliability and availability of WT depend on age, size, weather, etc., and
the application of proper CM Techniques (CMTs) increase them.

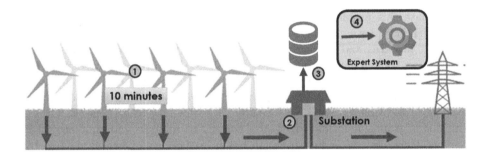

Fig. 2. CEP Wind Farm Environment.

Figure 2 shows the steps to perform CM. It includes the data acquisition
using sensors (1–2), information processing (3–4) and status retrieval (output in
4). The information considered in CM includes not only the current information,
but also the historical information gathered about the system status in the past.
Fault diagnosis can imply corrective maintenance to fix a detected fault, but the
objective of CM is focused on preventing the fault. This diagnosis should provide
information regarding the location and severity of the fault. Furthermore, an
early detection providing warning signs is desirable in these systems.

CMTs for WTs consider several subsystems to analyze different aspects. In
this work, we focus on the CM as a WT Global System, also known as Nonin-
trusive CMTs, and the aspect considered is the performance monitoring (PM).
The parameters considered in PM are capacity factors, power, wind velocity,
rotor speed, and blade angle, which are compared with those specified by the
manufacturer to analyze the WT efficiency. These parameters can be used to
determine significant deviations [19]. Sensors are in charge to measure most of
these parameters, which are integrated in a SCADA system.

The trend in WT industry is the implementation of Intelligent Machine Health Management (IMHM), which is considered the fourth-generation maintenance strategy. The ultimate goal is to have autonomous intelligent systems that can make decisions without human intervention.

3.2 Complex Event Processing

Users can analyze and correlate large volumes of data using CEP to detect situations of interest in a real time domain. The data considered by CEP is formatted as events, and the situations of interest are detected using the so-called event patterns. Specific languages are used to specify the event patterns, which are known as Event Processing Languages (EPLs) [7]. Patterns are processed by CEP engines, where they are deployed. Once a situation is detected it becomes as a new complex event, which can be used again as a building block for more elaborated patterns.

However, in practice, domain experts are not aware with these technologies. To alleviate this problem, Boubeta-Puig et al. [4] presented the MEdit4CEP tool, which provides a graphical modeling editor for specifying the CEP domain, the event patterns and the actions to be performed. The models created with this tool can then be automatically translated into the corresponding Esper EPL code [8]. In addition, Valero et al. presented the MEdit4CEP-BPCPN approach [18], in which the MEdit4CEP tool was extended using a further translation into Prioritized Colored Petri Nets with Black sequencing transitions (BPCPN), which is a graphical formalism that allows us to analyze and simulate CEP systems, thus allowing a semantic validation of the event patterns defined by the user using Colored Petri Nets (CPNs), and specifically using CPN Tools [1].

4 Approach

Figure 2 presents the WT farm scenario of the FDS proposed in this work, in which the information and electricity flow are shown. The steps followed in this FDS proposal are the following:

(1) Each WT releases a pulse of information every ten minutes.
(2) The stream information flows to the substations.
(3) This information is pre-processed as a flow of simple events.
(4) These simple events are analyzed using a CEP engine to detect the situations of interest and produce the corresponding complex events, which provide us with real time alerts, thus allowing the implementation of a preventive maintenance protocol.

The power produced by a WT in a wind farm presents a stable trend over time under similar wind conditions. As an illustration, in a wind farm with five neighbor turbines disposed in different places, we will have that some of them will produce more electricity than others, but this difference is kept over time as we have seen in [13,21]. This is caused because of different variables, such as altitude, atmospheric pressure, environmental temperature and other factors related to the position of a WT within the wind farm. But we can observe that abnormal behaviors imply fast drops in the produced power. In this situation, the likelihood of a failure is high. To detect these anomalies, a diary production ranking is considered to analyze position changes. Thus, we use CEP to detect a rush change in the production by analyzing this ranking, and these event patterns are encoded in Esper EPL.

Figure 3 depicts the detailed diagram[2] of this proposal. The data streams produced by the sensors in the WTs are gathered by the SCADA system. Specifically, we consider the following information for the simple events produced by the sensors: the WT identifier, the measurement time, the wind speed and the electricity produced. The CEP module consists of a set of patterns and the CEP engine. The engine analyzes the information using the patterns and detects sudden changes in the daily ranking. In this figure, an alert for WT 2 has been raised to inform the maintenance team that a maintenance for this WT must be scheduled.

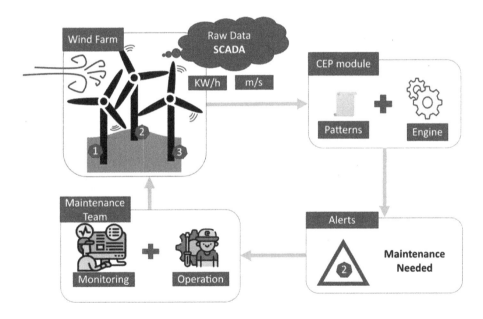

Fig. 3. General diagram of FDS4WTCM-CEP design

Next, we present the EPL schema and patterns encoded for this approach. Figure 4a shows a graphical view of the schema used, as produced by the MEdit4CEP tool, and Fig. 4b its corresponding EPL code, as produced by the tool. According to this schema, as indicated above, simple events consist of the

---

[2] This diagram has been designed using images from Flaticon.com.

identifier (id) of each WT, the time-stamp (time), the number of wind turbines in the wind farm (numAeros), the average wind speed (wind, in m/s) and the throughput power (power) produced (KW/h). This EPL schema contains enough information to detect a sudden ranking change.

Specifically, three event patterns are defined for this purpose. The first pattern is `RankingProd` (Fig. 5a), which produces a descending list with the average daily production for each WT. A 1-day batch window of Esper EPL is used to produce this information. The corresponding EPL code is shown in Fig. 5b. The second pattern is `Scale` (Fig. 6a), which takes as input the complex events produced by the first pattern and produces the WT ranking in the power production. It generates the daily position of each WT in the ranking list. Figure 6b shows the corresponding EPL code. Finally, the third pattern is `Transition` (Fig. 7), which uses this information (complex events produced by pattern *Scale*) to detect sudden changes in the ranking with respect to the previous day. It computes the position difference in the ranking with respect to the previous day.

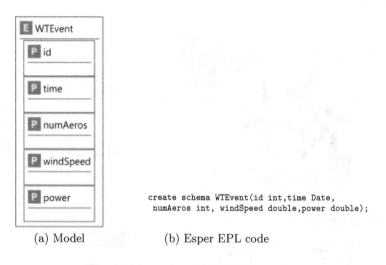

```
create schema WTEvent(id int,time Date,
    numAeros int, windSpeed double,power double);
```

(a) Model   (b) Esper EPL code

Fig. 4. Schema for a WT simple event.

5   Use Case

We consider a real WT farm that consists of 30 WTs, GE 1.5 MW. and the production data is gathered from its SCADA system. The maintenance of this farm is provided by Ingeteam Corporación S.A., which is a company established in 24 countries, with more than 4,000 employees, leader in renewable energy generation (wind, solar, and hydroelectric) and their provisioning of operation and maintenance includes more than 18 GW of maintained capacity. The WT 1.5 MW series of General Electric consists of three upwind blades and a horizontal axis wind turbine with a rated capacity of 1.5-MW. The rotor operates

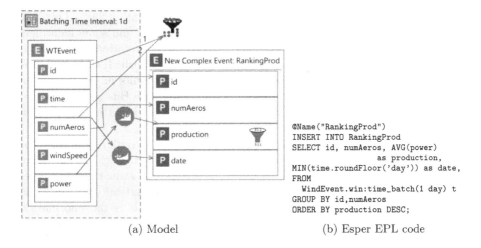

(a) Model

```
@Name("RankingProd")
INSERT INTO RankingProd
SELECT id, numAeros, AVG(power)
                   as production,
MIN(time.roundFloor('day')) as date,
FROM
   WindEvent.win:time_batch(1 day) t
GROUP BY id,numAeros
ORDER BY production DESC;
```

(b) Esper EPL code

Fig. 5. Pattern RankingProd.

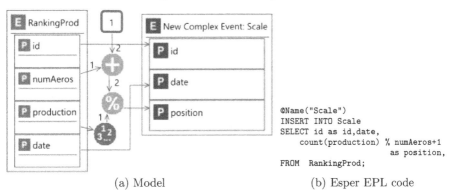

(a) Model

```
@Name("Scale")
INSERT INTO Scale
SELECT id as id,date,
      count(production) % numAeros+1
                        as position,
FROM  RankingProd;
```

(b) Esper EPL code

Fig. 6. Pattern Scale.

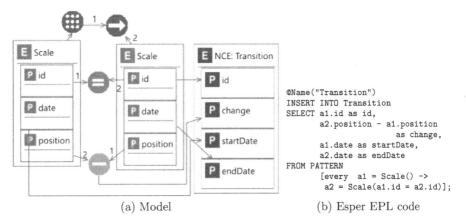

(a) Model

```
@Name("Transition")
INSERT INTO Transition
SELECT a1.id as id,
       a2.position - a1.position
                      as change,
       a1.date as startDate,
       a2.date as endDate
FROM PATTERN
     [every  a1 = Scale() ->
      a2 = Scale(a1.id = a2.id)];
```

(b) Esper EPL code

Fig. 7. Pattern Transition.

in an upwind configuration at 10 to 20 revolutions per minute (rpm). This WT produces an information flow consisting on 70 parameters and some identifiers and time stamps every ten minutes. This includes the information we have considered as simple events, but other information is also published, such as alarms and warnings. These data has been injected as simple events in an instance of Mulesoft Anypoint provided by MuleSoft®, an Enterprise Bus Service (ESB). We have installed a CEP engine in the ESB, where the EPL patterns have been deployed.

To illustrate the proposal, we consider a sample of the ranking obtained for 4 WTs (from the 30 WTs of the farm), which is shown in Fig. 8. This figure shows the ranking for these 4 WTs during 9 days, from January 4th to January 12th. The position in the ranking is shown in the vertical axis from 1 to 30, while the dates are shown in the horizontal axis. This is the information provided by the pattern Scale. As we can see, each WT basically maintains its position in the ranking over time, but some exceptions are observed. From January 5th to January 6th, there is a sudden change in the ranking for WT 4, while the other turbines remain in similar positions. This WT has experienced a drastic change from position 14th to 27th in the ranking. Therefore, the pattern Transition computes a difference of 13 positions (change attribute is –13). The operation and maintenance team should then study this drastic change and plan the appropriate actions. We assume that maintenance actions have been performed on January 7th, and a regular power production has been retrieved again on January 8th. A similiar drastic change in the ranking can be observed for WTs 1 and 2 on January 9th, but in this case the cause is a change in the wind configuration and the power production recovers in the following journey for both of them. A similar situation can be observed for the four WTs in the last days of the sample.

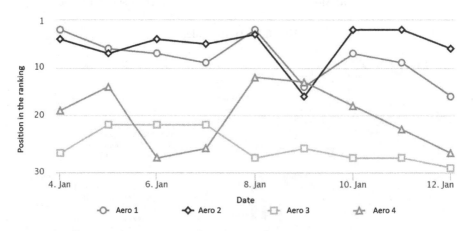

Fig. 8. Rankings from the 4/1/2018 to 12/1/2018

6   Conclusions and Future Work

This work presents the design of an FDS in the context of WTCM using CEP technologies to analyze the data streams of a wind farm. Sensors are installed in WTs to monitor their status via the SCADA system. We demonstrate that the use of CEP technologies can provide real-time analysis and detect abnormal WT behaviors.

Specifically, we have shown how the power produced by WTs can be used to detect exhaustion problems. Maintenance team can take advantage of this technique helping them in the decision-making process to plan WT maintenance. A CEP engine with real-time capabilities is used here to detect these situation of interest using a set of EPL patterns. These patterns can detect sudden changes in the power production using a daily ranking. The daily ranking is analyzed to study the changes with respect to the previous day.

This information can be analyzed by the operation and maintenance team to help them determine the WT status and implement preventive measures to increase the WT lifespan. As an illustration, the data obtained from a real use case of a farm with 30 WTs has been analyzed, showing the WT rankings and how the FDS would inform of the relevant changes in the ranking positions.

As future work, we intend to develop and deploy this proposal in a real wind farm system by implementing a SOA 2.0 architecture. New patterns can then be considered in this development to consider other relevant information provided by the WT sensors and the SCADA system. In addition, the techniques developed in this work can be combined with other technologies such as Fuzzy Logic [5].

Acknowledgement. Thanks to Ingeteam Power Technology S.A for its collaboration, specially regarding the data used in this work. This work was envisioned and performed during Enrique Brazález's intership at this company.

References

1. CPN Tools. http://www.cpntools.org/. Accessed 15 Mar 2023
2. Ait-Alla, A., Lütjen, M., Lewandowski, M., Freitag, M., Thoben, K.D.: Real-time fault detection for advanced maintenance of sustainable technical systems. Procedia CIRP **41**, 295–300 (2016). https://doi.org/10.1016/j.procir.2016.01.015
3. Betz, A.: Wind-Energie und Ihre Ausnutzung durch Windmühlen. Göttingen, Germany (1926)
4. Boubeta-Puig, J., Ortiz, G., Medina-Bulo, I.: MEdit4CEP: a model-driven solution for real-time decision making in SOA 2.0. Knowl.-Based Syst. **89**, 97–112 (2015). https://doi.org/10.1016/j.knosys.2015.06.021
5. Brazález, E., Maciá, H., Díaz, G., Baeza-Romero, M., Valero, E., Valero, V.: Fume: an air quality decision support system for cities based on cep technology and fuzzy logic. Appl. Soft Comput. **129**, 109536 (2022). https://doi.org/10.1016/j.asoc.2022.109536
6. Chandy, K., Schulte, W.: Event Processing: Designing IT Systems for Agile Companies, 1st edn. McGraw-Hill Inc., New York (2010)

7. Cugola, G., Margara, A.: Processing flows of information: from data stream to complex event processing. ACM Comput. Surv. **44**(3), 1–62 (2012). https://doi.org/10.1145/2187671.2187677

8. EsperTech: Esper EPL Online. http://esper-epl-tryout.appspot.com/epltryout/index.html. Accessed 15 Mar 2023

9. Fischer, K., Besnard, F., Bertling, L.: Reliability-centered maintenance for wind turbines based on statistical analysis and practical experience. IEEE Trans. Energy Conv. **27**(1), 184–195 (2012). https://doi.org/10.1109/TEC.2011.2176129

10. Khodabakhsh, A., Ari, I., Bakir, M., Ercan, A.O.: Multivariate sensor data analysis for oil refineries and multi-mode identification of system behavior in real-time. IEEE Access **6**, 63489–64405 (2018). https://doi.org/10.1109/ACCESS.2018.2877097

11. Luckham, D.C.: The Power of Events: An Introduction to Complex Event Processing in Distributed Enterprise Systems. Addison-Wesley Longman Publishing Co., Inc., Boston (2001)

12. Ma, J., Yuan, Y.: Application of SCADA data in wind turbine fault detection - a review. Sensor Rev. **43**(1), 1–11 (2023). https://doi.org/10.1108/SR-06-2022-0255

13. Marti-Puig, P., Cusidó, J., Lozano, F.J., Serra-Serra, M., Caiafa, C.F., Solé-Casals, J.: Detection of wind turbine failures through cross-information between neighbouring turbines. Appl. Sci. (Switzerland) **12**(19), 9491 (2022). https://doi.org/10.3390/app12199491

14. Mehdi, G., et al.: Model-based approach to automated calculation of key performance indicators for industrial turbines. In: Proceedings of the Annual Conference of the Prognostics and Health Management Society, PHM, pp. 632–639 (2015). https://doi.org/10.36001/phmconf.2015.v7i1.2599

15. Méndez, M., Merayo, M.G., Núñez, M.: Long-term traffic flow forecasting using a hybrid cnn-bilstm model. Eng. Appl. Artif. Intell. **121**, 106041 (2023). https://doi.org/10.1016/j.engappai.2023.106041

16. Pandit, R., AstolfC;, D., Hong, J., Infield, D., Santos, M.: SCADA data for wind turbine data-driven condition/performance monitoring: a review on state-of-art, challenges and future trends. Wind Eng. **I**(20), 0309524X221124031 (2022). https://doi.org/10.1177/0309524X221124031

17. Tchakoua, P., Wamkeue, R., Ouhrouche, M., Slaoui-Hasnaoui, F., Tameghe, T.A., Ekemb, G.: Wind turbine condition monitoring: state-of-the-art review, new trends, and future challenges. Energies **7**(4), 2595–2630 (2014). https://doi.org/10.3390/en7042595

18. Valero, V., Díaz, G., Boubeta-Puig, J., Macià, H., Brazález, E.: A compositional approach for complex event pattern modeling and transformation to colored petri nets with black sequencing transitions. IEEE Trans. Softw. Eng. **48**(7), 2584–2605 (2022). https://doi.org/10.1109/TSE.2021.3065584

19. Verbruggen, T.W.: Wind Turbine Operation and Maintenance based on Condition Monitoring WT-omega. Final report. Technical Report ECN-C-03-047 (2003). https://www.osti.gov/etdeweb/biblio/20376548

20. Vidal, Y.: Artificial intelligence for wind turbine condition monitoring. Energies **16**(4), 1632 (2023). https://doi.org/10.3390/en16041632

21. Wen, W., Liu, Y., Sun, R., Liu, Y.: Research on anomaly detection of wind farm scada wind speed data. Energies **15**(16), 5869 (2022). https://doi.org/10.3390/en15165869

22. Yang, W., Court, R., Jiang, J.: Wind turbine condition monitoring by the approach of scada data analysis. Renew. Energy **53**, 365–376 (2013). https://doi.org/10.1016/j.renene.2012.11.030

Halyomorpha Halys Detection in Orchard from UAV Images Using Convolutional Neural Networks

Alexandru Dinca[1]([✉]), Dan Popescu[1], Cristina Maria Pinotti[2], Loretta Ichim[1], Lorenzo Palazzetti[3], and Nicoleta Angelescu[4]

[1] University POLITEHNICA Bucharest, Bucharest, Romania
marius.dinca1411@stud.acs.upb.ro, {dan.popescu, loretta.ichim}@upb.ro
[2] University of Perugia, Perugia, Italy
cristina.pinotti@unipg.it
[3] University of Florence, Florence, Italy
lorenzo.palazzetti@unifi.it
[4] Valahia University of Targoviste, Targoviste, Romania
nicoleta.angelescu@valahia.ro

Abstract. Halyomorpha Halys, commonly known as the brown marmorated stink bug, is an invasive insect that causes significant damage in orchards. Neural networks have the potential to improve insect pest detection and classification in modern agriculture, which can lead to better pest management. The detection of these insects in orchards using drones imposes special problems because the images are taken from a limited distance and the foliage of the trees makes detection difficult. In this article, we studied the possibility of detecting the respective insects using the latest generation YOLOv8 neural networks and compared the results with the well-known YOLOv5 network. The results were obviously better for YOLOv8 (accuracy of 94.55%). However, satisfactory results were also obtained in the case of YOLOv5 (accuracy of 90.91%).

Keywords: Convolutional Neural Networks · Harmful Insects · Pest Detection · Orchard

1 Introduction

The insect pest causes damage to crop, livestock, forests, or other natural resources, causing economic losses or ecological imbalances. Halyomorpha Halys (HH), commonly known as the brown marmorated stink bug, is a species of insect in the family Pentatomidae. It is native to East Asia, including China, Japan, and Korea, but has become an invasive species in many parts of the world, including North America and Europe [1]. Following this context there are many studies available for HH and its effects as an insect pest. They provide valuable information on the behavior, biology, and management of the brown marmorated stink bug, as well as its impact on various crops and industries [2–5].

I. Rojas et al. (Eds.): IWANN 2023, LNCS 14135, pp. 315–326, 2023.
https://doi.org/10.1007/978-3-031-43078-7_26

HHs feed on a variety of plants, including fruits, vegetables, and ornamental plants, and can cause significant damage to crop [2]. Some of the main crops that are affected by this invasive species include apples, grapes, peaches, and soybeans [3]. The invasive nature of the brown marmorated stink bug has caused concern among farmers and homeowners, and efforts are being made to control its spread and limit its impact on agriculture and the environment [4].

Regarding insect pests, effective management strategies include technological, biological, and chemical methods to control their populations and reduce their impact on ecosystems and agriculture [5]. Deep learning and neural networks are modern ways of detecting insect pests in the field. In recent years, there has been a growing interest in using machine learning techniques to automate pest detection in agriculture [6]. However, it is important to note that these models require significant amounts of high-quality training data and computational resources to develop and deploy [7–12]. Additionally, the accuracy of the model's predictions can be affected by factors such as lighting conditions, camera angles, and the diversity of insect pests in the field. Therefore, careful calibration and validation of the model are necessary to ensure reliable pest detection.

Deep learning has become popular in modern agriculture due to its ability to analyze and process large amounts of data efficiently. With the increasing availability of sensors and data collection devices, agriculture has become a data-rich industry. Deep learning algorithms can process this data and extract valuable insights that can help farmers make informed decisions. For precision agriculture, deep learning algorithms can analyze various environmental and crop-related data [6]. On the other hand, deep learning algorithms can analyze images of crops taken from drones, digital cameras, or other devices and identify patterns of growth, disease, and pest infestation [7]. This information can help farmers make timely decisions about crop management. To detect insect pests using deep learning and neural networks, the need is to train the models using large datasets of images of both healthy and infested crops [8, 9, 11]. The model would learn to recognize patterns in the images that distinguish healthy plants from infested ones. Once trained, the model can then be used to automatically detect pest infestations in new images of crops taken in the field [12].

Neural networks have the potential to improve insect pest detection and classification in modern agriculture, which can lead to better pest management and higher crop yields. One common approach is to use image classification techniques based on convolutional neural networks (CNNs) to identify and classify insect pests. This involves training a neural network on a large dataset of images of crops and insects, where the images are labeled with information about the presence of insect pests. The neural network learns to recognize the patterns and features of the insects in the images and can then classify new images as containing a specific insect pest or not.

This paper is organized in four sections, presenting the details that were the basis of the study. Considering the presented introduction section, the rest of the sections are organized as follows: Sect. 2, Materials and Methods, presents the data set used to develop the present work in relation to the object detection task for HH, the models used are presented. On the other hand, the hardware and software part used to implement neural networks is also noted. Section 3 presents the experimental results, performances, and corresponding discussions. Finally, a short section of conclusions is presented.

2   Material and Methods

2.1   Dataset Used

To develop and train accurate neural networks for pest detection, a substantial amount of data is required, which can be a challenge in some agricultural contexts. A proprietary dataset was created for this work using real images taken in the orchard field. The dataset tracks the presence of the HHs insect pest in the images. A custom dataset was designed and implemented for the purpose of YOLO (You Only Look Once) family to HH detection. Transfer learning applied to a custom dataset with images in the field and fine-tuning hyperparameters were followed for this study. The data set is proprietary, and no external or public image databases were queried or included in this study.

Because this work relies on an object detection task, LabelImg [13] was the choice to create the labeled data to train the YOLOv5 and YOLOv8 models to detect object classes in our case. The labels were exported to YOLO format, using a .txt file to describe the manually bounding box labeled objects in each image, with one object in each row. A dataset partition was implemented to group images in each directory for training, validation, and testing. The created dataset consists of 312 images divided as follows: 70% for training (218 images), 20% for validation (62 images), and finally 10% for testing (32 images). For the data set created, it is important to mention that the images may contain one or more instances of the insect of interest.

The images in the dataset depict a real context represented by various orchard images taken using a precision camera drone. The images taken from the drone went through a pre-processing step before being fed into the training part of the network. An example of images from the training set can be viewed in Fig. 1 below. In this way, manual identification of regions of interest was pursued and image patches of size 640 × 640 pixels were gathered to create the mentioned dataset. The image size followed is the one from the official YOLOv8 documentation.

Fig. 1. Example of training images.

2.2 Neural Networks Used

Pre-trained models from YOLO object detection family were used for training and evaluation for HH detection using digital images in the field. YOLO is an object detection algorithm developed in [14]. It was first introduced in 2015 and has since then become one of the most popular and widely used object detection algorithms in computer vision. The YOLO algorithm is based on a deep convolutional neural network (CNN) architecture that uses a single pass to perform object detection. Unlike other object detection algorithms that rely on region proposals, YOLO uses a grid of cells to divide the image into smaller regions and predicts the object class and location for each cell. The algorithm is trained on large datasets, such as the COCO (Common Objects in Context) dataset and has been designed to be fast and accurate. Today, YOLO is used in a wide range of applications, including autonomous vehicles, security systems, and video analytics. The YOLO algorithm has been integrated into many popular computer vision frameworks, such as OpenCV, TensorFlow, or PyTorch, and has become a standard benchmark for object detection in computer vision. The key feature of the YOLO family is its ability to perform object detection in real-time on GPU hardware.

YOLOv5 is a cutting-edge deep learning-based object recognition method based on the YOLO family of object detection models [15]. The structure of the YOLOv5 model can be broken down into several main components: input processing, backbone network, neck network, and detection head. The input image is first preprocessed by resizing it to a fixed size and then normalized by subtracting the mean pixel value and dividing it by the standard deviation. The backbone network is responsible for extracting features from the input image. YOLOv5 uses a CSPNet backbone network, which consists of a series of convolutional layers that gradually reduce the spatial resolution of the feature maps while increasing their depth. Next, the neck network is placed for combining the features extracted by the backbone and producing a set of feature maps with rich spatial information. In the case of YOLOv5 the spatial pyramid pooling (SPP) module is used, pooling features at multiple scales and capturing objects of different sizes. The detection head is the next component for YOLOv5, using a multi-scale anchor-based detection head. This one predicts a set of bounding boxes and class probabilities for each anchor at multiple scales. Using non-maximum suppression (NMS) the predicted boxes are filtered and refined to compute the final detections.

The current state-of-the-art (SOTA) release, YOLOv8, is an upgraded version from the YOLO family that has improved accuracy and speed compared to the previous versions [16]. YOLOv8 has a more complex architecture that includes multiple neural network layers and skip connections to better model object relationships and features. It has improved anchor box design and data augmentation techniques, resulting in better accuracy. At the time of the development of the present study, there was no official published paper for the new YOLOv8. The ideas developed and the analysis of the YOLOv8 architecture were based on the online documentation made available by the team that implemented the new version and the research of the open-source code [16].

As a quick summary, developed by Ultralytics, YOLOv8 offers a boost in performance and flexibility over its predecessors and still follows a wide area in object detection and image classification, and segmentation tasks. Following the Ultralytics release, all YOLOv8 models could be used as pre-trained models under different availability – YOLOv8n, YOLOv8s, YOLOv8m, YOLOv8l, YOLOv8x. The models used for detection and segmentation were tested and evaluated over MS COCO (Microsoft Common Objects in Context) dataset, while the models used for classification are pre-trained on ImageNet data. The extensibility of YOLOv8 is an important feature.

The current SOTA is built as a framework that supports all prior YOLO versions, having the option to move between them and evaluate their performance. To note the extensive changes starting from the previous versions and the novelty part, YOLOV8 is described using a new head with anchor-free detection and introducing a new backbone and loss function. The anchorless addition will not follow any anchor box offset and will detect the center of the object. Also, this addition will have a big impact on Non-Maximum Suppression (NMS). Closing the mosaic augmentation was also implemented for the last epochs in the training steps. The PyTorch integration and a well-documented CLI could also describe the innovations in this YOLOv8 release.

For this study, YOLOv8 was used as the main model for HH detection using images from a pear orchard. The choice to implement also the YOLOv5 model was made to describe a comparison area with the new version in the YOLO family - YOLOv8.

As a programming language, Python version 3.10 was used for the implementation. PyTorch was considered as deep-learning framework in this context with Miniconda and Jupyter Notebook to define and organize the code. The hardware part was defined by a system that aimed to integrate a GPU component for training and evaluating the used models. In this regard, the system features an NVIDIA RTX 2080Ti GPU with 11GB memory and CUDA V11.7. The other components of the system were represented by 128GB of RAM and an Intel Core i9-11900K processor. The operating system was represented by Ubuntu V22.

3   Experimental Results and Discussions

Training experiments are based on several model settings using the YOLOv8 and YOLOv5 architectures: the number of epochs was 300, the models used were YOLOv8m and YOLOv5m, and the input image size for object detection was as each model suggests 640×640 pixels. The custom class for object detection to highlight the HH pest was denoted with the acronym: HH. A configuration YAML file was added to define the paths for each directory involved in the training and evaluation part. The images in the test and validation set are different from the ones the model analyzes in the training phase so that the evaluation is robust and based on new contexts and information. Both models were trained and evaluated using the same dataset. An example with validation images is shown in Fig. 2.

For both models used, training and validation accuracy and loss plots are shown in Fig. 3 for YOLOv8m and in Fig. 4 for YOLOv5m to describe the model performance. Following the object detection task, the final output for each model is a list of bounding boxes and associated class labels, along with their confidence scores. Analyzing the training results, we can note a great accuracy for the YOLOv8m model using transfer learning for object detection regarding the HH pest from real orchard context images.

In the same scenario for HH detection, Fig. 5 shows the computed precision-recall curve obtained for YOLOv8m and Fig. 6 for YOLOv5m. For HH object detection, this

Fig. 2. Example images from the validation dataset.

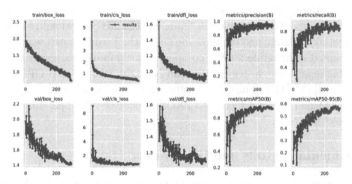

Fig. 3. YOLOv8m training and validation results.

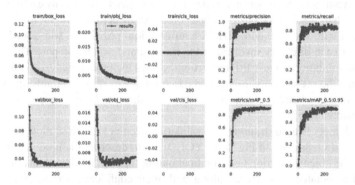

Fig. 4. YOLOv5m training and validation results.

is a valuable tool for evaluating algorithm performance and making decisions about the threshold for predictions. In practice, it is commonly used to evaluate the performance of an object detection algorithm and to make decisions about the appropriate threshold for predictions. The precision-recall curve is plotted as a scatter plot with precision on the y-axis and recall on the x-axis. Each point on the curve represents a different threshold for predictions, and the curve is created by connecting the points in order of increasing the threshold. Precision is the proportion of true positive predictions out of all positive predictions, while recall is the proportion of true positive predictions out of all true positive cases. As the threshold increases, precision will typically increase while recall will decrease. The ideal curve would have a slope of 1 and achieve a precision of 1 and a recall of 1 at the highest threshold. However, in practice, the precision-recall curve will often be an S-shaped curve, to understand the balance between false positive and false negative errors.

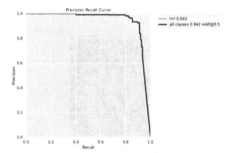   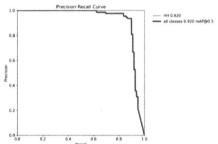

Fig. 5. The precision-recall curve obtained for HH pest detection using YOLOv8m.   **Fig. 6.** The precision-recall curve obtained for HH pest detection using YOLOv5m.

Examples of the prediction part resulting from network training and analysis are shown in Fig. 7 for YOLOv8m and in Fig. 8 for YOLOv5m. It can be seen in the attached prediction figure for YOLOv5m that the model failed to identify the insect for row one, image number two, and in image number three the model identified a spot on the leaf as an HH-type insect. The same reference images were used for both models to illustrate the area of comparison for the obtained performances and predictions.

For both models, we aimed to attach the confusion matrix resulting from the training and evaluation processes to calculate the performance indicators. The figures with representative confusion matrices are attached in Fig. 9 and Fig. 10 for YOLOv8m and YOLOv5m, respectively in the case of learning and validation only with images containing HH and in Fig. 11 and Fig. 12 in the case of testing with images containing HH and images without HH (background).

For testing the network various images were considered. These contain images in which the insect of interest (HH) is present in one or more instances, images with a complex background that do not feature this insect or any other insects, and selected images containing other types of insects to test the ability used models trained for HH detection and rejection of other insects or artifacts at the digital image level. In the case of the models in the present study, it is observed that they manage with high accuracy to

Fig. 7. Examples with HH predictions generated by YOLOv8m.

Fig. 8. Examples with HH predictions generated by YOLOv5m.

identify the insect of interest of the HH class and avoid the detection of artifacts from the background area or representations of other insects.

Testing the models was also done by attaching images in which the insect of interest is partially visible, illustrating various positions of it or being partially visible because it can be covered by various elements (flowers, leaves, branches) or located under various lighting areas (sun rays, shading or insects present at the base of the fruits). Representative example types were also attached to the training and validation datasets. The results are notable in these cases, the models used to be able to locate the insect of interest with great precision. Examples of this type are attached in Fig. 13a. For YOLOv8m and Fig. 13b. For YOLOv5m. We note in the first instance in Fig. 13b. – second row that the YOLOv5m network fails to correctly identify the partially visible insect in the top left. At the same time, the models used can successfully identify the presence of several

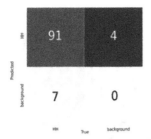

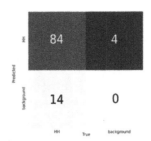

Fig. 9. The confusion matrix obtained for HH pest detection using YOLOv8m after training and validation.

Fig. 10. The confusion matrix obtained for HH pest detection using YOLOv5m after training and validation.

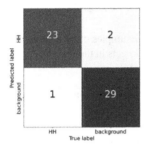

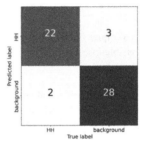

Fig. 11. YOLOv8m confusion matrix obtained using the test dataset.

Fig. 12. YOLOv5m confusion matrix obtained using the test dataset.

insects at the image level. These are exemplified in Fig. 14a for YOLOv8m and Fig. 14b for YOLOv5m. From these examples, the trend of increased accuracy can be observed in the case of the newly developed network of the YOLO family, namely v8, although we can note the performance gradually resulting from the optimization of the v5 model.

On the other hand, the model is not perfect and sometimes manages to identify areas in the images that are not represented by the insect of interest, defining parts of the background area that are represented by the presence of spots on leaves or spots on fruits. Examples of this type are indicated in the previously attached confusion matrices (Fig. 11 and Fig. 12), where the problem of false positive and false negative detections is observed.

To evaluate the performance and robustness of the models used, the test phase aimed to evaluate the networks based on images that represent real contexts, from the orchard level, in which the detection of insects of interest of the HH type is aimed. The images illustrate both areas of complex background where the insect of interest is not present, as well as images illustrating the presence of HH class insects in poses where the insect is difficult to spot, partially hidden, or even difficult to see due to differences in brightness or even of the blur that can be present from the image acquisition area. The test dataset consisted of 31 images with complex backgrounds and 19 images where the HH insect is visible. For images where the insect is visible, the number of insect instances that have

Fig. 13. Examples of HH predictions with partially visible insects.

Fig. 14. Examples of predictions with multiple insects in the image.

been attached in this test dataset is 24. In this case, the insect may be visible in one or more instances at the image level.

To test the network, it was followed to pass the trained and validated models through the new data set to calculate accuracies and prediction areas, and finally a confusion matrix to be populated with the false and true predictions. Since the resulting confusion matrices from the validation dataset of both networks do not provide the necessary indices, it was aimed to create new instances of the confusion matrices based on the test dataset. After testing, the confusion matrices are attached in Fig. 11 for the YOLOv8m model and in Fig. 12 for the YOLOv5m model. Following the results obtained, increased performance can be seen in the case of the new YOLOv8 model. However, satisfactory results were also obtained in the case of YOLOv5, opening a research and development direction for both models. The statistical result indicators were as follows:

- YOLOv8m: Precision = 0.957, Recall = 0.928, mAP50 = 0.942 (training and validation), Accuracy = 94.55% (test).
- YOLOv5m: Precision = 0.954, Recall = 0.857, mAP50 = 0.920 (training and validation), Accuracy = 90.91% (test).

Looking from the perspective of an explanatory analysis to describe the behavior of the YOLO models in relation to the identification of the HH insect, one observes the tendency of the models to detect the object represented by the insect of interest when it is present on leaves or even on fruits. In these cases, the model identification accuracy tends to the maximum. For these image areas, the insect is depicted surrounded by a uniform background, its features being distinctly different.

It is observed that most false detections intervene when the insect is present on the branches or is in occlusion areas being covered by background areas. In these cases, the model tends to direct its attention to the leafy area, as previously mentioned, but the insect being visible in another context, in areas of shading, occlusion or in areas close to branches, the accuracy values start to decrease for such examples. The pixels representing the branch areas appear to have similar colors to those of the HH insects, and occlusions or various shading areas decrease the algorithm's ability to detect the full features of the insect. For these marginal cases, image enhancement and pre-processing techniques can bring out the characteristics of the insect much better, so that it is better differentiated from the background area.

To describe the novelty and the research directions, the implementations presented in this study represent a strong point, with notable results in research on the automatic detection and HH identification, at the level of digital images illustrating real contexts and which are based on techniques that include convolutional neural networks and related software modules. The results obtained based on neural network models bring to the fore the advantages and popularity of these automatic identification techniques. Although there are some studies based on the HH identification at the level of digital images and using deep learning techniques, from our knowledge there are no studies that focus on the identification of this insect of interest at the level of real context by using representative images or data sets in cases of various types of crops that are affected by the presence and massive spread of this insect pest. In this sense, due to the highly invasive characteristics of this population of harmful insects, we emphasize an important and necessary direction of research that can be developed along the way in the precise identification of this type of pest using various techniques and developing the recent studies attached to this topic.

4   Conclusions

In this study, a novel implementation using YOLOv8 was used for HH detection. The new version of YOLOv8 opened the road regarding the state of the art in real-time object detection and improved the accuracy of its predecessors. It is worth noting that the YOLO architecture has undergone several iterations, with each new version introducing improvements in terms of accuracy, speed, and robustness. Each new version of YOLO has built upon the previous versions and has added new features and techniques to improve object detection performance. Although the number of images was relatively small, good results were obtained through transfer learning. Unfortunately, no comparison articles were found for the investigated application.

Acknowledgment. This work was supported by HALY.ID project. HALY.ID is part of ERA-NET Co-fund ICT-AGRI-FOOD, with funding provided by national sources [Funding agency UEFISCDI, project number 202/2020, within PNCDI III] and co-funding by the European Union's Horizon 2020 research and innovation program, Grant Agreement number 862665 ERA-NET ICT-AGRI-FOOD (HALY-ID 862671).

References

1. Haye, T., Weber, D.C.: Special issue on the brown marmorated stink bug, Halyomorpha halys: an emerging pest of global concern. J. Pest Sci. **90**, 987–988 (2017)
2. Ivancic, T., Grohar, M.C., Jakopic, J., Veberic, R., Hudina, M.: Effect of Brown Marmorated Stink Bug (Halyomorpha halys Stål.) Infestation on the Phenolic Response and Quality of Olive Fruits (Olea europaea L.). Agronomy 12 (2022)

3. Aigner, B.L., Kuhar, T.P., Herbert, D.A., Brewster, C.C., Hogue, J.W., Aigner, J.D.: Brown Marmorated Stink Bug (Hemiptera: Pentatomidae) infestations in tree borders and subsequent patterns of abundance in soybean fields. J. Econ. Entomol. **110**(2), 487–490 (2017)
4. Rice, K.B., et al.: Biology, ecology, and management of brown marmorated stink bug (Hemiptera: Pentatomidae). J. Integr. Pest Manag. **5**(3), A1–A13 (2014)
5. Elahe, P., et al.: Population genomic insights into invasion success in a polyphagous agricultural pest, Halyomorpha halys. Molecular Ecol. **32**(1), 138–151 (2023)
6. Li, W., Zheng, T., Yang, Z., Li, M., Sun, C., Yang, X.: Classification and detection of insects from field images using deep learning for smart pest management: a systematic review. Ecol. Inform. **66**, 101460 (2021), ISSN 1574–9541
7. Ayan, E., Erbay, H., Varçın, F.: Crop pest classification with a genetic algorithm-based weighted ensemble of deep convolutional neural networks. Comput. Electron. Agric. (179), 105809 (2020)
8. Bereciartua-Pérez, A., et al.: Insect counting through deep learning-based density maps estimation. Comput. Electron. Agric. **197**, 106933 (2022)
9. Rustia, D.J., et al.: Automatic greenhouse insect pest detection and recognition based on a cascaded deep learning classification method. J. Appl. Entomol. 1–17 (2020)
10. Teng, Y., Zhang, J., Dong, S., Zheng, S., Liu, L.: MSR-RCNN: a multi-class crop pest detection network based on a multi-scale super-resolution feature enhancement module. Front. Plant Sci. **13**, 810546 (2022)
11. Zhichao, S., Dang, H., Liu, Z., Zhou, X.: Detection and identification of stored-grain insects using deep learning: a more effective neural network. IEEE Access **8**, 163703–163714 (2020)
12. Nanni, L., Manfè, A., Maguolo, G., Lumini, A., Brahnam S.: High performing ensemble of convolutional neural networks for insect pest image detection. Ecol. Inform. 67 (2022)
13. Tzutalin. LabelImg. Git code. https://github.com/tzutalin/labelImg (2015)
14. Redmon, J., Divvala, S., Girshick, R., Farhadi, A.: You only look once: Unified, real-time object detection, arXiv:1506.02640 (2015)
15. Jocher, G.: YOLOv5 by Ultralytics (Version 7.0) [Computer software]. https://doi.org/10.5281/zenodo.3908559 (2020)
16. Jocher, G., Chaurasia, A., Qiu, J.: YOLO by Ultralytics (Version 8.0.0) [Computer software]. https://github.com/ultralytics/ultralytics (2023)

Chemistry-Wise Augmentations for Molecule Graph Self-supervised Representation Learning

Evgeniia Ondar[1] and Ilya Makarov[2,3]($\boxtimes$)

[1] Zelinsky Institute of Organic Chemistry, Moscow, Russia
[2] Artificial Intelligence Research Institute (AIRI), Moscow, Russia
[3] AI Center, NUST MISiS, Moscow, Russia
iamakarov@misis.ru

Abstract. In molecular property prediction tasks, graph neural networks have become a widely used tool. Recently, self-supervised learning frameworks, especially contrastive learning, gathered growing attention for the potential to learn molecular representations that generalize to the meaningful chemical space. Unlike supervised, self-supervised learning can directly leverage extensive unlabeled data, which significantly reduces the effort to acquire molecular property labels through costly and time-consuming simulations or experiments. However, most of them do not take into account the unique cheminformatics (e.g., molecular fingerprints) and multi-level molecular graph structures (e.g., functional groups).

In toxicity prediction tasks the molecule substructure can be crucial. Structure alerts (e.g. toxicophores) are studied pretty well and proven to be responsible for different types of toxicity. In this work, we propose chemistry-wise augmentations for a contrastive learning framework. Two augmentations were implemented: (1) toxicophore subgraph removal, and (2) toxicophore subgraph saving. This approach does not violate chemical principles while pushing the model to learn the toxicity-dependent parts of a molecule.

Experiments showed that novel augmentations are more efficient than the random subgraph masking approach usually used in molecular contrastive learning. The performance comparison with other GNN-based frameworks is carried out as well.

Keywords: Graph Neural Networks · Deep Learning, Chemistry

1 Introduction

Molecular property prediction is a challenging task in cheminformatics. In particular, the toxicity prediction of organic molecules is a hot topic in the field of drug discovery.

I. Makarov—The work of Ilya Makarov was made in the framework of the strategic project "Digital Business" within the Strategic Academic Leadership Program "Priority 2030" at NUST MISiS.

It is known, that toxicophores are responsible for different types of toxicity, such as genotoxicity, carcinogenicity, mutagenicity, etc. Toxicophores are chemical functional groups that can show destructive behavior to vital parts of cell. In drug discovery, it can be utilized in different phases of drug optimization: supporting early risk, hazard assessments, the design, synthesis, and ranking of chemicals [23].

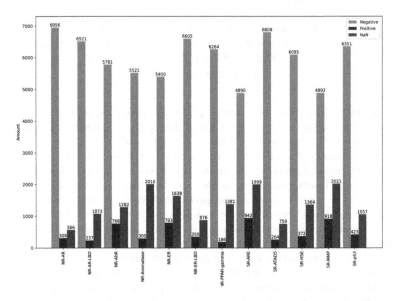

Fig. 1. Tox21 dataset classes histogram plot. Red bars denote the number of non-toxic compounds. Green bars denote the number of toxic compounds. Blue bars denote the number of compounds with no information on toxicity provided. (Color figure online)

To address the problem of rapid, accurate, and cost-efficient toxicity assessment, the Tox21 data challenge was held [5]. Using chemical structure data from 12 of the cellular assays on nuclear receptor signaling and stress response pathways, a Tox21 dataset was created. As shown in Fig. 1, the dataset is imbalanced, and a small amount of data points for each assay is not enough to use neural networks.

A large number of molecular representations and models aimed at predicting molecular properties have been developed so far. Graph representations [21] and graph neural networks [13–16] take a special place among them, since the information on the geometry of the molecule and atomic environment is crucial [29].

Benefiting from a large number of available molecule data [6], self-supervised molecular representation learning has also been developed. The graph-level pre-training showed to be effective by the use of contrastive learning method [2, 11, 12, 17, 26]. Generally, the contrastive method generates new views of an instance from different augmentation. Then, the aim is to maximize the agreement of two jointly sampled positive pairs against the agreement of two independently sampled negative pairs.

You et al. developed GraphCL (graph contrastive learning) framework [34] for learning unsupervised representations of graph data. They showed that data augmentations are crucial in graph contrastive learning, and composing different augmentations benefits more.

Wang et al. [28] extended it to the MolCLR (Molecular Contrastive Learning of Representations) framework based on molecule-wise augmentations and GIN (graph isomorphism network), GCN (graph convolutional network) backbone.

In this work, we propose a novel molecular augmentation strategy based on known toxicophores structure for the contrastive learning framework MolCLR.

2   Related Work

The molecular property prediction task from the toxicity prediction point-view can be solved by the use of many approaches. In this work, we will discuss fingerprint-based GNN methods and contrastive learning GNN methods [16].

2.1   Fingerprint-Based Methods

Molecular fingerprints (descriptors) are some properties of a molecule. It can be physical property (melting/boiling point, molar mass, etc.), count of specific atom type/hybridization type in the molecule, or more complicated as MACCS key. Fingerprints are pretty informative and easy to obtain, so widely used to learn molecular representation.

Xiong et al. developed Attentive FP which is a graph attention mechanism with attentive layers for atom and molecule embeddings [31]. Extension to MPNN has been done by Withnall et al. by the introduction of attention and edge memory schemes [30]. In [3], message passing neural networks (MPNN) as a framework for chemical property prediction was proposed. It consists of a message passing algorithm and aggregation procedure to compute a function of their entire input graph. Yang et al. introduced a directed message passing neural network (D-MPNN), which uses messages associated with directed edges instead of nodes used in MPNN, thereby preventing repeated message passing from the same node [33]. Another approach based on graph neural networks are proposed to solve molecular property prediction task as well [29].

As for now, the discussion was moved beyond descriptor-based approaches. One of the developing methods is a self-supervised contrastive learning.

2.2   Contrastive Methods

Contrastive learning methods are used on many tasks [4,20,27] but yet poorly applied in molecular property prediction.

Li et al. [8] proposed a graph contrastive learning framework (GeomGCL) utilizing the geometry of the molecule across 2D and 3D structures to learn chemical semantics. They maximized consistency between 2D and 3D positive pairs compared to negative ones. Also, contrastive loss with spatial regularization was introduced.

InfoGraph by Sun et al. [24] manages to maximize the mutual information between the representations of the graph and its substructures to guide molecular representation learning. To alleviate the problem of random corruption on molecular graphs which may alter the chemical semantics, MoCL adopts the domain knowledge-driven contrastive learning framework at both local- and global-level to preserve the semantics of graphs in the augmentation process [25].

Fang et al. proposed the chemical element knowledge graph (KG) usage for molecular contrastive learning [1]. Chemical element KG describes the relations between elements and their chemical attributes. Also, a knowledge contrastive learning framework (KCL) consisting of knowledge-guided graph augmentation, knowledge-aware graph representation, and contrastive objective was introduced.

3   Problem Statement

For a contrastive learning part, the problem statement is the following. Given molecule graphs G_n constructed from SMILES, we augment them into positive and negative pairs G_i', G_j', where $i = 2n - 1$ and $j = 2n$. Having pairs of augmented graph molecules, we minimize an NT-Xent (the normalized temperature-scaled cross-entropy loss) contrastive loss function for a positive pair of examples (i, j) [34]:

$$l(i, j) = -log \frac{e^{(sim(z_i, z_j)/\tau)}}{\sum_{k=1}^{2N} \mathbf{1}_{[k \neq i]} \cdot e^{(sim(z_i, z_k)/\tau)}},$$

where $\mathbf{1}_{[k \neq i]} \in {0, 1}$ is an indicator function, τ is a temperature parameter, and $sim(z_i, z_j) = z_i^T z_j / ||z_i|| ||z_j||$ is cosine similarity. The final loss is computed across all positive pairs, both (i, j) and (j, i), in a mini-batch.

For a fine-tuning part the problem is a multi-class classification task [18]. We minimize cross-entropy loss having observation probabilities $p_{o,c}$:

$$Cross - entropy\ loss = \sum_{c=1}^{M} \mathbf{1} \cdot log(p_{o,c}),$$

where M is a number of classes, $\mathbf{1}$ is an indicator function (if class label c is the correct classification for observation o), and $p_{o,c}$ is a predicted probability of observation o in class c derived from predictions by use of sigmoid function.

4   Methods

The model was build upon MolCLR framework [28]. Latent representations from positive augmented molecule graph pairs are contrasted with representations from negative pairs. The pipeline consisted of: data augmentation, graph isomorphism network (GIN) feature extractor, non-linear projection head, and the NT-Xent contrastive loss.

4.1   Chemistry-Wise Molecule Graph Augmentations

Three types of molecule graph augmentation were implemented: random subgraph removal, toxicophore substructure removal, and toxicophore substructure saving.

Random subgraph removal consisted of node and edge masking. First, a random atom was picked, and the atom and its neighbors were considered to be masked. Then, edges between chosen atoms were masked as well. These steps were executed till 25% of atoms and 25% of bonds were removed. This augmentation resulted in two correlated molecules with different subgraphs removed. Frequently, the random subgraph removal augmentation is used in graph contrastive learning methods, so we used it as a baseline approach.

To augment with toxicophore substructure removal, the search for such a subgraph was implemented. When the toxicophore was detected, it was removed from the molecular graph. When no toxicophore was observed for the given molecule graph, it was augmented with random subgraph removal described above. This augmentation resulted in two correlated molecules: first one had no toxicophore subgraph, and the other was obtained by random subgraph removal.

To augment with toxicophore substructure saving, the search of such a subgraph was implemented as well. If a toxicophore subgraph was found, the removal of atoms and bounds not included in the toxicophore was carried out. Otherwise, random subgraph removal was used. Thus, one molecule was augmented into two molecules: the first was with a randomly deleted subgraph having saved toxicophore substructure, and the second was with a randomly deleted subgraph.

4.2   Neural Network Structure

The graph of molecule G_n was defined as $G = (V, E)$, where node $v \in V$ represents an atom and edge $e \in E$ represents the chemical bond between atoms. Each graph was built from molecule SMILES s_n (Simplified Molecular Input Line Entry System). Using molecule graph augmentation strategies described below, G_n was transformed into two molecule graphs: G'_i and G'_j, where $i = 2n - 1$ and $j = 2n$. Two molecule graphs were considered to be positive pair if augmented from the same molecule, and negative pair otherwise. The feature extractor was modeled by graph isomorphism network (GIN) with average pooling. It mapped the graphs into the molecular representations $h_i, h_j \in \mathbb{R}$.

A non-linear projection head was modeled by a multilayer perceptron with one hidden layer, which mapped the representations h_i and h_j into latent vectors z_i and z_j, respectively. NT-Xent loss was applied to the $2N$ latent vectors to maximize the agreement of positive pairs and minimize the agreement of negative ones. The framework was pre-trained on the 100K unlabeled data from PubChem [6].

The pre-trained GIN model was fine-tuned on Tox21 dataset for toxicity prediction. The fine-tuned model consisted of a GIN backbone as the model on pre-train and an MLP head for features mapping into the predictions. The GIN backbone was initialized by parameter sharing from the pre-trained model and the MLP head was initialized randomly.

5   Computational Experiment

5.1   Metrics

ROC-AUC (area under the receiver operating characteristic curve) was used to evaluate the model result on each of the 12 classes. It is a plot of the false positive rate on the x-axis vs the true positive rate on the y-axis. The points are calculated for different thresholds $T \in [0, 1]$, which determines whether the observation is classified correctly.

5.2   Training Settings

At the first stage of training the model, we split train 0.95 and validation 0.05 from a dataset of 100k structures from the pubcam. On the subsequent classification of tox21, we used train 0.8, validation 0.1, and test 0.1 split. The splitting was carried out as a scaffold split: all molecules were divided into structural types by smiles (MurckoScaffold was generated for each, and matching ones were grouped). Then they were divided into train, validation, and test so that they include different scaffolds.

5.3   Evaluation and Results

The pre-train and fine-tune models were trained on GPU GeForce RTX 2060. Configured parameters for both models are presented in Table 1.

Table 1. Used model parameters.

Parameter	Pre-train model	Fine-tune model
batch size	512	32
epochs	50	100
initial learning rate	0.0005	0.0005
weight decay	$1 \cdot 10^{-5}$	$1 \cdot 10^{-6}$
GIN number of layers	5	5
Embedding dimension	300	300
Feature dimension	512	512
Pool	mean	mean

Prediction results on each assay from Tox21 dataset are provided in Fig. 2, and the comparison with other graph neural frameworks is in Table 2.

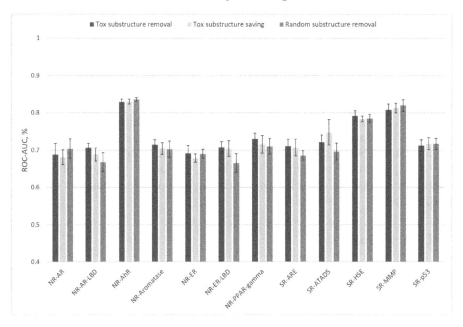

Fig. 2. Mean ROC-AUC values on Tox21 dataset with different augmentations. Error bars represent the standard deviation. Results obtained from 5 runs.

Augmentation with random substructure removal was implemented as a baseline. As can be seen in Fig. 2, toxicophore-oriented augmentations showed better results on 7 assays out of 12: NR-AR-LBD, NR-Aromatase, NR-ER-LBD, NR-PPAR-gamma, SR-ARE, SR-ATAD5, and SR-HSE.

Also, toxicophore substructure removal showed to be more efficient than toxicophore substructure saving augmentation.

Compared to other GNN-based models (Table 2), our model showed average performance.

Table 2. Performance of different models on Tox21 dataset. Mean and standard deviation of ROC-AUC are provided.

Model	ROC-AUC, %
SchNet [22]	77.2 ± 2.3
N-gram [9]	76.9 ± 2.7
GCN [7]	70.9 ± 2.6
GIN [32]	74.0 ± 0.8
MGCN [10]	63.4 ± 4.2
D-MPNN [33]	68.9 ± 1.3
$Ours_{tox\ removed}$	73.4 ± 1.6
$Ours_{tox\ saved}$	73.1 ± 1.7
$Ours_{random\ removed}$	71.3 ± 1.8

6   Discussion

In this work, the problem of molecule graph augmentation in the molecular contrastive learning framework is addressed. In solving the toxicity prediction task, the chemical meaning of subgraphs is taken into account by use of a novel augmentation strategy.

There is no essential difference between the usage of toxicophore substructure removal/saving augmentations according to the overall result on Tox21 dataset. Probably, since both approaches lead to learning the toxic subgraph anyways, they are equally effective. On the other hand, it can be the result of the pre-training on a 100k set of molecules which can be not large enough to identify the difference.

7   Conclusion

The discussed framework gains from pre-training on unlabeled data which is easy to obtain [4]. Novel augmentations of molecule graphs with chemistry-wise node and edge masking are proposed. These showed to be more efficient than the random subgraph masking approach usually used. The performance comparison with other GNN-based frameworks is carried out as well.

8   Future Work

As a future work, we plan to enlarge the pre-train dataset from 100k of molecules to 10 million. It should make the molecular representation learning more meaningful. The improvement of augmentation politics can be carried out as well [19]. E.g., Lipinski's rule of 5 and its extensions could be involved to identify toxic substructure. Also, other datasets will be used in the fine-tuning step to evaluate the model flexibility on different molecular property tasks.

References

1. Fang, Y., et al.: Molecular contrastive learning with chemical element knowledge graph. In: Proceedings of the AAAI Conference on Artificial Intelligence. vol. 36 (4), pp. 3968–3976 (2022)
2. Gerasimova, O., Makarov, I.: Higher school of economics co-authorship network study. In: Proceedings of the 2nd IEEE International Conference on Computer Applications and Information Security (ICCAIS 2019), pp. 1–4. King Saud University, IEEE, New York (2019). https://doi.org/10.1109/CAIS.2019.8769556
3. Gilmer, J., Schoenholz, S.S., Riley, P.F., Vinyals, O., Dahl, G.E.: Neural message passing for quantum chemistry. In: International Conference on Machine Learning, pp. 1263–1272. PMLR (2017)
4. Grachev, A.M., Ignatov, D.I., Savchenko, A.V.: Neural networks compression for language modeling. In: Shankar, B.U., Ghosh, K., Mandal, D.P., Ray, S.S., Zhang, David, Pal, S.K. (eds.) PReMI 2017. LNCS, vol. 10597, pp. 351–357. Springer, Cham (2017). https://doi.org/10.1007/978-3-319-69900-4_44

5. https://tripod.nih.gov/tox21/challenge/

6. Kim, S., et al.: PubChem 2019 update: improved access to chemical data. Nucleic Acids Res. **47**(D1), D1102–D1109 (2019). https://doi.org/10.1093/nar/gky1033

7. Kipf, T.N., Welling, M.: Semi-supervised classification with graph convolutional networks. CoRR abs/1609.02907, https://arxiv.org/abs/1609.02907 (2016)

8. Li, S., Zhou, J., Xu, T., Dou, D., Xiong, H.: GeomGCL: geometric graph contrastive learning for molecular property prediction. In: Proceedings of the AAAI Conference on Artificial Intelligence, vol. 36 (4), pp. 4541–4549 (2022)

9. Liu, S., Chandereng, T., Liang, Y.: N-gram graph, a novel molecule representation. CoRR abs/1806.09206, https://arxiv.org/abs/1806.09206 (2018)

10. Lu, C., Liu, Q., Wang, C., Huang, Z., Lin, P., He, L.: Molecular property prediction: a multilevel quantum interactions modeling perspective. In: Proceedings of the AAAI Conference on Artificial Intelligence, vol. 33, pp. 1052–1060 (2019). https://doi.org/10.1609/aaai.v33i01.33011052

11. Makarov, I., Gerasimova, O.: Link prediction regression for weighted co-authorship networks. In: Proceedings of the 15th International Work-Conference on Artificial Neural Networks (IWANN 2019), pp. 667–677. Universitat Politecnica de Catalunya, Springer, Berlin (2019). https://doi.org/10.1007/978-3-030-20518-8_55

12. Makarov, I., Gerasimova, O.: Predicting collaborations in co-authorship network. In: Proceedings of the 14th IEEE International Workshop on Semantic and Social Media Adaptation and Personalization (SMAP 2019), pp. 1–6. Cyprus University of Technology, IEEE, New York (2019). https://doi.org/10.1109/SMAP.2019.8864887

13. Makarov, I., Kiselev, D., Nikitinsky, N., Subelj, L.: Survey on graph embeddings and their applications to machine learning problems on graphs. PeerJ Comput. Sci. **7**, e357 (2021). https://doi.org/10.7717/peerj-cs.357

14. Makarov, I., Korovina, K., Kiselev, D.: JONNEE: joint network nodes and edges embedding. IEEE Access **9**, 144646–144659 (2021). https://doi.org/10.1109/ACCESS.2021.3122100

15. Makarov, I., Makarov, M., Kiselev, D.: Fusion of text and graph information for machine learning problems on networks. PeerJ Comput. Sci. **7**(e526), 1–26 (2021). https://doi.org/10.7717/peerj-cs.526

16. Makarov, I., et al.: Temporal network embedding framework with causal anonymous walks representations. PeerJ Comput. Sci. **8**(e858), 1–27 (2022). https://doi.org/10.7717/peerj-cs.858

17. Makarov, I., Savostyanov, D., Litvyakov, B., Ignatov, D.I.: Predicting winning team and probabilistic ratings in "dota 2" and "counter-strike: Global offensive" video games. In: Proceedings of the 6th International Conference on Analysis of Images, Social Networks and Texts (AIST 2017), pp. 183–196. LNCS, Polytechnic University, Springer, Berlin (2017). https://doi.org/10.1007/978-3-319-73013-4_17

18. Savchenko, A.V.: Fast inference in convolutional neural networks based on sequential three-way decisions. Inf. Sci. **560**, 370–385 (2021)

19. Savchenko, A.V., Belova, N.S.: Statistical testing of segment homogeneity in classification of piecewise-regular objects. Int. J. Appl. Math. Comput. Sci. **25**(4), 915–925 (2015)

20. Savchenko, A.V., Belova, N.S.: Unconstrained face identification using maximum likelihood of distances between deep off-the-shelf features. Expert Syst. Appl. **108**, 170–182 (2018)

21. Savchenko, A.V., Savchenko, L.V.: Towards the creation of reliable voice control system based on a fuzzy approach. Pattern Recogn. Lett. **65**, 145–151 (2015)

22. Schütt, K., Kindermans, P.J., Sauceda Felix, H.E., Chmiela, S., Tkatchenko, A., Müller, K.R.: SchNet: a continuous-filter convolutional neural network for modeling quantum interactions. In: Advances in Neural Information Processing Systems, vol. 30 (2017)

23. Singh, P.K., Negi, A., Gupta, P.K., Chauhan, M., Kumar, R.: Toxicophore exploration as a screening technology for drug design and discovery: techniques, scope and limitations. Arch. Toxicol. **90**(8), 1785–1802 (2016). https://doi.org/10.1007/s00204-015-1587-5

24. Sun, F.Y., Hoffmann, J., Verma, V., Tang, J.: InfoGraph: unsupervised and Semi-supervised Graph-level Representation Learning Via Mutual Information Maximization. arXiv preprint arXiv:1908.01000 (2019)

25. Sun, M., Xing, J., Wang, H., Chen, B., Zhou, J.: MoCL: data-driven molecular fingerprint via knowledge-aware contrastive learning from molecular graph. In: Proceedings of the 27th ACM SIGKDD Conference on Knowledge Discovery and Data Mining, pp. 3585–3594. ACM, New York (2021). https://doi.org/10.1145/3447548.3467186

26. Tikhomirova, K., Makarov, I.: Community detection based on the nodes role in a network: the telegram platform case. In: Proceedings of the 9th International Conference on Analysis of Images, Social Networks and Texts (AIST 2020), pp. 294–302. LNCS, Skoltech, Springer, Berlin (2020). https://doi.org/10.1007/978-3-030-72610-2_22

27. Wang, X., Liu, N., Han, H., Shi, C.: Self-supervised heterogeneous graph neural network with co-contrastive learning. In: Proceedings of the 27th ACM SIGKDD Conference on Knowledge Discovery and Data Mining, pp. 1726–1736 (2021)

28. Wang, Y., Wang, J., Cao, Z., Barati Farimani, A.: Molecular contrastive learning of representations via graph neural networks. Nature Mach. Intell. **4**(3), 279–287 (2022). https://doi.org/10.1038/s42256-022-00447-x

29. Wieder, O., et al.: A compact review of molecular property prediction with graph neural networks. Drug Discov. Today Technol. **37**, 1–12 (2020). https://doi.org/10.1016/j.ddtec.2020.11.009

30. Withnall, M., Lindelöf, E., Engkvist, O., Chen, H.: Building attention and edge message passing neural networks for bioactivity and physical-chemical property prediction. J. Cheminformatics **12**(1), 1 (2020). https://doi.org/10.1186/s13321-019-0407-y

31. Xiong, Z., et al.: Pushing the boundaries of molecular representation for drug discovery with the graph attention mechanism. J. Med. Chem. **63**(16), 8749–8760 (2020). https://doi.org/10.1021/acs.jmedchem.9b00959

32. Xu, K., Hu, W., Leskovec, J., Jegelka, S.: How powerful are graph neural networks?. https://arxiv.org/abs/1810.00826 (2018)

33. Yang, K., et al.: Analyzing learned molecular representations for property prediction. J. Chem. Inf. Model. **59**(8), 3370–3388 (2019). https://doi.org/10.1021/acs.jcim.9b00237

34. Yuning, Y., Tianlong, Ch., Yongduo, S., Ting, C., Zhangyang, W., Shen, Y.: Graph contrastive learning with augmentations. In: Lin, H.L., Ranzato, M., Hadsell, R., Balcan, M.F., H. (eds.) Advances in Neural Information Processing Systems, pp. 5812–5823. Curran Associates, Inc. (2020)

Interaction with Neural Systems in Both Health and Disease

Machine Learning Models for Depression Detection Using the Concept of Perceived Control

Prosper Azaglo[1], Pepijn van de Ven[1], Rachel M. Msetfi[2], and John Nelson[1(✉)]

[1] Department of Electronic and Computer Engineering, University of Limerick, Limerick, Ireland
{prosper.azaglo,pepijn.vandeven,john.nelson}@ul.ie
[2] Office of the Vice President Research and Innovation, Maynooth University, Maynooth, Co., Kildare, Ireland
rachel.msetfi@mu.ie

Abstract. In this paper, machine learning techniques are used to detect and predict the mental health status of individuals based on the concept of Perceived Control using a mobile app. Perceived control has long been established to have a strong link with an individual's mental health. Individuals with a high level of perceived control seem to have good mental health while those with low levels of perceived control usually suffer from depression, anxiety and stress.

In the proposed method, an individual's measure of perceived control is solicited by allowing them to download and install an android app called the Judgement App. The users then participate in an experiment, where they perform a number of trials and make a judgement after 8 trials. The data generated is then analysed and used to train supervised machine learning models to predict whether an individual is suffering from depression or not. Data generated for internal and external perceived control were of both tabular and time-series types. The data is labelled by the subject's Beck Depressive Inventory (BDI-II) score, which is performed by the individual answering the 21-questions before the experiment begins. Due to the imbalanced nature of the data available, Synthetic Minority Oversampling Technique (SMOTE) and some of its variants were used to process the training data before being used to train ML algorithms. Simple evaluation criteria consisting of Precision, Recall, F1-score and overall model efficiency were used.

The evaluation was completed by analyzing 274 samples from 140 participants. Out of the 274 samples, 53 were labelled as mildly depressed and 221 as non-depressed.

Keywords: Perceived Control · Internal Control · External Control · contingency · trial · judgement · depression

1 Introduction

The problem of mental health affects a very significant portion of the world's population. The World Health Organization (WHO) declared depression as the leading cause of ill health and disability worldwide in 2015: more than 300 million people live with

© The Author(s), under exclusive license to Springer Nature Switzerland AG 2023
I. Rojas et al. (Eds.): IWANN 2023, LNCS 14135, pp. 339–351, 2023.
https://doi.org/10.1007/978-3-031-43078-7_28

it. The resultant adverse effects on the economic, social, and educational sectors are quite enormous [1–3]. Being able to diagnose depression at an earlier stage helps the professionals to treat them on time and it improves the patient's quality of life. One way of achieving this is to use a reliable and easy manner of detecting the condition so that appropriate clinical attention can be given. The typical clinical manifestations and symptoms include but are not limited to cognitive and emotional disorders and loss of confidence in life. Although various methods have been introduced to diagnose depression, the most commonly used method is through consulting with experts. However, this method can have a number of logistical challenges. For example, monitoring the illness trajectory frequently requires patients to travel to a clinical centre within its limited hours of operation. This can be quite burdensome for patients whose mental illness is quite serious. Also, these methods are highly resource-intensive because they require one-to-one interactions with a trained clinician, and, thus, their large-scale dissemination is challenging [2].

Given the high prevalence of depression and its suicide risk, finding new methods for diagnosis and treatment becomes more and more critical. There is growing interest in using data generated by automatic human behaviour for computer-aided diagnosis of mental health based on behavioural information such as facial expressions, speech prosody and daily routine activities because of convincing evidence that depression and related mental health disorders are associated with changes in patterns of behavior [3, 7–12].

In this paper, an attempt is made to monitor, detect and predict the mental health status of individuals based upon the concept of perceived control using a mobile app and machine learning techniques. Perceived control has a very important link to the mental health status of an individual. It involves the concept which measures the degree to which an individual thinks they have control over certain events around them or in their life [23].

The associated "illusory" Perception of Control is the idea where people believe they have more control over events than they actually do [21]. Patrick, Skinner, and Connell defined these perceptions of control as "a person's belief that there is a connection between their behaviour and specific possible outcomes, and such the belief that they are capable of desired results and preventing undesired results" [4].

These perceptions of control can be used to detect depression in individuals. For example, Alloy and Abramson [18] found that non-depressed people are more likely than depressed people to think that their actions are contingent on their outcomes when in reality they are not.

Perceived control in contingency learning is generally related to a good psychological well-being while a low level of perceived control is believed to be a cause or effect of depression [22]. According to locus of control theory [5], an individual's perception of control is either internal or external. Internal perception of control is linked to mental healthiness. Individuals with an internal perception of control perceive they can control their actions to achieve a desired outcome. Whereas, External Perception of Control (no control) is associated with an individual having no control over outcomes, which is linked to depression. Skinner [6] has linked these internal perceptions of control to positive emotions such as joy and has shown that external perceptions of control are linked to negative emotions such as anxiety and fear.

Measurements of perceived control usually take place in a well-controlled laboratory environment where participants are made to perform some contingency judgement after they have carried out an experiment or an action requiring an outcome [21, 22]. One of the earliest attempts at measuring perception of control in a non-laboratory setting is in [21] where a mobile phone app is used to examine and enhance perceived control in 106 participants who are either mildly depressed or non-depressed. Given the widespread adoption of smartphones makes them very suitable personal device for collection of user data, including perceived control data in the normal context of daily life. The perceived control data can be used, given appropriate modelling, to obtain the depression status and hence the mental health status of the individual.

2 Literature Review

Artificial intelligence and machine learning are being increasingly used to analyse, detect and predict most of these mental health ailments and they are increasingly gathering pace because of improving reliability in the early detection of mental health conditions [7]. A review of some AI and ML techniques used have been summarised below, to give an insight into the different methods adopted. Overall, the feature extraction and the evaluation techniques used in their approaches, to a large extent, informed the approach used in this project.

In [7] a system was designed for automatic depression detection using visual and vocal expressions. This uses facial expression images, voice audio captured using a mobile phone. VGG-Face (Visual Geometry Group for Facial recognition), a type of pre-trained Convolutional Neural Network (CNN) for facial recognition, is adapted to form what is called Feature Dynamic History Histogram (FDHH) which is used to capture the dynamic temporal movement on the deep feature space. Partial least squares (PLS) and linear regression (LR) are used to model the mapping between the dynamic features and the depression scales. The model was able to achieve a Mean Absolute Error (MAE) and Root Mean Square Error (RMSE) values of 6.14 and 7.43, respectively.

A speech analysis library for unobtrusive monitoring of mood and stress on mobile phones was designed in [8]. Voice audio, was collected through mobile phone via the AMMON (Affective and Mental health MONitor) software. The AMMON performs feature extraction on the voice audio. The extracted features include pitch, energy, and Low Level Descriptors (LLDs) including Zero-crossing-rate, Root Mean Square (RMS) frame energy, harmonics-to-noise ratio (HNR), Mel-Frequency Cepstral Coefficients (MFCC) 1- 12. AMMON Recognition or classification uses linear Support Vector Machines (SVM) to recognise emotions based on the feature vectors with a speech library as a reference. The AMMON library is based on the InterSpeech Emotion Challenge 2009 speech library database. For recognition of positive and negative emotion clips, the AMMON implementation without glottal timings had an efficiency of 75.51% with openSMILE, which is an open-source audio feature extraction achieving the same result. For recognition of stressed and neutral speech, AMMON achieves 84.43% with openSMILE 83.18%.

In [9], heart rate, skin temperature, and galvanic skin response (GSR), where changes in sweat gland activity are reflective of the intensity of one's emotional state, were collected with the help of a monitoring patch with a mobile app interface. The data gathered with the above 3 parameters were fed into an SVM algorithm for a two-class classification, with stressed and non-stressed labels for training. Up to 91.26% overall accuracy was obtained during the testing phase.

The authors in [10] carried out forecasting of depressed mood based on self-reported histories via Recurrent Neural Networks. The behavioural logs are collected through the users input on a mobile application. These include mood at each time of the day, actions at a particular time of the day (go to work, work at home, sleeping time, wake up time, etc.). A total of 2,382 individuals were involved in the trial over a period of 22 months. The collected data is pre-processed into a time series data, involving activities at a particular time of the day.

The depression detection and estimation in [11] was carried out with the help of the Audio Visual Emotion Challenge 2016 (AVEC2016) dataset. The data was obtained from 9 individuals to form the training set, 9 individuals for the development set and also 9 for the test set. The system uses a 2-level hierarchical Deep Convolutional Generative Adversarial Network (DCGAN) to generate audio features, which is then used to augment the training data for depression estimation.

An Emotion Sensing for Mental Health with the aid of Galvanic Skin Response (GSR) obtained through wearables was implemented in [13]. Feature vectors extracted from the raw GSR signal which included the mean, standard deviation, mean of absolute value of the first difference of the raw signal, mean of absolute value of the first difference of the normalised signal, mean of absolute value of the second difference of the raw signal, and mean of absolute value of the second difference of the normalised signal. The Sequential Floating Forward Selection (SFFS) technique is applied to find the optimal feature subspace combination to maximise the inter-class distance in emotion classification. The classification model used is the k-NN which achieved up to 82.86% accuracy.

Tackling Mental Health by Integrating Unobtrusive Multimodal Sensing was implemented in [14] using social media stream activities (Likes, Comments, Shares) and a close-up video of the subject while engaged in these activities. Computer vision algorithms including Cascade Classifier and Tracking-Learning-Detection were used to monitor real-time psycho-behavioural signals through the obtained video, including the heart rate, eye blink rate, pupil variations, head movements, and facial expressions of the users. The mood and emotion of the subject are also extracted from the social media activities by the subject as a prelude to assessing the effects of social contacts and context within such media. The social media content is analysed using an NLP tool, Sentiment 140, to discover the sentiment polarity. Logistic regression and Support Vector Machines are used for modeling and prediction and the system is able to achieve an AUC_ROC of up to 95%.

Unobtrusive monitoring of behaviour and movement patterns to detect clinical depression severity level via smartphone [15] was implemented using the smartphone acceleration sensor signal, geographical location information from a GPS sensor. A feature selection method was applied to the extracted features for optimal selection. An

SVM was used to classify the depression severity level among individuals (absence, moderate, severe) with an accuracy of 87.2% in severe depression cases, which outperformed other classification models including the K- Nearest Neighbour (KNN) and Artificial Neural Network (ANN).

Unobtrusive monitoring to detect depression for elderly with chronic illnesses using infra-red motion detection sensor for activity detection was implemented in [16]. 12 features are extracted from the infrared sensor signals. They include mean, mode and number of: washroom visits during the day, washroom visits during the night, times subject was out, times of sleep. These were carried out for various period lengths ranging from 1 to 7 days. Dimensionality reduction is carried out using PCA resulting in 8 features which are then used with SVM, ANN, C4.5 Decision Tree and Bayesian Network to classify the severity of depression. The system achieved up to 96% accuracy on normal condition and mild depression, although using a limited dataset of 20 elderly participants.

Using Wearable Sensors in addition to Mobile Phones to detect Stress in [17] was implemented using a wrist sensor for acceleration, skin conductance and mobile phone to record call usage, SMS, location and screen on/off status. Surveys are also used to collect stress, mood, sleep, tiredness, general health, alcohol or caffeinated beverage intake and electronics usage. Linear correlation analysis was applied to the features/data and a determination was made on which features were significantly correlated with the self-reported perceived stress scale ratings. These were labelled according to the perceived stress scale (PSS) ratings and fed to the SVM, k-NN classifiers. Post survey, skin conductance and accelerometer sensor data, and mobile phone usage showed up to 87.5% accuracy, for a limited 5 days of data collection for 18 participants.

3 Methodology

Data for this project was collected through an experiment with the use of an android application designed for that purpose. The recruitment and the design of the experiment are described as follows.

3.1 Recruitment

Participants consisted of people who volunteered through social media platforms through an advertisement and fulfilled the inclusion criteria: (1) access to an Android mobile phone and (2) over the age of 17. Participants completed the Beck Depression Inventory (BDI) [19] consisting of 21 questions after downloading the app.

On the basis of their BDI scores, participants were categorised as members of the low BDI group (BDI $\leq$ 9, representing non-depression) or high BDI group (BDI > 10, representing depression). A second categorization was made for multi-class classification with the BDI score ranges summarised in Tables 1 and 2.

Table 1. Summary of Dataset

CLASS	NUMBER OF SAMPLES	BDI RANGE	DESCRIPTON
0	221	0–9	NON-DEPRESSED
1	53	≥ 10	DEPRESSED

Table 2. Class Summary of the Multiclass Case

BDI RANGE	DESCRIPTION
0–9	NON-DEPRESSED
10–18	MILD DEPRESSION
19–29	MODERATE DEPRESSION
30–63	SEVERE DEPRESSION

3.2 Perception of Control Task

The task is implemented using an app developed using an Android library [24] which is able to send messages and notifications to user's mobile device. The app requires network connectivity with the server initially in order for the experimental condition settings to be downloaded to the phone. Once this was complete, the app functioned independently and did not require continuous connectivity. Incremental data upload to the server was programmed to take place as soon as connectivity was available, both during the task and once the task was finished.

After downloading the app, the user is made to confirm they are of legal age of 17 years and above in order to commence after which the user is asked to complete a BDI-21 questionnaire. The experiment settings are then downloaded and scheduled. A short message pops up on the screen informing the user to be ready to start receiving notifications. In this experiment, a trial is a single instance of the experiment. It involves the participant responding to a notification from the judgement app and taking an action which would result in an outcome. Finally, a judgement refers to the participant's rating of how much control they have over the outcome, which is referred to as the internal control, and their rating of how much external factors were involved in the outcome. A single judgement rating can also be referred to as a sample.

The experiment is as follows. The user is alerted randomly when the scheduled trial is due and has two minutes to complete. The user accesses the alert and clicked and a round blue ball appears on the screen, see Fig. 1 (a). The action is whether the user either chooses to click or not to click on the ball within 3 s. After that, the outcome is that a 'boing' sound may or may not be heard within a few seconds, and is determined by probability settings. The trial repeats again after the inter-trial interval elapses. After 8 trials, the user is presented with a screen to rate their control over the outcome of the auditory sound, see Fig. 1 (b). A second screen asks to rate how much control external factors had over the outcome of the sound, see Fig. 1 (c).

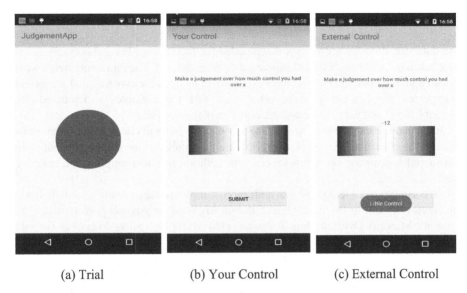

(a) Trial   (b) Your Control   (c) External Control

Fig. 1. Application Screen Shots during Trial and Judgements

In total, 5 of these judgements or ratings are expected to be completed, with a total of 40 trials in the process. The associated data and the user behaviour are sent securely to a server for further processing and to be used for machine learning models. Table 3 below is a summary of the data collected during the perceived control experiment.

Table 3. Summary of Data Collected During the Experiment

Data	Description
Age	This is the age of the participant
Gender	This is the gender of the participant
Internal Perceived Control	This value provides the participant's rating of how much influence they believe their input/actions contributed to the outcome. A total of 5 of such ratings are generated when the experiment is fully completed.
External Perceived Control	This value provides the participant's rating of how much influence they believe external factors contributed to the outcome. A total of 5 of such ratings are generated when the experiment is fully completed.
BDI	This provides the Beck Depressive Inventory score of the participant.

4   Data Pre-processing and Modeling

The data stored on the server was downloaded for processing. The unnecessary fields were taken out and rows that had missing data were deleted. The remaining fields were plotted and outliers removed to improve data quality. Given the experimental group and as expected, the data classes were imbalanced, with few examples of depressed (the minority class) and many of non-depressed (the majority class).

Imbalanced data problems in machine learning are generally dealt with using resampling of the class data. Below is a description of a number of the techniques that were considered to improve the minority class, in addition to undersampling the majority class.

Oversampling is a technique that adds more samples to the minority class to balance the classes present in the data. One of the mostly used oversampling methods is the Synthetic Minority Oversampling Technique (SMOTE) [20]. Others do exist as variants of the SMOTE and brief description is given below.

SMOTE is an improvement on duplicating examples from the minority class by synthesizing new examples from the minority class. This is done by having examples that are close in the feature space selected, and having a line drawn between the examples in the feature space and then selecting a new sample at a point along that line.

Borderline-SMOTE: An extension to smote involves selecting those instances of the minority class that are misclassified, such as with a k-NN classification model. It can then oversample just those difficult instances, providing more resolution only where it may be required.

Borderline-SMOTE SVM: An alternative of borderline-smote where an SVM algorithm is used instead of a k-NN to identify misclassified examples on the decision boundary. An SVM is used to locate the decision boundary defined by the support vectors and examples in the minority class that are close to the support vectors become the focus for generating synthetic examples.

Adaptive Synthetic Sampling (ADASYN): This method is quite similar to the smote approach, but the differences are described as follows. It involves generating synthetic samples inversely proportional to the density of the examples in the minority class. That is, generate more synthetic examples in regions of the feature space where the density of minority examples is low, and fewer or none where the density is high.

The concept of ADASYN is to produce the right number of synthetic alternatives for each observation belonging to the minority class. The concept of "appropriate number" here depends on how hard it is to learn the original observation. In particular, an observation from the minority class is quite difficult to learn if many samples from the majority class with features similar to that observation share the same feature space.

4.1   Classification Models

After all the preliminary data cleaning and processing, a portion of the data was taken to train a number of ML algorithms. After the training, a test set is used to predict if participants are mildly depressed or not. To carry out this binary classification, models were trained and tested based on the algorithms listed below in Table 4. These were selected because they produced the best results.

Table 4. The Machine Learning algorithms

- Support Vector Machines (SVM)
- K- Nearest Neighbour (KNN)
- Decision Tree (DT)
- Random Forest (RF)

- Naïve Bayes (NB)
- Gradient Boosting (GB)
- XGBoost (XGB)

These models consume data generated by the single valued data pairs of the internal judgement rating and external judgement ratings and a combination of other features. The performance metrics were measured using a 5-fold cross validation technique. The models were trained and tested on a computing system equipped with an 8th Generation Intel® Core™ i7-8565U **Processor** (8MB Cache, up to 4.6 GHz, 4 cores) and equipped with a RAM of 16 GB LPDDR3 at 2133 MHz. The prediction models were implemented in Python using *scikit-learn, scipy* and *imblearn* packages.

4.2 Model Hyper-Parameters

A number of hyper-parameters were considered and tuned for the models implemented as summarized in Table 5. The various ML model performances were assessed using simple metrics including Precision, Recall, F1-Score, and overall model Efficiency.

Table 5. Summary of ML Model Hyper-parameters

Model	Hyper-Parameters
SVM	Kernel types, *rbf*, *sigmoid*, *linear* and *poly*. C, the regularization parameter (varied between 1 and 10000 in steps of 100). Degree varied in the range 2 to 6 for the poly kernel. Gamma values of 1e-3 and 1e-4 for the rbf kernel. GridsearchCV used to determine the optimal performance parameters
K-NN	k, representing the number of neighbours or most similar samples considered to classify a sample: (3 to 8 in increments of 1),
DT	Different values for the *criterion* parameter, namely gini and entropy, and for the *splitter* parameter, namely best and random. Values of parameters *min_samples_split* (1 to 5) and *max_depth* (2 to 6)
RF	As RF is an ensemble model using DT, most of its parameters are as for DT. Similarly tested values for the *criterion*, *splitter*, *max_depth* and *min_samples_split* parameters. *n_estimators* (values in the range 1 to 10)
XGB	Tested different *booster* choices, including gbtree and gblinear, along with different *eval_metric* choices, such as logloss and error. Apart from these parameters, the *min_child_weight* and *max_depth* parameters were varied in the range 1 to 5 and 3 to 10, respectively, in increments of 1

5 Results

The obtained user data after going through the preprocessing stage is split into training and test portions. A train test split ratio of 70/30 was used, yielding a train set of 154 and 38 samples for the majority and minority classes, respectively; and the test set had 67 and 15 samples, respectively.

The training data was then subjected to the Synthetic Minority Oversampling Technique (SMOTE) and other variants of SMOTE to deal with the issue of class imbalance. After SMOTE was applied, both the minority and majority classes had 154 samples each. The results are displayed in Fig. 2 and Table 6. The best results were achieved with a combination of Extreme Gradient Boosting and the SMOTE technique. The confusion matrix indicates the model without the SMOTE is able to correctly classify 64 out of 67 samples for non-depressed individuals while only able to correctly classify 9 out of 15 samples in the mildly depressed case, while misclassifying 6 samples in the process.

However, in the case where SMOTE was applied to the training data, the model was able to classify 13 out of 15 samples correctly, a significant improvement from the initial model where SMOTE was not considered. The SMOTE & Extreme Gradient Boosting outperforms the case where only the Extreme Gradient Boosting was used with a Precision and Recall of 81% and 87% respectively. Adaptive Synthetic Minority Oversampling Technique, a variant of SMOTE, produced similar results to SMOTE.

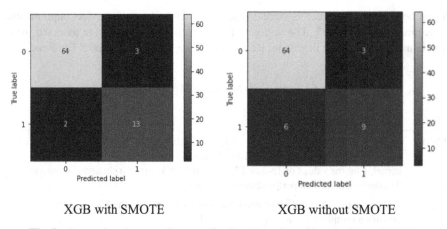

XGB with SMOTE   XGB without SMOTE

Fig. 2. Comparison between Extreme Gradient Boosting with and without SMOTE

Also, taking a look at the performance of the other models in Table 7 shows a general deficit in being able to correctly classify samples from the minority class and this is clearly indicated by the Recall metric. The best recall value registered is around 67% which is quite low. Employing the SMOTE technique, which helped generate more samples for training improved that to 87% with the Extreme Gradient model. All the above results were reported on the test data.

In an attempt to carry out the same analysis to determine the extent of seriousness of depression an individual was experiencing, the BDI range for the classes were adjusted as in Table 2 giving rise to a multiclass problem. However, the three minority classes contain very little data samples and the performance is hampered.

Table 6. Classification Report for XGBoost with and without SMOTE

SMOTE	CLASS	PRECISION	RECALL	F1-SCORE	SUPPORT
SMOTE/ XGB	0	0.97	0.96	0.96	67
	1	0.81	0.87	0.84	15
No SMOTE	0	0.91	0.96	0.93	67
	1	0.75	0.6	0.67	15
ADASYN	0	0.97	0.96	0.96	67
	1	0.81	0.87	0.84	15

Table 7. Comparison between SMOTE and non-SMOTE for Other Models

ALGORITHM	SMOTE	CLASS	PRECIS-ION	RECALL	F1-SCORE
Decision Tree	SMOTE	0	0.93	0.94	0.93
		1	0.71	0.67	0.69
	No SMOTE	0	0.91	0.96	0.93
		1	0.75	0.6	0.67
Gradient Boosting	SMOTE	0	0.97	0.93	0.95
		1	0.72	0.87	0.79
	No SMOTE	0	0.92	0.91	0.92
		1	0.62	0.67	0.65
Random Forest	SMOTE	0	0.9	0.78	0.83
		1	0.38	0.6	0.46
	No SMOTE	0	0.82	1	0.9
		1	0	0	0
KNN	SMOTE	0	0.86	0.93	0.89
		1	0.5	0.33	0.4
	No SMOTE	0	0.86	0.82	0.84
		1	0.33	0.4	0.36

6 Conclusion and Future Work

Mental health issues are a major concern for the wider society. With the advent of Covid-19, the proportion of the world's population affected is likely to rise significantly. Using the concept of perception of control becomes useful, given the use of a relatively unobtrusive and simple way of collecting data from individuals. The purpose of the experiment detailed in this paper was to establish whether depressive state can be predicted from perceived control data by machine learning models.

The current small number of samples from the minority class presented challenges. The initial imbalanced data set resulted in poor model performances. Currently, further trials are being run to increase the population size, and in particular the minority class. In addition, work is ongoing to classify more accurately the degree of seriousness of depression using multi-class classification and by adopting more complex models.

This is the first research to the best of our knowledge of establishing the link between the concept of Perceived Control and mental health using machine learning techniques. Even with the current imbalanced data set, the ML techniques have shown great promise of accurately classifying a large number of individuals/samples from the experiments already performed. It was shown that the performance is significantly improved when oversampling techniques were used on the training data to generate more samples in the minority class.

References

1. Ringeval, F., et al.: AVEC 2019 workshop and challenge: state-of-mind, detecting depression with AI, and cross-cultural affect recognition. In: Proceedings of the 9th International on Audio/visual Emotion Challenge and Workshop (2019)
2. Abdullah, S., Choudhury, T.: Sensing technologies for monitoring serious mental illnesses. IEEE MultiMedia, 25(1), 61–75 (2018)
3. World Health Organization: Depression and Other Common Mental Disorders: Global Health Estimates. Technical Report. World Health Organization, 2017. Licence: CC BY-NC-SA 3.0 IGO
4. Patrick, B.C., Skinner, E.A., Connell, J.P.: What motivates children's behavior and emotion? joint effects of perceived control and autonomy in the academic domain. J. Pers. Soc. Psychol. 65, 10 (1993)
5. Rotter, J.B.: Generalized expectancies for internal versus external control of reinforcement. Psychol. Monogr. Gen. Appl. 80(1), 1–28 (1966)
6. Skinner, E.A.: Perceived control, motivation, & coping. Sage Publications (1995)
7. Jan, A., Meng, H., Gaus, Y.F.B.A., Zhang, F.: Artificial intelligent system for automatic depression level analysis through visual and vocal expressions. IEEE Trans. Cogn. Dev. Syst. 10(3), 668–680 (2018)
8. Chang, K., Fisher, D., Canny, J., Hartmann, B.: How's my mood and stress? an efficient speech analysis library for unobtrusive monitoring on mobile phones. In: 6th International ICST Conference on Body Area Networks, pp. 71–77, June 2012
9. Akbar, F., Mark, G., Pavlidis, I., Gutierrez-Osuna, R.: An empirical study comparing unobtrusive physiological sensors for stress detection in computer work. Sensors (2019)
10. Yoshihiko, S., Xu, Y., Pentland, A.: DeepMood: forecasting depressed mood based on self-reported histories via recurrent neural networks. In: Proceedings of the 26th International Conference on World Wide Web, pp. 715–724 (2017)

11. Yang, L.: Multi-modal depression detection and estimation. In: 8th International Conference on Affective Computing and Intelligent Interaction Workshops, pp. 529–536 (2019)
12. Ben-Zeev, D., Scherer, E.A., Wang, R., Xie, H., Campbell, A.T.: Next-generation psychiatric assessment: using smartphone sensors to monitor behavior and mental health. Psychiatr. Rehabil. J. **38**(3), 218–226 (2015)
13. Guo, R., Li, S., He, L., Gao, W., Qi, H., Owens, G.: Pervasive and unobtrusive emotion sensing for human mental health. In: 7th International Conference on Pervasive Computing Technologies for Healthcare and Workshops, pp. 436–439 (2013)
14. Zhou, D., et al.: Tackling mental health by integrating unobtrusive multimodal sensing. In: Proceedings of the 29th AAAI Conference on Artificial Intelligence, AAAI 2015, pp. 1401–1408 (2015)
15. Masud, M., Mamun, M., Thapa, K., Lee, D., Griffiths, M., Yang, S.: Unobtrusive monitoring of behavior and movement patterns to detect clinical depression severity level via smartphone. J. Biomed. Inform. **103**, 103371 (2020)
16. Kim, J.Y., Liu, N., Tan, H.X., Chu, C.H.: Unobtrusive monitoring to detect depression for elderly with chronic illnesses. IEEE Sens. J. **17**, 5694–5704 (2017)
17. Sano, A., Picard, R.: Stress Recognition Using Wearable Sensors and Mobile Phones. Humaine Association Conference on Affective Computing and Intelligent Interaction (2013)
18. Alloy, L.B., Abramson, L.Y.: Judgment of contingency in depressed and nondepressed students: sadder but wiser? J. Exp. Psychol. Gen. **108**, 441 (1979)
19. Beck, A.T., Ward, C.H., Mendelson, M., Mock, J., Erbaugh, J.: An inventory for measuring depression. Arch. Gen. Psychiatry **4**, 561–571 (1961)
20. Chawla, N.V., Bowyer, K.W., Hall, L.O., Kegelmeyer, W.P.: SMOTE: synthetic minority over-sampling technique. J. Artif. Intell. Res. **16**, 321–357
21. Msetfi, R., O'Sullivan, D., Walsh, A., Nelson, J., Van de Ven, P.: Using mobile phones to examine and enhance perceptions of control in mildly depressed and non-depressed volunteers: intervention study. JMIR Mhealth Uhealth **6**(11), e10114 (2018)
22. Msetfi, R.M, Kornbrot D.E., Matute, H., Murphy, R.A.: The relationship Between Mood State and control in contingency learning: effects of individualist and collectivist values. Front Psychol. **29**(6), 1430 (2015)
23. Wallston, K.A., Wallston, B.S., Smith, S., Dobbins, C.J.: Perceived control and health. Curr. Psychol. Res. Rev. **6**, 5–25 (1987)
24. van de Ven, P., et al.: ULTEMAT: a mobile framework for smart ecological momentary assessments and interventions. Internet Interv. **9**, 74–81 (2017)

ECG Hearbeat Classification Based on Multi-scale Convolutional Neural Networks

Ondrej Rozinek[✉][ORCID] and Petr Dolezel[ORCID]

University of Pardubice, Studentska 95, 532 10 Pardubice, Czech Republic
ondrej.rozinek@gmail.com, petr.dolezel@upce.cz
https://fei.upce.cz/

Abstract. Clinical applications require automating ECG signal processing and classification. This paper investigates the impact of multiscale input filtering techniques and feature map blocks on the performance of CNN models for ECG classification. We conducted an ablation study using the AbnormalHeartbeat dataset, with 606 instances of ECG time series divided into five classes. We compared five multiscale input filtering techniques and four multiscale feature map blocks against a base model and non-multiscale input. Results showed that the combination of mean filter for multiscale input and residual connections for multiscale block achieved the highest accuracy of 64.47%. Residual connections were consistently effective across different filtering techniques, highlighting their potential to enhance CNN model performance for ECG classification. These findings can guide the design of future CNN models for ECG classification tasks, with further experimentation needed for optimal combinations in specific applications.

Keywords: ECG classification · deep learning · multiscale CNN · convolutional neural networks

1 Introduction

CNNs (Convolutional Neural Networks) have been shown to be state of the art in many computer vision applications due to their ability to learn complex patterns in images with high accuracy [10,12,25]. CNNs achieve this by using convolutional layers that extract and combine local features from the input image, followed by fully connected layers that use the learned features to make a prediction [13]. Additionally, CNNs can be trained end-to-end using backpropagation, which allows them to automatically learn the optimal set of weights that minimize the prediction error [23]. This makes CNNs highly effective for a wide range

The work has been supported by the SGS grant no. SGS_2023_016 at the Faculty of Electrical Engineering and Informatics, University of Pardubice, Czech Republic. This support is very gratefully acknowledged.

I. Rojas et al. (Eds.): IWANN 2023, LNCS 14135, pp. 352–363, 2023.
https://doi.org/10.1007/978-3-031-43078-7_29

of image recognition tasks, including object detection [21], image classification [25], and semantic segmentation [17].

CNNs have demonstrated significant improvements over traditional machine learning approaches in many computer vision tasks, especially in object recognition tasks such as the ImageNet Large Scale Visual Recognition Challenge (ILSVRC) [24]. The winning entries in the ILSVRC from 2012 to 2015 were all CNN-based models [10,12,25,27]. Moreover, CNN-based models have set new records in various other image recognition tasks, including face recognition [28], image captioning [29], and visual question answering [2].

Overall, the effectiveness of CNNs can be attributed to their ability to learn and extract features from images in an end-to-end manner, which allows them to handle complex visual patterns and large datasets. With continued improvements in model architecture, optimization techniques, and hardware, CNNs are expected to continue to push the state of the art in computer vision applications.

2   Multiscale CNN and Related Work

Multiscale CNNs process signals at multiple resolutions, capturing fine-grained details and larger structural features. They can handle input data that varies in scale, leading to improved accuracy and reduced overfitting. Three types of multiscale CNNs are: (1) downsampled inputs, (2) multiple receptive field inputs, and (3) feature pyramid inputs [6,8,15,16,22,26].

Inception network [26] uses convolutional layers with different filter sizes for processing images at multiple scales. U-Net [22] is designed for biomedical image segmentation tasks. Multiscale CNNs for time series classification [6] use convolutional layers at multiple scales to extract features. Pyramid Scene Parsing Network (PSPNet) [15] is for semantic segmentation of high-resolution images. Feature Pyramid Network (FPN) [16] is for object detection. Dual Attention Network (DANet) [8] uses self-attention mechanisms for image segmentation.

Relevant scientific competitions include ILSVRC, Kaggle, COCO Detection Challenge, VisDA, and Robust Vision Challenge.

3   CNN Classification of Electrocardiogram

Electrocardiograms (ECGs) are non-invasive tests that record the heart's electrical activity, with the PQRST complex being key for diagnosing cardiac conditions. Abnormalities in this complex can indicate arrhythmias, ischemia, or infarction. Research in ECG analysis focuses on developing algorithms to better detect and diagnose these abnormalities, often using machine learning models.

ECG classification is challenging due to signal variability, large dataset requirements, and interpretability. Convolutional neural networks (CNNs) dominate state-of-the-art methods, with transfer learning and data augmentation techniques improving performance. Publicly available datasets, like PhysioNet/CinC Challenge datasets [1], facilitate research. Recent studies propose hybrid deep learning models, such as CNN-RNNs [14] and convolutional attention models [18], achieving high accuracy in atrial fibrillation detection and PTB Diagnostic ECG Database tasks.

Further research is needed to address challenges in designing accurate, interpretable ECG classification models for clinical use.

4 Methodology

4.1 Multiscale Inputs

Our methodology utilizes multiscale input filtering techniques to preprocess a set of 1D electrocardiogram (ECG) signals. The primary goal of our methodology is to evaluate the impact of different multiscale filtering techniques on the performance of convolutional neural network (CNN) architectures for ECG classification. To ensure a fair comparison, all methods described in this section will have the same input and output formats, with the input being a preprocessed ECG signal and the output being the predicted class label. The input matrix $X \in \mathbb{R}^{N \times M}$ consists of separate signals in each row. We aim to develop a multiscale representation of these signals by filtering and downsampling them. The filtered and downsampled matrix $y(j) \in Y_{downsampled}$ is computed for each downsampling factor f_i using different filtering techniques.

In this work, we will compare five different multiscale input filtering techniques (subsampling, mean filter, Gaussian filter, bilateral filter and wavelet-based downsampling and denoising method).

Additionally, we will include the original (non-multiscale) 1D signal as a baseline for comparison. This "Original" method will not apply any filtering or downsampling to the input ECG signals, preserving the raw data for the CNN architecture. By including the original signal, we can evaluate the performance improvements, if any, that the multiscale filtering techniques provide over the raw input data.

Subsampling. Subsampling is performed by selecting every f_i-th sample from the input signal x:

$$y(j) = x(j \cdot f_i), \tag{1}$$

where j is the index within the downsampled signal. This process is applied for various downsampling factors $f_1, f_2, \ldots, f_i, \ldots f_n$ to create a multiscale representation of the input signal.

Mean Filter. The Mean Filter technique averages neighboring samples within the input signal $x \in \mathbb{R}^M$:

$$y(j) = \frac{1}{f_i} \sum_{n=0}^{f_i-1} x(j \cdot f_i + n), \tag{2}$$

where j is the index within the downsampled signal. This process is repeated for different downsampling factors to obtain a multiscale representation of the input signal.

Gaussian Filter. The Gaussian filter smooths the input signal x_i with a Gaussian function:

$$y(j) = \frac{1}{\sqrt{2\pi}\sigma} \sum_{n=-\infty}^{\infty} x_i(j+n) e^{-\frac{n^2}{2\sigma^2}}, \tag{3}$$

where σ is the standard deviation of the Gaussian filter. The filtered signal is then downsampled with a step size equal to the downsampling factor f_i. This process is executed for various downsampling factors to create a multiscale representation of the input signals.

Bilateral Filter. The 1D Bilateral filter [19] is a non-linear, edge-preserving filter that smooths signals while preserving sharp edges. Given an input signal $x \in \mathbb{R}^M$, the output signal y is obtained by applying the 1D bilateral filter to each point j:

$$y(j) = \frac{1}{W} \sum_{n=-\lfloor \frac{k}{2} \rfloor}^{\lfloor \frac{k}{2} \rfloor} x(j+n) \cdot g_c(x(j+n) - x(j)) \cdot g_s(n), \tag{4}$$

where W is the normalization term, k is the kernel size, $g_c(\cdot)$ is the color Gaussian function with standard deviation σ_c, and $g_s(\cdot)$ is the spatial Gaussian function with standard deviation σ_s. The color Gaussian function and the spatial Gaussian function are defined as follows:

$$g_c(\Delta x) = e^{-\frac{(\Delta x)^2}{2\sigma_c^2}}, \tag{5}$$

$$g_s(n) = e^{-\frac{n^2}{2\sigma_s^2}}. \tag{6}$$

Wavelet-Based Downsampling and Denoising Method. The wavelet-based method for downsampling and denoising [7] time series data is built upon the discrete wavelet transform (DWT), thresholding techniques, and signal reconstruction.

Given a time series data X, the DWT decomposes the signal into a set of wavelet coefficients as follows:

$$X = A_n + D_n + D_{n-1} + \cdots + D_1, \tag{7}$$

where A_n is the approximation coefficients at level n, and D_i represents the detail coefficients at level i. The DWT is performed using a chosen wavelet function, such as the Daubechies wavelet (e.g., 'db4'), which offers a balance between smoothness and compact support.

The wavelet coefficients are thresholded to denoise the data. The threshold value (T) is computed as:

$$T = k \cdot \text{median}(|D_n|), \tag{8}$$

where k is the downsample factor, and $|D_n|$ is the absolute value of the detail coefficients at the highest level of decomposition. The threshold is applied using either a soft or hard thresholding mode. Soft thresholding is defined as:

$$Y_i = \text{sign}(D_i) \cdot \max(0, |D_i| - T), \tag{9}$$

where Y_i is the thresholded coefficient, and $\text{sign}(D_i)$ represents the sign of the detail coefficient D_i.

Hard thresholding is defined as:

$$Y_i = D_i \cdot I_{(|D_i|>T)}, \tag{10}$$

where $I_{(|D_i|>T)}$ is an indicator function that equals 1 if the condition is true and 0 otherwise. In our experiment we use only soft thresholding.

The denoised signal is reconstructed using the inverse discrete wavelet transform (IDWT) as follows:

$$X_{\text{denoised}} = \text{IDWT}(A_n, Y_n, Y_{n-1}, \dots, Y_1), \tag{11}$$

where $x(j) \in X_{\text{denoised}}$ represents the denoised signal, and Y_i corresponds to the thresholded detail coefficients at level i.

Finally, the subsampling is performed by selecting every f_i-th sample from the input signal x:

$$y(j) = x(j \cdot f_i), \tag{12}$$

where j is the index within the downsampled signal. This process is applied for various downsampling factors $f_1, f_2, \dots, f_i, \dots f_n$ to create a multiscale representation of the input signal.

4.2   Multiscale Feature Maps

In this section, we explore the effects of incorporating different multiscale feature map blocks in our CNN models for ECG classification. We investigate four distinct types of multiscale blocks, each designed to capture different aspects of the input data. These blocks are compared against a base model that serves as a foundation for our ablation study.

Through the ablation study, we aim to quantify the contribution of the each block to the overall performance of our model by removing it and comparing the results with and without the block.

Base Model. We present a base model that serves as a starting point for our ablation study. The base model consists of three 1-D convolutional layers with different filter sizes (3, 5, and 7) and a hyperbolic tangent (tanh) activation function. No pooling or residual connections are employed in this base model. The base model serves as a foundation for further investigation and comparison with other more complex models that incorporate techniques such as multiscale blocks, residual connections, and different types of convolutions.

Multiscale Block Inception Convolution. In this ablation study, we investigate the impact of a used block known as the multiscale block inception convolution (further Inception Convolution) on the performance of our model. The Inception Convolution is inspired by the inception module introduced by [26] in their seminal work on the GoogLeNet architecture. The inception module aims to capture various spatial and channel-wise patterns within the input tensor by employing different filter sizes in parallel.

The Inception Convolution consists of three convolutional layers with different filter sizes (1, 3, and 5) and a max-pooling layer with a pool size of 3. The output feature maps of these layers are concatenated to form a combined feature map.

Multiscale Block with Depthwise Separable Convolution. We investigate the impact of another used block known as the multiscale block with depthwise separable convolution [11] (further Depthwise Convolution) on the performance of our model. The Depthwise Convolution employs depthwise separable convolutions, which factorize a standard convolution into a depthwise convolution followed by a point-wise convolution, thus reducing the number of parameters and computational cost.

The Depthwise Convolution consists of three depthwise separable convolutional layers with different filter sizes (3, 5, and 7). The resulting feature maps from these layers are merged also together to create a unified feature map.

Multiscale Block with Dilated Convolution. We examine the influence of an alternative building block, known as the multiscale block with dilated convolution [30] (Dilated Convolution), on our model's performance. The Dilated Convolution employs dilated convolutions, which introduce a dilation factor to increase the receptive field of the convolutional layers without increasing the number of parameters or computational cost.

The Dilated Convolution consists of three convolutional layers with different filter sizes (3, 5, and 7) and a dilation rate of 2. The output feature maps of these layers are concatenated to form a combined feature map.

Multiscale Block with Residual Connections. We explore the effect of incorporating an additional block known as the multiscale block with residual connections [10] (also referred to as Residual Connections) on our model's performance.

The Residual Connections employs residual connections, which are a technique to mitigate the vanishing gradient problem in deep networks by allowing the gradients to flow through skip connections, thus improving the model's training and performance.

The Residual Connections consists of three pairs of convolutional layers with different filter sizes (3, 5, and 7) and a residual connection for each pair. The resulting feature maps from these layers are combined to create a unified feature map.

4.3 Training Details

We use the Adam optimizer for the training of the models, minimizing a categorical cross entropy loss function. The models are trained for a maximum of 10,000 epochs with early stopping, which monitors the validation loss and has a patience of 10 epochs. We use a validation split of 0.2, meaning that 20% of the training dataset is reserved for validation purposes.

In our methodology, we employ the Glorot Uniform initializer (also known as Xavier Uniform initializer) for weight initialization in our 1D convolutional layers. The Glorot Uniform initializer is designed to maintain a specific variance in the activations of the neurons, which helps avoid vanishing or exploding gradients during training.

For each model, we carry out 10 training sessions to mitigate the stochasticity of the training experiment. In each session, the dataset is randomly split into training and validation subsets, and the best model validated on the validation set is saved at the end of the training session. The performance of the trained models is then evaluated on the test set.

5   Experimental Results

In our ablation study, we used the AbnormalHeartbeat dataset for the task of classifying heartbeat recordings into one of five classes. This dataset contains a total of 606 instances, with each instance being a time series of length 3,053. The time series represent the change in amplitude over time during an examination of patients suffering from common arrhythmias. The dataset is divided into training and testing sets, each containing 303 instances.

The AbnormalHeartbeat dataset was obtained from a combination of sources, including the iStethoscope Pro iPhone app and clinical trials using the digital stethoscope DigiScope. All instances were resampled to 4,000 Hz and truncated to the shortest instance length. The original data can be found at the provided link, and the original paper is by [3].

The dataset is composed of five classes:

- **Artifact** (40 cases): These are recordings that contain noise or other artifacts that can obscure the true nature of the heartbeat. They are not indicative of any particular disease but are important to recognize in order to distinguish them from pathological cases [9].

- **ExtraStole** (46 cases): This class corresponds to recordings of heartbeats with premature ventricular contractions (PVCs) or extra systoles. PVCs are extra heartbeats that disrupt the normal rhythm of the heart and can be caused by various factors such as stress, caffeine, and heart diseases [31].
- **Murmur** (129 cases): Murmurs are abnormal heart sounds caused by turbulent blood flow across the heart valves. They can be indicative of various heart conditions, such as valve stenosis, regurgitation, or congenital heart defects [4].
- **Normal** (351 cases): This class consists of recordings of normal heartbeats, which exhibit a regular rhythm and no abnormal sounds. A normal heartbeat typically includes two main sounds: the first (S1) and second (S2) heart sounds, caused by the closure of the atrioventricular and semilunar valves, respectively [5].
- **ExtraHLS** (40 cases): This class corresponds to recordings with extra heart lung sounds, which are additional heart sounds that can be indicative of certain cardiac conditions, such as heart failure or pericarditis.

Algorithm 1. Ablation study of multiscale inputs and multiscale feature map blocks

---

$DownFactors \leftarrow [8, 4, 2, 1]$
$Results \leftarrow []$
$Reps \leftarrow 10$
for each $InputFunc$ in $InputFuncs$ **do**
    for each $MultiFunc$ in $MultiFuncs$ **do**
        for i in $Reps$ **do**
            $TrainData \leftarrow InputFunc(X\_train, DownFactors)$
            $TestData \leftarrow InputFunc(X\_test, DownFactors)$
            $Model \leftarrow main\_model(MultiFunc)$
            Compile $Model$ with loss, optimizer, and metrics
            Define early stopping callback
            Fit $Model$ with $TrainData$, y_train and validation split
            $Score \leftarrow Model.evaluate(TestData, y\_test)$
        Compute accuracy and loss
        Store results in $Results$

As shown in Algorithm 1, we conducted an ablation study to compare the performance of different multiscale input filtering techniques and multiscale feature map blocks. The default train-test split was created through a random partition. In our ablation study, we investigated the impact of incorporating different multiscale feature map blocks in our CNN models for ECG classification. We analyzed four distinct types of multiscale blocks against a base model. By comparing the performance of each model, we aimed to determine the best multiscale input block and feature map block for ECG classification.

Table 1. Experimental results sorted by accuracy evaluated over testing set (descending)

Multiscale Inputs	Multiscale Block	Max Accuracy (%)
Mean Filter	Residual Connections	64.47
Gaussian Filter	Residual Connections	63.49
Wavelet Denoising	Residual Connections	63.49
Bilateral Filter	Residual Connections	63.16
Gaussian Filter	Base Model	63.16
Bilateral Filter	Base Model	62.17
Subsampling	Residual Connections	62.17
Subsampling	Base Model	61.84
Bilateral Filter	Inception Convolution	61.51
Gaussian Filter	Inception Convolution	61.18
Original	Base Model	61.18
Mean Filter	Base Model	60.86
Gaussian Filter	Depthwise Convolution	60.86
Mean Filter	Dilated Convolution	60.53
Gaussian Filter	Dilated Convolution	60.53
Subsampling	Depthwise Convolution	60.53
Mean Filter	Depthwise Convolution	60.20
Original	Residual Connections	59.87
Subsampling	Dilated Convolution	59.87
Bilateral Filter	Depthwise Convolution	59.54
Subsampling	Inception Convolution	59.54
Wavelet Denoising	Dilated Convolution	59.87
Wavelet Denoising	Base Model	60.86
Mean Filter	Inception Convolution	58.22
Bilateral Filter	Dilated Convolution	58.88
Original	Inception Convolution	57.89
Original	Depthwise Convolution	57.89
Original	Dilated Convolution	57.89
Wavelet Denoising	Inception Convolution	57.89
Wavelet Denoising	Depthwise Convolution	57.89

Please refer to the experimental results table for a detailed comparison of the performance of each model in terms of accuracy. In Table 1 we can identify the best multiscale input block and feature map block for this task.

6   Discussion and Conclusion

In this study, we investigated the impact of different multiscale input filtering techniques and multiscale feature map blocks on the performance of CNN architectures for ECG classification. We evaluated five multiscale input filtering techniques (subsampling, mean filter, Gaussian filter, bilateral filter, and wavelet-based downsampling and denoising method) and four multiscale feature map blocks (base model, depthwise convolution, inception convolution, and different filters) using the AbnormalHeartbeat dataset.

Our experimental results showed that the combination of the mean filter for multiscale input and the residual connections for multiscale block provided the best performance with an accuracy of 64.47%. The residual connections consistently achieved higher accuracy across different multiscale input filtering techniques, highlighting the effectiveness of using residual connections in the multiscale blocks for enhancing the performance of CNN models for ECG classification.

It is worth noting that the results are specific to the dataset and problem at hand. Therefore, further experimentation with different datasets and ECG classification tasks may yield different optimal combinations of multiscale input filtering techniques and multiscale feature map blocks. Additionally, future research can explore other filtering techniques and multiscale block designs to further improve the performance of CNN models for ECG classification. An experiment was also performed on this data in the paper [20]

In conclusion, our study demonstrated the importance of multiscale analysis in ECG classification and provided insights into the effectiveness of different multiscale input filtering techniques and multiscale feature map blocks in CNN architectures. Our findings can be useful for designing effective CNN models for ECG classification, which can have practical applications in the field of cardiology, such as in automated ECG diagnosis systems for early detection of arrhythmias and other cardiac conditions.

Acknowledgment. We would like to thank the software technology company Rozinet s.r.o. (www.rozinet.net) without whose support, this article would not have been possible to create by providing technical resources and space for research.

References

1. Physionet/cinc challenge (2017). https://physionet.org/content/challenge-2017/. Accessed 08 Mar 2023
2. Antol, S., et al.: VQA: visual question answering. In: Proceedings of the IEEE International Conference on Computer Vision, pp. 2425–2433 (2015)
3. Bentley, P., Nordehn, G., Coimbra, M., Mannor, S., Getz, R.: The pascal classifying heart sounds challenge 2011 (chsc2011) results (2011)
4. Bonow, R.O., Otto, C.M.: Valvular Heart Disease: A Companion to Braunwald's Heart Disease. Elsevier (2020)
5. Boron, W.F., Boulpaep, E.L.: Medical physiology E-book. Elsevier Health Sciences (2016)

6. Cui, Z., Chen, W., Chen, Y.: Multi-scale convolutional neural networks for time series classification. arXiv preprint arXiv:1603.06995 (2016)
7. Donoho, D.L.: De-noising by soft-thresholding. IEEE Trans. Inf. Theory **41**(3), 613–627 (1995)
8. Fu, J., et al.: Dual attention network for scene segmentation. IEEE Trans. Image Process. **29**, 7873–7887 (2020)
9. Grady, D.: Manual of Medical-Surgical Nursing Care: Nursing Interventions and Collaborative Management. Elsevier Health Sciences (2010)
10. He, K., Zhang, X., Ren, S., Sun, J.: Deep residual learning for image recognition. In: Proceedings of the IEEE Conference on Computer Vision and Pattern Recognition, pp. 770–778 (2016)
11. Howard, A.G., et al.: MobileNets: efficient convolutional neural networks for mobile vision applications. arXiv preprint arXiv:1704.04861 (2017)
12. Krizhevsky, A., Sutskever, I., Hinton, G.E.: ImageNet classification with deep convolutional neural networks. In: Advances in neural information processing systems, pp. 1097–1105 (2012)
13. LeCun, Y., Bottou, L., Bengio, Y., Haffner, P.: Gradient-based learning applied to document recognition. Proc. IEEE **86**(11), 2278–2324 (1998)
14. Li, X., Li, F., Zhou, X., et al.: A hybrid CNN-RNN model for atrial fibrillation detection from ECG recordings. In: Proceedings of the 2018 Computing in Cardiology Conference, pp. 1–4 (2018)
15. Li, Z., Peng, C., Yu, G., Zhang, X., Deng, Y.: Pyramid scene parsing network. In: Proceedings of the IEEE conference on computer vision and pattern recognition, pp. 2881–2890 (2017)
16. Lin, T.Y., Goyal, P., Girshick, R., He, K., Dollar, P.: Focal loss for dense object detection. In: Proceedings of the IEEE International Conference on Computer Vision, pp. 2980–2988 (2017)
17. Long, J., Shelhamer, E., Darrell, T.: Fully convolutional networks for semantic segmentation. In: Proceedings of the IEEE Conference on Computer Vision and Pattern Recognition, pp. 3431–3440 (2015)
18. Ozturk, S., Yildirim, O., Kocadag, I., et al.: A novel 1D convolutional attention model for ECG classification. J. Med. Syst. **44**(7), 1–10 (2020)
19. Paris, S., et al.: Bilateral filtering: theory and applications. Found. Trends Comput. Graph. Vision **4**(1), 1–73 (2009)
20. Raza, A., Mehmood, A., Ullah, S., Ahmad, M., Choi, G.S., On, B.W.: Heartbeat sound signal classification using deep learning. Sensors **19**(21), 4819 (2019)
21. Ren, S., He, K., Girshick, R., Sun, J.: Faster R-CNN: towards real-time object detection with region proposal networks. In: Advances in neural information processing systems, pp. 91–99 (2015)
22. Ronneberger, O., Fischer, P., Brox, T.: U-Net: convolutional networks for biomedical image segmentation. In: Navab, N., Hornegger, J., Wells, W.M., Frangi, A.F. (eds.) MICCAI 2015. LNCS, vol. 9351, pp. 234–241. Springer, Cham (2015). https://doi.org/10.1007/978-3-319-24574-4_28
23. Rumelhart, D.E., Hinton, G.E., Williams, R.J.: Learning representations by back-propagating errors. Nature **323**(6088), 533–536 (1986)
24. Russakovsky, O., et al.: ImageNet large scale visual recognition challenge. Int. J. Comput. Vision **115**(3), 211–252 (2015)
25. Simonyan, K., Zisserman, A.: Very deep convolutional networks for large-scale image recognition. arXiv preprint arXiv:1409.1556 (2014)
26. Szegedy, C., et al.: Going deeper with convolutions. In: Proceedings of the IEEE Conference on Computer Vision and Pattern Recognition, pp. 1–9 (2014)

27. Szegedy, C., Vanhoucke, V., Ioffe, S., Shlens, J., Wojna, Z.: Going deeper with convolutions. In: Proceedings of the IEEE Conference on Computer Vision and Pattern Recognition, pp. 1–9 (2015)
28. Taigman, Y., Yang, M., Ranzato, M., Wolf, L.: DeepFace: closing the gap to human-level performance in face verification. In: Proceedings of the IEEE Conference on Computer Vision and Pattern Recognition, pp. 1701–1708 (2014)
29. Xu, K., et al.: Show, attend and tell: neural image caption generation with visual attention. In: International Conference on Machine Learning, pp. 2048–2057 (2015)
30. Yu, F., Koltun, V.: Multi-scale context aggregation by dilated convolutions. arXiv preprint arXiv:1511.07122 (2015)
31. Zipes, D.P., Jalife, J.: Cardiac Electrophysiology: From Cell to Bedside. Elsevier Health Sciences (2018)

Fast Convolutional Analysis
of Task-Based fMRI Data for ADHD
Detection

Federica Colonnese, Francecso Di Luzio, Antonello Rosato,
and Massimo Panella[✉]

Department of Information Engineering, Electronics and Telecommunications
University of Rome "La Sapienza", 00184 Rome, Italy
{antonello.rosato,massimo.panella}@uniroma1.it

Abstract. Among the most common neurodevelopmental disorders,
Attention Deficit Hyperactivity Disorder (ADHD) is a complex and challenging one to identify. This happens because there are no objective medical techniques, and diagnoses are made only on interviews and a set of
symptoms tests evaluated by psychiatrists. However, in recent years, the
use of Deep Learning techniques has emerged as a promising solution
for accurately classifying ADHD, using non-intrusive advanced imaging
techniques such as fMRI data. Specifically, task-based fMRI data can
be especially valuable for identifying ADHD, as it enables researchers to
examine the functional activity of the brain during tasks that involve
working memory, which is known to be affected in individuals with this
disorder. The presented paper introduces 3D-ADHD, a 3D Convolutional Neural Network-based approach that uses individual time instant
of working memory task-based fMRI data for real-time ADHD classification. The proposed model not only achieves remarkable accuracy, but is
also very fast in the inference phase. This makes 3D-ADHD suitable for
real-time classification and integration into medical systems, including
computer-aided diagnosis one. As a result, this model has great potential for clinical use and can significantly contribute to the diagnosis of
ADHD. This work has the potential to advance the accurate and efficient
detection of ADHD with real-time diagnoses, opening new opportunities
for research on faster detection and early treatment of this dysfunction.

Keywords: ADHD Detection · Deep Learning · Task-based fMRI
Data · Fast Detection

1 Introduction

Nowadays neuroimaging has become a valuable tool in studying neuropsychiatric
disorders. The use of deep learning (DL) and machine learning (ML) in medical
image analysis has enabled researchers to identify complex patterns and features
that are difficult for humans to detect. In fact, ML and DL are now widely used
in neuroimaging to identify clinical signs of neurodevelopmental disorders and
develop automated diagnostic systems [21]. ML and DL have shown promising

I. Rojas et al. (Eds.): IWANN 2023, LNCS 14135, pp. 364–375, 2023.
https://doi.org/10.1007/978-3-031-43078-7_30

results analyzing medical data such as electroencephalogram (EEG), Magnetic Resonance Imaging (MRI), and functional MRI (fMRI) [3,12]. fMRI is a non-invasive imaging technique that is preferred for studying the human brain [7]. There are two approaches to fMRI data analysis: task-based and resting-state. Task-based fMRI investigates how specific brain regions or networks are involved in different cognitive processes, while resting-state fMRI measures brain activity when the patient is not focused on any particular task. Deep learning algorithms have been used to analyze fMRI data to identify patterns of brain activity that are associated with specific cognitive processes, emotions, and neurological disorders [23]. Convolutional neural networks (CNNs) are commonly used for analyzing fMRI images to extract features like regional activation patterns. They can identify these patterns across different subjects and conditions, making them useful for detecting neuropsychiatric disorders such as schizophrenia [24]. Other DL technologies have shown remarkable accuracy and efficiency in tackling difficult tasks like gesture recognition [14,15], neurorehabilitation [11], and patterns' detection of Autism Spectrum Disorder and Alzheimer's disease [6,8,19].

Among those studies, growing attention has been given to Attention Deficit Hyperactivity Disorder (ADHD) which is one of the most common neurodevelopmental disorders, affecting 4-12 % of young children and 4-5 % of adults [22]. When it comes to ADHD, and more in general to neurodevelopmental disorder, the main problem is that the standard diagnostic procedures (defined by The Diagnostic and Statistical Manual of Mental Disorders (DSM-5) [1]) are based only on interviews and a set of symptom tests evaluated by psychiatrists. ADHD is frequently misdiagnosed because its symptoms can be similar to other conditions, and healthcare professionals may lack adequate training, also the stigma surrounding ADHD may result in under-reporting or underdiagnosis. Because of what just said, accurate diagnosis can be challenging to achieve due to the subjective nature of these diagnostic methods [5]. Evaluating ADHD using DNN in order to create automated diagnosis tools could assist the diagnosis process. It is important to note, however, that automated diagnosis tools should not be used to replace standard diagnostic procedures. Instead, they could serve as a useful screening tool, allowing clinicians to identify individuals who may require further assessment. This could help to reduce the time and resources as are necessary for diagnosis, and improve access to early treatments for those who need it.

Extensive brain imaging studies have provided evidence supporting the notion that ADHD is associated with structural and functional brain differences: numerous regions have been implicated in the pathophysiology of ADHD, revealing how it exhibits distinct neurological responses to the same stimuli when compared to neurotypical individuals [10,18]. The use of DNN for detecting ADHD from resting-state fMRI data has been explored by [16] with FCNet: a deep learning technique that has been introduced as a means for classifying ADHD. The method employs a CNN that extracts functional brain connectivity directly from fMRI time-series signals overcoming the problem that traditional methods might not be able to capture the latent characteristics of raw time-series signals. The method then resorts to traditional ML approaches for feature extraction and an SVM classifier for classifying labels. Also in [17] the same authors proposed

DeepFMRI. The DeepFMRI method is composed of three consecutive networks: a feature extractor, a functional connectivity network, and a classification network that achieves a classification accuracy of 73.1%. The model is specifically created to receive preprocessed fMRI time series signals as input and produce a diagnosis. One of the main advantages is that DeepFMRI is an end-to-end deep learning model and this means that it can learn the features and classify ADHD patients in a single framework. This approach avoids the need for manual feature extraction, which can be time-consuming and error-prone. But this also means that is a complex model that requires significant computational resources for training and inference. Also task-based fMRI has been used to train DL models. For example, [4] employs visuospatial working memory (VSWM) fMRI data. The authors conducted the analysis on data from 20 boys with ADHD and 20 typically developed, using brain activity from 16 regions of interest that were significantly activated or deactivated during VSWM tasks, identified using an independent univariate analysis. They reduced the number of input variables for the classifier by applying sparse principal component analysis. Then, the classification model for ADHD was trained using logistic regression, resulting in good accuracy. Also, [20] utilized a transfer learning approach, specifically a ResNet-50 pre-trained 2D-Convolutional Neural Network (CNN), to classify ADHD and healthy children automatically, using as input task-based fMRIs recorded from 20 children without neuropsychiatric conditions and 19 children with ADHD while doing a spatial attention task. The results showed that the ResNet-50 architecture achieved a remarkable overall classification accuracy with 10-k cross-validation.

The employment of DNNs for this specific problem is constantly evolving, and there is an increasing demand for an approach that can accelerate and support the entire diagnosis procedure. This paper introduces a method for detecting ADHD that employs a DNN based on three-dimensional convolutional layers. The classification accuracy achieved by this approach exceeds the current state-of-the-art results. The dataset utilized in this study is comprised of preprocessed task-based fMRIs extracted from the dataset proposed in [13]. The main key contributions of this research can be summarized as follows:

- The model focuses on utilizing fMRI recordings during working memory tasks, whereas most previous studies on ADHD detection through ML and DL solutions have relied on resting-state fMRI. This is important because individuals with ADHD tend to exhibit deficits in working memory compared to those without the condition, as has been demonstrated in previous research [9].
- It provides valuable insights into the potential utility of task-based fMRI recordings in the diagnosis of ADHD, which could lead to improved treatment options for individuals with this condition.
- Lastly, the input of the DNN is made of single-time steps of the fMRI. This allows for the classification of brain activity to be performed instantaneously rather than across the entire fMRI scan. By breaking down the data into individual time steps, the DNN is able to analyze neural patterns presented

in each moment, providing a more detailed and accurate understanding of the brain's activity.

This whole approach not only increases the speed and efficiency of the classification process but also offers the potential for more precise and targeted diagnoses and treatments in the field of neuroscience.

In the following paragraphs, we will provide a detailed explanation of the adopted dataset and the training procedure, along with an accurate description of the achieved accuracy results.

2   Problem Statement

Through our research, we aim to contribute to the state-of-the-art by focusing specifically on working memory task-based fMRI data.

The dataset used in this paper pertains to the one proposed in [13]. The original dataset is composed of task-based fMRI, MRI scans and also scores obtained from a series of standardized measures aimed at assessing: cognitive ability, ADHD symptoms questionnaires, and reading skills. To conduct this study, we have exclusively utilized fMRIs which are a total of 543 obtained from 79 children, including 35 with ADHD and 44 without ADHD. FMRI scans are organized according to the Brain Imaging Data Structure (BIDS) and were acquired while participants completed a series of 8 different tasks specifically designed to investigate working memory, monetary reward, and feedback processing in both typically developing children and those diagnosed with ADHD. Going into details, the eight different working memory tasks varied in the domain (i.e. spatial, verbal), reward amount (i.e. large, small), and feedback (i.e. immediate, delayed). Participants had to make judgments in two different domains depending on the task. In the verbal working memory task (V), they had to determine if the letter on the screen was the same as the one presented n letters back. In the spatial working memory task (S), they had to judge if the letter on the screen was in the same position as the one presented n letters back. To respond, participants used a right-handed button box with two buttons. The tasks also differed in the amount of reward offered. Participants were informed that they would earn .02 or .25 for every correct response in the small (S) and large (L) reward tasks, respectively. Additionally, the tasks had two types of feedback times: immediate (I) and delayed (D). In the immediate feedback tasks, a green or red fixation square would appear after each letter presentation to indicate a correct or incorrect response. In the delayed feedback tasks, participants would see a black fixation square after each letter presentation, and at the end of each experimental block, they would be informed of their percentage of correct responses.

Slices were acquired interleaved from bottom to top with even slices acquired first. Details regarding parameters applied when acquiring fMRI can be found in Table 1, where TR (Time Repetition) is the amount of time that passes between consecutive acquired brain volumes, TE (Time Echo) is the time between the excitation of protons in the brain and the measurement of the signal produced by the protons as they relax back to their original state, and Voxel Size refers to the volume of a three-dimensional pixel (called 'voxel') that represents a small volume of brain tissue.

Table 1. Parameters used for fMRI acquisition

Parameter	Value
TR	2000 ms
TE	20 ms
Matrix size	128 × 120
Slice thickness	3 mm (0.48 mm gap)
Number of slices	32
Number of volumes for each task	139

Since the fMRI in the original dataset presented in [13] are made of raw data, the fMRIs used in this paper underwent a series of preprocessing steps. To do that, we used the latest version of SPM (Statistical Parametric Mapping), a software package for fMRI analysis run in Matlab ® [2]. The preprocessing steps employed in this paper are standard procedures when dealing with this type of data and we decided to not use raw fMRIs due to their high dimensionality, which can make it difficult to train DL models, and also because of their limited generalization capability. We will provide a brief explanation of why and how we applied these preprocessing steps, both to ensure the replicability of the process and also to enable readers to better understand the differences. The purpose of the preprocessing we decided to apply was twofold: firstly, to distinguish real brain signals from noise caused by factors like head movements and physiological processes; and secondly, to standardize the fMRI scans by aligning them with a common space. This is important because it reduces variability between individuals and allows for meaningful group analysis, cross-study comparisons, and the generalization of findings to the population level.

The preprocessing steps employed are summarized in Fig. 1 and briefly explained in the following lines.

In the preprocessing step we decided to leave most of the default values as they are. This was done because in SPM these defaults have been calculated to produce the best results for a wide range of image field of views, voxel sizes, and so on. Realignment: the objective of this step is to remove any motion artifacts in the scans. This is achieved by aligning all scan volumes to ensure that the brain is in the same position. The amount of misalignment of each volume with a reference volume is assessed. The first volume is chosen as the reference volume for this purpose. Slice timing: the purpose of this step is to rectify any discrepancies in image acquisition time across slices. In this paper interleaved slice acquisition was used: it acquires every other slice, and then fills in the gaps on the second pass. Coregistration: it aligns the structural (MRI) and functional (fMRI) data to improve the translation into a standardized space. Segmentation: automatic segmentation of SPM was used in order to segment the anatomical image to extract various brain regions for subsequent use. Normalization: it adjusts the anatomical image to a standardized space using a nonlinear transformation algorithm. Here, in the writing options we changed the voxel resolution of the images that are warped. The default of $2 \times 2 \times 2$ will create higher-resolution images, but the files would take up too much space, so in order to create smaller files with lower resolution we changed this to $3 \times 3 \times 3$. Smoothing: it applies a spatial filter to the image to reduce noise and improve the signal-to-noise ratio. The choice of the smoothing kernel size should balance improving the signal-to-noise ratio with preserving spatial resolution.

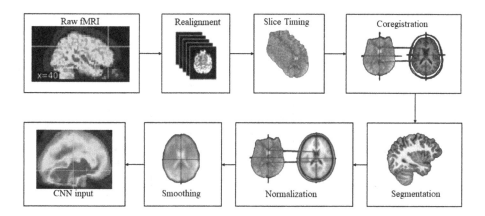

Fig. 1. Summarized representation of the preprocessing steps employed

3 Proposed Methodology

One of the most important contributions of the approach herein presented is that the spatial coordinates of the task-based fMRIs are sufficient to model, train and implement a fast and precise classification model for ADHD detection, as will be proven in the following.

3.1 Data Adaptation

Each data sample in the dataset is an fMRI exam which can be represented by a tensor with dimensions: $T \times X \times Y \times Z$ where:

- T is the number of time steps of each fMRI;
- X, Y, Z are the spatial coordinates of the slices of the fMRI exam.

Each data instance is labeled with a single label which assumes the categorical value of 0 if the exam related to that data sample is associated with a typical patient, and 1 if it is the exam of a patient suffering from ADHD. To increase the number of samples in the dataset and to streamline the training procedure we employed every single time step of the exam as a single data sample, replicating the label related to that instance.

Hence, the final dataset can be represented by a four-dimensional tensor $\mathbf{X}$ whit this dimensionality: $N \times X \times Y \times Z$ where N is the number of samples in the dataset. In this sense, the dataset is handled as a 3D-images database, and the implemented network, which is based on three-dimensional convolutional layers, reflects this aspect.

3.2 Model Design

The proposed architecture, which will be denoted in the following as '3D-ADHD' is illustrated in Fig. 2.

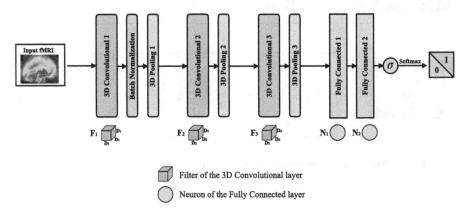

Fig. 2. Graphical representation of the proposed 3D-ADHD model architecture for ADHD detection

The involved layers are described in the following:

- *3-D Convolutional layer 1.* It is a one-channel three-dimensional convolutional layer receiving the i-th sample x_i. It applies F_1 different convolutional filters. Each filter is associated with a $D_1 \times D_1 \times D_1$ nonlinear kernel which uses a Rectified Linear Unit (ReLU) activation function.
- *Batch normalization layer.* It normalizes the inputs from the first convolutional layer using the means and standard deviations of the batches of inputs seen during training, obtaining for each input a corresponding output tensor to be fed to the successive pooling layer.
- *Average Pooling layer 1.* It applies a 3D average pooling over the output of the batch normalization layer. The pooling kernel dimension is fixed to 2 and the stride of the same is fixed to 1.

- *3-D Convolutional layer 2*. It is a second three-dimensional convolutional layer. In this case, each of the F_2 filters is associated with a $D_2 \times D_2 \times D_2$ nonlinear kernel which uses a ReLU activation function.
- *Average Pooling layer 2*. In this case, it is used to apply three-dimensional average pooling directly over the output of the previous convolutional layer. Also in this case the pooling kernel dimension is fixed to 2 and the stride of the same is fixed to 1.
- *3-D Convolutional layer 3*. This is the last 3D convolutional layer of the proposed architecture. It applies F_3 filters, each associated with a $D_3 \times D_3 \times D_3$ nonlinear kernel which uses a ReLU activation function.
- *Average Pooling layer 3*. It is the last pooling layer and it is used to end the feature extraction process before passing the data to the fully connected layers. The pooling kernel dimension is fixed to 2 and the stride of the same is fixed to 1.
- *Flatten layer*. This is utilized to compress the tensor linked with the standardized characteristic maps into a one-dimensional column vector, which serves as the input for the following dense layer.
- *Fully Connected layer 1*. It is the typical feed-forward layer of the network with N_1 neurons connecting the flattened features to the succeeding final layers. Also for this layer, the employed activation function is the ReLU.
- *Fully Connected layer 2*. It is the second feed-forward layer characterized this time by N_2 neurons with ReLU activation function.
- *Fully Connected layer 3*. It is the last layer, consisting of one only neuron with a sigmoid activation function for the prediction of the probability of belonging to the ADHD class.

4   Results

To evaluate and assess the performance of the proposed methodology, we perform a comparison with the state-of-the-art models for ADHD detection with task-based fMRI data. As benchmark papers for our study, we have selected two models that utilize task-based fMRI data, both of which were mentioned in Sect. 1. The first model proposed in [4], which will be denoted in the following as 'LRM' (Logistic Regression Method), uses visuospatial working memory (VSWM) fMRI data from which they extracted 16 brain regions of interest through independent univariate analysis. Then they reduced the number of input variables using independent univariate analysis and employed Logistic Regression as a classifier. The second benchmark proposed presented in [20] will be denoted as 'TLM' (Transfer Learning Method). Here the authors proposed a transfer learning approach on the ResNet-50 type pre-trained 2D-Convolutional Neural Network on fMRI data recorded while children were doing spatial-attention tasks. The model presented in this paper is trained 5000 samples, while the test set is composed of 8900 samples. In this case, the number of samples in the test set is greater than the number of samples in the training set. This is primarily due to two different factors: firstly training the network with fewer samples increases the speed of the training procedure; furthermore, the accuracy rate of the classification procedure is high enough. In this sense, increasing the number of training samples will only result in a more demanding training procedure from the computational point of view. The model has been trained using the ADAM algorithm with the learning rate set to 0.001 and mini-batch size 128, selected as default values. The additional hyperparameters and training options have been set through the employment of a grid search procedure performed to avoid overfitting, increasing the generalization capabilities of the proposed approach. The optimal configuration of the hyperparameters found with the grid-search procedure, referring to Fig. 2, is stated in the following for reproducibility tasks. The dimension of the filters of the 3D Convolutional layers is $D_1 = D_2 = D_3 = 3$; the number of filters for the 3D Convolutional layers is $F_1 = 4$, $F_2 = 8$ and $F_3 = 16$ and the number of neurons of the Fully Connected layers is $N_1 = 16$ and $N_2 = 32$. All the experiments were performed using Python and the Pytorch® back-end on a machine equipped with an AMD Ryzen^{TM} 7 5800X 8-core CPU at 3.80 GHz, 64 GB of RAM, and an NVIDIA® GeForce^{TM} RTX 3080 Ti GPU at 1.365 GHz and 12288 MB of GDDR6X RAM using the GPU for training, testing and for the hyperparameters optimization procedure.

Table 2. Classification Performance

Network	Training	Test
3D-ADHD	1.00	1.00
TLM	0.9344	0.9345
LRM	-	0.925

Considering that the initial value of the network parameters is selected randomly, 10 runs are performed for each test and the average and standard deviation values of the error over the different runs are reported, in order to highlight the robustness of the approach. The error measure employed to evaluate the performance of the implemented model is the accuracy, the ratio of well-classified samples over the number of total samples in the training/test set. The loss function used for all the optimization procedures is the binary cross entropy BC, which is defined by Eq. 1:

$$BC = -\frac{1}{S} \sum_{i=1}^{S} \left[y_i \log(\hat{y}_i) + (1 - y_i) \log(1 - \hat{y}_i) \right], \tag{1}$$

where S is the number of fMRIs in the training set, y_i is the actual binary label (either 1 or 0) of the i-th sample and $\hat{y}_i$ is the probability estimated by the adopted model that the i-th sample is representative of ADHD.

The numerical results obtained with the proposed approach are summarized in Table 2. A graphical representation of the numerical results is also displayed in Fig. 3, where the confusion matrix achieved by the model on the test set is proposed. The confusion matrix is also useful to reveal that the test set is not evenly distributed in regards to the number of samples belonging to the two distinct classes and this is true also for the training set. Starting from the numerical results presented in Table 2 it can be seen that the method proposed herein not only outperforms the state-of-the-art methods used for benchmarking purposes, but reaches the perfect accuracy both on the training set and on the test set. The goodness of the presented results, highlighted also by the confusion matrix in Fig. 3, demonstrates the robustness of the proposed method and the reliability of the approach presented in this paper. In fact, it must be underlined once again that the accuracy is computed with 10 different partitions of the dataset, using 10 different seeds for the random initialization of the network parameters. However, the classification accuracy equal to 1 is probably motivated by the employment of a single data set that presents for every sample the same environmental conditions and the same data-generation procedure. Using three-dimensional tensors as input to the network is quite expensive from a computational point of view, leading to an important training time of 1085.607 seconds. Despite this, the inference time for a single sample is practically insignificant, 0.001 seconds and, considering that the period between the acquisition of two different time steps of the fMRIs is $2s$, as soon as the single time step is acquired, it can be passed to the network for the classification. This fact enhances the possibility to perform several classification procedures for ADHD detection during the single task-based fMRI exam, providing a useful tool to be used in real-time in the medical field. The high accuracy reached with this method and the practically insignificant inference time needed, when framed in the medical domain of task-based fMRI, can represent a powerful resource for ADHD detection. In particular, considering that people suffering from this disorder have often issues with working memory-related tasks. Hence, this work paves the way for the construction of neural models trained with task-based fMRI data which can be more efficient in analyzing the different areas of the brain highlighted while the patients are performing certain actions.

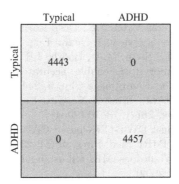

Fig. 3. Confusion matrix obtained with 3D-ADHD on the test set.

5   Conclusions

Considering the emphasis on investigating and uncovering new and valuable applications for deep learning methods in the medical setting, it is of great interest to find new solutions which are tailored to real-world problems. In this paper, we focused on the challenging and crucial problem of ADHD detection from task-based fMRI data, proposing a deep 3D Convolution based neural architecture trained on real-world data. A key element of the dissertation proposal centers around the idea that creating models using fMRI data from task-based training can be particularly beneficial for individuals with ADHD who struggle with working memory. In fact, this type of exam can help highlight the specific challenges faced by these patients. Through experimental analysis, the advantageous aspects of the suggested structure have been validated, with a high level of confidence in reporting its accuracy benefits when compared to the state-of-the-art deep neural models. Moreover, the speed of the inference procedure allows the real-time employment of this algorithm on new data samples during the acquisition of the fMRI, enriching the usability of the model. As there are numerous deep neural models available for addressing this complex problem and the medical field is rapidly expanding, we believe that our work can pave the way for studying even more intricate solutions to be applied to various task-based fMRI datasets for ADHD detection. Moreover, from the clinical point of view, given the easiness of implementation and the fast optimization procedure, such an accurate methodology could represent a useful and helpful tool for assisting neurologists in the detection of this particular neurodevelopmental disorder.

References

1. American Psychiatric Association, D., Association, A.P., et al.: Diagnostic and statistical manual of mental disorders: DSM-5, vol. 5. American psychiatric association Washington, DC (2013)
2. Ashburner, J., et al.: Spm12 manual. Wellcome Trust Centre for Neuroimaging, London, UK 2464(4) (2014)
3. Craik, A., He, Y., Contreras-Vidal, J.L.: Deep learning for electroencephalogram (EEG) classification tasks: a review. J. Neural Eng. **16**(3), 031001 (2019)
4. Hammer, R., Cooke, G.E., Stein, M.A., Booth, J.R.: Functional neuroimaging of visuospatial working memory tasks enables accurate detection of attention deficit and hyperactivity disorder. NeuroImage Clin. **9**, 244–252 (2015)
5. Hartnett, D.N., Nelson, J.M., Rinn, A.N.: Gifted or ADHD? the possibilities of misdiagnosis. Roeper Rev. **26**(2), 73–76 (2004)
6. Haweel, R., et al.: A robust DWT-CNN-based CAD system for early diagnosis of autism using task-based fMRI. Med. Phys. **48**(5), 2315–2326 (2021)
7. Heeger, D.J., Ress, D.: What does fMRI tell us about neuronal activity? Nat. Rev. Neurosci. **3**(2), 142–151 (2002)
8. Heinsfeld, A.S., Franco, A.R., Craddock, R.C., Buchweitz, A., Meneguzzi, F.: Identification of autism spectrum disorder using deep learning and the abide dataset. NeuroImage Clin. **17**, 16–23 (2018)

9. Kofler, M.J., Rapport, M.D., Bolden, J., Sarver, D.E., Raiker, J.S., Alderson, R.M.: Working memory deficits and social problems in children with ADHD. J. Abnorm. Child Psychol. **39**, 805–817 (2011)
10. Konrad, K., Eickhoff, S.B.: Is the ADHD brain wired differently? a review on structural and functional connectivity in attention deficit hyperactivity disorder. Hum. Brain Mapp. **31**(6), 904–916 (2010)
11. Liparulo, L., Zhang, Z., Panella, M., Gu, X., Fang, Q.: A novel fuzzy approach for automatic Brunnstrom stage classification using surface electromyography. Medical & Biological Engineering & Computing **55**(8), 1367–1378 (2017)
12. Lundervold, A.S., Lundervold, A.: An overview of deep learning in medical imaging focusing on MRI. Z. Med. Phys. **29**(2), 102–127 (2019)
13. Lytle, M.N., Hammer, R., Booth, J.R.: A neuroimaging dataset on working memory and reward processing in children with and without ADHD. Data Brief **31**, 105801 (2020)
14. Maisto, M., Panella, M., Liparulo, L., Proietti, A.: An accurate algorithm for the identification of fingertips using an RGB-D camera. IEEE J. Emerg. Sel. Top. Circuits Syst. **3**(2), 272–283 (2013)
15. Panella, M., Altilio, R., Panella: A smartphone-based application using machine learning for gesture recognition: using feature extraction and template matching via Hu image moments to recognize gestures. IEEE Consum. Electron. Mag. **8**(1), 25–29 (2019)
16. Riaz, A., et al.: FCNet: a convolutional neural network for calculating functional connectivity from functional MRI. In: Wu, G., Laurienti, P., Bonilha, L., Munsell, B.C. (eds.) CNI 2017. LNCS, vol. 10511, pp. 70–78. Springer, Cham (2017). https://doi.org/10.1007/978-3-319-67159-8_9
17. Riaz, A., Asad, M., Alonso, E., Slabaugh, G.: DeepFMRI: end-to-end deep learning for functional connectivity and classification of ADHD using fMRI. J. Neurosci. Methods **335**, 108506 (2020)
18. Salmi, J., et al.: Out of focus-brain attention control deficits in adult ADHD. Brain Res. **1692**, 12–22 (2018)
19. Sarraf, S., Tofighi, G.: Classification of alzheimer's disease using fMRI data and deep learning convolutional neural networks. arXiv preprint arXiv:1603.08631 (2016)
20. Uyulan, C., Erguzel, T.T., Turk, O., Farhad, S., Metin, B., Tarhan, N.: A class activation map-based interpretable transfer learning model for automated detection of ADHD from fMRI data. Clin. EEG Neurosci. **54**(2), 151–159 (2023)
21. Vieira, S., Pinaya, W.H., Mechelli, A.: Using deep learning to investigate the neuroimaging correlates of psychiatric and neurological disorders: methods and applications. Neurosci. Biobehav. Rev. **74**, 58–75 (2017)
22. Wilens, T.E., Spencer, T.J.: Understanding attention-deficit/hyperactivity disorder from childhood to adulthood. Postgrad. Med. **122**(5), 97–109 (2010)
23. Yin, W., Li, L., Wu, F.X.: Deep learning for brain disorder diagnosis based on fMRI images. Neurocomputing **469**, 332–345 (2022)
24. Zheng, J., Wei, X., Wang, J., Lin, H., Pan, H., Shi, Y., et al.: Diagnosis of schizophrenia based on deep learning using fMRI. Comput. Math. Methods Med. **2021**, 8437260 (2021)

On Stability Assessment Using the WalkIT Smart Rollator

Manuel Fernandez-Carmona[1]([✉])(iD), Joaquin Ballesteros[2](iD),
Jesús M. Gómez-de-Gabriel[3](iD), and Cristina Urdiales[1](iD)

[1] Ingeniería de Sistemas Integrados Group, University of Málaga, Málaga, Spain
{mfcarmona,acurdiales}@uma.es
[2] ITIS Software, Universidad de Málaga, Málaga, Spain
jballesteros@uma.es
[3] Robotics and Mechatronics Group, University of Málaga, Málaga, Spain
jesus.gomez@uma.es

Abstract. Stability loss may lead to serious injury. Fall risk is often assessed using wearable sensors and/or external motion capture devices. However, these approaches may not be valid for all environments. As rollators are widely used to improve stability and provide physical support, in this work we propose a stability assessment method to proactively predict balance using only onboard rollator sensors. We use Machine Learning to predict how a person's gait is affecting their stability. Prediction can be used to proactively provide warnings or, if the rollator allows it, to physically affect gait to reduce fall risk. The method has been tested by volunteers using a smart rollator. Best results in terms of instability prediction were obtained using Regression Trees.

Keywords: Fall risk · Stability Assessment · Smart Rollator · Machine Learning · Prediction

1 Introduction

Population over 65 years in the EU-27 is expected to increase from 90.5 million at the start of 2019 to reach 129.8 million by 2050 [1]. Elder population is more likely to develop disabilities that may affect their independence [9]. For instance, in Spain, the ratio of dependency of the population older than 64 years was 30.46% in 2021 [6]. This increase will heavily impact healthcare systems, which need to adapt to cope with the increasing demand. The best option is to prevent disability when possible and many approaches rely on technology to do so.

Most these systems rely on monitoring, combined with technologies such as Artificial Intelligence, Data Mining, Home-based Healthcare, and Robotics. Robotics are specifically interesting when people require physical support, as is

This work has been partially supported by projects PID2021-127221OB-I00 (Plan Estatal de Investigación Científica, Técnica y de Innovación 2021–2023) and UMA20-FEDERJA-052 (Programa Operativo FEDER Andalucía 2014–2020).

I. Rojas et al. (Eds.): IWANN 2023, LNCS 14135, pp. 376–387, 2023.
https://doi.org/10.1007/978-3-031-43078-7_31

the case with mobility, which has a major impact on dependency. Mobility assistive devices, such as rollators, canes, or wheelchairs could play an important role in monitoring since they are employed by users in a fairly continuous way. Smart assistive devices include sensors, actuators, and a processing unit to monitor and assist users.

Stability is a parameter of interest in ambulation, as falls may result in major injuries and a higher degree of dependency. Manual stability assessment is performed via supervised medical tests. For example, the Tinetti Mobility Test [19] scores a person's gait and balance in 17 different tasks. Manual assessment requires supervision by clinicians, so it can only be obtained in controlled situations rather than in everyday life. Besides, scales only provide general assessment, rather than fall risk in specific situations, maneuvers, and postures. Automatic assessment extracts information from sensors and, thus, may provide continuous feedback to warn about imminent fall. These methods may require wearable sensors, ambient sensors and/or onboard sensors. Wearable sensors-based approaches attach sensors to specific limbs, joints, or body parts [17,21]. Ambient sensors-based approaches distribute sensors in the environment to capture a person's activity or posture [7]. Finally, sensors can be placed onboard assistive devices [2,5]. Wearable sensors are usually easy to attach and affordable but might require personal calibration and some might not be practical for everyday use. Ambient sensors provide high accuracy, but are frequently expensive and constrained to specific areas where they are installed. Onboard sensors provide less accuracy than ambient sensors, but: i) are attached to the device in a transparent way to the user; ii) monitoring can be performed anytime, anywhere as long as the device is used; and iii) if the device can provide physical help, motion can be altered to improve stability.

A common approach for onboard based fall risk assessment is measuring spatiotemporal gait parameters, such as walking speed [14] or stride-to-stride variability [10], but it needs to be considered that fear of falling may alter these parameters [15], so it may be necessary to use additional information. Posture, captured by range sensors on board can also be used as a fall risk estimator, for example, by calculating the projection of the person's center of mass [4] or their base of support [11], or by detecting unusual postures/positions that may lead to balance disorders [18]. However, posture alone does not account for the weight supported on the device for balance estimation [12,20].

This work presents a novel method for stability assessment based on a smart rollator including sensors for handlebar support, feet position, and odometry, which is described in Sect. 2. Although the technique can be extended to any other rollator with this sensor configuration, in this work we use a low-cost Open modular rollator set proposed by the authors in [8] (WalkIT). Its main advantages are that it can be deployed on any commercial conventional rollator in a simple way and that it relies on cheap, off-the-shelf sensors and electronics. Thus, both the device and the method can be easily reproduced. In order to work with explainable data, we have adapted the balance assessment method in [16] to rollator users. This method relies on projection of a person's center of mass (CoM) on feet/ground contact surface during gait. Subsection 2.1 presents the extension of this approach to rollator users, and Subsect. 2.2 shows how all required parameters can be extracted from the onboard sensors of a smart rollator. The main novelty of our approach is that, rather than evaluating whether a pose is stable or not, we predict how the next maneuver will affect the user's stability using the rollator odometry. Since gait differs from one person to another, we rely on Machine Learning (ML) techniques to predict future CoM. Then, we determine how far the predicted CoM projection is with respect to the area of maximum stability of the rollator user. Our prediction methodology is described in Subsect. 2.3. The method has been tested by volunteering rollator users in a care centre facility, as presented in Sect. 3. Finally, Sect. 4 outlines our conclusions and Future work lines.

2   Methodology

The proposed methodology is based on rollator on-board sensors, namely, strain gauges to estimate weight bearing on handlebars, laser to estimate feet position and speed, and wheel odometry. It consists of two stages. First, using estimated weight support on user's limbs, the user's CoM is calculated and projected on the ground. If the projection falls within a so-called safe area referred to the rollator frame, stability is deemed appropriate. The further the CoM is from the centroid of the safe area, the lower the user's stability is. Rather than working with punctual stability values, the next location of the CoM projection is predicted depending on the user's gait and ongoing maneuver, which is extrapolated from the device odometry. As people may perform the same maneuver in different ways, it is necessary to create a prediction model for each user. Prediction allows us to proactively warn users of destabilizing maneuvers.

2.1   Theoretical Background

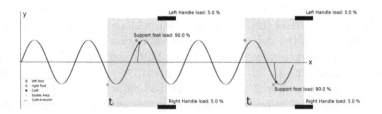

Fig. 1. Walker user CoM evolution over a straight path

In [22], authors describe the *Margin of Stability (MoS)* as a stability metric for unassisted gait, where stability depends on how far the projection of a person's CoM falls on the projected contact segment of the next foot on the ground or base of support (BoS). A rollator improves MoS as it expands BoS with four support points, so the contact segment becomes a surface. In rollators, a user's gait is considered stable when they are at an appropriate distance from the frame when their handlebar height is correctly selected. Authors in [3] analyze correct user posture when using a walker. They suggest that by holding the device at its proper height[1], the patient's elbow is naturally flexed at a 15-to-30° angle. That implicitly limits the maximum rollator-to-user safe distance in the x axis. In this angle range, users can support weight safely on the rollator. However, if the elbow flexion is larger than 15°, support is limited and if they are closer to the rollator, the CoM oscillation area is reduced.

Since shoulder location and upper limb lengths are tabulated for people depending on height and sex, the ideal BoS can be calculated to keep the desired elbow flexing range. Our dynamic safe BoS is the area imposed by that limit, plus half the stride of each specific user. Ideally, the person's CoM is always contained within that BoS. However, unstable gait/posture or wrong rollator settings may bring the CoM outside of BoS.

Figure 1 shows an ideal contralateral gait for a rollator user who shifts weight symmetrically from right to left side. The plot includes weight supported on left and right feet and rollator handlebars, plus BoS at two different time instants (t_i, t_j). Ideally, the CoM follows a sinusoid function centered on the rollator middle axis, peaking towards the supporting feet. This plot is synthetic: real users rarely produce a clean sinusoidal function.

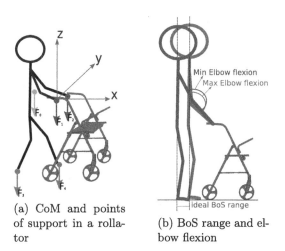

(a) CoM and points of support in a rollator

(b) BoS range and elbow flexion

Fig. 2. CoM and BoS definition.

---

[1] The handlebar height (H_h) in a walker is the same as a cane and can be estimated with the user height (u_h) as $H_h = user_h * 0.45 + 0.087$ [13].

2.2   CoM Calculation Using a Smart Rollator

Figure 2a presents all forces used to obtain the CoM position. Our reference point is the right handle location and the x axis is aligned with the advance direction. Force at the CoM (F_0) (user's weight) is by definition equal to the sum of right and handle support forces (F_1, F_2) and forces applied on right and left foot (F_3, F_4), as summarized on Eq. 1.

$$F_0 = F_1 + F_2 + F_3 + F_4 \tag{1}$$

Figure 3 shows XZ and YZ projections of Fig. 2a. These projections will be used to obtain the three spatial components of the CoM. Specifically, CoM equations can be expressed in this frame of reference as:

$$F_0 x_0 = F_3 x_3 + F_4 x_4$$
$$F_0 y_0 = F_2 y_2 + F_3 y_3 + F_4 y_4 \tag{2}$$
$$F_0 z_0 = F_3 z_3 + F_4 z_4$$

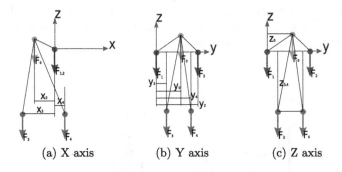

(a) X axis   (b) Y axis   (c) Z axis

Fig. 3. Center of mass projection in different axes.

Where the spatial coordinates are measured from the right handle position to: CoM position (x_0, y_0, z_0), left handle position (x_2, y_2, z_2), right foot position (x_3, y_3, z_3) and to left foot position (x_4, y_4, z_4).

When one foot is on air (single support), all weight is supported on the other foot at (x_s, y_s). We can define a support force (F_s) as:

$$F_s = F_0 - F_1 - F_2$$

Using F_s definition, Eq. 2 can be rewritten in single support cases as:

$$F_0 x_0 = F_s x_s$$
$$F_0 y_0 = F_2 y_2 + F_s y_s$$
$$F_0 z_0 = F_s z_s \tag{3}$$

Using the safe BoS defined in the previous subsection, the closer the CoM is to the BoS boundaries, the less stable the current posture is. Figure 4 shows three different poses: In Fig. 4a, the left foot is supporting most of the body weight, but the right handlebar is supporting more weight than the left and the CoM is closer to the BoS central area. This is a typical safe contralateral gait. Figure 4b shows a low-stability configuration: most weight is supported on the left foot, but significant weight is supported on the left handlebar at the same time, meaning the body is canting to the left. Figure 4c is a dangerous pose: feet are too far from the rollator frame, so the elbows flexing angle is too wide and the CoM is outside of the BoS area.

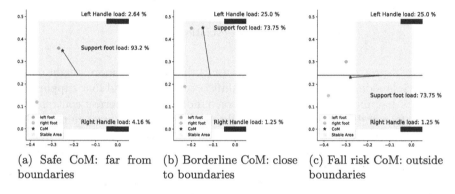

(a) Safe CoM: far from boundaries   (b) Borderline CoM: close to boundaries   (c) Fall risk CoM: outside boundaries

Fig. 4. Center of Mass (CoM and Base of Stability (BoS).

Figure 5 shows data from a real rollator user in a care facility. This person moves asymmetrically, keeping mostly to the left side of the rollator. Figure 6 shows the evolution of all parameters involved in the calculation of the CoM during a 10 m Test, namely load of the handlebars, relative speed of the feet and estimated load on each foot. As commented, it is assumed that weight must be distributed among all limbs offering support. The foot on the ground is the one presenting the lowest speed with respect to the rollator frame, and feet speeds are extracted from the onboard laser sensor.

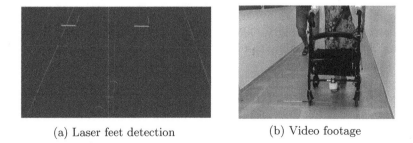

(a) Laser feet detection   (b) Video footage

Fig. 5. User asymmetries in gait.

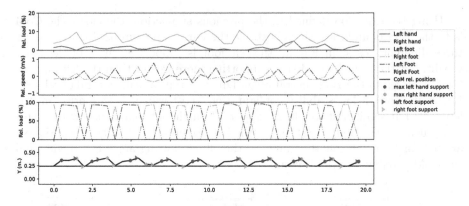

Fig. 6. Center of mass evolution and forces

The plot provides some meaningful data: i) they supports more weight on the rollator right side; ii) there is usually a shift between hand and foot support, i.e. they steps first and then supports weight on handlebars (in a perfect contralateral gait, left foot and right hand peaks would align and in ipsilateral gait, left foot and left hand peaks would align) and iii) their CoM is always shifted to the left side of the rollator central axis: for an asymmetry this significant, the user is walking on one side of the rollator at all times, as observed in Fig. 5. In brief, this user's stability is always subpar so any maneuver affecting it, like sharp steering, implies fall risk.

2.3   CoM Prediction Using Machine Learning

CoM prediction allows us to determine whether current gait is increasing or decreasing stability in order to generate warnings or, if the rollator allows it, alter motion to reduce fall risk. The CoM position can not be simply extrapolated from the rollator odometry, due to user's gait specifics, so we rely on regression techniques using weight support on limbs along with odometry as input instance. Regression requires data from a given person using the rollator, preferably in a path involving different maneuvers. Rather than binary classification of maneuvers as dangerous, i.e. producing a CoM configuration outside of the user's BoS or not, our predictors returns the expected CoM location in the near future. Distance from CoM to the BoS central region provides a quantitative estimation of stability. CoM information is useful to determine stability evolution. Then, thresholding can be used to determine if the predicted CoM is outside of the BoS.

3   Experiments and Results

3.1   Test Volunteers

The proposed method has been tested with volunteers presenting some degree of physical dependency at *Macrosad Arroyo de la Miel* nursing home. Tests were approved by an Ethical Committee, and all volunteers gave consent. During tests, they were asked to follow a corridor, perform a U-Turn at the end of the corridor and return to their departure point. Corridors in the *Macrosad Arroyo de la Miel* nursing home facilities are adapted for wheelchair and rollator navigation and no other persons were allowed in the corridor during tests, except for medical staff and a researcher walking behind the user in case of emergency (see Fig. 5b). The smart rollator was used uniquely in monitoring mode.

Gathered data was stored and processed to obtain three models per volunteer, using Linear Regression (LR), Support Vector Regression (SVR) and Regression Trees (RT), since the best classifier for a particular task is task-dependent. Observation of people's gait led us to believe that the lineal model could be unfit for this task, but it was still tested for comparison.

To illustrate results, we present models from three specific volunteers to account for three different conditions (healthy, recovered, and current disabilities):

- Person A: A 43 years old male with a height of 1.74 m, a weight of 94 kg and no previous conditions, nor experience using rollators. This person is a reference (healthy) user. It was required that he put weight on the handlebars, but it was expected that his gait differed from rollator users.
- Person B: A 38 years old male with a height of 1.84 m, a weight of 110 kg and recovered from severe leg polytrauma. He has wide experience with rollators, although he does not need one anymore.
- Person C: A 82 years old female, with a height of 1.5 m, a weight of 61 kg and several chronic conditions such as hypertension, polyarthritis, generalized arthrosis, and polyarthralgia. She uses rollators on a daily basis.

3.2 The WalkIT Architecture

One of the main drawbacks of many existing smart rollators is that they are usually blackbox systems or difficult to replicate. Furthermore, rollator users are reluctant to use any device but the one they are used to. The authors proposed in [8] a modular architecture to turn any commercial rollator into a smart device[2] using cheap, off-the-shelf components. The WalkIT architecture is meant to be deployed on a need basis. In this work, only monitoring modules are included in the rollator frame. All specifications, 3D printable pieces, and a basic ROS2 control system are provided under an Open Source license. The most important constraint in building WalkIt has been to have the smallest possible impact on the base commercial rollator, so that attached modules do not affect usability and user acceptance is as high as possible.

3.3 CoM Calculation

Gathered data during tests is used to feed our regression models. In this work, a random partition of 75% of data is used to train our models and the remaining 25% is used to evaluate them.

---

[2] https://github.com/TaISLab/WalKit/tree/humble.

During test it was observed that Person A (healthy) beared little weight on the handlebars. This was expected: even though it was explicitly requested that he did so, he does not need any physical support. Person B (recovered) beared far more weight than A and C. However, it is interesting to note that Person C (disabilities) beared weight asymmetrically, usually leaning to her right side. Person C also presented larger speed variation than the other two volunteers, possibly due to balance issues. The most relevant parameter for balance assessment is the CoM position. For example, for Person A CoM was symmetrical and consistent. Person B still favored a side, even though his CoM only showed little oscillations. Finally, Person C presented larger CoM oscillations with respect to the rollator axis and significant asymmetries.

3.4   CoM Estimation

Table 1 shows CoM estimation errors using all three regression methods for our three volunteers. Error is measured as the module of the difference between test data and predicted CoM position. No method has been finetuned in this work.

LR returns the largest average error and variance for all users, as expected, due to non linearity in users' gait. Lineal prediction average error is close to other methods' for patient C, but even then, it presents a larger variance.

SVR and RT errors are closely matched in all three scenarios, with almost the same average error. However, SVR has the lowest error variance, meaning that the error magnitude is similar despite the situation and chosen maneuver. RT may be affected by overfitting problems, which could explain its bigger variance for most users. This could be solved by better tuning of the model or by moving to Random Forest techniques. In brief, SVR seems to be the best performing method regarding CoM prediction. However, since we are interested in stability assessment rather than CoM prediction, it is necessary to evaluate whether the models are fit for fall risk assessment or not, as shown in next subsection.

Table 1. Estimators error distribution

Patient	Patient A			Patient B			Patient C		
Estimator	LR	SVR	RT	LR	SVR	RT	LR	SVR	RT
Mean error	0.14	0.06	0.07	0.17	0.06	0.09	0.09	0.09	0.11
Std. dev	0.11	0.05	0.07	0.08	0.05	0.08	0.07	0.05	0.09

3.5   Fall Risk Assessment

Using predicted CoM values, fall risk can be obtained depending on whether these values fall within their corresponding user's BoS. Table 2 shows results for all models in terms of their corresponding confusion matrix.

Results must be compared per user, as models are calculated for each individual. As a whole, RT returns the best results in all cases in terms of precision,

Table 2. Results for three different prediction methods for selected volunteers

Patient	Patient A			Patient B			Patient C		
Estimator	LR	SVR	RT	LR	SVR	RT	LR	SVR	RT
True Positives (TP)	1496	1532	1567	465	526	522	238	200	237
True Negatives (TN)	6	27	84	1	1	14	0	11	18
False Positives (FP)	97	76	19	17	17	4	26	15	8
False Negatives (FN)	84	48	13	61	0	4	3	41	4
Precision	93.91	95.27	98.8	96.47	96.87	99.24	90.15	93.02	96.73
Recall	94.68	96.96	99.18	88.4	100.0	99.24	98.76	82.99	98.34
F-Score	94.3	96.11	98.99	92.26	98.41	99.24	94.26	87.72	97.53

but not necessarily in recall. Although precision is above 90% for all models, it needs to be noted that the method returns lower values for Person C, as it is harder to predict her maneuvers. This issue could be solved by additional training, to account for each possible maneuver.

In general, results are consistent with the previous example. RT has a higher TP and TN rate than LR and SVR, although results may be more or less similar depending on the user, but it yields better results than the other two models in terms of FP and FN. It is interesting to note that RT precision is higher than 95% for all three models, despite the user's condition, i.e. it works correctly despite the amount of weight on the handlebars and gait asymmetry. Although in the previous section, SVR was slightly better in terms of prediction error than RT, these results show that, in fact, RT is better for stability assessment for these users. This happens because RT predictions tend to better follow the general direction of CoM, even if they are not as close to real points as in SVR predictions. In conclusion, SVR predictions are more accurate than RT for CoM estimation, but RT predictions are safer for stability assessment, i.e. to check if predicted CoM will be contained in the BoS or not.

4   Conclusions and Future Work

This work has presented a novel stability assessment method for smart rollator users, valid for any device capable of monitoring weight on handlebars, wheel odometry, and distance to the user's feet. Specifically, the method has been deployed on a Walk-IT rollator, a low cost, open modular design proposed by the authors to transform a commercial rollator into a smart one. The new method is based on the prediction of the projection of each user's CoM onto their BoS. Stability depends on the distance of their predicted CoM to the central area of their BoS. Commands can be classified into safe or not depending on whether the CoM projection on the ground falls inside the BoS. The method has been tested by three volunteers with different tipology for comparison: a healthy one, a recovered one, and a person with a disability. We have tested three different prediction techniques: Lineal Regression, Support Vector Regression and Regression Trees. Lineal Regression consistently returns the poorest results for all tested users. SVR and RT present similar results on average, although variation is larger in RT. These results can be explained by RT tendency to overfit. Despite this larger variation, assessment of whether the predicted CoM falls inside the BoS or not provides better results for RT, meaning that RT estimation may be less precise, but errors are typically less significant than in the other methods. Our results prove that assessment must be adapted to each specific user, not just in terms of height, weight, and sex, but specifically considering their gait traits. The main advantage of the proposed technique is that it is valid for a low cost sensorized rollator, using only onboard sensors, so it can be used anywhere, anytime. Furthermore, it analyses stability in a continuous, predictive way, so warnings or even gait adaptation can be proactively performed.

Future work will consist of extending tests to a significant number of volunteers and, once results are sufficiently validated, on physically affecting gait when necessary using selective braking in the smart rollator.

Acknowledgments. This work has been partially supported by projects PID2021-127221OB-I00 (Plan Estatal de Investigación Científica, Técnica y de Innovación 2021–2023) and UMA20-FEDERJA-052 (Programa Operativo FEDER Andalucía 2014-2020). The authors would also like to thank *Macrosad Arroyo de la Miel* nursing home and the volunteers in our tests for their time and help.

References

1. Amt, E.K.S., Corselli-Nordblad, L., Strandell, H., of the European Union, E.C.S.O.: Ageing Europe: Looking at the Lives of Older People in the EU : 2020 Edition. Statistical books / Eurostat, Publications Office of the European Union (2020)
2. Ballesteros, J., Urdiales, C., Martinez, A.B., Tirado, M.: Automatic assessment of a rollator-user's condition during rehabilitation using the i-walker platform. IEEE Trans. Neural Syst. Rehabil. Eng. **25**(11), 2009–2017 (2017)
3. Bradley, S.M., Hernandez, C.R.: Geriatric assistive devices. Am. Family Phys. **84**(4), 405–411 (2011)
4. Chalvatzaki, G., Koutras, P., Hadfield, J., Papageorgiou, X.S., Tzafestas, C.S., Maragos, P.: Lstm-based network for human gait stability prediction in an intelligent robotic rollator. In: 2019 International Conference on Robotics and Automation (ICRA), pp. 4225–4232. IEEE (2019)
5. Chou, H.C., Han, K.Y.: Developing a smart walking cane with remote electrocardiogram and fall detection. J. Intell. Fuzzy Syst. **40**(4), 8073–8086 (2021)
6. Instituto Nacional de Estadística: Tasa de dependencia de la población mayor de 64 años, por comunidad autónoma, instituto nacional de estadística. https://www.ine.es/jaxiT3/Datos.htm?t=1455. Accessed 30 Jan 2023
7. Ezatzadeh, S., Keyvanpour, M.R., Shojaedini, S.V.: A human fall detection framework based on multi-camera fusion. J. Exp. Theor. Artif. Intell. **34**(6), 905–924 (2022)

8. Fernandez-Carmona, M., Ballesteros, J., Díaz-Boladeras, M., Parra-Llanas, X., Urdiales, C., Gómez-de Gabriel, J.M.: Walk-it: an open-source modular low-cost smart rollator. Sensors **22**(6), 2086 (2022)
9. Grammenos, S., et al.: European comparative data on Europe 2020 & People with Disabilities. Academic Network of European Disability experts, Leeds (2013)
10. Hausdorff, J.M.: Gait dynamics, fractals and falls: finding meaning in the stride-to-stride fluctuations of human walking. Human Mov. Sci. **26**(4), 555–589 (2007)
11. Hirata, Y., Komatsuda, S., Kosuge, K.: Fall prevention control of passive intelligent walker based on human model. In: 2008 IEEE/RSJ International Conference on Intelligent Robots and Systems, pp. 1222–1228. IEEE (2008)
12. Hirata, Y., Muraki, A., Kosuge, K.: Motion control of intelligent walker based on renew of estimation parameters for user state. In: 2006 IEEE/RSJ International Conference on Intelligent Robots and Systems, pp. 1050–1055. IEEE (2006)
13. Kumar, R., Roe, M.C., Scremin, O.U.: Methods for estimating the proper length of a cane. Arch. Phys. Med. Rehabil. **76**(12), 1173–1175 (1995)
14. Kyrdalen, I.L., Thingstad, P., Sandvik, L., Ormstad, H.: Associations between gait speed and well-known fall risk factors among community-dwelling older adults. Physiother. Res. Int. **24**(1), e1743 (2019)
15. Makino, K.: Fear of falling and gait parameters in older adults with and without fall history. Geriat. Gerontol. Int. **17**(12), 2455–2459 (2017)
16. McAndrew Young, P.M., Wilken, J.M., Dingwell, J.B.: Dynamic margins of stability during human walking in destabilizing environments. J. Biomech. **45**(6), 1053–1059 (2012)
17. Qian, Z., et al.: Development of a real-time wearable fall detection system in the context of internet of things. IEEE Internet Things J. **9**(21), 21999–22007 (2022)
18. Rubenstein, L.Z.: Falls in older people: epidemiology, risk factors and strategies for prevention. Age Ageing **35**(suppl_2), ii37-ii41 (2006)
19. Tinetti, M.E.: Performance-oriented assessment of mobility problems in elderly patients. J. Am. Geriat. Soc. (1986)
20. Tung, J.Y., Gage, W.H., Poupart, P., McIlroy, W.E.: Upper limb contributions to frontal plane balance control in rollator-assisted walking. Assist. Technol. **26**(1), 15–21 (2014)
21. Van Schooten, K.S., et al.: Daily-life gait quality as predictor of falls in older people: a 1-year prospective cohort study. PLoS One **11**(7), e0158623 (2016)
22. Watson, F., Fino, P.C., Thornton, M., Heracleous, C., Loureiro, R., Leong, J.J.H.: Use of the margin of stability to quantify stability in pathologic gait - a qualitative systematic review. BMC Musculoskel. Disord. **22**(1), 597 (2021)

Data Analysis and Generation in the ENVELLINT Longitudinal Study to Determine Loss of Functionality in Elderly People

John Nelson[1]([✉]), Jordi Ollé[2], Xavier Parra[3], Carlos Pérez-López[4],
Oscar Macho-Pérez[4], Marta Arroyo-Huidobro[4], and Andreu Català[3]

[1] Department of Electronic and Computer Engineering, University of Limerick,
Limerick, Ireland
john.nelson@ul.ie
[2] Conceptos Claros, Barcelona, Spain
[3] CETpD-UPC, Technical University of Catalonia, Vilanova i la Geltrú, Spain
[4] Consorci Sanitari Alt Penedes-Garraf, Sant Pere de Ribes, Spain

Abstract. This paper presents the initial data analysis and modelling for detecting health changes from data gathered on a low-cost smartphone used during normal daily activities. The work is part of the ENVELLINT project, where one of the main objectives is to explore if it is possible to evaluate the functional aspects of frailty indices automatically using smartphones.

The project involves both longitudinal and cross-sectional studies involving elderly participants. In the longitudinal study a comprehensive set of sensor, application and other smartphone data is gathered over lengthy periods for each participant, together with extensive medical assessments. The purpose is to provide a comprehensive data set for investigating frailty and health changes. The larger cross-sectional study, which included only the medical assessments, was necessary to gather more medical related health and frailty data, and to balance project costs.

The analysis work to date has involved data and feature engineering to identify, extract and select the most useful features. Insights are given for the potential use of the location and application usage features.

A core aspect, given the expense and the limited number of participants in the longitudinal study, is to explore the use of synthetic data generation to leverage the real data from both studies. Generative Adversarial Network and Gaussian Copula models have been investigated to create a larger representative dataset of longitudinal participants. Initial results and insights show generated synthetic data that closely mirrors the real data, especially using Gaussian Copula.

Keywords: Smartphone · Frailty · ADLs · Medical Assessment · Sensors · Data Generation · CTGAN · Gaussian Copula

I. Rojas et al. (Eds.): IWANN 2023, LNCS 14135, pp. 388–399, 2023.
https://doi.org/10.1007/978-3-031-43078-7_32

1   Introduction

In OECD countries the percentage of the population over 65 years old continues to rise and is estimated to surpass 26% by 2050 [1]. Changing demographics though are leading to an aging population where it is estimated that by 2050 a third of the population will be over 65 years in various southern European countries. More relevant is that the population increase has been particularly rapid among the oldest group – people aged 80 and over.

A key challenge is to identify risk markers for the normal and natural deterioration of health, both physical and mental, based upon low-cost monitoring of Activities of Daily Living (ADLs), and to proactively manage any detected changes in functionality, frailty and dependency. The ENVELLINT pilot project, funded by the Catalonia FEDER program, is tasked with investigating and developing an unsupervised monitoring system, using primarily the smartphone, to identify and diagnose health changes arising from aging. The long-term objective is to initiate appropriate interventions. Thereby preventing more serious issues such as falls and hospitalizations.

The project involves the comprehensive use of medical assessments as baselines to determine the participants' actual physical and medical health, combined with the gathering of extensive smartphone sensor data over extended periods (minimum two weeks). Two studies were designed and performed during the project. The first one was a cross-sectional study dedicated to collect the medical assessment Case Report Forms (CRF) and to measure the isometric force of the hand to access grip strength. The second one was a longitudinal study, more complex and comprehensive, with two periods of data gathering, separated by a one-year washout period. For the longitudinal pilot, smartphones and inertial sensors were the main devices used to gather data.

The challenge is to infer possible changes in the person's underlying health, by processing only the smartphone sensor data and logs. For example, for physical health a change could be detected by reduced activity or by spending more extended periods at home; for mental health, changes could identified from the reduced use of smartphone applications, such as the phone call app or social media apps.

The data analysis and artificial intelligence challenge is to process the various smartphone log files and the extensive sensor information gathered per participant (in the longitudinal study around ten gigabytes each). The challenge is to extract appropriate health indicators to identify and monitor changes in functional health. This requires extensive modelling, initially using supervised machine learning, through the extraction of appropriate features from both the raw sensor data and the smartphone logs and mapping the features to subsets of the comprehensively labelled medical assessment data.

However, the costs of gathering such longitudinal data for 30 participants in the target age group are very expensive due to the required technical, medical and administrative time and overheads. Therefore, the project included a much larger cross-sectional study of over 100 participants, who completed only the medical assessments as for the longitudinal pilot. Currently, the actual completed longitudinal cases to date are significantly lower due to dropouts and Covid impacts.

Hence, a key question arose as to whether the longitudinal and cross-sectional data gathered to date could be used to generate a more comprehensive synthetic data set for creating models to determine changing health and frailty. Consequently, various techniques were considered to generate new data with similar statistics to the measured data: Synthetic Data Augmentation for Tabular Data (SMOTE) [2], Conditional Tabular Generative Adversarial Network (CTGAN) [3], and Gaussian Copula (GC) [4].

The rest of the paper is as follows. Section 2 describes the studies, the sensor data and the medical data gathered. Section 3 covers the data preparation and initial insights. In Sect. 4, the data generation models adopted are introduced, and their evaluation is presented. The final Sect. 6 draws conclusions and describes the next steps.

2 Data Gathered

The longitudinal study period for each participant was organized, for project reference purposes, as a sequential set of days supported by a comprehensive set of paper-based Case Report Forms (CRF). The CRFs are used for both medical and technical reporting, supported by detailed inclusion and exclusion criteria for participation.

Although the target monitoring device was a user's smartphone, concerns about its adoption in previous projects and the state-of-the-art algorithms resulted in two one-week periods of baseline monitoring of the inertial sensors. An additional proven waist mounted sensor [5] was worn by the participant in conjunction with the smartphone. As the mobile placement is flexible, e.g. in a pocket or bag or in the hand, this presents challenges to estimate a participant's movements using algorithms and supervised learning techniques. The fixed position of the waist sensor gives a gold standard acquisition with proven algorithms for extraction of the ambulatory movements and ADLs.

By ensuring that both the smartphone and waist sensor were simultaneously used, the data provided allows the benchmarking of the developed algorithms and models for the smartphone derived features.

2.1 Sensor Data

In the longitudinal part, the main monitoring device is the user smartphone. An application was run to capture sensor readings of the Accelerometers, to log the Application Usage, the Bluetooth Usage and the Location/GPS readings. The sensor sample frequency of 100 Hz resulted in the order of GBs of data capture files per monitoring period and per participant. The proven waist mounted sensor [5], provides more extensive and accurate measurement [6, 7] with a set of three 3-axes Accelerometers, a Gyroscope, Magnetometer and Barometer.

2.2 Medical Assessment

For each participant sociodemographic data, pathological antecedents (diseases), and a comprehensive list of medications and the doses were captured. Comprehensive medical assessments for ADLs, physical and mental health were adopted. The responses were gathered by a qualified geriatrician and the answers handwritten on the appropriate CRF form, and later scanned. These included indices such as Fried Frailty Index [8], Índice frágil-VIG [9], MEC Scale [10], Barthel ADL Scale [11], FI-CGA [12], Lawton-Brody Instrumental ADL Scale [13], Yesavage Geriatric Depression Scale [14], and Charlson Comorbidity Index [15].

The responses were then entered into a structured Excel form with additional checks and summarisations, and later extracted into tables for adoption in modelling.

The comprehensive set of assessments consisting of 178 questions and test results, despite some overlaps, provides an extensive set of usable labels for the physical and mental health state. Many of these indices in normal usage result in a cumulative score which is then compared against various thresholds as a crude indicator of overall health and independence. However, capturing the responses to the individual questions provides considerable opportunities to develop better and more accurate predictive models.

In addition, a separate Dynamometer was used to capture hand grip strength [16].

3 Data Preparation

It was decided to use MATLAB for initial data preparation based on an extensive set of analysis and visualization tools created for prior projects. The data from the waist sensor was as anticipated with regular sampling and complete data stored in the file-based database over the week duration. However, the smartphone sensor data had significant issues arise during the data logging phase. Frequently there were brief outages in the sensor readings logged, abrupt file closures and irregular sampling of sensor data.

To maximize the data extracted, a significant number of repairs to the raw data was required, supported mainly by using exception handling in MATLAB to detect and where possible correct the anomaly. A key objective, given the large number of files and sensor readings, was to avoid manually correcting any of the raw data files by identifying and addressing each anomaly during data processing. This objective was fulfilled and has resulted in successful preprocessing of all the available participant data.

An optimized and effective MATLAB application, with a simple GUI front end was developed to clean, merge, and store the raw data for later input to Python based notebooks for more detailed processing and modelling. The GUI application allows participant data to be selected, summarized, and if necessary visualized, significantly improving the ability to navigate through and understand the large datasets. Furthermore, in the earlier stages it allowed the team to have improved understanding and better insights into the data and to focus on handling the various data anomalies.

By adopting fine granular control of the file reading in MATLAB, through import options and using byte data types, the file processing time was reduced by orders of magnitude to around a second per 25 MB file on a medium specified laptop.

4 Initial Insights

A subset of eight longitudinal study participants is being thoroughly analyzed. The preliminary focus is on investigating the inertial and location data to determine changes in physical health, and to analyze the app usage as a predictor of changes in social interactions and mental health, all of which are potential predictors of frailty. Features such as energy expenditure, activities, step counts, and distance walked are extractable from the motion sensors. These features combined with the location information provide an improved feature vector for more improved and insightful modelling.

4.1 Location

The location data frequently showed daily and weekly similarities, which can be used to identify repeated patterns. Fig. 1 (a) shows a static plot of positions on a map giving an insight into the locality and distances travelled. In addition, specific locations can be identified, and the speed of movement detected. It is clear by adding the time dimension to positions as shown in Fig. 1 (b), in this case for a single week, that this participant had a daily repeating mobility pattern; most days taking a regular walk, with one exceptional car journey on the last day (deduced from the speed of movement).

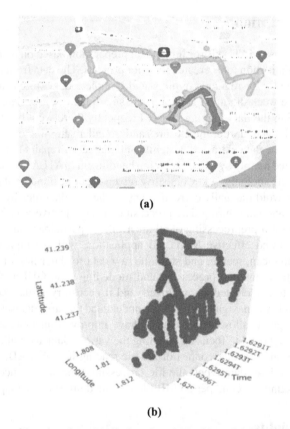

(a)

(b)

Fig. 1. Extraction of location data, (a) displayed on Map (blurred deliberately) and (b) versus time to show and identify repeated behavior.

The location data was further processed to deduce additional features such as time spent at home, activity time, time in a vehicle, and more generally time spent outside and inside, visiting and stopping at identifiable places or specific buildings e.g. a park, a restaurant or a church. Such features are potential key indicators of current health and more importantly for this project potentially usable for detecting changing health, when observed over extended periods.

4.2 Application Usage

It is possible to use the Android UsageStats and/or Activity Events to log information about the smartphone application usage time. However, the resultant logged information requires significant data cleaning to extract useful information due to duplicate records, overlapping records, reset timers etc.

In Fig. 2 an extract of the application usage during a 4-week period shows that this octogenarian user spent significant time using WhatsApp, or at least with WhatsApp active in the foreground, with average usage of nearly 3 h per day. In addition, they also spent considerable time on voice calls making around 3 to 4 calls a day, many lasting over an hour. Finally, they used the browser (Chrome) for extended periods.

	sum	count
App		
com_whatsapp	4804.5510	4.9
com_google_android_dialer	3381.2939	3.6
com_android_chrome	2492.7511	3.0

Fig. 2. Extract of Application sum of usage in minutes and average usage count over 4 weeks.

Other features include the number of different apps used, the types of app used (such as social media e.g. WhatsApp, cognitive e.g. chrome, organizational e.g. calendar app etc.), the usage rate and the patterns of usage, which are indicators of the level of engagement of the smartphone user.

The analysis of the app features for a specific participant over the longitudinal study was explored for the early detection of potential health degradation. The extraction of various features of the total app usage and the categorized app usages for the first and second observation periods frequently showed marked differences.

To illustrate, for one participant based on daily usage over extended periods of 2 to 3 weeks, the total app usage and frequency of usage in the first and second periods showed significant change. Furthermore, a deeper analysis of the categorized app usage showed increased time using social apps, but significant reductions in the use of cognitive and organization apps.

App usage alone may be insufficient to classify frailty. However, the classification can be improved by combining the usage with additional features for the same participant, such as energy expenditure. For this participant, in parallel with the app feature changes from the first to the second observation periods, the movement energy feature was reduced by 60%.

5 Data Generation

Starting from the completed datasets from the 8 longitudinal participants, various methods to generate similar data were investigated. The basic process started by creating a table of the features of interest. The initial focus was on two distinct feature subsets:

(a) the per day extracted features as follows: 12 features related to application usage and the average energy expended feature, calculated using signal magnitude area per second.
(b) the CRF captured medical features using the indices: Barthel, Lawton & Brody, Fragile Index VIG; and Age and Sex.

Using these 13 per day and 5 medical features, various generator models were trained and evaluated.

The raw data, dominated by the inertial sensor readings, was more than 60 GB for the 8 participants, which was reduced to a tabular dataset with a total of 151 day-based observations, where 128 were fully complete with the above 18 features. Location features were omitted as they are currently undergoing further validation.

Two generator models were selected and developed and evaluated for synthetic data generation: Conditional Tabular Generative Adversarial Network (CTGAN) [3] and Gaussian Copula (GC) [4]. The target of 1000 generated observations was selected.

The basic statistics remained consistent for both models, the means and standard deviations of the real and fake data from the Gaussian Copula (GC) method are shown below in Fig. 3.

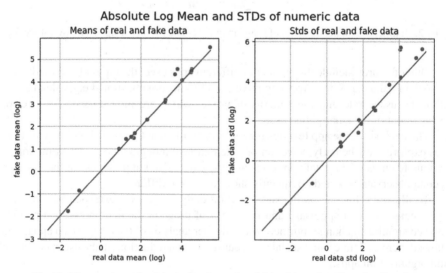

Fig. 3. Mean and standard deviation for real and fake data using Gaussian Copula.

However, the correlations differed significantly between CTGAN and GC, see Fig. 4 for the correlation matrix plots for the original data and fake data generation. Visually

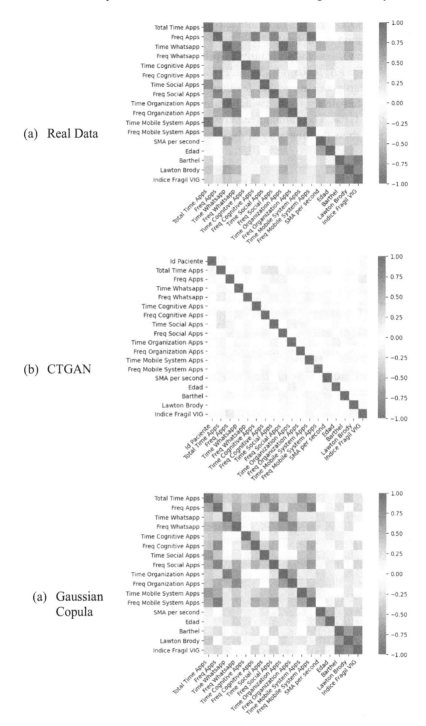

Fig. 4. Real data and Generated Data Correlation Matrices.

it can be seen that GC results in significantly improved correlations, whereas CTGAN had poor correlation throughout.

Figure 5 below shows the probability distributions for real and fake data for two of the features: *Total Time Apps Usage* and the *Frequency of Apps Usage*, indicating visually the strong similarity of data. Other features were similarly distributed, including the categorial male/female sex feature which maintained the approximately two thirds male ratio, to within 1.1%. The cumulative sums curves shown in Fig. 6 for the same two features further show the similarity of the data generated.

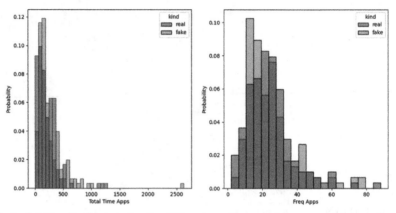

Fig. 5. Probability distributions for *Total Time Apps Usage* and *Frequency of Apps Usage* for Real and Fake data.

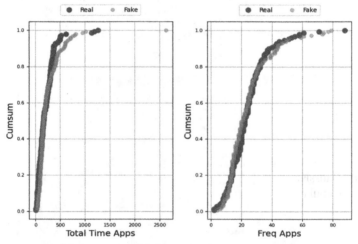

Fig. 6. Cumulative sums for *Total Time Apps Usage* and *Frequency of Apps Usage* for Real and Fake data.

An analysis was completed using the Synthetic Data Value (SDV) Metrics evaluation, SDMetrics [17], in particular from the Quality and Diagnostic Reports, which detail overall quality, column shapes and correlations and present a diagnosis summary. For both CTGAN and GC, the resultant Diagnostic Result was SUCCESS with the synthetic data satisfying the following criteria: over 90% for both coverage of the numerical ranges and categories present in the real data; over 90% of synthetic rows not being copies of the real data and for following over 90% of the min/max boundaries set by the real data. Nevertheless, GC significantly outperformed CTGAN, see Table 1 below for the Overall Quality Score and the column shapes and column pair trends properties.

Table 1. SDMetrics evaluation: Overall Scores and Properties.

Scores/Property	CTGAN	GC
Overall Quality Score	76.9%	87.3%
Column Shapes	71.3%	83.0%
Column Pair Trends	82.6%	91.7%

The deeper analysis of the individual features uses the Kolmogorov-Smirnov (KS) statistic to give the maximum difference between the cumulative density functions for the real and fake data. SDMetrics reports the KS complements, as summarised in Table 2 for both CTGAN and GC. For nearly all features, GC outperformed CTGAN.

Table 2. SDMetrics evaluation: Individual feature KS complement Quality Scores.

Feature	KS complement	
	CTGAN	GC
Total Time Apps	0.479	0.872
Freq Apps	0.630	0.910
Time Whatsapp	0.743	0.703
Freq Whatsapp	0.642	0.843
Time Cognitive Apps	0.600	0.870
Freq Cognitive Apps	0.804	0.956
Time Social Apps	0.635	0.937
Freq Social Apps	0.737	0.914
Time Organization Apps	0.762	0.779
Freq Organization Apps	0.786	0.889
Time Mobile System Apps	0.839	0.905
Freq Mobile System Apps	0.807	0.879
SMA per second	0.532	0.659

(continued)

Table 2. (*continued*)

Feature	KS complement	
	CTGAN	GC
Edad	0.572	0.607
Barthel	0.762	0.808
Lawton Brody	0.801	0.799
Indice Fragil VIG	0.721	0.696
Sexo	0.846	0.915

Using the trained model, a subset of 76 of the cross-sectional observations were selected and combined with the real longitudinal data to successfully generate an extended longitudinal data set of 379 observations, with similar quality scores and properties. This extended dataset is being used to develop and evaluate new models for determining frailty.

6   Conclusion and Next Steps

The initial data analysis and modelling in the ENVELLINT pilot study has shown significant opportunities for detecting changes in both physical and mental health using only the smartphone sensors and the application logs. As currently many of the younger elders already use one, the smartphone provides a universal and low-cost way, with appropriate medical monitoring, to enable proactive intervention on deteriorating or changing health.

Although only a pilot has been completed targeting 30 longitudinal and 100 cross-sectional participants, the comprehensive medical and sensor data gathered from the existing participants is proving valuable and already has enabled much deeper analysis and modelling. The analysis of the energy expenditure, location and application data and the extraction of relevant features is aiding a deeper understanding of their combined potential use for detecting health changes due to aging.

The use of data generation has shown significant potential to address the high-cost limitations of the real data set necessary to allow the development and evaluation of frailty models. In particular, the Gaussian Copula model generated synthetic longitudinal data that most closely mirrored the real data and work is ongoing to further expand this initial synthetic dataset.

A resultant project output is the objective to significantly reduce the amount of data stored on and transmitted from the smartphone. This requires identifying through further modelling the ambulatory and activity features most useful for predicting frailty and health degradation. The smartphone would then locally extract these features from the inertial sensor measurements and eliminate their store and forward.

Acknowledgements. This work was partially supported by the Catalonia FEDER program, resolution GAH/815/2018 under the project, PECT Garraf : Envelliment actiu i saludable i dependència.

References

1. OECD: Health at a Glance 2021 (Ageing and Long Term Care: Demographic trends): OECD Indicators, OECD Publishing, Paris (2021), https://doi.org/10.1787/ae3016b9-en
2. Chawla, N.V., Bowyer, K.W., Hall, L.O., Kegelmeyer, W.P.: SMOTE: synthetic minority over-sampling technique. J. Artif. Intell. Res. **16**, 321–357 (2002)
3. Xu, L., Skoularidou, M., Cuesta-Infante, A., Veeramachaneni, K.: Modeling tabular data using conditional GAN. In: Proceedings of the 33rd International Conference on Neural In-formation Processing Systems, Article 659, pp. 7335–7345. Curran Associates Inc., Red Hook, NY, USA (2019)
4. Kamthe, S., Assefa, S., Deisenroth, M.: Copula flows for synthetic data generation. arXiv preprint arXiv:2101.00598 (2021)
5. Rodríguez-Martín, D., et al.: A waist-worn inertial measurement unit for long-term monitoring of Parkinson's disease patients. Sensors **17**(4), 827 (2017)
6. Rodríguez-Martín, D., Samà, A., Pérez-López, C., et al.: Posture transition analysis with barometers: contribution to accelerometer-based algorithms. Neural Comput. Appl. **32**, 335–349 (2020). https://doi.org/10.1007/s00521-018-3759-8
7. Pérez-López, C., et al.: Dopaminergic-induced dyskinesia assessment based on a single belt-worn accelerometer. Artif. Intell. Med. **67**, 47–56 (2016). https://doi.org/10.1016/j.artmed.2016.01.001
8. Fried, L.P., et al.: Frailty in older adults: evidence for a phenotype. J. Gerontol Ser. A Biol. Sci. Med. Sci. **56**, M146–M157 (2001). https://doi.org/10.1093/gerona/56.3.M146
9. Amblàs-Novellas, J., Martori, J.C., Brunet, N.M., Oller, R., Gómez-Batiste, X., Panicot, J.E.: Revista Espa-ñola de Geriatría y Gerontología Índice frágil-VIG: dis no y evaluación de un índice de fragilidad basado en la Valoración Integral Geriátrica, Rev Esp Geriatr Gerontol. (2017). https://doi.org/10.1016/j.regg.2016.09.003
10. Lobo, A., Ezquerra, J., Gómez Burgada, F., Sala, J.M., Seva Díaz, A.: El Mini-Examen Cog-noscitivo (un test sencillo, práctico, para detectar alteraciones intelectuales en pacientes médi-cos)., Actas Luso. Esp. Neurol. Psiquiatr. Cienc. Afines. (1979)
11. Mahoney, F., Barthel, D.: Functional evaluation: the Barthel index. Md State Med J. **14**, 61–65 (1965)
12. Rockwood, K., Rockwood, M.R.H., Mitnitski, A.: Physiological redundancy in older adults in relation to the change with age in the slope of a frailty index. J. Am. Geriatr. Soc. **58**, 318–323 (2010). https://doi.org/10.1111/j.1532-5415.2009.02667.x
13. Lawton, M.P., Brody, E.M.: Assessment of older people: self-maintaining and instrumental V1.1 – 24/12/2019 20 activities of daily living. Gerontologist **9**, 179–86 (1969). http://www.ncbi.nlm.nih.gov/pubmed/5349366
14. Martínez de la Iglesia, J., Onís Vilches, M.C., Dueñas Herrero, R., Aguado Taberné, C., Albert Colomer, C., Arias Blanco, M.C.: Abreviar lo breve.aproximación a versiones ultracortas del cuestionario de Yesavage para el cribado de la depresión. Atención Primaria **35**(1), 14–21 (2005). https://doi.org/10.1157/13071040
15. Soler, P.A., Mellinas, G.P., Sánchez, E.M., Jiménez, E.L.: Evaluación de la comorbilidad en la población anciana: utilidad y validez de los instrumentos de medida. Rev. Esp. Geriatr. Gerontol. **45**(4), 219–228 (2010). https://doi.org/10.1016/j.regg.2009.10.009
16. Pérez, E., et al.: Frailty Level Prediction in Older Age Using Hand Grip Strength Functions Over Time. In: Rojas, I., Joya, G., Català, A. (eds.) Advances in Computational Intelligence: 16th International Work-Conference on Artificial Neural Networks, IWANN 2021, Virtual Event, June 16–18, 2021, Proceedings, Part II, pp. 356–366. Springer International Publishing, Cham (2021). https://doi.org/10.1007/978-3-030-85099-9_29
17. Synthetic Data Metrics. Version 0.9.3. DataCebo, Inc. (2023). https://docs.sdv.dev/sdmetrics/

Machine Learning for 4.0 Industry Solutions

Prediction of Transportation Orders in Logistics Based on LSTM: Cargo Taxi

Tomasz Grzejszczak[1,2]([✉]) [iD], Adam Gałuszka[1,2] [iD], Jarosław Śmieja[1,2] [iD], Marek Harasny[2], and Maciej Zalwert[2]

[1] Silesian University of Technology, Gliwice, Poland
{tomasz.grzejszczak,adam.galuszka,jaroslaw.smieja}@polsl.pl
[2] Giełda Papierów Wartościowych w Warszawie (GPW), Warsaw, Poland
{marek.harasny,maciej.zalwert}@gpw.pl

Abstract. This work presents the application of LSTM neural network in prediction of transportation orders. In case of logistic transport, the empty return routs can be minimized by matching new orders in the vicinity of drop off of actual order. This is a similar approach to taxi, thus the approach is named cargo-taxi. To find the new orders in the vicinity of carrier, an LSTM network is used to predict the next towns that the carrier would visit basing on his actual route and archival routs that the network was trained on. This research focus on proof of concept, the way of constructing the training data, and the research of parameters influence on time of training and prediction.

Keywords: LSTM · cargo-taxi · matching in logistics · planing

1 Introduction

Standard handling of orders by logistics operators consist of matching clients with carriers. A matched order consist of a price and pick-up - drop-off addresses. Carrier needs to deliver package from one place to another and that is basically end of his order, thus one can return to base. On the other hand, there is an ongoing field of optimization in this topic with use of dynamic pricing and planing. An iconic example of such approach in Uber [6].

The main idea that we wish to implement into cargo logistics field is inspired from taxi, thus we call it cargo-taxi. A taxi delivers client and does not return to base. Taxi driver can wait for or search for a new order within a region that he actually is in. Noreover, minimizing the number of empty routs, the idea of cargo-taxi in logistics reduces the carbon footprint.

One can assume that a tractor and trailer of maximum weight 40 tonnes and a payload of 25 tonnes consumes 37 l/100 km when fully loaded or 25 l/100 km when empty in return trip. With well organised logistics, the return trips can be reduced to 25% [3].

Thus the main aspect in reducing carbon footprint is the well organized logistics. The research in logistics is aimed at planning algorithms [2]. However,

I. Rojas et al. (Eds.): IWANN 2023, LNCS 14135, pp. 403–410, 2023.
https://doi.org/10.1007/978-3-031-43078-7_33

the planing algorithms usually are off-line or the on-line parts are mostly reactive systems for weather or traffic conditions [7].

In order to substitute the well trained human logistic operator, the dynamic on-line planning can be implemented with use of neural networks and artificial intelligence. Deep learning, mainly recursive networks can be applied in route planning [5]. In this research we would like to use Long short-term memory (LSTM) network. LSTM is an efficient tool in many different fields including statistics, linguistics, medicine, transportation, computer science and others [4].

The purpose of the work is to study and develop artificial intelligence algorithms for determining paths based on data from the Graph Base. The Graph Base is a special way of storing data, an alternative to the sequential database [1]. However, for the purpose of studying algorithms, the key point is that the data on which the calculations will be carried out is in the nature of a graph.

1.1   Aim of the Research

The presented research connects few aspects, each shown on the Fig. 1. First is the preparation of data in form of the graph. The graph can be prepared in multiple sizes determined by parameter N. This map is used to create a list of orders and simulates data obtained from logistic operator. Any number L of orders can be created basing on probability represented by graph edges. This data is used in LSTM neural network for training. The output of the network predicts the next edges that will be most probably visited by the carrier basing on the previous edges that he was in. Thus the network learns the map basing on the order list, not knowing the map.

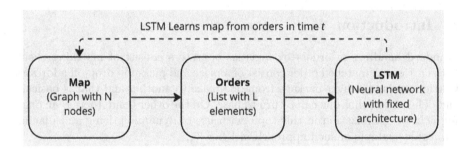

Fig. 1. data flowchart

In real life there is no map and the list of archival orders are provided by logistic operator or are dynamically gathered while the system operates. The described project is in research phase, thus no real data is available yet.

In this research, the influence of order list on LSTM times t of training and prediction are measured. The order list is controlled directly by parameter L and indirectly by map size N, that determines number of available towns in LSTM dictionary.

2   Materials and Methods

The task of predicting a transportation orders in logistics is performed using Long short-term memory (LSTM) artificial neural network, that is used in the fields of artificial intelligence and deep learning. Usually in case of neural networks, the key aspect is a dataset. Thus, this chapter focus on both, construction of LSTM neural network and preparation of data set, that is a list of orders created on the map, that is represented in form of a graph.

2.1   LSTM

The core of this research is an LSTM network. The network architecture was implemented with use of Keras library for Python as a sequential model. There is an LSTM layer, and two dense layers: relu and softmax, with Adam optimizer. The network is perfect for word prediction basing on the text that it was trained on. During the prediction, a sentence is passed as an input and the output is the next word from dictionary.

In order to prepare a dataset for training, each word in the text needs to be mapped in a dictionary. Thus, the input sentence is divided into words and each word has its index. Thus, the output is an index to a word in dictionary. Moreover, concatenating the predicted word into the input sentence and predicting one more time gives the second predicted word. This procedure can be repeted several times to predict next few words.

This approach can be used in route prediction. In case a carrier makes a route through several towns, passing this order list into LSTM can predict the most probable town that the carrier will visit next. This information is useful to narrow the list of orders that the carrier can take next.

2.2   Map

Because of the need to study various scenarios, not always available in archival collections, the need arose to develop a method for generating synthetic data. Therefore, a method was developed to create a random graph that mimics a map with delivery points and routes with a certain delivery cost between them. The developed method of generating a graph has several parameters, but the most important is the ability to set the number of nodes N. An example of a simple graph with 30 nodes is presented in Fig. 4. This figure contains $N = 30$ nodes, which are delivery points. The nodes on the plot are numbered from left to right. The nodes are connected by vertices with a random value. Each node is connected to several nearest neighbors and the value of a node is determined by multiplying the distance by a random variable. The randomisation simulates the country map with roads between towns and the cost of traveling from town to town.

The vertex values correspond to the probability of selecting a given node in the order. The cost of moving through the graph is determined by the inverse of the vertex values. These values can be read from the figure based on the color

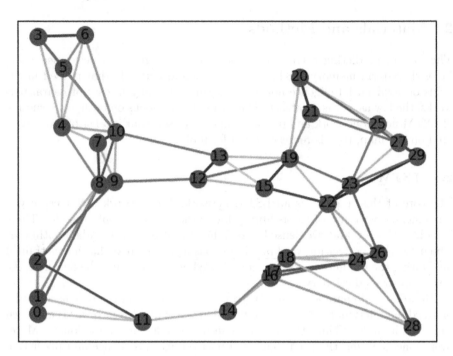

Fig. 2. Map in form of a graph with $N = 30$

scheme. The vertices were given a color scheme based on the "jet" color space, which is presented in Fig. 3. Intense red is used to indicate the most frequently selected routes with high probability, while "cold" blue is used to indicate the reluctant routes. For example, between node 12 and 13 there is a very good connection with a low cost. Similarly, between node 13 and 19. However, the direct connection between 12 and 19 is not very likely to be chosen. A similar situation in reality can occur if a highway is routed between cities 12, 13 and 19, which is not at all the closest route, but is the fastest. The alternative national road between 12 and 19 runs through built-up areas, so this route is reluctantly chosen, even though it is shorter.

A graph with 30 nodes is easy to illustrate, but the work will conduct research on much larger graphs. Figure 3 contains a random graph with 1000 nodes. Its readability is significantly limited, but it is possible to perform the order list on such graph.

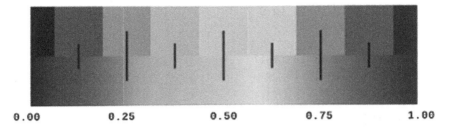

0.00   0.25   0.50   0.75   1.00

Fig. 3. Definition of edge color

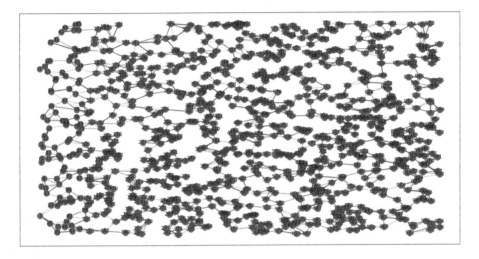

Fig. 4. Map in form of a graph with $N = 1000$

2.3   Orders

Another aspect important for algorithm verification is the creation of a list of orders. For this purpose, a script was created that generates random archival orders on an existing graph. To do this, a random node is selected and then a random route is run based on the probability of the node. The parameters of the order generation script are the order length range (that is, the number of nodes visited) and the number of orders. Below is a part of the file that was created after generating a random list of $L = 10$ orders with lengths ranging from 3 to 10, based on the graph in Fig. 2.

```
orders
22 19 12 13 10 9 7 4 5 3 6
21 25 27 29 22
6 4 3 5 10 9
23 27 21 19 15 12
1 8 9 12 13 10 5 3
```

```
22 29 23 25 21 19 15 13 12 9 7
11 14 16 17 18 28 26
25 20 27
15 19 22 26 28 24 18 16 14 17
8 10 9 2 1 0 11 14 16 18
```

3 Training and Prediction

At this moment, there is an order list with L orders that consist of routs passing thought different towns on the map. Town names are just consecutive numbers, but the algorithm will work in the same way if those names would be real town names. Each town is indexed in dictionary, thus the dictionary consist of N elements. Those parameters are influencing the size and training time of the neural network.

It is worth noticing that the network is trained on the order list, but not on the map. The map is unknown, because in real example there is no map. System does not know whether there is a direct connection between towns. This information is learned basing on order list and whether there was a transport order that was routed between those towns.

Training the network on the exemplary order list that comes from Fig. 2, and providing 252119 as an input sequence will predict that the next town that will be visited would be 15. This information was learned from orders 4, 6. One can note that town 19 is also in orders 1 and 9, but in this case those orders had different predecessor town.

3.1 Training Parameters

The test of training times is important because this solution should be applied on large scale. A check of the network's training time and prediction time is performed to see the impact of the parameters. This analysis will help select appropriate data for the training process to determine the target response time. The tests were not performed on a computing server in order not to block computing power for the other research being conducted in the research team, but on a personal PC with one GPU, so the times presented may differ significantly from those achieved on a dedicated computing server. Thus, the times presented in Table 1 are for trending and selecting the appropriate data structure for the final network model.

The data presented in Table 1 was used to determine the effect of parameters on calculation time. The data is presented in Fig. 5. Parameter analysis shows a large effect of the number of generated orders on the network training time. This phenomenon is due to the fact that the model is trained on a significantly larger set. On the other hand, the size of the graph also has an impact, but it is not so significant. A set of tests was performed where one parameter was held constant, while the other was changed. This determined the area in which the model's training time changes. The blue area is significantly larger, because for

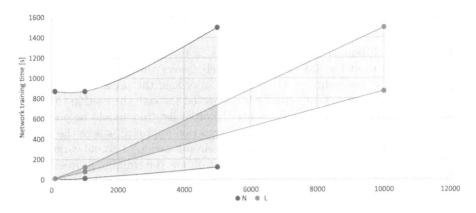

Fig. 5. Influence of parameters N and L on training time t

a fixed graph size, modifying the number of orders has a large effect on time. Alternatively, for a fixed value of the number of orders, the size of the graph has little effect, so the orange area is smaller.

Table 1. Tests on influence of parameters on training times

N	L	training time	prediction time	accuracy
1000	1000	79.96	14.12	0.8265
1000	10 000	871.07	13.76	0.4098
5 000	1000	121.58	15.69	0.9821
100	100	13.34	13.26	0.92
100	10 000	872.51	13.76	0.5817
1000	100	13.44	13.48	0.9954

The next parameters worth analyzing from Table 1 are prediction time and efficiency. However, the analysis does not yield valuable conclusions. The prediction time for 200,000 orders is at a constant level and does not depend on the size of the data used for training. As for efficiency, the number of orders has a large impact on efficiency. This is due to the random nature of the data used for training. The more data is used, the more the distribution of data resembles a normal distribution, so the prediction approaches a choice of 50:50. However, this is a good value, because looking at the nature of the graph, about 3–4 routes can be selected from a given point. An untrained neural network does not take these relationships into account and may suggest going to a city that is not directly connected to the current one, which is a significant error. Such a choice would reduce the efficiency to a value of a few to several percent.

4   Conclusions

The main conclusion of the presented work is that the network was able to learn the map from the list of orders, which proves the concept and can be applicable in the real solution. Giving a list of orders, the prediction output of the network was a town that actually is connected through the road in the map. This information can significantly reduce the order list that should be presented to carrier during the transportation in a Cargo-taxi approach, where the carrier can continue delivering new cargos from the drop location of actual cargo.

The tests of training time gives the insight of how the training process would behave in real life, depending on the structure of input data. In Poland, there are 979 towns and 43 784 post codes, making the training process a real challenge. If the system should work on the whole Europe, the database is even bigger. However, the biggest influence is on the order list, which can be reduced to only a subset, but still with L proportionally smaller than N there is a risk that a particular town will not be included in the dictionary and the network will not train properly and will omit orders from this town.

Moreover this solution has a potential of further development and can help reduce empty routs.

Acknowledgement. The work has been supported by NCBiR grant No. POIR.01.01.01-00-0142/21 "Development of the Polish Digital Logistics Operator - PCOL" in 2023. The work of AG, JS, TG was supported in part by the Silesian University of Technology (SUT) through the subsidy for maintaining and developing the research potential grant (BK) in 2023.

References

1. Cook, D., Holder, L.: Graph-based data mining. IEEE Intell. Syst. Appl. **15**(2), 32–41 (2000). https://doi.org/10.1109/5254.850825
2. Kolhe, P., Christensen, H.: Planning in logistics: a survey. In: Proceedings of the 10th Performance Metrics for Intelligent Systems Workshop, pp. 48–53 (2010)
3. Rizet, C., Browne, M., Cornelis, E., Leonardi, J.: Assessing carbon footprint and energy efficiency in competing supply chains: review-case studies and benchmarking. Transp. Res. Part D: Transp. Environ. **17**(4), 293–300 (2012)
4. Smagulova, K., James, A.P.: A survey on LSTM memristive neural network architectures and applications. Eur. Phys. J. Spec. Top. **228**(10), 2313–2324 (2019)
5. Teng, S.: Route planning method for cross-border e-commerce logistics of agricultural products based on recurrent neural network. Soft. Comput. **25**(18), 12107–12116 (2021)
6. Yan, C., Zhu, H., Korolko, N., Woodard, D.: Dynamic pricing and matching in ride-hailing platforms. Naval Res. Logist. (NRL) **67**(8), 705–724 (2020)
7. Zolfpour-Arokhlo, M., Selamat, A., Hashim, S.Z.M.: Route planning model of multi-agent system for a supply chain management. Expert Syst. Appl. **40**(5), 1505–1518 (2013)

X-ELM: A Fast Explainability Approach for Extreme Learning Machines

Brandon Warner[1], Edward Ratner[1(✉)], and Amaury Lendasse[2,3]

[1] Verseon Corp., Fremont, CA, USA
eratner@verseon.com
[2] ILT Department, University of Houston, Houston, USA
[3] Arcada University of Applied Sciences, Helsinki, Finland

Abstract. In recent years, Explainable Artificial Intelligence (XAI) has emerged as one of the key specializations in Machine Learning (ML) research. XAI has gained significant interest in the last decade due in part to the reluctance of sensitive domains to adopt the use of "black box" models (i.e., models that use ambiguous or obfuscated reasoning). This motivation has led to the rediscovery of Shapley values [1], a method originally applied to coalitional game theory to optimally distribute the "payout" (i.e., importance) of the "players" (i.e., features) of a model. More recently, Lundberg and Lee developed a sophisticated methodology to approximate Shapley values by computing SHAP (SHapley Additive exPlanations) [2] values. SHAP uses the coefficients from local linear models created for every sample in a test set, providing robust sample-level, model-agnostic explainability. Calculating global SHAP values is computationally expensive, requiring models to be built for every sample in a test or validation set. To address these tractability concerns, we propose eXplainable Extreme Learning Machine (X-ELM) values, which can be computed using coefficients of ELM parameters to wholistically evaluate the global importance of each feature in a dataset using a single ELM ensemble model. We compare the extracted ELM coefficients to values extracted using SHAP methods to show that our approach yields values comparable to the state-of-the-art (SOTA) game theoretic approaches at a dramatically lower computational cost.

Keywords: Machine Learning · Extreme Learning Machine · Variable Importance · Feature Importance · Explainable Artificial Intelligence · XAI · Interpretability · Explainability · Comprehensibility · Black-box

1 Introduction

Being able to interpret a model's decisions is increasingly vital in real-world artificial intelligence (AI) applications. This has led to the emergence of a sub-field of AI research, explainable artificial intelligence (XAI). Two main approaches to building explainable models have come to the forefront of the industry. The first is to simply use inherently interpretable models (e.g., linear/logistic regression, decision trees). While these models are more interpretable than more complex models, their lack of complexity generally

© The Author(s), under exclusive license to Springer Nature Switzerland AG 2023
I. Rojas et al. (Eds.): IWANN 2023, LNCS 14135, pp. 411–422, 2023.
https://doi.org/10.1007/978-3-031-43078-7_34

leads to poorer predictive performance in high-dimensional space. The second approach is to pair complex "black box" models with model-agnostic, interpretable models. Some of the most widely adopted model-agnostic approaches include Shapley values [1] and SHAP values [2], among others [5–9]. Shapley values measure a feature's average contribution to a model's predictive performance as a whole. SHAP values build on Shapley values by incorporating coefficients from local linear models introduced in LIME [5] to approximate Shapley values.

Extreme Learning Machines (ELMs) have been studied for the selection of the optimal number of features based on accuracy, speed, and compression ratios [3, 10–13] due to their extremely fast training times and high generalization capabilities [15–21]. Unlike many other ML algorithms, ELMs do not require backpropagation, allowing for analysis of the inherent structure of the model. To our knowledge, this is the first study that investigates whether ELM coefficients can serve as a sound strategy to approximate game theoretic Shapley values, providing insights into the inherent interpretability of ELMs.

Our proposed method, X-ELM, uses the coefficients of an ensemble of ELMs to measure the contribution of each feature to a model's neural structure, effectively approximating Shapley values. To best measure, whether our method can serve as a comparable approach to SOTA approaches, we benchmark feature contribution values made by our method to those made by SHAP. Additionally, we note the amount of time required to compute each of the implementations. The findings in the given experiments show that not only does X-ELM provide a robust approximation of game theoretic approaches but does so requiring orders of magnitude less computing time compared to SHAP values.

2 Methods

We benchmarked the two model explanation methodologies using two binary classification datasets. We chose to limit our experiments to binary classification datasets to confirm that our methodology performs well before progressing with more complex multiclass models (Table 1).

Table 1. Data sets used in benchmarking.

Name	Number of features	Number of samples
Pima Indians Diabetes [20]	8	768
Wisconsin Breast Cancer [21]	9	699
Electrical Grid Stability [22]	13	10,000

2.1 ELM

Extreme Learning Machines (ELMs) are a form of feedforward neural networks that employ a single layer of randomly generated hidden neurons. ELMs may use any activation function that is an infinitely differentiable function (i.e., monotonic) and is bounded from -1 to 1 (e.g., hyperbolic tangent). ELMs have illustrated learning speeds thousands of times faster than other feedforward neural network algorithms that rely on backpropagation while also obtaining greater generalization performance [14–21]. The output of an ELM may be denoted as

$$y_k = \sum_{j=1}^{m} \beta_{j,k} \, g\left(\sum_{i=1}^{n} w_{i,j} x_i + b_j\right) \tag{1}$$

where x_i represents the input to the network and y_k as the k'th output. n, m, and k represent the number of neurons in the input, hidden, and output layers, respectively [4, 15]. $w_{i,j}$ and $\beta_{j,k}$ represent the corresponding weights for the input and output neurons. b_j represents the bias values used by the neurons in the hidden layer. Lastly, $g(.)$ denotes the activation function employed for each neuron. Equation (1) may then be written as

$$\mathbf{H}\beta = y \tag{2}$$

where $\mathbf{H}$ is the hidden layer output matrix. The output neurons' weights, $\beta 1 \cdots m, 1 \cdots k$, are then computed with the generalized Moore–Penrose pseudoinverse matrix as

$$\beta^* = \mathbf{H}^+ y, \tag{3}$$

where $\mathbf{H}^+$ is the generalized Moore–Penrose pseudoinverse matrix of H, and y is the resultant output of the system.

2.2 X-ELM

As elucidated in [4], Eq. (1) may be written in a compact form as

$$y_k = \sum_{i=1}^{n} \propto_{1,k} g(x_1) + \psi_k. \tag{4}$$

The coefficients $\alpha_{i,k} = \beta_{1,k} w_i, 1 + \cdots + \beta_{j,k} w_{i,j} + \beta_{m,k} w_{i,m}$ are then used to measure the j'th feature contribution for the k'th output in the system, which may also be expressed in matrix form as

$$y_k = \begin{bmatrix} \beta_{1,k} \\ \beta_{2,k} \\ \beta_{3,k} \\ \cdots \\ \beta_{m,k} \end{bmatrix}^T \begin{bmatrix} w_{1,1} & w_{2,1} & \cdots & w_{n,1} \\ w_{2,k} & w_{2,2} & \cdots & w_{n,2} \\ w_{3,k} & w_{2,3} & \cdots & w_{n,3} \\ \cdots & \cdots & \cdots & \cdots \\ w_{1,m} & w_{2,m} & \cdots & w_{n,m} \end{bmatrix} g \begin{bmatrix} x_1 \\ x_2 \\ x_3 \\ \cdots \\ x_n \end{bmatrix} + \psi_k \tag{5}$$

The set of coefficients for every k'th output in the system may then be expressed in matrix form as

$$
\left[\alpha_{1,k}, \alpha_{2,k}, \ldots \alpha_{i,k}, , , , \alpha_{n,k}\right] = \begin{bmatrix} \beta_{1,k} \\ \beta_{2,k} \\ \beta_{3,k} \\ \vdots \\ \beta_{m,k} \end{bmatrix}^T \begin{bmatrix} w_{1,1} & w_{2,1} & \cdots & w_{n,1} \\ w_{2,k} & w_{2,2} & \cdots & w_{n,2} \\ w_{3,k} & w_{2,3} & \cdots & w_{n,3} \\ \ddots & \ddots & \vdots & \ddots \\ w_{1,m} & w_{2,m} & \cdots & w_{n,m} \end{bmatrix}. \tag{6}
$$

Lastly, coefficients $\alpha_{1,k}, \alpha_{2,k}, \ldots \alpha_{i,k}, , , , \alpha_{n,k}$ are calculated for each feature in the model using the system

$$
y_k = \left[\alpha_{1,k}, \alpha_{2,k}, \ldots \alpha_{i,k}, , , , \alpha_{n,k}\right] g \begin{pmatrix} x_1 \\ x_2 \\ x_3 \\ \vdots \\ x_n \end{pmatrix} + \psi_k. \tag{7}
$$

The resultant coefficients are what we refer to as X-ELM values. We compute these values for each model in a 400-model ensemble of ELMs. We then compute the weighted average of these values over all models, weighing them by each Model's weight in the ensemble. We compare the resulting values to the other two methodologies. As shown in Sect. 3, X-ELM values effectively approximate the theoretically optimal Shapley values in a fraction of the time taken by other methods.

2.3 SHAP (SHapley Additive exPlanations)

In recent years, SHAP [2] has become one of the foremost approaches used in explaining ML models' reasoning. SHAP values can be computed in a variety of ways, including using both model-agnostic and model-specific approaches. In our experiment, we employ the model-agnostic Kernel SHAP method to explain output made by and ensemble of ELM models. Unlike other XAI approaches, Kernel SHAP values can be calculated for a single prediction in a test set using coefficients of local surrogate (LIME) models [5]. Like LIME, SHAP employs explainable linear models for each prediction, using coefficients from the sample-level models to estimate theoretically optimal Shapley values in coalition. SHAP improves LIME in a few ways. First, SHAP authors avoid the need for heuristics in choosing model parameters (i.e., loss function, weighting kernel, and regularization term) by providing a rigorously proven theorem to choose optimal parameters, recovering Shapley values in the process. Consequentially, SHAP computes approximated Shapley values using weighted linear regression, while LIME employs a simplified input mapping strategy. This enables SHAP to provide greater sample efficiency and interpretability than classical Shapley estimations. While this approach provides valuable sample-level granularity, calculating model-wide SHAP values may be intractable in many circumstances due to combinatorial time and compute constraints resulting from training coalitional models on every test point.

3   Results

For each of the data sets below, we used the sample 80/20 train/test splits to provide a comparable comparison between the two methodologies. We initially planned on using 5-fold cross-validation but SHAPs vast compute requirements made this intractable. Instead, we selected a random subset of 350 background training samples to use for the SHAP portion of the experiments. Furthermore, we chose to explain 50 samples for each test set using the 350 background samples. While this represents a small sample of the total data in some of the experiments, using more data would require weeks of compute time to explain the models. We believe this sampling introduces some noise into the SHAP estimates, however, we believe our conclusions remain unaffected. The time reported for the X-ELM method is equal to the amount of time it took to train the ELM ensemble model since we use only coefficients used in training to measure feature importance. All experiments were computed using the same Windows machine with an AMD Ryzen 9 5950X 16-Core Processor and 64 GB of RAM.

3.1   Pima Indians Diabetes Data Set

The Pima Indians data set [22] was originally collected and made public by the National Institute of Diabetes and Digestive and Kidney Diseases for the purpose of developing methods to diagnostically predict whether patients are diabetic or not. The predictor variables include BMI, insulin level, age, and more. The values obtained using the two methodologies are shown in Table 2 below. Figure 1 visualizes the importance values obtained by SHAP and X-ELM.

Table 2. Normalized feature values computed for each class by each method.

Name	Time	Feature 1	Feature 2	Feature 3	Feature 4	Feature 5	Feature 6	Feature 7
X-ELM	75 seconds	0.4319	0.0467	0.036	0.0461	0.1604	0.1225	0.1563
SHAP	4 days 3 hours	0.3171	0.0722	0.1128	0.1166	0.1638	0.1314	0.0862

Notably, both SHAP and X-ELM agree that glucose is the most influential feature in the model. Both methods also agree on the relative impact of the following three most influential features. Overall, X-ELM and SHAP provided analogous explanations of the models, with a Pearson correlation coefficient between the two normalized feature importance vectors of 0.84. There is, however, a vast separation in the time taken to compute the two methodologies. While it took 75 seconds to produce X-ELM values, it took over 4 days to produce SHAP values on the same data - over 5000 times longer than X-ELM. X-ELM can provide nearly instantaneous explainability because the requisite information is acquired from the model's coefficients in training.

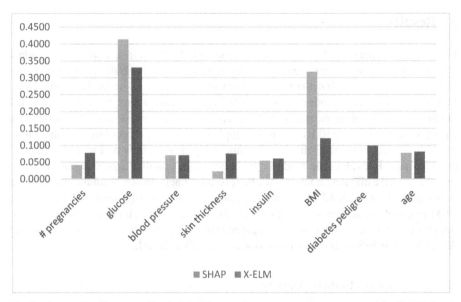

Fig. 1. Comparing the normalized global feature importance values computed by X-ELM and SHAP using the Pima Indians data set

3.2 Wisconsin Breast Cancer Data Set

The Wisconsin breast cancer (original) data set was collected by Dr. William H. Wolberg at the University of Wisconsin Hospitals system in Madison, Wisconsin in 1991 to develop predictive models for identifying malignant cancers. The features in this data set include clump thickness, uniformity of cell size, uniformity of cell shape, marginal adhesion, single epithelial cell size, bare nuclei, bland chromatin, normal nucleoli, and mitoses. The global feature contribution values for the three methods are given in Table 3 below.

Once again, both SHAP and X-ELM methods agree that the first feature, clump thickness, is the most influential variable in the model. Additionally, both methods agree that feature 5, clump area, is the least influential variable in the model. The Pearson correlation coefficient of the two normalized importance vectors is 0.73. Again, the notable difference in the methodologies is in the amount of time needed to produce the model explanations. Using this data set, X-ELM provided robust explainability over 5,000 times faster than SHAP (Fig. 2).

Table 3. Normalized feature values computed by each method.

Name	Time	Feature 1	Feature 2	Feature 3	Feature 4	Feature 5	Feature 6	Feature 7	Feature 8	Feature 9
SHAP	5 days 11 hours	0.2488	0.0768	0.1522	0.1015	0.0081	0.1832	0.1021	0.0725	0.0548
X-ELM	85 seconds	0.1707	0.0833	0.1022	0.1038	0.0637	0.1198	0.0742	0.0782	0.1371

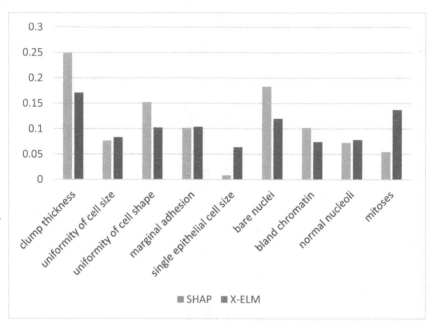

Fig. 2. Comparing the normalized global feature importance values computed by X-ELM and SHAP using the Wisconsin breast cancer data set

3.3 Electrical Grid Stability Simulated Data Set

The electrical grid stability simulated data set [24] was developed by Vadim Azamasov and colleagues for the purpose of automating electrical grid control systems. The variables in this data set include tau[x] (reaction time of producer), p[x] (nominal power consumed/produced), and g[x] (gamma coefficient proportional to price elasticity). There are three tau[x] variables (tau2, tau3, and tau4). There are four p[x] variables (p1, p2, p3, and p4). The output of this data set is the binary class of stable or unstable (Table 4).

Again, there is significant agreement between SHAP and X-ELM values. The Pearson coefficient of the two feature importance vectors is 0.95. Both methods place overwhelming importance on the 'stab' feature in the models. The 'stab' feature is defined as "the maximal real part of the characteristic equation root" [24]. The significant agreement between the two methodologies further illustrates X-ELM's ability to accurately approximate Shapley values (Fig. 3).

Table 4. Normalized feature values computed by each method.

Name	Time	Feature 1	Feature 2	Feature 3	Feature 4	Feature 5	Feature 6	Feature 7	Feature 8	Feature 9	Feature 10	Feature 11	Feature 12	Feature 13
SHAP	6 days 10 hours	0.0982	0.0757	0.0904	0.1118	0.0031	0.0036	0.0026	0.0039	0.0009	0.0040	0.0055	0.0047	0.5956
X-ELM	103 seconds	0.0396	0.0287	0.0314	0.0339	0.0815	0.0445	0.404	0.0461	0.0177	0.016	0.0149	0.0179	0.5415

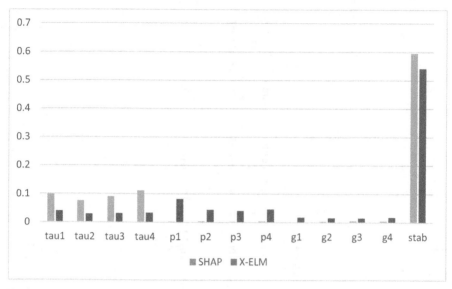

Fig. 3. Comparing the normalized global feature importance values computed by X-ELM and SHAP using the electrical grid stability data set

4 Conclusion

While both methods provide robust explainability, X-ELM does so thousands of times faster than SHAP. Unless one is using an inherently explainable model like XG-Boost, using SHAP to provide global, model-wide explainability is likely not going to be tractable due to SHAP's combinatorial compute requirements. SHAP, however, has the advantage of being able to explain individual predictions. This suggests that SHAP's model-agnostic approach could be used concurrently with X-ELM values to provide varying levels of granularity when explaining models. In future research, we will build on this benchmarking analysis by including more complex multi-class data sets.

References

1. Shapley, L.: A Value for N-Person Games. Contributions to the Theory of Games, vol. 2, no. 28, pp. 307–317 (1953)
2. Lundberg, S.M., Lee, S.: A unified approach to interpreting model predictions. In: Advances in Neural Information Processing Systems, vol. 30 (2017)
3. Štrumbelj, E., Kononenko, I.: Explaining prediction models and individual predictions with feature contributions. Knowl. Inf. Syst. **41**(3), 647–665 (2014)
4. Ertuğrul, Ö.F., Tağluk, M.E.: A fast feature selection approach based on extreme learning machine and coefficient of variation. Turk. J. Electr. Eng. Comput. Sci. **25**(4), 3409–3420 (2017)

5. Marco T.R., Singh, S., Guestrin, C.: Why should I trust you?: Explaining the predictions of any classifier. In: Proceedings of the 22nd ACM SIGKDD International Conference on Knowledge Discovery and Data Mining, pp. 1135–1144. ACM (2016)
6. Shrikumar, A., Greenside, P., Shcherbina, A., Kundaje, A.: Not just a black box: learning important features through propagating activation differences. arXiv preprint:arXiv:1605. 01713 (2016)
7. Datta, A., Shayak, S., Yair, Z.: Algorithmic transparency via quantitative input influence: theory and experiments with learning systems. In: 2016 IEEE Symposium on Security and Privacy (SP), pp. 598-617. IEEE (2016)
8. Lipovetsky, S., Conklin, M.: Analysis of regression in game theory approach. In: Applied Stochastic Models in Business and Industry, vol. 17, no. 4, pp. 319–330 (2001)
9. Bach, S., Binder, A., Montavon, G., Klauschen, G., Müller, K.R., Samek, W.: On pixel-wise explanations for non-linear classifier decisions by layer-wise relevance propagation. PloS one **10**(7), e0130140 (2015)
10. Benoît, F., Van Heeswijk, M., Miche, Y., Verleysen, M., Lendasse, A.: Feature selection for nonlinear models with extreme learning machines. Neurocomputing **102**, 111–124 (2013)
11. Termenon, M., Graña, M., Barrós-Loscertales, A., Ávila, C.: Extreme learning machines for feature selection and classification of cocaine dependent patients on structural MRI data. Neural Process. Lett. **38**, 375–387 (2013)
12. Salcedo-Sanz, S., Pastor-Sánchez, A., Prieto, L., Blanco-Aguilera, A., García-Herrera, R.: Feature selection in wind speed prediction systems based on a hybrid coral reefs optimization–extreme learning machine approach. Energy Convers. Manag. **87**, 10–18 (2014)
13. Guillén, A., et al.: Fast feature selection in a GPU cluster using the delta test. Entropy **16**(2), 854–869 (2014)
14. Huang, G., Zhu, Q., Siew, C.: Extreme learning machine: theory and applications. Neurocomputing **70**(1–3), 489–501 (2006)
15. Huang, G., Zhu, Q., Siew, C.: Extreme learning machine: a new learning scheme of feed-forward neural networks. In: 2004 IEEE International Joint Conference on Neural Networks (IEEE Cat. No. 04CH37541), vol. 2, pp. 985-990. IEEE (2004)
16. Huang, G., Wang, D.H., Lan, Y.: Extreme learning machines: a survey. Int. J. Mach. Learn. Cybern. **2**, 107–122 (2011)
17. Khan, K.,Ratner, E., Ludwig, R., Lendasse, A.: Feature bagging and extreme learning machines: machine learning with severe memory constraints. In: 2020 International Joint Conference on Neural Networks (IJCNN), Glasgow, UK, pp. 1–7 (2020)
18. Li, Z., Ratner, K., Ratner, E., Khan, K., Bjork, K..-M.., Lendasse, A..: A novel ELM ensemble for time series prediction. In: Cao, J.., Vong, C..M.., Miche, Y.., Lendasse, A.. (eds.) ELM 2018. PALO, vol. 11, pp. 283–291. Springer, Cham (2020). https://doi.org/10.1007/978-3-030-23307-5_31
19. Warner, B., Ratner, E., Lendasse, A.: Edammo's extreme AutoML technology – benchmarks and analysis. In: Björk, KM. (ed.) Proceedings of ELM 2021. ELM 2021. Proceedings in Adaptation, Learning and Optimization, vol. 16. Springer, Cham. (2023). https://doi.org/10. 1007/978-3-031-21678-7_15
20. Carolus Khan, K., Ratner, E., Douglas, C., Lendasse, A.: A novel methodology for object detection in highly cluttered images. In: Björk, KM. (ed.) Proceedings of ELM 2021. ELM 2021. Proceedings in Adaptation, Learning and Optimization, vol. 16. Springer, Cham. (2023). https://doi.org/10.1007/978-3-031-21678-7_2

21. Ratner, E., Garcia, E., Warner, B., Douglas, C., Lendasse, A.: Extreme AutoML – analysis of classification, regression and NLP performance. To be submitted, (2023)
22. Smith, J.W., Everhart, J.E., Dickson, W.C., Knowler, W.C., Johannes, R.S.: Using the ADAP learning algorithm to forecast the onset of diabetes mellitus. In: Proceedings of the Annual Symposium on Computer Application in Medical Care, p. 261. American Medical Informatics Association (1988)
23. Wolberg, W.H., Mangasarian, O.L.: Multisurface method of pattern separation for medical diagnosis applied to breast cytology. Proc. Nat. Acad. Sci. **87**(23), 9193–9196 (1990)
24. Arzamasov, V., Böhm, K., Jochem, P.: Towards concise models of grid stability. In: 2018 IEEE International Conference on Communications, Control, and Computing Technologies for Smart Grids (SmartGridComm), pp. 1–6. IEEE (2018)

Replacing Goniophotometer with Camera and U-Net with Hypercolumn Rescale Block

Marek Vajgl$^{(\boxtimes)}$ and Petr Hurtik

Centre of Excellence IT4Innovation, Institute for Research and Applications of Fuzzy Modelling, University of Ostrava, 30. dubna 22, Ostrava, Czech Republic
{marek.vajgl,petr.hurtik}@osu.cz
http://irafm.osu.cz

Abstract. We deal with replacing a costly and slow goniophotometer device with a standard, inexpensive, and fast camera in the task of evaluating an illuminated area by a car headlamp. This solution is novel, has not yet been solved, and has the potential to speed up the process of prototyping headlamps. The difficulties lie in the significantly different resolutions of the two devices and in the disparity between intensities captured by the camera and goniophotometer due to the nonlinear behavior of the light. We propose to capture images by a camera with various exposure times and handle them as a multispectral image. The image is processed by U-Net architecture where we replaced the standard decoder with a Hypercolumn rescale block. The proposed scheme produces a mean absolute percentage difference between the real goniophotometer and our solution of less than 0.5%.

Keywords: Goniophotometer · U-Net · Asymmetric segmentation · Car headlamp

1 Introduction

Research and recent experiments in the field of neural networks (NN) have allowed their application in many areas and tasks [1] with the ability to replace manual or time-consuming procedures with fast solutions with sufficient precision. One of the main areas with massive applications of AI and NN is the automotive industry. The use of AI in conjunction with autonomous vehicles is a very common example [18]; however, NN can be applied in manufacturing to make production faster and more effective [24,30].

One of the areas is the development of car headlamps (or lights in general). The design and development must ensure the correct direction of light rays and luminosity of light beams produced by the light emitter. Those properties of prototypes and of the final designed headlight are evaluated on the specific device - a goniophotometer (GM). It measures the intensities of particular rays,

© The Author(s), under exclusive license to Springer Nature Switzerland AG 2023
I. Rojas et al. (Eds.): IWANN 2023, LNCS 14135, pp. 423–434, 2023.
https://doi.org/10.1007/978-3-031-43078-7_35

and the obtained data are compared with the legal requirements for headlamps valid for a particular local market.

The process of collecting the required data using a GM is expensive due to the cost of the device and is time-consuming because the information is captured piece by piece and needs precise orientation and adjustment during the capturing process. However, the obtained intensities are precise, as it is a physics-based measurement.

Our motivation is to replace the GM with a standard camera. A standard camera is inexpensive and capable of capturing the entire scene in a fraction of a second. It is obvious that replacing an expensive and slow device with a cheap and fast one may increase the speed of prototyping of the headlamps. However, the camera is much less precise, is affected by noise, has a limited dynamic range, captures the surface of the projection wall as well, and is more sensitive to ambient light conditions. Therefore, a simple replacement of the GM with a camera is insufficient. An additional post-processing must be made to eliminate camera issues. In this paper, we propose an efficient postprocessing based on NN. For a comparison of the output produced by the GM and the camera, see Fig. 1.

Fig. 1. An example of the same car light captured by the two different devices. Left: goniophotometer, right: camera. There are visible auxiliary markings and a hole (behind it is placed GM) on the image taken by camera.

2   Related Work

Based on our best knowledge, the simulation of a goniophotometer is a new way of research that has not been described yet in the literature. The topic of simulating light measurement was tackled by estimating incoming light from all directions to a 3D point in the RGB image [23], or by developing a software GM device to estimate LED Luminaires Luminous Intensity Distribution [20]. Unfortunately, these attempts cannot simulate the GM for a general light projection on a wall. If we look at the problem from the point of view of GM evaluation or simulation, there is not published research yet. The connections with our research can be found in a simulation of physics-based processes, hyperspectral image processing, and image segmentation.

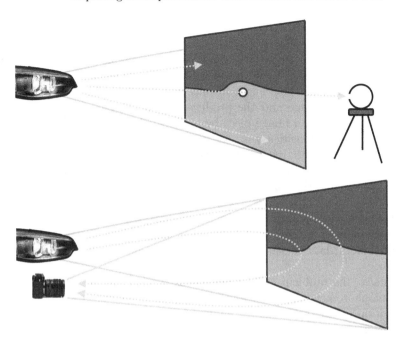

Fig. 2. Headlamp light evaluation. Upper: Current solution using screen and Goniophotometer. The GM captures the intensities of the light ray passing through the hole in the wall directly. Lower: Proposed idea using a white wall and a standard camera. The camera captures the reflection made by the headlamp on the wall.

The first attempts to simulate physical devices using an NN were made by estimating fluid mechanics [27] using a shallow neural network or by simulating physics-based models [6]. The replacement of partial equations with an NN was then regularly improved [4,14,31,32] with the finding that it can be carried out under restricted scenario and conditions. For us, that indicates a possibility of replacing the GM with an NN, because is a simpler task than a fluid simulation.

Hyperspectral image processing [5,11,28] is usually connected to satellite imaging, where it benefits from multiple sensors that capture the same area. In our task, we have a single sensor (camera) but with various settings of the exposure time, producing multiple images with different light intensities. Motivated by the heperspectral image processing, we fuse these images into a multichanneled input that is processed by the neural network at once.

Finally, image segmentation is a well-solved task, where [21] proposed an encoder/decoder based fully convolution architecture U-Net, which has evolved into uncountable versions [10,19,33]. Although there is also research leading to *asymmetric* U-Net [12,29], the term asymmetric is related to the complexity of encoder/decoder and not to the resolution. From this point of view, our research where the input has a higher resolution than the output is novel. Note that the input has approximately 8000× more pixels than the output.

3   Problem Formulation

There are two devices: a standard camera and a goniophotometer (GM). There is a significant disparity between the two devices. The camera is cheap and capable of taking the whole image at once in a fraction of a second, but it has a limited contrast range and the produced image contains noise. The GM is expensive, capturing an image piece-by-piece, so the capturing process is long, but the captured intensities are precise. The motivation is to use a standard camera with its benefits, but improve the produced captured intensity precision to be on par with the GM. For an illustration of the difference between the GM capture process and the camera, see Fig. 2. For a comparison of the output, see Fig. 1.

We are going to use a standard camera to capture a large area (a wall) lighted by the headlight. The main task is the transformation of imprecise and noise-affected camera data to simulate the goniophotometer output with sufficient precision.

Formally, the problem can be defined as follows. Let $g : W_g \times H_g \rightarrow L_g$ be a goniophotometer function and $c : W_c \times H_c \rightarrow L_c$ be a camera function. The solution of the task relies on finding f such that $f : c \mapsto g$. Although the formulation seems simple, we will show in this paper that the practical solution may be tricky and domain-specific.

The first issue is that $L_g \neq L_c$, that is, $L_g = \{0, 1, \ldots, 32767\}$ and $L_c = \{0, 1, \ldots, 255\}$. The solution is to normalize the ranges, so $L_g', L_c' \in [0, 1]$. The second issue is that the data resolution differs significantly, that is, $H_g \ll H_c$ and $W_g \ll W_c$. Therefore, the unification of the domains is more complicated and its solution is one of the contributions of this paper; a detailed description is given in Sect. 5. Note that simple resampling techniques (Bicubic, Lanczos, etc.) bring distortion into the image in the case of upscaling and lead to omitting important context information in the case of downsampling, so are not directly applicable to our task.

4   Analysis of a Model-Driven Solution

Let us recall that we are finding f such that $f : c \mapsto g$, where c is the camera function and g is the GM function. Model-driven is a "classical" way in which we take input (c), define transformation rules (our f), and obtain the required output (g). Here, we discuss two transformation rules.

The first rule we propose is $f(c) = \alpha c$ where $\alpha \in (-\infty, \infty)$. The particular value of α is given by

$$\alpha = \frac{1}{WH} \sum_{x,y \in W \times H} \frac{g(x,y)}{c(x,y)},$$

where the necessary condition is that c and g are resized into the same domain of $W \times H$. The advantage of such a solution is that it is spatial-shift invariant, with the disadvantage that it cannot capture the non-linearity in intensity. The approach in which we determined α using couple (c_1, g_1) and then applied it to c_2 to obtain g_2 is presented in Fig. 3. The figure shows the difference between the ground truth (given by the GM measurements) and the computed g_2. It is obvious that the auxiliary marks on the wall and the center hole for GM reflect the light in a local manner, so the proposed global rule fails here. This proves the fact that the whole transformation cannot be described using a single constant only.

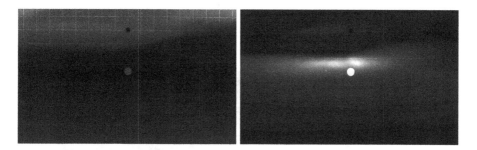

Fig. 3. Left: model-driven solution based on single constant; right: model driven solution based on matrix. The solution based on constant is not capable to describe local differences while the solution based on matrix is sensitive to spatial distortions.

The second rule is motivated by the failure of the first one and is given by $f(c) = A \odot c$, where $\odot$ is a Hadamard product, and $A : W \times H \to (-\infty, \infty)$. The computation is on the same basis as the first rule, i.e., $A(x, y) = \frac{g(x,y)}{c(x,y)}$. The main advantage is that A handles the local distortions caused by the auxiliary marks and by the wall itself. The output visualized in Fig. 3 confirms the claim. On the other hand, a disparity in the projected light can be seen. The hypothesis is that the GM and camera capture the light in different ways, and the relationship between them is not-linear. The verification and confirmation of the hypothesis is shown in Fig. 4.

5 The Proposed Data-Driven Solution

An alternative to the 'model-driven solution' is the attitude-based 'data-driven' approach using deep learning. In contrast to the model-driven solution, the input to the data-driven one is the image together with the required output (label), and the output of the solution is a set of rules that can be used to map the input to the label.

5.1 Preliminaries

The process of mapping one image to another is similar to the already known task of dense classification, i.e., image segmentation. Thus, we propose to use an

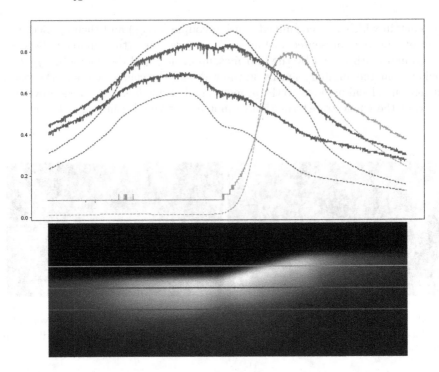

Fig. 4. Upper: Comparison of GM-data (dashed) vs. camera-data (solid) intensities at chosen lines of interest. Lower: Visualisation of lines of interest in GM-data image.

established technique of image segmentation based on a deep neural network and modify it for our task so it is able to manage the discrepancy between a large camera output in comparison with a small GM output. Common approaches such as U-Net [21], PAN [15], Deeplab [3], MANet [16], or HR-Net [25] are based on the encoder-decoder scheme in which the input is sequentially mapped into compressed latent representation and then decoded into the desired output. The important remark here is that these schemes are symmetric; that is, the input resolution is equal to the output resolution. To resolve this, we propose an asymmetric HR-B U-Net; a U-Net with Hypercolumn rescale block.

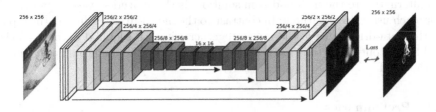

Fig. 5. U-Net scheme. Image taken from deepsense.ai

Formally, the standard U-Net [21] performs $U : \mathbb{R}^{h \times w \times 3} \to \mathbb{R}^{h \times w \times c}$, where we consider a standard raster color image of size $h \times w \times 3$ to be the input and the task is a classification of each pixel into one of the c classes. The inner structure consists of a sequence of convolutional, batch normalization, aggregation, and activation layers, see Fig. 5. The simple encoder presented in the original paper [21] is usually replaced by a backbone known from the image classification task such as ResNet [8], EfficientNet [26], or Swin [17], to name a few. It reflects the importance of proper coding of input information. The decoding from latent representation into the output is simpler compared to the encoding part, and thus usually, only a few convolution + batch normalization + activation blocks with upsampling are used. The only requirement for the backbone is that it must be hierarchical to realize the skip connections between the encoder and the decoder.

That is, at the same time, the main issue of U-Net. Let us assume that U consists of a fixed number of L layers. Taking into account the convolutional layer (with/without stride), batch normalization layer, activation layer, or upscaling layer (standard or transpose), we can write $L : \mathbb{R}^{h_1 \times w_1 \times c_1} \to \mathbb{R}^{h_2 \times w_2 \times c_2}$, where c_1 and c_2 can be arbitrary, but for h_2 must hold $h_2 \in \{1/sh_1, h_1, sh_1\}$ where $s \in \mathbb{N}^+$ is stride and the same for w_2. Violation of the condition leads to the impossibility of applying the skip connection due to the input shape mismatch. When the stride (or the pooling) is $s = 2$, and d is the number of scales of the encoder, it must hold $h = \alpha 2^d$ where $\alpha \in \mathbb{N}^+$ is an arbitrary constant, and h is the height of the input image; the same applies for w. If the condition is broken, the input image must be padded to make the condition valid. The problem arises when the output resolution is not equal to the input image and, moreover, it is not d times naturally divisible by s. That is our case. Its solution is described in the following section.

5.2 The Realization

We propose to use a common encoder part with a custom decoder part to fulfill our needs.

Encoding. The proposed approach starts with an arbitrary backbone where the only requirement is that it is hierarchical, i.e., it includes several scales from which residual outputs can be extracted. We suppose R residual outputs, where each of them is followed by a 'convolution + batch normalization + activation' block (CBA, for short). In our case, the resolution of the GM image, i.e., the resolution of our desired output, is roughly similar to the resolution of a feature map produced by the deepest block. In this case, a standard solution is to use only the last 'residual' output and build a decoder on top of it.

Decoding. We propose to reuse information captured by the scales with higher resolution available through the particular residual outputs and built from them Hypercolumn [7]. To build it, it is necessary to unify the resolution of the partial feature maps, thus downsampling by the factor of 2^r where $r \in \{0, 1, \ldots, R-1\}$ is the scale index, sorted in descending order (the last scale has index 0). The obtained Hypercolumn is the input to the following bilinear resize (BR) lambda layer, where the purpose of BR is to have the final output of the network in the identical resolution as the GM has. Because the BR is not differentiable and does not include trainable parameters, it must be followed by a CBA block to suppress artifacts made by the interpolation itself. For a visualization of the entire scheme, see Fig. 6.

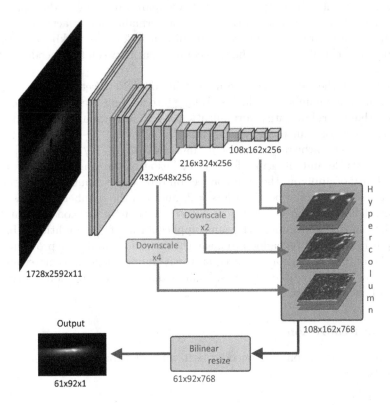

Fig. 6. Scheme of the proposed approach.

Loss Function. The standard loss functions such as MSE or crossentropy minimizes the global difference without connection to the scale. Thus, we propose to use MSE and modified MAPE in compound loss function ℓ defined as

$$\ell(y, \hat{y}) = \psi(y, \hat{y}) + \alpha \frac{|y - \hat{y}| + s}{y + s},$$

where $y, \hat{y}$ is the ground truth with prediction, ψ stands for binary cross-entropy, α is the loss balancing constant, and s is the smoothing constant enriching the original MAPE to avoid numerical instability for low values of y.

6   Experimental Verification

Data. We have obtained two types of car headlamps, and for each of them, two pieces with different configurations (left and right; note that the left and right light beams are not symmetric). Using a camera, we have captured each of the projections of a car hadlamp with 11 images, where they differ by shooter setting. The resolution of these images is 2592×1728 pixels. Furthermore, each projection has been captured simultaneously by GM, where the output resolution is 92×61 pixels. For training, we use the left car lights, and for testing, we use the right ones. Due to the non-linear intensity nature of the data (see Fig. 4), we omitted intensity augmentations as same as augmentations that use internally an interpolation. Thus, we are restricted to flips and shifts only. Note that there is a widespread incorrect assumption that convolution neural networks are invariant to shifts, and thus, this augmentation is useless; in fact, they are equinvariant [2]; therefore, the augmentation of shift is beneficial.

Training Setting. The network was trained with Adam [13] optimizer with $\alpha = 1 \cdot 10^{-4}$ and the decrease to half after three consecutive not decreasing training losses. The used hardware, GPU RTX2080Ti with 11GB VRAM was able to process a batch of two images at max. The model was trained for 100 epochs of 300 iterations each, and the final trained model was the subject of evaluation on the test set.

Evaluation. During the research, we trained the network in various configurations and examined the impact of the proposed steps; the results are shown in Table 1. The error given in the table represents MAPE and is measured on the test set, and the obtained results support the fact that a physical device can be replaced by a simulation given by the neural network model, as the GM itself has an error of approximately 5%. For an illustration of the visual difference, see Fig. 7. Three important milestones can be seen in Table 1: increasing the capacity of the network, adjusting the loss function w.r.t. the solved problem, and the addition of trainable layers after non-trainable bilinear interpolation. Based on this behavior, there is a hypothesis that a further increase of parameters in the backbone, e.g., by utilizing EfficientNet, may lead to another error decrease. However, as the resolution of the input is enormously large, the increased capacity of the network leads to increased memory requirements on the graphics card.

7   Summary

This paper proposes a solution to make the car headlamp prototyping process more effective. The task was to replace the expensive device, the goniophotometer, with a standard camera. To ensure that data provided by standard-camera-approach can substitute goniophotometer, we have modified the U-Net architecture to be able to process significantly different resolution of its input and

Table 1. The error measured on the test set for the modifications of the baseline. The MAPE is in percentage between the predictions and the ground truth labels.

Setting	MAPE [%]	
	mean	max
Baseline, binary crossentropy loss, VGG-16 [22] backbone	11.4	453.5
Backbone replaced by SE-ResNet-34 [9]	6.0	252.4
Binary crossentropy loss replaced by percentual loss	2.8	100.0
Combination of percentual and binary crossentropy loss	2.5	100.0
Added Hypercolumn block	2.2	100.0
Added convolution block after interpolation	0.4	23.7

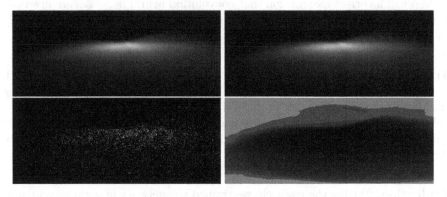

Fig. 7. The prediction on test set. Top row: ground truth (left) and the prediction (right). Bottom row: visualization of the error between ground truth and prediction; MAE (left) and MAPE (right). Note, both images illustrating the error were normalized to increase the visibility.

output. Namely, we have combined a U-Net encoding part with Hypercolumn decoding part, including U-Net skip connections to achieve better result quality.

The proposed solution achieves precision comparable to that of a goniophotometer solution and under the standard goniophotometer's measurement error. Therefore, we have proven that the solution is capable of replacing the original expensive and time-consuming approach. The completed solution is being implemented as a practical application in cooperation with Plastic Omnium company.

Acknowledgement. The work is supported by ERDF/ESF "Centre for the development of Artificial Intelligence Methods for the Automotive Industry of the region" (No. CZ.02.1.01/0.0/0.0/17_049/0008414).

References

1. Alam, M., Samad, M.D., Vidyaratne, L., Glandon, A., Iftekharuddin, K.M.: Survey on deep neural networks in speech and vision systems. Neurocomputing **417**, 302–321 (2020)
2. Azulay, A., Weiss, Y.: Why do deep convolutional networks generalize so poorly to small image transformations? arXiv preprint arXiv:1805.12177 (2018)
3. Chen, L.C., Papandreou, G., Schroff, F., Adam, H.: Rethinking atrous convolution for semantic image segmentation. arXiv preprint arXiv:1706.05587 (2017)
4. Cheng, C., Zhang, G.T.: Deep learning method based on physics informed neural network with resnet block for solving fluid flow problems. Water **13**(4), 423 (2021). https://doi.org/10.3390/w13040423, https://www.mdpi.com/2073-4441/13/4/423
5. Ghamisi, P., et al.: Advances in hyperspectral image and signal processing: a comprehensive overview of the state of the art. IEEE Geosci. Remote Sens. Mag. **5**(4), 37–78 (2017)
6. Grzeszczuk, R., Terzopoulos, D., Hinton, G.: Neuroanimator: fast neural network emulation and control of physics-based models. In: Proceedings of the 25th Annual Conference on Computer Graphics and Interactive Techniques, SIGGRAPH 1998, pp. 9–20. Association for Computing Machinery, New York (1998). https://doi.org/10.1145/280814.280816
7. Hariharan, B., Arbeláez, P., Girshick, R., Malik, J.: Hypercolumns for object segmentation and fine-grained localization. In: Proceedings of the IEEE Conference on Computer Vision and Pattern Recognition, pp. 447–456 (2015)
8. He, K., Zhang, X., Ren, S., Sun, J.: Deep residual learning for image recognition. In: Proceedings of the IEEE Conference on Computer Vision and Pattern Recognition, pp. 770–778 (2016)
9. Hu, J., Shen, L., Sun, G.: Squeeze-and-excitation networks. In: Proceedings of the IEEE Conference on Computer Vision and Pattern Recognition, pp. 7132–7141 (2018)
10. Hurtik, P., Ozana, S.: Dragonflies segmentation with u-net based on cascaded resnext cells. Neural Comput. Appl. **33**, 4567–4578 (2021)
11. Imani, M., Ghassemian, H.: An overview on spectral and spatial information fusion for hyperspectral image classification: current trends and challenges. Inf. Fusion **59**, 59–83 (2020)
12. Kazerouni, I.A., Dooly, G., Toal, D.: Ghost-unet: an asymmetric encoder-decoder architecture for semantic segmentation from scratch. IEEE Access **9**, 97457–97465 (2021)
13. Kingma, D.P., Ba, J.: Adam: a method for stochastic optimization. arXiv preprint arXiv:1412.6980 (2014)
14. Kumar, A., Ridha, S., Narahari, M., Ilyas, S.U.: Physics-guided deep neural network to characterize non-newtonian fluid flow for optimal use of energy resources. Expert Syst. Appl. **183**, 115409 (2021). https://doi.org/10.1016/j.eswa.2021.115409. https://www.sciencedirect.com/science/article/pii/S0957417421008307
15. Li, H., Xiong, P., An, J., Wang, L.: Pyramid attention network for semantic segmentation. arXiv preprint arXiv:1805.10180 (2018)
16. Li, R., et al.: Multiattention network for semantic segmentation of fine-resolution remote sensing images. IEEE Trans. Geosci. Remote Sens. **60**, 1–13 (2021)
17. Liu, Z., et al.: Swin transformer: hierarchical vision transformer using shifted windows. In: Proceedings of the IEEE/CVF International Conference on Computer Vision, pp. 10012–10022 (2021)

18. Mozaffari, S., Al-Jarrah, O.Y., Dianati, M., Jennings, P., Mouzakitis, A.: Deep learning-based vehicle behavior prediction for autonomous driving applications: a review. IEEE Trans. Intell. Transp. Syst. **23**(1), 33–47 (2020)
19. Nabiee, S., Harding, M., Hersh, J., Bagherzadeh, N.: Hybrid u-net: semantic segmentation of high-resolution satellite images to detect war destruction. Mach. Learn. Appl. **9**, 100381 (2022)
20. Novak, T., Valicek, P., Mainus, P., Becak, P., Latal, J., Martinek, R.: Possibilities of software goniophotometer usage for led luminaires luminous intensity distribution curves modelling - case study. In: 2022 22nd International Scientific Conference on Electric Power Engineering (EPE), pp. 1–5 (2022). https://doi.org/10.1109/EPE54603.2022.9814141
21. Ronneberger, O., Fischer, P., Brox, T.: U-Net: convolutional networks for biomedical image segmentation. In: Navab, N., Hornegger, J., Wells, W.M., Frangi, A.F. (eds.) MICCAI 2015. LNCS, vol. 9351, pp. 234–241. Springer, Cham (2015). https://doi.org/10.1007/978-3-319-24574-4_28
22. Simonyan, K., Zisserman, A.: Very deep convolutional networks for large-scale image recognition. arXiv preprint arXiv:1409.1556 (2014)
23. Song, S., Funkhouser, T.: Neural illumination: lighting prediction for indoor environments (2019)
24. Staar, B., Lütjen, M., Freitag, M.: Anomaly detection with convolutional neural networks for industrial surface inspection. Procedia CIRP **79**, 484–489 (2019)
25. Sun, K., et al.: High-resolution representations for labeling pixels and regions. arXiv preprint arXiv:1904.04514 (2019)
26. Tan, M., Le, Q.: Efficientnet: rethinking model scaling for convolutional neural networks. In: International Conference on Machine Learning, pp. 6105–6114. PMLR (2019)
27. Teo, C., Lim, K., Hong, G., Yeo, M.: A neural net approach in analyzing photograph in piv. In: Conference Proceedings 1991 IEEE International Conference on Systems, Man, and Cybernetics, vol. 3, pp. 1535–1538 (1991). https://doi.org/10.1109/ICSMC.1991.169906
28. Wang, C., et al.: A review of deep learning used in the hyperspectral image analysis for agriculture. Artif. Intell. Rev. **54**(7), 5205–5253 (2021). https://doi.org/10.1007/s10462-021-10018-y
29. Wang, S., et al.: Stacked dilated convolutions and asymmetric architecture for u-net-based medical image segmentation. Comput. Biol. Med. **148**, 105891 (2022)
30. Wang, T., Chen, Y., Qiao, M., Snoussi, H.: A fast and robust convolutional neural network-based defect detection model in product quality control. Int. J. Adv. Manuf. Technol. **94**, 3465–3471 (2018)
31. Wessels, H., Weißenfels, C., Wriggers, P.: The neural particle method - an updated lagrangian physics informed neural network for computational fluid dynamics. Comput. Methods Appl. Mech. Eng. **368**, 113127 (2020). https://doi.org/10.1016/j.cma.2020.113127. https://www.sciencedirect.com/science/article/pii/S0045782520303121
32. Zhang, Y., Ban, X., Du, F., Di, W.: Fluidsnet: end-to-end learning for lagrangian fluid simulation. Expert Syst. Appl. **152**, 113410 (2020). https://doi.org/10.1016/j.eswa.2020.113410. https://www.sciencedirect.com/science/article/pii/S0957417420302347
33. Zhou, Z., Rahman Siddiquee, M.M., Tajbakhsh, N., Liang, J.: UNet++: a nested U-net architecture for medical image segmentation. In: Stoyanov, D., et al. (eds.) DLMIA/ML-CDS -2018. LNCS, vol. 11045, pp. 3–11. Springer, Cham (2018). https://doi.org/10.1007/978-3-030-00889-5_1

Unsupervised Clustering at the Service of Automatic Anomaly Detection in Industry 4.0

Dylan Molinié$^{(\boxtimes)}$ ⓘ, Kurosh Madani, and Véronique Amarger

LISSI Laboratory EA 3956, Université Paris-Est Créteil,
Sénart-FB Institute of Technology,
Campus de Sénart, 36-37 Rue Georges Charpak, 77567 Lieusaint, France
{dylan.molinie,madani,amarger}@u-pec.fr

Abstract. Industrial processes are among the most complex systems, for they are dynamic, nonlinear and comprise many interdependent parts. In the scope of the contemporaneous fourth industrial revolution, the Industry 4.0, the trend is to integrate Artificial Intelligence and hyper-connectivity to intelligently exploit any available resources of a system. A key issue for system management is the control of anomalies, which may cause severe failures if not corrected rapidly, or affect product quality; defining and knowing how to handle them is thus of major importance. This paper proposes to apply Machine Learning-based unsupervised clustering to industrial data to automatically identify the anomalies historically encountered; this is expected to help understand the system and define a framework for diagnosis of failures. Results show that unsupervised clustering is able to detect salient groups of data, which can therefore be classified as anomalies by comparing them to the regular system's behaviors, obtained using another round of unsupervised clustering.

Keywords: Machine Learning · Industry 4.0 · Automatic diagnosis · Anomaly detection · Unsupervised clustering · Data Mining

1 Introduction

With the global trend of production optimization and resource usage reduction, the industrial sector is mutating in depth: more and more projects emerge to propose new, innovative ways to handle these keys issues [14]. To that purpose, one may consider two fashions to operate [13]: 1) Always keep a vigilant eye on the systems to master them and control how they evolve; 2) Build predictive, automatic models to estimate their future evolution and react accordingly.

The first point corresponds to what is often done in most industries: senior, highly experienced technicians regularly check the state of the processes, and

This paper has received funding from the European Union's Horizon 2020 research and innovation program under grant agreement No 869886 (project HyperCOG).

mechanically operate the actuators so as to keep them in their operating areas. Constant monitoring is commonly used, but it is expensive, and when a senior operator leaves or retires, his or her knowledge does as well, knowledge which may be hard to replace. Likewise, due to systems' intrinsic complexity, a small change in a process may get worse if not handled properly, and cause severe failures by snowball effect; unfortunately, such small changes might not be visible before it is too late to prevent them from affecting the other processes, which can thus only be handled afterwards, when their negative effects become noticeable [4].

Although real-time analysis by an expert is the actual "standard" way to monitor industrial systems, new possibilities emerge with the development of Artificial Intelligence [1]. Nowadays, the industrial sector is turning to the integration of Machine Learning, and Artificial Intelligence more broadly, alongside highly connected systems. Such combination provides the basis of a new industrial revolution, known as Industry 4.0, which aims to globally improve system management, by better exploiting the production units, while proposing smart, cognition-oriented tools to handle and control them [5].

In that scope, a common problem one may encounter with any industrial system is process drift, i.e. when it evolves toward an unwanted direction [18]. Such events are detrimental to the industrial efficiency and product quality, and, as such, operators strive to prevent them from occurring, or, if not possible, strive to correct them as soon as possible. They are aptly assimilated to anomalies, in comparison to the normal way a system was crafted to behave; identifying, handling and correcting these anomalies is a cornerstone of system management, quality assessment, and is a key to help improve production chains [15].

There might be several ways to define and point out such anomalies, but intuition would assimilate them as any system drift, i.e. when any sensor reaches an unwanted range of values. Assuming a system behaves "normally"most of the time (i.e. in the way it was crafted for), most of its sensors' states (data) should take very similar values, contained within a restricted interval, whilst the abnormal events should not. Consequently, one may identify and isolate them by searching for salient regions in the feature space, where some data would form local, compact groups, apart from the largest ones (the regular "behaviors").

In the context of anomaly detection and assisted diagnosis, the first step consists in defining the different anomalies the system may encounter; to that purpose, assuming they form local groups of data, isolated from one another, one may consider using data-driven unsupervised clustering to isolate and classify the anomalies a system may encounter. Indeed, these Machine Learning-based algorithms aim to automatically regroup data sharing similarities into compact groups. With that definition of anomaly, unsupervised clustering seems to be the most appropriate way to automatically isolate and point out the events of a system [11]. Then, an averaging, higher-level unsupervised clustering may help identify the behaviors of the system (ground truth), to which the clusters can be compared to so as to classify them as either an anomaly or a true behavior.

This paper is organized as follows: first, an overview of the already-existing techniques for anomaly detection is made, then, the proposed methodology is described in detail and is applied to a didactic example, before extending it to real industrial data, collected within the European project HyperCOG (introduced in Sect. 4); the last section finally concludes this paper.

2   State-of-the-Art

When dealing with dynamic systems, such as industrial ones, fault diagnosis is of major importance, for an error should be corrected before it propagates and affects the other systems. With dynamic processes, the notion of anomaly can be defined as a state when some sensors take abnormal values, outside their nominal ranges. Such event can occur when a system is wrongly controlled, or even just by itself, especially with chemical compounds which may react differently than expected, due to the inertia of the reactions. This last example is called a "drift"; they are very common in real industrial, dynamic systems, and may be the commonest sort of anomalies one may encounter.

With that definition, the simplest way to identify the anomalies encountered by a system is certainly by comparing them to its regular behaviors (that one may call modes): an anomaly is therefore whatever differs from them.

In that scope, [3] proposed to use a Petri Net to build a dynamic model, which learns by example through time. Starting from a manual net, the model adds the different states the system passes through over time as new nodes. Although relevant, this approach relies on a manual initialization, which may be case-dependent, and the learning stage is ambiguous, since it does not really propose a way to distinguish between regular behaviors and abnormal anomalies.

In the same vein, [16] proposed HyBUTLA, which trades the Petri Net of [3] for a more classic finite-state automaton, but extended to the domain of hybrid systems (i.e. containing both analog and digital processes).

The main drawback of the approaches based on models is that it is generally difficult to identify the regular modes (behaviors) of a system, and to isolate the anomalies then. As a consequence, [2] proposed to apply unsupervised clustering to classify the data of a sensor through time, and then apply more classic statistical distribution functions to propose a model to the sensor, which can then be used to follow its temporal evolution. This tool was later integrated to a dedicated interface, named SCADA method [10].

Some works preferred using Deep Learning for anomaly detection [15,17], but, in that domain, it requires many labeled and classified examples for the training, which is prohibitive in an unknown context such as the proposed one.

Finally, a last mention on the work of Latham et al. is worth being made [8]: they proposed Machine Learning-based tools to pre-process data so as to make them ready for more complex, higher level modeling and anomaly diagnosis-oriented approaches, such as applying a CNN to classify time-series images for anomaly detection in a steel industry context [7].

The main drawback with all these works is that they generally rely on much knowledge, for instance an initial automaton of the system, or many manually-labeled data, making these approaches poorly appropriate for unknown systems, in blind contexts where the purpose is to automatically detect and classify the anomalies a system may encounter. For that reason, it would be relevant to use tools from the domain of Data Mining to address the problem of anomaly detection, especially in the context of industrial process' drifts, for they generally require very little, if not no, information on the system to perform.

3   Proposed Methodology

The proposed methodology consists in using unsupervised clustering to isolate the anomalies of an unknown, real industrial system. It operates in two steps: a first clustering of the historical data, and a classification of these groups by comparing them to the regular system's behaviors, identified and delimited by using another round of higher-level clustering.

3.1   Clustering Algorithms

Clustering consists in gathering data into compact and homogeneous groups within which they share similarities, but not between the different groups. The purpose of clustering is essentially to find a partition of the feature space minimizing the dissimilarity between the data of a same cluster, for every cluster, whilst also maximizing it between the data of different clusters.

Clustering algorithms can be regrouped into two categories: the supervised and the unsupervised ones. The former require the labels of the classes, i.e. the data must be labeled upstream by any means, and the purpose is to find the best borders delimiting the classes. The latter do no require such knowledge, they perform very generally, assuming nothing on the data or classes; they are more generic, but are also more random, for they draw their own conclusions, without possibility to confirm them. Since the proposed methodology operates in a blind, Data Mining-oriented context, only the second category can be considered.

One of the simplest unsupervised clustering algorithm is the K-Means (KM) [9]: it draws K points, assimilates them as the clusters' barycenters, labels any data with the same class as that of the closest barycenter, updates the barycenters as the means of the so-built clusters, and redoes that procedure until satisfying any criterion, typically when few enough data change of category from a round to the next. The K-Means is one of the most widespread algorithms due to its simplicity: although naive, it is intuitive. Generally sufficient to get an idea on the regions of the feature space, it generally provides weak partitionings, highly sensitive to initialization, and is only able to propose linear borders.

Based on a similar idea, the Self-Organizing Maps (SOMs) replace the notion of mean of the K-Means by the concept of attraction [6]. Linked within a grid (referred to as "map"here), the barycenters are attracted by the samples: when a training data is drawn, the closest the barycenter, the more attracted it is. This attraction takes the shape of an update of position, which is propagated within the whole grid, but which diminishes as the further away the barycenters get. Doing so allows to greatly speed up the learning stage, since every data updates the whole grid, not only one cluster; moreover, by linking the barycenters to one another, their final topology respects that of the database. Also, a great advantage of the SOMs is that they are able to deal with even nonlinearly separable datasets, by using nonlinear learning functions (typically Gaussian-like).

Notice that, thus far, the clusters' representative feature vectors have been referred to as "barycenters" to create a parallel with Physics, but it is more rigorous to call them "patterns"; a pattern is alike a true barycenter, but adapted to the data by learning. With the K-Means, the pattern of a cluster is its true barycenter, i.e. the mean of its data, which contribute all equally; for a SOM, this pattern is generally not equivalent to the barycenter, since it is built by attraction during the training stage, thus the latest data used for training typically weight a little more. Moreover, is called "region of influence" of a cluster the surrounding area within which the nearest pattern is that of the considered cluster.

Since the purpose of this paper is to identify and classify the anomalies (drifts) of dynamic, industrial processes, assuming no previous information, the K-Means and the SOMs are two good candidates for that task. As a consequence, both will be considered for a simple, first version of anomaly detection round.

3.2   Clustering for Behavioral Identification and Anomaly Detection

The second step of the proposed methodology is to classify the clusters issued by the K-Means or the SOMs; indeed, these algorithms partition the whole database, and there is no distinction between the clusters found: they can either be regular states or true anomalies. Therefore, a way to distinguish between both is needed; this task is often done manually, but it may be automatized by comparing the clusters to the regular behaviors (modes) of the system.

That domain of behavioral identification was investigated in some previous works [11,13], which resulted in proposing a clustering method based on the SOMs, the Bi-Level Self-Organizing Maps (BSOMs), specifically crafted to automatically identify the behaviors of an unknown system, only using its historical data [12]. It relies on a two-step clustering: first, the dataset is clustered N times, providing several, possibly different partitionings, which are then all projected onto a new SOM, averaging the different maps obtained in a data-driven fashion.

In [12], it was showed that this method achieves more accurate results than both the K-Means and the SOMs in the identification of the behaviors of an unknown system. As a consequence, in this paper, this approach will be used to identify the regular behaviors of industrial systems, to which can be compared the clusters obtained in the first stage to automatically classify them.

The motivation of using an averaging clustering such as BSOM for behavioral identification is to estimate the typical state vectors corresponding to the different modes of a system; indeed, despite the presence of anomalies, one can assume that they are greatly rarer than the regular values, thus, using clustering allows to point out the main regions of the feature space (corresponding to the real system's standard ways to behave), and using an averaging clustering diminishes the impact of the outliers (the anomalies); as a consequence, the BSOM's patterns should be representative of these modes' typical state vectors.

3.3 Hybridization of Clustering for Anomaly Classification

Since the BSOMs are good in identifying the behaviors of an unknown industrial system, by using its historical data only, they can be used to craft a reference, similar to a ground truth, of the system, using a somewhat statistical approach.

Meanwhile, since the manifestation of an anomaly is assumed to be a local, compact group of data, standing out from the others, it should be quite easily isolated by more classic clustering, such as the K-Means or SOMs. Unfortunately, these algorithms cannot distinguish between anomalies or true behaviors. Therefore, in order to classify the clusters issued in the first stage, they can be compared to the estimated behaviors issued in the second stage. To do so, one may compare the erratic clusters (K-Means/SOMs) to the averaged ones (BSOM): if a cluster is in the middle of a behavior, it is actually a part of it; else, it is an anomaly, as depicted on the flowchart of Fig. 1.

Finally, to compare the clusters to the estimated behaviors, one can compare their representative feature vectors, i.e. their patterns. As a consequence, for a given cluster, the first step consists in searching for the closest behavior, by computing the Euclidean distances between its pattern and that of all the behaviors, and then selecting that with the lowest distance. Once identified, the pattern of this behavior can be compared to that of the considered cluster to be classified; this comparison can be a vectorial ratio, which indicates how close the cluster's pattern is from the corresponding behavior's pattern. Eventually, since an anomaly generally occurs for only a few sensors, the maximal value among all the dimensions can be considered: if high enough, it means that the cluster greatly differs from its closest behavior, thus it can be classified as an anomaly.

Notice that ratios were preferred to absolute differences to ease readability. Indeed, since data are normalized between 0 and 1, patterns' features should take any value close to them (but may be smaller or greater, due to the learning), thus an absolute difference would result in scores typically comprised between -2 and $+2$; on the contrary, using a ratio provides scores comprised between $-\infty$ and $+\infty$, which may be easier to read and distinguish.

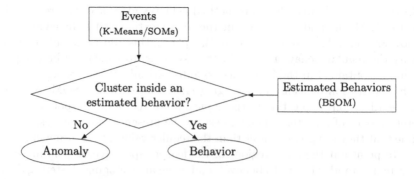

Fig. 1. The two-level methodology for anomaly classification.

In order to describe quantitatively the comparison stage, one should consider the following notations: d will refer to a distance, such as the Euclidean distance, $\mathcal{C} = \{C_i\}_{i\in[1,N_C]}$ is the set of the N_C clusters C_i issued by the K-Means or SOM, $\mathcal{B} = \{B_j\}_{j\in[1,N_B]}$ is the set of the N_B clusters B_j issued by the BSOM, and $\mathcal{P}_{\mathcal{C}} = \{p_{C_i}\}_{i\in[1,N_C]}$ and $\mathcal{P}_{\mathcal{B}} = \{p_{B_j}\}_{j\in[1,N_B]}$ are the sets of the patterns of the clusters from sets $\mathcal{C}$ and $\mathcal{B}$, respectively.

The first step is to build the matrix $\mathcal{M} \in \mathbb{R}^{N_C \times N_B}$ of all the distances between any cluster of the K-Means or SOM (set $\mathcal{C}$) and any behavior of the BSOM (set $\mathcal{B}$), as expressed by (1).

$$\mathcal{M} = \big\{\mathrm{d}(p_{C_i}, p_{B_j})\big\}_{i\in[1,N_C],\ j\in[1,N_B]} \tag{1}$$

Then, once this matrix is built, the second step consists in finding the closest behavior B_j for any cluster C_i; this value corresponds to the minimal distance of every row of matrix $\mathcal{M}$. This set of distances of given by (2).

$$\mathcal{P} = \{p_i\}_{i\in[1,N_C]} = \left\{ \operatorname*{argmin}_{j\in[1,N_B]} \big(\mathrm{d}\left(p_{C_i}, p_{B_j}\right) \big) \right\}_{i\in[1,N_C]} \tag{2}$$

Third, the set of ratios $\mathcal{R}$ is given by (3).

$$\mathcal{R} = \left\{ \frac{p_{C_i}}{p_i} \right\}_{i\in[1,N_C]} \tag{3}$$

If the data of a cluster are in the middle of that of an identified behavior, thus the cluster's and behavior's patterns should be very close, therefore their ratio should be close to 1. As a consequence, to classify the clusters as either a regular behavior or a true anomaly, one may select the extremal ratios of $\mathcal{R}$. Notice that

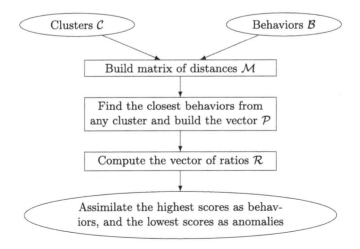

Fig. 2. The flowchart of the anomaly and behavior classification stage.

the patterns are vectors, therefore so are the ratios; to obtain a simple score, one may consider searching for the minimum, mean or maximum value among the vector. Indeed, an anomaly may occur in a single dimension, therefore only one component of the ratio may be high (or low); as a consequence, it is relevant to only consider the extremal values, i.e. the minimum and maximum. The overall methodology is summarized as the flowchart on Fig. 2.

3.4   Didactic Example

Before applying the proposed methodology to real industrial data, it is relevant to assess it first on a didactic example. To that end, a 1-dimensional set comprising 2500 samples is generated over time, which represents a system which periodically takes two distinct states (e.g. an on/off procedure), which are locally disturbed. The data are generated following two distinct uniform distributions, to which a white noise is added; then, some offset values are locally added to simulate local events. The database is depicted on the leftmost graph of Fig. 3.

Then, the database is partitioned using the clustering algorithms introduced in Sect. 3.1; the purpose of this stage is to isolate the events (the local offsets) from the true regular states. The clusters issued are represented on the two graphs in the middle of Fig. 3; notice that the patterns are scalars in the feature space, resulting in constant values in the time space. With respect to the original dataset and its six distinct groups, the number of objective clusters was set to nine for both methods; the K-Means issued nine clusters, whilst the SOM contained three empty clusters, which have been removed. In both cases, the main regions of the feature space were correctly identified, especially the green clusters (the orange group in the original dataset), but the K-Means split the main behaviors (the blue groups in the original dataset) into several pieces, due to the proximity of their respective patterns. Nonetheless, with this very didactic example, the SOM proved able to correctly separate the groups, and the K-Means to be less accurate, failing in merging the closest data as one unique cluster.

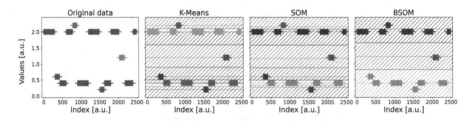

Fig. 3. The didactic 1-dimensional dataset and the corresponding clustered versions. The original database is represented on the leftmost graph, where the blue crosses are the regular states' data, and the anomalies are all other colored groups. The three other graphs are the clusters issued by every clustering method introduced in Sect. 3, one color per cluster. For every cluster, its region of influence is represented as a hatched area of the same color, with its pattern represented as a horizontal dotted line. (Color figure online)

The second stage of the proposed methodology consists in automatically searching for the regions of the regular states (the behaviors); indeed, a behavior should be assimilated to a region of influence, i.e. a trend, a typical feature vector representative of that regular state. For instance, with the original dataset, the upper blue group contains an anomaly as of the green cluster; yet, this anomaly is fairly close to the main group, thus the objective would be to merge both so as to obtain a true region whose typical representative feature vector would be contained within the blue data, but which would also incorporate the green ones. Therefore, a BSOM is applied to the database in order to estimate the areas of the regular behaviors: the issued partition is represented on the rightmost graph of Fig. 3. The map consists in ten 3×3 SOMs, but resulting in only three nonempty clusters (the others have been removed). On the graph, the two main regions of the feature space have been identified as of the blue and orange clusters, and an anomaly has been isolated as of the green cluster. As a result, following the methodology introduced in Sect. 3.3, each of these groups will be assimilated as an estimate of the typical regular states of the system, which is true for the blue and orange ones, but not for the green one, which is a true anomaly there.

Finally, the clusters must be classified as either regular states or as anomalies; to that end, the different steps of the flowchart of Fig. 2 are computed, and the ratios $\mathcal{R}$ are provided within Table 1. The color squares in columns "Clusters"and "Closest Mode" correspond respectively to the colors used to represent the clusters issued by the K-Means/SOM and the BSOM on Fig. 3.

In the table, assuming the estimated behaviors correspond to the true regular states, the patterns' ratios can be assimilated to scores: the closer to 1, the better. Therefore, the clusters 5 and 6 of both the K-Means and SOM are classified as anomalies, which is actually true, since the purple and brown clusters correspond respectively to the red and purple groups of data in the original database, which are both anomalies. However, the red clusters, corresponding to the green group

Table 1. Ratios between the clusters' patterns and the estimated behaviors' patterns.

Clusters		K-Means			SOM	
		Closest Mode	Patterns' ratio		Closest Mode	Patterns' ratio
Cluster 1		1	0.956		1	0.961
Cluster 2		2	0.978		2	0.988
Cluster 3		3	0.976		3	0.955
Cluster 4		1	1.057		1	1.035
Cluster 5		2	1.483		2	1.473
Cluster 6		2	0.494		2	0.653
Cluster 7		1	0.958		–	–
Cluster 8		2	1.002		–	–
Cluster 9		1	0.962		–	–

in the original database, have been classified as a behavior, which is wrong here. Moreover, the green clusters, corresponding to the original orange group has been correctly identified and isolated, but, since it has been assimilated as a behavior by the BSOM, so it is for the clusters. Finally, regarding the additional clusters issued by the K-Means (clusters 7 to 9), they have been aptly incorporated to the closest behaviors (orange behavior for cluster 8 and blue behavior for clusters 7 and 9), which would allow to confidently merge them into the correct behavior.

With this didactic example, the proposed methodology proved able to automatically identify the real regular states of the dataset, but assimilated an anomaly as a behavior due to its isolation, and the methodology allowed to correctly classify most of the clusters issued by the K-Means and the SOM as either a true anomaly or a true behavior, even though some errors occurred.

4   Results

4.1   Presentation of the Dataset

This paper takes place in the context of the European project HyperCOG, whose main objective is the study of the feasibility of the hyper-connected cognitive plant with an Industry 4.0 orientation: a Cyber-Physical System is designed, within which intelligent, Machine Learning-based and cognition-oriented tools are integrated for automatic intelligent control and system management.

One of the industrial partners of that project is a cement plant situated in Turkey, belonging to the group ÇimSA. They produce white Portland cement and want to integrate high-level tools based on Artificial Intelligence to ameliorate their management of their processes and smooth their production by identifying, predicting and eventually correcting the drifts of their systems.

They provided a scenarized dataset containing some events (assimilated to anomalies), as of a six-day long recording of eighteen sensors, at a sampling rate of one data every five minutes, for a total of 1440 samples. In this study, for confidentiality concerns, the sensors' tags and units are removed; this does not impact the results, since the proposed methodology does not require that information to perform. Besides, the data are normalized between 0 and 1.

An event is any period when any expert took some action to handle a sensor being out of range; this led to five annotated anomalies, summarized in Table 2. The eighteen sensors are represented against time on Fig. 4. The regular data are represented as blue crosses, whilst the events are highlighted as colored crosses. For every column, the plots share the same abscissa axis (representing the time in minutes), and for every row, the plots share the same ordinate axis (representing the normalized data, with arbitrary unit and no dimension).

Although the database comprises eighteen sensors, in all the following, only the sensors 5, 12 and 14 will be depicted, for the events are easier to distinguish with them. Notice that the input signals of the clustering algorithms introduced in Sect. 3 are every sample of every sensor: they perform in the 18-dimensional feature space spanned by the basis consisting of all these eighteen sensors.

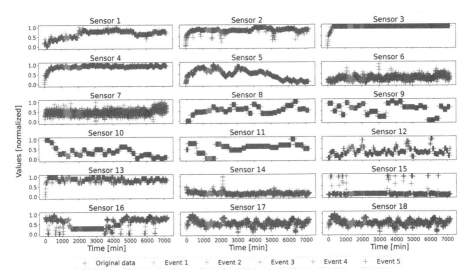

Fig. 4. The eighteen sensors and the corresponding events (all colors but blue). (Color figure online)

Table 2. List of the events to be identified.

Events	Starting Time	Ending Time		Color Scheme
Event 1	2022-09-10 19:40:00	2022-09-11 01:00:00		Orange
Event 2	2022-09-11 07:45:00	2022-09-11 11:45:00		Green
Event 3	2022-09-12 18:00:00	2022-09-12 19:50:00		Red
Event 4	2022-09-13 04:10:00	2022-09-13 07:25:00		Purple
Event 5	2022-09-14 03:50:00	2022-09-14 07:35:00		Brown

4.2 Anomaly Detection with Unsupervised Clustering

The first stage of the proposed methodology is to apply unsupervised clustering to a scenarized dataset, in order to isolate the salient groups (the anomalies) from the main trend of the system (the behaviors). To that end, the database is clustered using both the K-Means and the SOMs; since there are a handful of true regular states, plus five events to be identified, the number of objective classes for both methods should be around a dozen. Notice that a higher number could be considered, but this would result in a finer partitioning, with very small groups, which may be difficult to classify without an additional step of post-processing, step that the proposed methodology aims to avoid to simplify the procedure; moreover, anomalies generally form local, compact groups of data, therefore there would be no true benefit in splitting them into smaller pieces.

As a consequence, the K-Means will be trained with $K = 15$ and the SOM will consist of a 4×4 nodes grid. The training is made using any of the eighteen sensors of the database; the three selected in Sect. 4.1 are depicted on Fig. 5, where the

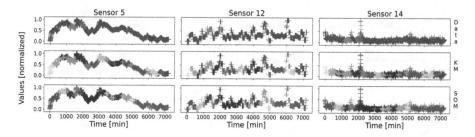

Fig. 5. The three selected sensors clustered using the K-Means and the SOMs.

original data are on the uppermost row of plots, with the events represented as colored crosses (with respect to Table 2), and the clustered database using the K-Means and a SOM are represented on the middle and lowermost rows, respectively, where every cluster is represented by a unique color. Notice that the colors are not consistent across the rows, since the clusters of both methods can not match each other.

The clusters issued by both methods generally achieved to isolate every event but Event 3 (red), even though some inertia can be noticed. For instance, Event 1 (orange in the uppermost row of Fig. 5) can be assimilated to the cluster in cyan of the K-Means (middle row), and to that in brown of the SOM (lowermost row); in both cases the clusters are a little larger than the real event's data, but the region is nevertheless correctly identified. For both methods, the correspondences between their respective clusters and the real events are summarized within Table 3; globally, all the events were correctly identified, except Event 3, which is too small and which does not comprise truly salient data, which was missed.

Table 3. Correspondence between the clusters and the events.

Events	Database	K-Means	SOMs
Event 1	Orange	Cyan	Brown
Event 2	Green	Brown	Dark Green
Event 3	Red	Missed	Missed
Event 4	Purple	Red	Khaki
Event 5	Brown	Orange	Orange

4.3   Behavioral Identification with BSOM

In order to classify the clusters obtained in Sect. 4.2, they may be compared to a ground truth; that piece of information is generally hard to obtain though, thus, with respect to [12,13], a BSOM is used to cluster the database, considering all its data, so as to identify and delineate the processes' behaviors. The identified system's behaviors (modes) are depicted on Fig. 6, obtained using a set of ten

SOMs of size 3×3. The BSOM comprises six nonempty clusters, which can be assimilated to the historical behaviors of the system; actually, the blue, red, green and orange ones are true modes, but the brown one is more a catch-all cluster (somehow outliers), and the purple one is a real anomaly, which was, ironically, isolated by the BSOM also, due to the great saliency of its data.

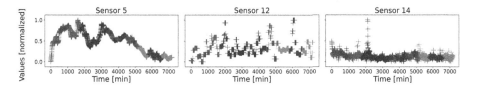

Fig. 6. The system's behaviors on the three selected sensors. (Color figure online)

4.4   Comparison of the Clusters with the Behaviors

Once the database is split into pieces by unsupervised clustering (identifying most of the anomalies) and that the behaviors' regions are estimated, both sets of results can be used to classify the clusters obtained in the first stage. The methodology described in Sect. 3.3 is therefore applied to all the clusters of both clustering methods, whose results are gathered within Table 4.

From this table, it is noticeable that the clusters corresponding to the anomalies (clusters 2, 4, 6 and 10 for the K-Means, and 2, 3, 6 and 9 for the SOM, with respect to Table 3) got globally higher ratios than the others. For instance, for the K-Means, they got the highest ratios, meaning they are the most dissimilar clusters from their respective closest behavior: they are true anomalies. This statement is a little less true for the SOM: three of these four clusters got high dissimilarity scores, but the last one got a quite low; this is due to the fact that the dark green cluster issued by the SOM spans across two regions of the feature space (corresponding actually to the Events 2 and 3), thus it becomes similar to a true mode of the system, whence the proximity to the green behavior. For the SOM again, some clusters got high scores, even though they are not expected to be true anomalies, such as cluster 8: this is due to Sensor 15, whose values are very high for this cluster, which may be a regular way to behave and thus which was not tagged as an anomaly by the CimSA's experts (for it is actually not).

Table 4. Ratios between the clusters' patterns and the behaviors' patterns. The ratios of the clusters corresponding to the events (cf. Table 3) are highlighted in bold.

Clusters	K-Means		SOM	
	Closest Mode	Patterns' ratio	Closest Mode	Patterns' ratio
Cluster 1	1	1.673	1	1.204
Cluster 2	5	**2.394**	5	**2.321**
Cluster 3	2	1.376	1	**1.346**
Cluster 4	1	**2.387**	3	1.363
Cluster 5	4	1.958	1	1.513
Cluster 6	1	**1.733**	4	**2.152**
Cluster 7	1	1.302	6	1.318
Cluster 8	1	1.180	5	2.779
Cluster 9	6	1.273	1	**2.357**
Cluster 10	4	**2.185**	2	1.954
Cluster 11	5	1.191	4	3.317
Cluster 12	3	1.328	1	1.843
Cluster 13	1	1.264	4	1.334
Cluster 14	2	1.373	2	1.650
Cluster 15	3	1.498	1	1.903

5   Conclusion

This paper introduced an automatic, Machine-Learning-based and data-driven methodology to address the problem of anomaly detection and identification of an unknown system, in an Industry 4.0 context. To that purpose, it proceeds in two steps: first, is partitions the feature space of the system with a standard unsupervised clustering method (e.g. K-Means or Self-Organizing Maps) so as to split it into pieces and isolate the anomalies; then, it partitions it anew with a higher level, averaging clustering algorithm (e.g. BSOM) so as to point out the behaviors of the system, which can serve as ground truth; finally, it compares the clusters issued by the first stage to the behaviors issued by the second stage, in order to classify the clusters as either anomalies or behaviors.

Applied to real industrial data, provided by a real cement plant (ÇimSA) in the form of a scenarized dataset comprising a handful of events (drifts) which occurred during the recording, the results showed that unsupervised clustering was globally able to isolate the event's data as distinct groups, that the averaging clustering proved able to propose a good estimate of the true, historical modes of the system, and that comparing the clusters to these behaviors allowed to classify most of them as true anomalies.

Although promising, the main weakness of this study is the relative smallness of the dataset: it comprises 1440 samples only, which is enough to get a first insight, but which may not be as representative as one would expect. Consequently, as future work, the authors are currently collecting more data and refining the notion of "events" and "anomalies" with their industrial partners to build larger, more representative and more meaningful datasets. Moreover, the authors plan to refine the approach by applying intelligent modeling, which is expected to help follow the evolution of the system and prevent anomalies.

Acknowledgements. The authors want to thank the ÇimSA team for their help in data collection and interpretation of results. The authors also want to express their gratitude to the EU Horizon 2020 Program for supporting this work.

References

1. Abdallah, M., et al.: Anomaly detection and inter-sensor transfer learning on smart manufacturing datasets. Sensors **23**(1) (2023). https://doi.org/10.3390/s23010486
2. Calvo-Bascones, P., Sanz-Bobi, M.A., Welte, T.M.: Anomaly detection method based on the deep knowledge behind behavior patterns in industrial components. application to a hydropower plant. Comput. Ind. **125**, 103376 (2021). https://doi.org/10.1016/j.compind.2020.103376
3. Dotoli, M., Pia Fanti, M., Mangini, A.M., Ukovich, W.: Identification of the unobservable behaviour of industrial automation systems by petri nets. Control Eng. Pract. **19**(9), 958–966 (2011). https://doi.org/10.1016/j.conengprac.2010.09.004, special Section: DCDS'09 – The 2nd IFAC Workshop on Dependable Control of Discrete Systems
4. Gupta, V., Mitra, R., Koenig, F., Kumar, M., Tiwari, M.K.: Predictive maintenance of baggage handling conveyors using IoT. Comput. Ind. Eng. **177**, 109033 (2023). https://doi.org/10.1016/j.cie.2023.109033
5. Huertos, F.J., Masenlle, M., Chicote, B., Ayuso, M.: Hyperconnected architecture for high cognitive production plants. Procedia CIRP **104**, 1692–1697 (2021). https://doi.org/10.1016/j.procir.2021.11.285, 54th CIRP CMS 2021 - Towards Digitalized Manufacturing 4.0
6. Kohonen, T.: Self-organized formation of topologically correct feature maps. Biol. Cybern. **43**(1), 59–69 (1982). https://doi.org/10.1007/BF00337288
7. Latham, S., Giannetti, C.: Pre-trained CNN for classification of time series images of anti-necking control in a hot strip mill. In: The 9th IIAE International Conference on Industrial Engineering 2021 (ICIAE2021), pp. 77–84 (2021)
8. Latham, S., Giannetti, C.: Root cause classification of temperature-related failure modes in a hot strip mill. In: Proceedings of the 3rd International Conference on Innovative Intelligent Industrial Production and Logistics - IN4PL, pp. 36–45. INSTICC, SciTePress (2022). https://doi.org/10.5220/0011380300003329
9. Lloyd, S.P.: Least squares quantization in PCM. IEEE Trans. Inf. Theory **28**, 129–136 (1982)
10. Maseda, F.J., López, I., Martija, I., Alkorta, P., Garrido, A.J., Garrido, I.: Sensors data analysis in supervisory control and data acquisition (Scada) systems to foresee failures with an undetermined origin. Sensors **21**(8) (2021). https://doi.org/10.3390/s21082762

11. Molinié, D., Madani, K., Amarger, C.: Identifying the behaviors of an industrial plant: application to industry 4.0. In: Proceedings of the 11th International Conference on Intelligent Data Acquisition and Advanced Computing Systems: Technology and Applications (IDAACS), vol. 2, pp. 802–807, September 2021. https://doi.org/10.1109/IDAACS53288.2021.9661018

12. Molinié, D., Madani, K.: BSOM: a two-level clustering method based on the efficient self-organizing maps. In: 2022 International Conference on Control, Automation and Diagnosis (ICCAD), pp. 1–6 (2022). https://doi.org/10.1109/ICCAD55197.2022.9853931

13. Molinié, D., Madani, K., Amarger, V.: Clustering at the disposal of industry 4.0: Automatic extraction of plant behaviors. Sensors **22**(8) (2022). https://doi.org/10.3390/s22082939

14. Pozzi, R., Rossi, T., Secchi, R.: Industry 4.0 technologies: critical success factors for implementation and improvements in manufacturing companies. Product. Plann. Control **34**(2), 139–158 (2023). https://doi.org/10.1080/09537287.2021.1891481

15. Ruiz-Moreno, S., Gallego, A.J., Sanchez, A.J., Camacho, E.F.: Deep learning-based fault detection and isolation in solar plants for highly dynamic days. In: 2022 International Conference on Control, Automation and Diagnosis (ICCAD) (2022). https://doi.org/10.1109/ICCAD55197.2022.9853987

16. Vodenčarević, A., Bürring, H.K., Niggemann, O., Maier, A.: Identifying behavior models for process plants. In: ETFA2011 (2011). https://doi.org/10.1109/ETFA.2011.6059080

17. Wang, H., Liu, X., Ma, L., Zhang, Y.: Anomaly detection for hydropower turbine unit based on variational modal decomposition and deep autoencoder. Energy Rep. **7**, 938–946 (2021). https://doi.org/10.1016/j.egyr.2021.09.179, 2021 International Conference on Energy Engineering and Power Systems

18. Yeshchenko, A., Di Ciccio, C., Mendling, J., Polyvyanyy, A.: Comprehensive process drift detection with visual analytics. In: Laender, A.H.F., Pernici, B., Lim, E.P., de Oliveira, J.P.M. (eds.) ER 2019. LNCS, vol. 11788, pp. 119–135. Springer, Cham (2019). https://doi.org/10.1007/978-3-030-33223-5_11, https://doi.org/10.48550/arXiv.1907.06386

Adversarial Attacks on Leakage Detectors in Water Distribution Networks

Paul Stahlhofen[(✉)], André Artelt[ⓘ], Luca Hermes[ⓘ], and Barbara Hammer[ⓘ]

Bielefeld University, Inspiration 1, 33615 Bielefeld, Germany
{pstahlhofen,aartelt,lhermes,bhammer}@techfak.uni-bielefeld.de

Abstract. Many Machine Learning models are vulnerable to adversarial attacks: One can specifically design inputs that cause the model to make a mistake. Our study focuses on adversarials in the security-critical domain of leakage detection in water distribution networks (WDNs). As model input in this application consists of sensor readings, standard adversarial methods face a challenge. They have to create new inputs that still comply with the underlying physics of the network. We propose a novel approach to construct adversarial attacks against Machine Learning based leakage detectors in WDNs. In contrast to existing studies, we use a hydraulic model to simulate leaks in the water network. The adversarial attacks are then constructed based on these simulations, which makes them intrinsically physics-constrained. The adversary maximizes water loss by finding the least sensitive point, that is, the point at which the largest possible undetected leak could occur. We provide a mathematical formulation of the least sensitive point problem together with a taxonomy of adversarials in WDNs, in order to relate our work to other possible approaches in the field. The problem is then solved using three different algorithmic approaches on two benchmark WDNs. Finally, we discuss the results and reflect on potentials to enhance model robustness based on knowledge about adversarial weaknesses.

1 Introduction

According to the EU guidelines on Trustworthy AI, robustness is a key requirement of Machine Learning (ML) based systems deployed in security-critical domains [5]. Judging and improving a models' robustness requires to understand its weakness so that appropriate countermeasure can be taken (see e.g. [17,22]). A structural weakness of many models is the existence of specifically designed inputs, so called adversarial inputs, that can cause the model to make wrong predictions with a high confidence. These adversarial inputs, first described in [18], have been used to expose a lack of robustness for many state of the art ML models [2,10,23].

In this paper we focus on an application domain where robustness is of utmost importance: Water Distribution Networks (WDNs). A high amount of annual water loss in water networks across the world [13] and an increased likelihood of droughts due to climate change [3] call for an improvement and robustification

I. Rojas et al. (Eds.): IWANN 2023, LNCS 14135, pp. 451–463, 2023.
https://doi.org/10.1007/978-3-031-43078-7_37

of water management. In the context of WDNs, ML based systems are used for various tasks [14] e.g. detection and localization of events such as leakages [19,20] and contaminations [24]. These can have severe consequences for human health [11]. Hence, studies to scrutinize the ML detection models are required.

Related Work. In order to gain insight into vulnerabilities of a water distribution network, the authors of [9] propose a methodology for building finite state processes to find a-priori week spots independent of any monitoring system. The authors of [12] construct physics-constrained adversarial attacks for ML models operating on cyber-physical systems. However, in their case study they only analyze the manipulation of sensor readings for flow sensors and do not consider pressure sensors which constitute the most common type of sensor in real world WDNs.

Our Contributions. In this work we make the following contributions:

- We propose a taxonomy for adversarial attacks against leakage detectors in WDNs.
- We investigate one particular type of attack in more detail: The search for the *least sensitive point* in a WDN, which is the location in the water network where the largest possible undetected leak could occur. We formalize this attack and propose three algorithmic approaches, which we empirically evaluate in a case study on two benchmark WDNs.

The remainder of this work is structured as follows: In Sect. 2, we introduce adversarial attacks in general, as well as some important concepts of leakage detection in WDNs. Following up on this, we give a taxonomy of adversarial attacks on leakage detectors and formalize the *least sensitive point problem* in Sect. 3. Next, in Sect. 4 we empirically evaluate our proposed methods for finding the least sensitive point in a WDN. Finally, Sect. 5 discusses the results before we conclude with a brief summary in Sect. 6.

2   Foundations

2.1   Adversarial Attacks

The concept of adversarial attacks was first described in the context of image classification in [18]: There exist small and imperceptible perturbations of the original image that lead the classifier to output a wrong prediction with high confidence. Based on the definition given in [8], this can be generalized to further application domains as follows:

Definition 1 (Adversarial Attack). *An adversarial attack is a procedure to create inputs to a Machine Learning model, that will cause the model to make a mistake.*

It is important to note that the procedure for constructing an adversarial does not necessarily operate in the input space directly. In particular, consider a ML model that uses measurements as inputs: These measurements reflect the state of a system monitored by the model. Procedures that change the system in order to create measurements that will cause the model to make a mistake are also considered as adversarial attacks here. These attacks have the advantage that they are intrinsically physics-constrained: When changing measurements directly, one has to make sure that the resulting measurements obey physical laws, as discussed in [12]. Making changes to the system itself will automatically result in measurements that reflect physical processes.

2.2   Leakage Detection in Water Distribution Networks

Before considering leakage detection, we have to define a water distribution network

Definition 2. *Water Distribution Network A water distribution network (WDN) is a graph $W = V, E$. The nodes V can be either pipe junctions (most common), tanks or reservoirs while the edges $E \subseteq V \times V$ are either pipes (most common), pumps or valves.*

Next, we describe our notion of a leak.

Definition 3. *Water Leak A water leak (often simply referred to as leak) is a node feature of a water distribution network. Given a WDN $W = (V, E)$ with a set of N nodes, water leaks at time t are given in a vector $a_t \in \mathbb{R}^N$, where each vector element $a_{n,t}$ gives the leak area at node n.*

This definition has two important implications: Firstly, a vector $a_t = 0$ corresponds to the normal state without any leak. Secondly, as a leak is simply defined by its area, the graph structure itself does not change depending on leaks.

Leakage detectors typically rely on measurements of hydraulic variables such as pressure and flow at certain points in a network. Measurements are taken at discrete timesteps and depend on various factors, including network topology, water consumption at network nodes and potential leakages. When creating an adversarial input to the detector, we are mainly interested in how the pressure values are influenced by leaks. All other factors like water consumption and network topology are assumed to be fixed.

Definition 4 (Measurement Function). *A measurement function maps a vector of leak areas $a_t \in \mathbb{R}^N$ to pressure measurements y_t at $S \leq N$ nodes equipped with sensors. It is parametrized also by the timestep itself.*

$$f_{measure} : \mathbb{R}^N \times \{1, \ldots, T\} \to \mathbb{R}^S$$
$$(a_t, t) \mapsto y_t$$

(1)

In our experiments, we obtain the pressure values using a simulator that computes the measurements based on a hydraulic model of the underlying water network. Given these measurements, we can now define the leakage detector as a binary predictor

Definition 5 (Leakage Detector). *A leakage detector f_{detect} is a function mapping pressure measurements y_t to a binary output, indicating whether a leak was detected or not.*

$$f_{detect} : \mathbb{R}^S \rightarrow \{0,1\}$$

$$y_t \mapsto \begin{cases} 1 & \textit{if a leak was detected} \\ 0 & \textit{otherwise} \end{cases} \tag{2}$$

3   Adversarials in Water Distribution Networks

3.1   Taxonomy of Adversarials in WDNs

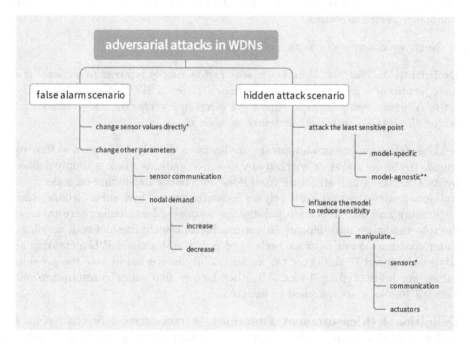

Fig. 1. Taxonomy showing different types of adversarial attacks in WDNs * Methods proposed in [12] ** Methods proposed here

There exist two different scenarios how a detector could be fooled: The first one would be an input causing the detector to predict a leak, even though no leak is actually present. We refer to this as the false alarm scenario; The second one would be an input causing the detector not to predict a leak, even though there is one. This case can be termed the hidden attack scenario.

In case of the false alarm scenario, there are again two different possibilities of attacking the system: An adversary might try to manipulate a few network elements (e.g. sensors, pipes, etc.) they have managed to access or the attacker makes external changes to the water demand at certain nodes. Note that, even if the adversary has managed to gain control over some parts of the network infrastructure, they are not interested in actually creating a leak or other damage in this scenario, but rather aims at causing the leakage detector to raise a false alarm. Thus, manipulating sensor values or disturbing the inter-sensor communication might be a promising approach. In particular, a large decrement of pressure measurements or the complete abortion of sensory recordings are likely to cause a wrong leakage prediction. In the attack scenario where the attacker can change node demands only, the success of the attack will depend on the number of nodes under adversary control and the extent to which the demand can be changed at these nodes. Generally, both a high increase in the demand e.g. by opening all water taps in an industrial complex as well as a high decrease constitute possible reasons for a prediction failure resulting in a false alarm.

The hidden attack scenario always involves an adversary who creates an actual leak in the physical network. In order to fool the leakage detector not to find the leak, there are again two possibilities: The attacker could try to manipulate the detector in order to make it less sensitive or they could place the leak at a point where the detector is not very sensitive anyway. In the manipulation case, sensors, programmable logic controllers (PLCs) or the sensory control and data acquisition system (SCADA) could be targets of the attack[1]. Otherwise, the adversary will have to search for points in the network where the detector has little sensitivity to leaks.

Because investigating all the aforementioned attack scenarios in detail is beyond the scope of this work, we focus on a particular type of hidden attack: Searching for locations in the network where the detector is the least sensitive to leaks.

3.2   The Least Sensitive Point Problem

If an adversary does not have access to sensors or other parts of the leakage detection system but still tries to create a leak that should not be detected, they have to find the location in the network where the largest undetected leak could occur. In other words, the adversary is looking for the location/point where the detector is least sensitive, which we formalize as follows:

Definition 6 (Least Sensitive Point). *Given a water distribution network* $W = (V, E)$ *with a set of* N *nodes* $V = \{v_n | n \in \{1, \ldots, N\}\}$ *and a number of*

---

[1] These constituents of a network monitoring system were used in the Battle of the Attack Detection Algorithms (BATADAL) [19]. A monitoring system might have additional entry points for an attack in practise.

$S \leq N$ *pressure sensors installed in it, a measurement function $f_{measure}$ and a leakage detector f_{detect}, the least sensitive point (LSP) $\text{lsp}(W)$ in this network W is defined as follows:*

$$\text{lsp}(W) = \arg\max_{\substack{v_n \in V \\ \alpha \in \mathbb{R}^+}} \max_{t \in \{1,...,T-K\}} \alpha$$

$$\text{s.t.} \quad f_{detect}\left(f_{measure}(\alpha e_n, t+k)\right) = 0 \quad \forall k \in \{0, \ldots, K\} \tag{3}$$

where K is a fixed time window length and e_n is the n-th canonical basis vector. The scalar $\alpha \in \mathbb{R}^+$ determines the leak area at node v_n.

In our analysis, we set the value of K equal to the leak duration which we fix at 3 h for all experiments.

We approach the least sensitive point problem (Definition 6) with three algorithmic methods for two different network topologies.

3.3 Algorithmic Approaches

We propose the following algorithmic approaches to find the least sensitive point (Definition 6):

- Bisection Search
- Basic Genetic Algorithm
- Genetic Algorithm with Spectral Embeddings

We empirically evaluate all three methods in a case study in Sect. 4.

Bisection Search. The Bisection Search is based on the fact that for the least sensitive point there has to exist some leak area α^*, such that a leak of that area remains undetected only at the LSP, while a leak of the same area would be detected at every other node. Thus, one can perform a simple line search over the leak area in order to find α^*. Bisection search can be costly, in particular when all timesteps in the simulation are viewed as potential starting times of the leak. However, it yields the benefit that it is guaranteed to find the least sensitive point in the given search space. To reduce the computational load, we implement a pruning of nodes and starting times based on intermediate results. Let α^i be the leak area after iteration i. If a leak of size α^i remained undetected for at least one node-time pair, we remove the following elements from the search space:

1. Every node for which a leak of size α^i was detected at every starting time
2. Every starting time for which a leak of size α^i was detected at every node

In practise, this lead to a strong reduction of the search space after the first iteration.

Genetic Algorithm. In the Basic Genetic Algorithm, we encode nodes and leak starting time as genes and optimize the following fitness function

$$Fitness(v_n, t) = \begin{array}{c} \max\limits_{\alpha \, in \, \mathbb{R}} \alpha \\ \text{s.t.} f_{detect}(f_{measure}(\alpha e_n, t)) = 0 \end{array} \tag{4}$$

where α is the leak area at node n. The largest undetected leak for each node-time pair is again computed via a line search. In order to avoid unnecessary area maximization, we use dynamic programming: The largest undetected leak area found so far is tested first for every new node-time pair. If this is detected, no area maximization was performed and the fitness is set to zero.

Genetic Algorithm with Spectral Embeddings. In order to take the network topology into account, we consider a second version of the Genetic Algorithm, using a spectral analysis of the network's graph Laplacian. Every node is assigned a four dimensional vector, containing elements of the 2nd through 5th eigenvector of the graph Laplacian for that node. Each vector element of the node embedding is then treated as a different gene by the Genetic Algorithm. After every recombination step, the nearest neighbour of the resulting embedding vector is returned as offspring.

None of the approaches used here makes any assumptions on the type of leakage detector. Hence, they are all model-agnostic in terms of the taxonomy (Fig. 1).

4 Case Study: LSP Search in Two Benchmark Networks

4.1 Leakage Detector

We implement a classic residual-based leakage detector [6,16] by means of a sensor specific linear model that calculates a pressure prediction $\hat{y}_{s,t}$ for sensor s and timestep t based on the pressure measurements of all other nodes [1]:

$$\hat{y}_{s,t} := \boldsymbol{w}_s^T \boldsymbol{y}_{-s,t} + b_s \tag{5}$$

where $\boldsymbol{y}_{-s,t}$ is a vector containing measurements from all sensors except s at time t and $(\boldsymbol{w}_s, b_s)$ are sensor specific weights learned using data. The residuals $r_{s,t} := |\hat{y}_{s,t} - y_{s,t}|$ determine the detector output. For the Hanoi network, we use a validation set to learn residual weights q_s for each sensor. An alarm is then raised if $\sum_s q_s r_s > 1$. In case of the L-Town network, lack of data makes the learning of residual weights impractical. Instead, we construct thresholds for each sensor by multiplying the maximum training error at the sensor location with a small constant. An alarm is raised if at least one sensor residual exceeds the corresponding threshold.

4.2   Water Networks

In order to obtain pressure values as input for the leakage detector, we used the hydraulic modelling software EPANET [15] which simulates flow and pressure values over time, given an hydraulic model of a water distribution network and water demands for each node and timestep. EPANET also allows to run simulations with leaks of arbitrary area, location, starting time and duration.

We first evaluate our proposed methodology on the Hanoi network [7]. This network consists of 31 junctions and one water reservoir from which water is entering the system through a pump. Realistic water demands at the network nodes are generated using the code provided with the LeakDB benchmark dataset [21]. Next, we extend our case study to the larger and more realistic L-Town network [20]. This network contains 782 junctions, receiving their water from two reservoirs and one tank which is used for intermediate storage. The authors of [20] provided realistic demand values along with the hydraulic model of the network.

For both networks we train the leakage detector on the first five days of the timeseries and search for the least sensitive point (Definition 6) during the two days afterwards. While detector training on realistic demands may not always be possible in advance, it is still reasonable to assume that the first few days produced by those demands are utilized to calibrate the detector for the network at hand. In particular, inferring thresholds from training errors on realistic data can help to avoid false alarms. For the Hanoi network, we conduct a second analysis with five training days and a search space of nine days afterwards. This is not done for L-Town as simulations over a longer timeseries are computationally demanding for larger networks.

5   Results and Discussion

5.1   Hanoi Network

For the Hanoi network, optimal solutions could be determined using the Bisection Search. First results show that the least sensitive point is always located close to the water source. This can be explained by the large water flow from the reservoir to the first nodes. A leakage event along a pipe with higher flow will lead to a less significant pressure drop when compared to a leak along a pipe with lower flow. As this fact is known to water utility administrators, we assume in follow up experiments that these nodes close to the reservoir might be subject to increased protection. We treat them as inaccessible for the adversary and remove them from the search space. In the 2-days dataset, the least sensitive point among the remaining nodes is located at node 10. For the 9-days dataset, it is found to be node 23. The results are visualized in Fig. 2.

To compare the performance between the Basic Genetic Algorithm and the extension with spectral embeddings we run five trials of each algorithm for both datasets. In all 20 cases, the respective algorithm is able to locate the least sensitive point correctly. This demonstrates the high accuracy of both methods

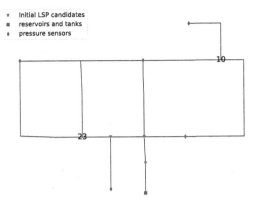

Fig. 2. Least Sensitive Point search on the Hanoi network: In initial analysis, points with the lowest detector sensitivity (marked by red triangles) are found close to the water reservoir. After excluding these nodes from the search space, the least sensitive point is found at node 10 for the 2-days dataset and at node 23 for the 9-days dataset. (Color figure online)

on small water networks. In order to gain evidence for this assumption, a larger number of trials could be conducted in future work.

5.2 L-Town Network

For the L-Town network, it is not computationaly feasible to find the least sensitive point globally with the Bisection Search, because the network is considerably larger than the Hanoi network and thus simulations are much more time consuming. In order to get a realistic estimate, we determine the LSP at three fixed leak starting times, 4, 25 and 28 h after the beginning of the timeseries. These starting times are assumed to have low overall demands based on the training data, which makes leaks more difficult to detect. During the line search we are able to observe intermediate results. For a given leak area, we find all nodes where a leak of that area starting at one of the fixed starting times is not detected. Similar to the Hanoi network, we observe an agreement between those intermediate steps for the different time trials, indicating that nodes with low sensitivity are located close to the water sources. Fig. 3 shows all nodes for which a leak with an area of 100 cm^2 is not detected in at least one of the start time trials as downward-pointing triangles. From the result we can also observe that the direction of water flow plays a major role in leakage detection: Even though there is a pressure sensor located directly next to the tank in the upper left area of the network, nodes on the other side of the tank are very vulnerable to large undetected leaks. This is because the sensor only measures the pressure of water flowing into the tank and does not support the detection of downstream leaks. In case of the L-Town network, pressure-based detection close to the water sources is also hampered by pressure reducing valves, located after each source. The valves smooth out incoming pressure values, making the detection of downstream leaks more difficult.

In order to compare the performance of our algorithmic approaches on the L-Town network, we first re-compute the least sensitive point for the same starting times used above after excluding the initial LSP candidates marked in Fig. 3 from the search space. All three runs of the Bisection Search agreed in node n387 as the new least sensitive point (upward-pointing triangle in Fig. 3). We then conduct five trials for each genetic algorithm. Results are shown in Fig. 4. The horizontal line inidicates that each run of the Bisection Search found the least sensitive point at a maximum undetected leak area of 80 cm². All genetic-algorithm results with a larger leak area identify n387 as the LSP, while all

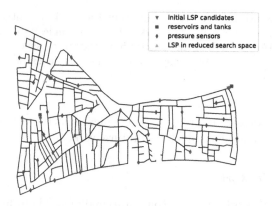

Fig. 3. Analysis of the L-Town network with Bisection Search for fixed starting times. Points with a low detector sensitivity are located close to the water sources. After excluding the initial LSP candidates from the search space, the least sensitive point is found at node n387 (upward-pointing triangle).

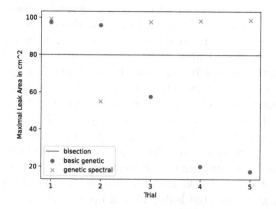

Fig. 4. Comparison of the three algorithmic approaches.

results with a smaller leak area fail to do so[2]. The comparison suggests that the genetic algorithm with spectral node embeddings performs well at finding the LSP, also for larger networks. A simple enhancement to reduce the uncertainty of a single trial would be to conduct multiple successive trials, keeping the result of the best one as global solution. This would still be computationally much cheaper than an exhaustive Bisection Search. In future work, a larger number of trials may be used to determine the junction accuracy of both algorithms.

6   Conclusion

In this work, we applied the concept of adversarial attacks to leakage detection in water distribution networks. For this purpose, we proposed a taxonomy of potential adversarials against a leakage detector. We then focused on a particular type of adversarial: Attacking the network at the least sensitive point. We formalized the least sensitive point problem and proposed three algorithmic approaches to solve it. In practice, knowing the least sensitive point and the vulnerability to attacks constitutes crucial information which enables practitioners to develop more robust methods using e.g. adversarial training [4] or an improvement of the robustness by introducing targeted sensors.

We empirically evaluated our proposed methods in a case study on two benchmark WDNs. For the genetic algorithm with spectral embeddings, experiments yielded promising results on both networks. This method allows much faster solutions when compared to the Bisection Search, due to a lower runtime complexity. The locations of the least sensitive point indicate that the detectors weak spots are highly dependent on network topology, in particular on the location of the water sources.

In future work, we plan to address the same problem with another leakage detection model to compare results between different leakage detectors. In order to increase the robustness, we will determine the effect of incremental sensor placement on the maximum undetected leak area. For this purpose, we will simulate sensors at the location where the least sensitive point has previously been detected and re-run the experiments with the new sensors in place. In this way, knowledge about adversarials can be helpful in practical applications.

Acknowledgements. We gratefully acknowledge funding from the European Research Council (ERC) under the ERC Synergy Grant Water-Futures (Grant agreement No. 951424).

References

1. Artelt, A., Vrachimis, S., Eliades, D., Polycarpou, M., Hammer, B.: One explanation to rule them all - ensemble consistent explanations. In: Workshop on XAI at IJCAI (2022). https://doi.org/10.48550/ARXIV.2205.08974, https://arxiv.org/abs/2205.08974

---

[2] In theory, it is possible that trials with a smaller leak area still find the same node, but this is not the case here.

2. Athalye, A., Engstrom, L., Ilyas, A., Kwok, K.: Synthesizing robust adversarial examples. CoRR abs/1707.07397 (2017). http://arxiv.org/abs/1707.07397
3. Ault, T.R.: On the essentials of drought in a changing climate. Science **368**(6488), 256–260 (2020). https://doi.org/10.1126/science.aaz5492, https://www.science.org/doi/abs/10.1126/science.aaz5492
4. Bai, T., Luo, J., Zhao, J., Wen, B., Wang, Q.: Recent advances in adversarial training for adversarial robustness. arXiv preprint arXiv:2102.01356 (2021)
5. Commission, E., Directorate-General for Communications Networks, C., Technology: Ethics Guidelines for Trustworthy AI. Publications Office (2019). https://doi.org/10.2759/346720
6. Eliades, D., Polycarpou, M.M.: Leakage fault detection in district metered areas of water distribution systems. J. Hydroinf. **14**(4), 992–1005 (2012)
7. Fujiwara, O., Khang, D.B.: A two-phase decomposition method for optimal design of looped water distribution networks. Water Resour. Res. **26**(4), 539–549 (1990)
8. Goodfellow, I., Papernot, N., Huang, S., Duan, Y., Abbeel, P., Clark, J.: Attacking machine learning with adversarial examples (2017). https://openai.com/blog/adversarial-example-research/
9. Karrenberg, C., Benavides, J., Berglund, E., Kang, E., Baugh, J.: Identifying cyber-physical vulnerabilities of water distribution systems using finite state processes. In: 2nd International Joint Conference on Water Distribution System Analysis and Computing and Control in the Water Industry, WDSA CCWI 2022 (2022)
10. Kurakin, A., Goodfellow, I.J., Bengio, S.: Adversarial examples in the physical world. CoRR abs/1607.02533 (2016). http://arxiv.org/abs/1607.02533
11. LeChevallier, M.W., Gullick, R.W., Karim, M.R., Friedman, M., Funk, J.E.: The potential for health risks from intrusion of contaminants into the distribution system from pressure transients. J. Water Health **1**(1), 3–14 (03 2003). https://doi.org/10.2166/wh.2003.0002
12. Li, J., Yang, Y., Sun, J.S., Tomsovic, K., Qi, H.: Conaml: constrained adversarial machine learning for cyber-physical systems (2020). https://doi.org/10.48550/ARXIV.2003.05631, https://arxiv.org/abs/2003.05631
13. Liemberger, R., Wyatt, A.: Quantifying the global non-revenue water problem. Water Supply **19**(3), 831–837 (07 2018). https://doi.org/10.2166/ws.2018.129
14. Ramotsoela, D.T., Hancke, G.P., Abu-Mahfouz, A.M.: Attack detection in water distribution systems using machine learning. HCIS **9**(1), 1–22 (2019). https://doi.org/10.1186/s13673-019-0175-8
15. Rossman, L.A., Woo, H., Tryby, M., Shang, F., Janke, R., Haxton, T.: EPANET 2.2. User Manual. U.S. Environmental Protection Agency, Washington D.C. (2020)
16. Santos-Ruiz, I., López-Estrada, F.R., Puig, V., Blesa, J.: Estimation of node pressures in water distribution networks by gaussian process regression. In: 2019 4th Conference on Control and Fault Tolerant Systems (SysTol), pp. 50–55. IEEE (2019)
17. Shafahi, A., et al.: Adversarial training for free! In: Wallach, H., Larochelle, H., Beygelzimer, A., d'Alché-Buc, F., Fox, E., Garnett, R. (eds.) Advances in Neural Information Processing Systems, vol. 32. Curran Associates, Inc. (2019). https://proceedings.neurips.cc/paper/2019/file/7503cfacd12053d309b6bed5c89de212-Paper.pdf
18. Szegedy, C., et al.: Intriguing properties of neural networks. In: International Conference on Learning Representations (2014). http://arxiv.org/abs/1312.6199
19. Taormina, R., et al.: Battle of the attack detection algorithms: disclosing cyber attacks on water distribution networks. J. Water Resourc. Plann. Manag. **144**(8), 04018048 (2018). https://doi.org/10.1061/(ASCE)WR.1943-5452.0000969

20. Vrachimis, S.G., et al.: Battle of the leakage detection and isolation methods. J. Water Resourc. Plann. Manag. **148**(12), 04022068 (2022). https://doi.org/10.1061/(ASCE)WR.1943-5452.0001601, https://ascelibrary.org/doi/10.1061/%28ASCE%29WR.1943-5452.0001601

21. Vrachimis, S.G., Kyriakou, M.S., Eliades, D.G., Polycarpu, M.M.: LeakDB: a benchmark dataset for leakage diagnosis in water distribution networks. In: WDSA/CCWI Joint Conference Proceedings, vol. 1, July 2018

22. Xie, C., Tan, M., Gong, B., Wang, J., Yuille, A.L., Le, Q.V.: Adversarial examples improve image recognition. In: Proceedings of the IEEE/CVF Conference on Computer Vision and Pattern Recognition (CVPR), June 2020

23. Xu, K., et al.: Adversarial t-shirt! Evading person detectors in a physical world (2019). https://doi.org/10.48550/ARXIV.1910.11099, https://arxiv.org/abs/1910.11099

24. Zhu, M., et al.: A review of the application of machine learning in water quality evaluation. Eco-Environ. Health (2022)

Neural Networks in Chemistry
and Material Characterization

Automatic Control of Class Weights in the Semantic Segmentation of Corrosion Compounds on Archaeological Artefacts

Ruxandra Stoean[1,2(✉)], Patricio García Báez[3], Carmen Paz Suárez Araujo[4], Nebojsa Bacanin[5], Miguel Atencia[6], and Catalin Stoean[1,2]

[1] Romanian Institute of Science and Technology, Cluj, Romania
{ruxandra.stoean,catalin.stoean}@inf.ucv.ro
[2] University of Craiova, Craiova, Romania
[3] Universidad de La Laguna, San Cristóbal de La Laguna, Spain
pgarcia@ull.es
[4] Universidad de Las Palmas de Gran Canaria, Las Palmas de Gran Canaria, Spain
carmenpaz.suarez@ulpgc.es
[5] Singidunum University, Belgrade, Serbia
nbacanin@singidunum.ac.rs
[6] Universidad de Málaga, Málaga, Spain
matencia@uma.es

Abstract. The semantic segmentation for irregularly and not uniformly disposed patterns becomes even more difficult when the occurrence of categories is imbalanced within the images. One example is represented by heavily corroded artefacts in archaeological digs. The current study therefore proposes a weighted loss function within a deep learning architecture for semantic segmentation of corrosion compounds from microscopy images of archaeological objects, where the values for the class weights are generated via genetic algorithms. The fitness evaluation of individuals is the estimation that a surrogate of the deep learner gives concerning the segmentation accuracy. The obtained class weight values are compared to a random search through the space of potential configurations and another automated means to compute them, in terms of resulting model accuracy.

Keywords: semantic segmentation · deep learning · class imbalance · weighting · evolutionary algorithms · surrogate models · archaeology

1 Introduction

Semantic segmentation is one computer vision task with numerous applications in the computational decision support for complex and sensitive fields. While the detection of human and animal beings, as well as general objects, from natural images, has reached outstanding performance, the delineation of more

I. Rojas et al. (Eds.): IWANN 2023, LNCS 14135, pp. 467–478, 2023.
https://doi.org/10.1007/978-3-031-43078-7_38

ambiguously formed shapes still poses a problem to the deep learning (DL) algorithms. Moreover, a second issue comes with the shortage of data available for analysis in practical scenarios. This leads to the situation when, while some classes that have to be distinguished are well represented in the data set, the others have very few examples from which the algorithm can learn the pattern.

The class imbalance problem is frequent in general in the classification process and a weighting procedure in the loss function is commonly employed to address it, either in a manual or in an automatic fashion. The open source machine learning libraries also offer in-built tools for an automatic calibration of class weights, based on standard formulas. However, these are general and thus not always able to capture entirely the particularities of the task at hand.

Under these circumstances, the current paper aims to introduce an automatic method to optimally adjust the class weighting by means of evolutionary algorithms (EA). However, since the repeated training of the DL model in order to evaluate every resulting potential solution would be highly resource-consuming, the behaviour of the deep architecture shall be simulated through a surrogate cheaper machine learning model. Surrogate-based optimization [13] has proven to be an efficient procedure, with various machine learning approaches offering a partial model of the original expensive phenomenon. Its use for hyperparameter tuning of DL models has also been exploited with success [3,12,18,21].

The real-world instance on which the surrogate-based class weighting will be performed comes from archaeology and envisages the intricate task of recognizing and delineating the corrosion compounds on recently excavated artefacts before proper restoration. Such objects are very old and therefore highly degraded from the long time contact with the soil and related components (temperature, water, minerals, salts, organisms), as well as from the sudden interaction with O_2 once excavated. The shape of the corrosion exhibition is thus irregular and variable, and necessitates a careful, laborious examination.

DL offers the possibility for the segmentation of corrosion compounds and a first inspection was made in [19]. There are 5 categories associated with this corrosion recognition task (4 compounds plus the clean metal). Nevertheless, the complexity of the shapes makes the semantic segmentation task difficult and a good detection accuracy is not possible for all the classes. Moreover, the issue grows significant due to the imbalance in the distribution of the classes.

The paper is structured as follows. A look at the related state of the art is provided in Sect. 2. Section 3 introduces the data set with the occurrence of each corrosion compound and the proposed method, together with the experimental setup. The results are discussed in Sect. 4. The conclusions and future directions are given in Sect. 5.

2 State of the Art

The detection of corrosion by means of DL can be in a first instance modeled through an easier, rougher classification formulation [15,16]. For the more complex semantic segmentation task, the reference DL architectures of U-Net and

Mask R-CNN [8] are employed as standard models. The problems usually come from civil engineering, where the presence of rust has to be effectively acknowledged on the resistance structures [6,8]. In the field of archaeology and cultural heritage preservation, the first attempt to segment the diverse corrosion products, specific to each material type of object manufacture, had been reached in [19] through a U-Net architecture. Nevertheless, it had been observed that the model focused on accurately detecting the clean metal part, which was larger defined and present in all images, and little emphasis was put on the corrosion products that were scarcer and more difficult to detect.

Hyperparameter optimization within DL is another issue connected to the present research. A proper tuning of the involved variables is directly related to the performance of the model. Various strategies based on heuristics have been proposed, ranging from the now already classical EA [10] to the newest entries of swarm intelligence [2]. Although the interest in the current work lies in the adjacent weighting of the classes, the problem is similar to that of DL parameter calibration: better chosen weights will lead to improved performance, and it all should be achieved within efficient computational restrictions. As pointed out in the introduction, surrogate-based optimization is ideal for such constraints.

The class imbalance problem has been a subject of interest in research works. Multi-class imbalance [4] is even more complicated, as there is a hierarchy of inequality [20]. The solutions come up at the data level and at the algorithmic stage. Sampling techniques are used for data driven changes, while the algorithmic modifications include cost targeted penalization, threshold adjustment for prediction probabilities and ensemble learning [4,20].

Literature examples where EA were used for class imbalance problems refer their usage at the algorithmic level to generate near balanced bags in bagging ensemble methods [14] and their application at the data level for generating synthetic examples for the minority class taking into account the distribution measure [1]. In the current paper, the optimal weight values are generated by an EA individual and included in the loss function of the U-Net.

3   Data and Methodology

The data, the methodological design and setup are described in the following.

3.1   Data

The current case scenario considers degraded artefacts made of copper. They are the property of the Oltenia Museum, Craiova, Romania. The task will be to spot 4 types of compounds on the microscopy image of their surface: two resulting from the corrosion in the presence of O_2, i.e. CuO (black color) and Cu_2O (red tone), and two other products caused by the interaction of the artefact with water and salts, i.e. $(CuOH)_2CO_3$ and $CuCl$ (green appearance). Additionally, a fifth category will be represented by the clean part of the object, where there is only raw Cu. A sample image is depicted in Fig. 1(a), with the manual annotations of the corresponding corrosion products given by labelled polygons, and performed with the support of the VGG Image Annotator tool [5].

(a)   (b)

Fig. 1. (a) An example of a copper artefact under the microscope and the corrosion compounds observable on its surface. The dark brown delineations (9, 10) are $CuCl$, the light brown (1, 3, 5, 6, 12, 13) represent Cu_2O, the blue ones (2, 7, 8, 11) denote CuO and the green markings (4, 14) are $(CuOH)_2CO_3$. (b) The top plot shows in percentages how many of the images in the training, validation and test sets contain pixels of each class. The bottom plot indicates the average coverage (%) over the entire image; this is computed considering for each class only the images where it occurs.

The images are split between training, validation and test sets as 131, 21 and 22, respectively. The images in the training set come from 20 objects, while the ones from the validation and test sets come from 4 and 3 distinct objects, respectively. Naturally, the separation of objects (and of the images accordingly) between the three sets assures there are no items appearing in any two sets at the same time.

Figure 1(b) illustrates the class occurrence in the training, validation and test sets. The top plot indicates that it is only clear Cu that appears in all the images, while the least represented component, both with respect to the number of images where it appears, but also as concerns the coverage area, is $CuCl$. The plots in Fig. 1(b) need to be analyzed only together, because the values of the bars from the second plot are calculated only for the percentage of images that are shown in the top plot for each corresponding class, i.e. only for the images that contain that corrosion compound. Accordingly, for instance, the CuO regions for validation and test sets in the second plot are 1.8% and 1.1% respectively, but in fact, the former is computed as an average from 47.6% of the pictures in validation (only these images contain such CuO pixels), while the 1.1% computed in the test set is computed from 95.5% of the test images.

It is therefore unarguably a very unbalanced pattern recognition task that demands a reconsideration of the learning within the model. Specifically, an EA will explore the search space of weight values for the different classes of the problem. The fitness evaluation will be powered by the estimation given by a surrogate model instead of running the convolutional model with the exploratory weights, in order to save computational time.

3.2   Design and Setup

The methodology involves several concepts that will be first briefly defined. The class weights are involved in the expression of the DL loss function, in order to balance the segmentation categories, and thus constitute important variables, whose values influence the accuracy of the model. Latin Hypercube Sampling (LHS) will be used to generate an initial random sample of values for the class weight parameters. Traditional regression models can be used in place of more expensive computational simulations (such as DL runs in this case) within the heuristic-based optimization of parameters, i.e. what is referred to as surrogate-based optimization. Among the nature-driven optimization techniques, EA are already a classical and reliable option. The heuristic will be employed to generate optimal configurations for the class weight parameters. The surrogates previously trained on the segmentation accuracy of initial LHS samples can further act on behalf of the DL technique, in evaluating these potential new variable values.

The flow of the proposed procedure is given in Fig. 2. The set of LHS configurations offers a random sweep over the space of values for the weights (Step 1). Each configuration is given to the DL model as the class weighting to be included in the loss function and the returning accuracy is appended as an output to the LHS samples (Step 2). A surrogate model is then assigned to this formed data set in order to learn the correspondence between LHS weight values and DL segmentation accuracy (Step 3). The surrogate will then be used as the fitness function when evaluating new individuals generated by an EA (Step 4).

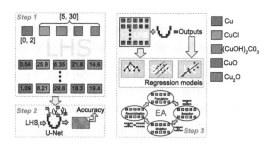

Fig. 2. The steps of proposed design.

There are 100 LHS configurations that are designed for the 5 weights that are controlling the categories of the classification problem: Cu, $CuCl$, $(CuOH)_2CO_3$, CuO and Cu_2O. The intervals for them are established as $[0, 2]$ for Cu and $[5, 30]$ for the Cu corrosion compounds. The intervals are inspired by the values calculated using the methodology in [9] and implemented under *scikit-learn* [11], which for this data set were equal to 0.21, 24.21, 8.9, 14.27 and 16.04, corresponding to the classes enunciated above. The values are computed based on the distributions of the samples (by samples it is referred to all the pixels in the images) to the classes from the training data. Through the LHS, a larger exploration of the search space is allowed.

The image pixel accuracy is employed for training the U-Net model that is utilized for the semantic segmentation task. The accuracy results reported will be calculated class-wise, due to the large imbalance in the occurrence of the corrosion labels. As Fig. 1(b) shows, all images contain pixels with Cu (or background, as we refer it interchangeably), but not all images contain the other compound classes. Thus, the pixel accuracy for each class in turn (ratio of correctly identified pixels of the class to total pixels of the class) is calculated and this value is averaged over the number of images corresponding to the label. Consequently, we will have a separate classification accuracy for each compound over the specific data set, be that validation or test. The final classification accuracy will be the average of the ones obtained for each class, including Cu.

There are thus 100 U-Net models that are trained using the different LHS configurations. The results from the validation set serve as reference points for the regression models that will be further used as surrogates. Three regression models are trained using all 100 LHS configurations and the validation DL output, in order to replace the more expensive model in evaluating new possible weight values. The three were chosen to be conceptually different, i.e. linear regression (LR), support vector machines regression (SVR) and random forests (RF). Each of them is further separately used as a fitness evaluation function for a genetic algorithm (GA) that explores the search landscape of the weights looking for a better value for fitness that should ideally correspond to the actual application of the U-Net model on the validation set.

The implementation of the GA uses the PyGAD library [7]. The same GA setup is used for each of the 3 surrogate models. An individual encodes the 5 weights for the classes with similar starting intervals as they are used for generating the LHS configurations. The GA setup options used are further detailed: the population size consisting of 10 individuals, 20 generations of evolution, tournament selection, single-point crossover and a mutation probability of 0.2. The latter corresponds to changing one gene per individual. There are 10 different runs completed for each of the 3 GA. When it comes to the evaluation of a newly obtained GA configuration, its accuracy value estimated by a surrogate regression model is attributed as the individual fitness.

4   Results and Discussion

This section presents the resulting outcomes, whose insights are further discussed. A comparison is made between those obtained for the class weights computed via the open source *scikit-learn* library (following the methodology derived from [9]) versus those when using LHS and surrogate-based optimization.

Figure 3 illustrates the connection between the generated LHS configurations and the validation accuracy that is obtained when these weights are used for training the U-Net models. The data is split into 3 intervals for each component for producing the box plots from Fig. 3. The *small* samples contain all LHS samples that have the values in $[0, 0.67)$ for Cu and $[5, 13.33)$ for the other classes, *medium* represents LHS with values for Cu in $[0.67, 1.33)$ and $[13.33, 21.67)$ for the others, and the highest rest of the configurations are in *large*.

Figure 4 illustrates the way the LHS samples are modeled by the 3 regression approaches in (a) and the feature importance given by the RF model in (b). Figure 4 (c) and (d) illustrate the histogram values of the GA genes and helps in understanding where the search is concentrated within the population (10 individuals) over the 20 generations of evolution.

Figures 5 and 6 provide various insights into the GA weight results obtained in the surrogate simulation, versus their success in the real DL application and also as compared to pre-calculated values as in [9]. All 4 plots in Fig. 5 show the results obtained by applying the U-Net using the best solutions discovered by the 10 repetitions of the GA using the 3 different regression models (SVR, LR and RF). The (a) plot shows side-by-side the outputs of the surrogate models (as they were embedded in the GA fitness function) and the actual results given by the U-Net when including the weights discovered by the GA using the 3 surrogates. The (b) plot depicts the DL validation accuracy for each corrosion class in turn from the configurations determined by the 10 GA runs.

Figure 6 (a) and (b) show through the box plots the weights that led to the top 10 results with respect to the validation accuracy. Additionally, the best weight configuration of the LHS, as well as of the regression (in particular, the SVR model led to the best DL accuracy on the validation set), are illustrated. Moreover, the calculated weights according to [9] are also pointed out in the graphic. The latter obviously appears in the same regions both in the LHS and regression sides in the plot. (c) and (d) show a comparison in DL results for validation and test sets. The dotted red lines indicate the result of the configuration with the computed weights following [9]. Out of the 10 best results determined by the weights found through GA and regression models, half are discovered by the SVR model, 3 by LR and 2 by RF. The results on the test set are obtained by applying the best configurations from validation.

Further, we discuss in detail the content of the resulting plots. Figure 3(a) does not only show how the 100 values are generated for each component in turn, but also indicates how the values are interconnected. Additionally, the colors express the quality of the validation accuracy when each specific setup is used. The weight of the Cu is the one that has a striking importance: the lines that depart from the smaller values of the parameter are particularly red, while towards the middle (between 0.5 and 1.5) the colors of the lines are closer to yellow and finally, going to the top, they are green. The other components do not have such a clear determination of the colors for the lines in any region. However, at a thorough analysis, it can be observed that lower values are discouraged for $CuCl$ and CuO, as there are fewer red lines at the bottom of the interval for them. The observations are also confirmed by the plots on the second row of the figure. As (b) illustrates, *small* values for Cu lead to the best overall accuracy

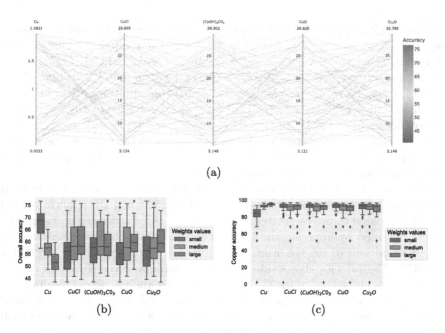

Fig. 3. (a) The connections between the 100 LHS configurations; the color corresponds to the accuracy. Box plots with intervals for class weights split into small, medium and large: (b) overall classification accuracy, (c) only Cu (background) accuracy.

result. On the contrary, as (c) exhibits, Cu *small* weights lead to the worst results for its class. This can be explained by the fact that the smaller weights values for Cu lead to a worse classification accuracy for Cu but, on the other hand, the quality of results improves for the other classes and thus the overall accuracy, which is computed as the mean over the classification accuracy of all classes (see Sect. 3.2), is enhanced. Smaller values for $CuCl$ and CuO indeed lead to a poorer overall accuracy ((b) and (a)).

The (a) plot from Fig. 4 is calculated from the 100 LHS configurations representing only the weight of Cu, as this is shown to be the most influential by the charts in Fig. 3. In fact, the RF model reveals that undoubtedly the Cu weights are the most important with respect to the validation accuracy (Fig. 4(b)).

The bottom of Fig. 4 shows the histograms for the 5 genes of the GA, for the entire population (c) and for the best solution (d). As noted in Sect. 3.2, the GA has a population of 10 individuals and runs for 20 iterations. The first row shows that the GA explores even beyond the initial intervals that are used for the initialization, i.e. gene 0 (Cu) and gene 2 (($CuOH)_2CO_3$). While values of almost 10 were also present in the population for the former (with the initialization in $[0, 2]$), values close to 0 were also explored for the latter (while the gene values were initialized in $[5, 30]$). However, the best individual (d) always had the values within the intervals considered for initialization.

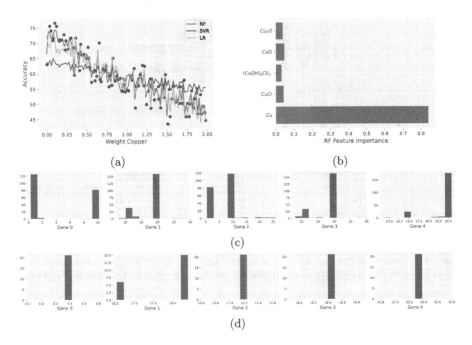

Fig. 4. (a) The training data with only the Cu weight and the accuracy, along with the solutions obtained from RF, SVR and LR. (b) The weight importance determined by RF. Histograms of the GA genes, obtained from all generations in one run when the fitness was given by SVR, for the entire population (c) and for the best solution (d). Genes 0 - 4 correspond to the weights for Cu, $CuCl$, $(CuOH)_2CO_3$, CuO and Cu_2O.

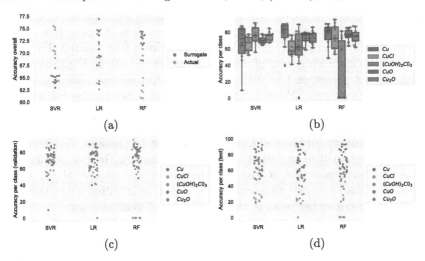

Fig. 5. Obtained results using the 3 different surrogate models. (a) The overall accuracy estimated from the surrogate vs the real DL outputs from the same inputs. (b) Box plots with the actual DL results per each class. The strip plots at the bottom show the real DL accuracy for each class on the validation (c) and test (d) sets.

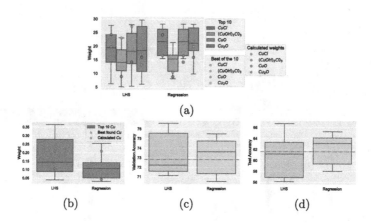

Fig. 6. Box plots with the weights that led to the top 10 validation results for the compounds (a) and Cu (b) during LHS and all surrogate models (dots indicate the best). DL results of the top 10 values for the validation set (c) and when applied to the test set (d). Pre-calculated weights as in [9] are also plotted by dots and red line.

Figure 5 illustrates various angles of the collected outputs. It can be observed from (a) that the SVR model proves to be *pessimistic*, meaning that the approximations of the model are below the actual outputs of the U-Net when using the weights discovered by the GA-SVR surrogate. On the contrary, RF is *optimistic*, as its results are above the actual U-Net values. LR has the results well interpolated with the actual values. While in (b) the GA-SVR surrogate has one very poor result for Cu (of 9.8%), there are 5 results when no pixels were identified at all for class $(CuOH)_2C0_3$, when using the weights given by the GA-RF surrogate. When checking the reason why this happens, it is observed that all these are cases when the weight for $(CuOH)_2C0_3$ is between 0.02 and 0.2. Thus, there are some runs in which gene 2 of the GA is very close to 0. In Fig. 4(d) the best results did not correspond to such a case, but that was for only one run and for SVR as a surrogate. What is more, there is also one outlier for which LR has a poor value for $(CuOH)_2C0_3$ and that happens as well for a low weight of that class, i.e. 0.2. The second row from Fig. 5 shows similar types of plots for the U-Net accuracy on the validation (c) and test (d) images, departing from the weights obtained in the 10 GA runs. Unfortunately, the test set has fewer images with any pixels of $CuCl$ and, in the ones where present, the areas were even smaller than for validation and a lot less than in training. Thus, the results for this class were rather weak in the test set for all configurations.

While the average on the validation set between the top 10 results given by LHS and surrogates-driven search in Fig. 6 are almost identical (73.22% and 73.18%, respectively), the median favors the surrogate regression (72.24% vs. 73.64%). However, on the test set, the results unarguably favor the search through surrogates, as the averages are at 60.75% and 62.17%, respectively. At the same time, if the configuration that leads to the most accurate validation result that is discovered through the surrogate-based GA model is applied on the test set, the result is 65.26%. If the same is done for the best LHS setting as found on the validation set, the test accuracy reaches only 62.73%. In comparison, the calculated weights through the approach in [9] lead to a test accuracy of 61.63%.

5 Conclusions

The paper discusses an appropriate class weighting for a semantic segmentation task where the pattern shapes are irregular and unbalanced. The best configuration is achieved by an EA backed by a surrogate regression model that gives the estimated accuracy of a DL with the loss function powered by the given values.

The results of the surrogate-based EA optimization (with SVR resulting as the best DL substitute) support the idea of evolving the optimal values for the class weighting in comparison to both a random LHS patameter search and the standard value of the *scikit-learn* package calculated for this purpose.

There are several directions for future work. The use of the Dice score as the segmentation metric could be a better alternative to handle class imbalance. The increase in the number of data images should also alleviate the issue. The consideration of an ensemble of surrogate models could also improve the performance. Also, relevant feature vectors for each compound could be extracted [17] and given to a standard neural network to view a parallel different processing.

Acknowledgement. This work was supported by a grant of the Romanian Ministry of Research and Innovation, CCCDI – UEFISCDI, project number 178PCE/2021, PN-III-P4-ID-PCE-2020-0788, *Object PErception and Reconstruction with deep neural Architectures (OPERA)*, within PNCDI III.

References

1. Arun, C., Lakshmi, C.: Genetic algorithm-based oversampling approach to prune the class imbalance issue in software defect prediction. Soft Comput. **26**(23), 12915–12931 (2022)
2. Bacanin, N., Stoean, R., Zivkovic, M., Petrovic, A., Rashid, T.A., Bezdan, T.: Performance of a novel chaotic firefly algorithm with enhanced exploration for tackling global optimization problems: application for dropout regularization. Mathematics **9**(21), 1 (2021)
3. Bartz-Beielstein, T., Chandrasekaran, S., Rehbach, F.: Case Study III: Tuning of Deep Neural Networks, pp. 235–269. Springer Nature Singapore, Singapore (2023)
4. Bi, J., Zhang, C.: An empirical comparison on state-of-the-art multi-class imbalance learning algorithms and a new diversified ensemble learning scheme. KBS **158**, 81–93 (2018)
5. Dutta, A., Zisserman, A.: The VIA annotation software for images, audio and video. In: 27th ACM International Conference on Multimedia, p. 4 (2019)
6. Forkan, A.R.M., et al.: Corrdetector: a framework for structural corrosion detection from drone images using ensemble deep learning. Expert Syst. Appl. **193**, 116461 (2022)

7. Gad, A.F.: PyGAD: An intuitive genetic algorithm python library. https://arxiv.org/abs/2106.06158 (2021)

8. Katsamenis, I., Protopapadakis, E., Doulamis, A., Doulamis, N., Voulodimos, A.: Pixel-level corrosion detection on metal constructions by fusion of deep learning semantic and contour segmentation. In: Bebis, G., et al. (eds.) ISVC 2020. LNCS, vol. 12509, pp. 160–169. Springer, Cham (2020). https://doi.org/10.1007/978-3-030-64556-4_13

9. King, G., Zeng, L.: Logistic regression in rare events data. Polit. Anal. **9**, 137–163 (2001)

10. Kumar, P., Batra, S., Raman, B.: Deep neural network hyper-parameter tuning through twofold genetic approach. Soft Comput. **25**(13), 8747–8771 (2021). https://doi.org/10.1007/s00500-021-05770-w

11. Pedregosa, F., et al.: Scikit-learn: machine learning in python. JMLR **12**, 2825–2830 (2011)

12. Postavaru, S., Stoean, R., Stoean, C., Caparros, G.J.: Adaptation of deep convolutional neural networks for cancer grading from histopathological images. In: Rojas, I., Joya, G., Catala, A. (eds.) IWANN 2017. LNCS, vol. 10306, pp. 38–49. Springer, Cham (2017). https://doi.org/10.1007/978-3-319-59147-6_4

13. Rehbach, F., Zaefferer, M., Naujoks, B., Bartz-Beielstein, T.: Expected improvement versus predicted value in surrogate-based optimization. In: The 2020 Genetic and Evolutionary Computation Conference, pp. 868–876 (2020)

14. Roshan, S.E., Asadi, S.: Improvement of bagging performance for classification of imbalanced datasets using evolutionary multi-objective optimization. Eng. Appl. Artif. Intell. **87**, 103319 (2020)

15. Samide, A., Stoean, C., Stoean, R.: Surface study of inhibitor films formed by polyvinyl alcohol and silver nanoparticles on stainless steel in hydrochloric acid solution using convolutional neural networks. Appl. Surf. Sci. **475**, 1–5 (2019)

16. Samide, A., Stoean, R., Stoean, C., Tutunaru, B., Grecu, R.: Investigation of polymer coatings formed by polyvinyl alcohol and silver nanoparticles on copper surface in acid medium by means of deep convolutional neural networks. Coatings **9**, 105 (2019)

17. Stoean, C., Stoean, R., Sandita, A., Ciobanu, D., Mesina, C., Gruia, C.L.: SVM-based cancer grading from histopathological images using morphological and topological features of glands and nuclei. In: De Pietro, G., Gallo, L., Howlett, R.J., Jain, L.C. (eds.) Intelligent Interactive Multimedia Systems and Services 2016. SIST, vol. 55, pp. 145–155. Springer, Cham (2016). https://doi.org/10.1007/978-3-319-39345-2_13

18. Stoean, R.: Analysis on the potential of an EA-surrogate modelling tandem for deep learning parametrization: an example for cancer classification from medical images. Neural Comput. Appl. **32**, 313–322 (2020)

19. Stoean, R., Bacanin, N., Stoean, C., Ionescu, L., Atencia, M., Joya, G.: Computational framework for the evaluation of the composition and degradation state of metal heritage assets by deep learning. J. Cult. Heritage (under review) (2023)

20. Tanha, J., Abdi, Y., Samadi, N., Razzaghi, N., Asadpour, M.: Boosting methods for multi-class imbalanced data classification: an experimental review. J. Big Data **7**(1), 1–47 (2020). https://doi.org/10.1186/s40537-020-00349-y

21. Zhang, M., Li, H., Pan, S., Lyu, J., Ling, S., Su, S.: Convolutional neural networks-based lung nodule classification: a surrogate-assisted evolutionary algorithm for hyperparameter optimization. IEEE Trans. Evol. Comput. **25**(5), 869–882 (2021)

Study on Semantic Inpainting Deep Learning Models for Artefacts with Traditional Motifs

Catalin Stoean[1,2(✉)], Nebojsa Bacanin[3], Zeev Volkovich[4], Leonard Ionescu[1,5], and Ruxandra Stoean[1,2]

[1] Romanian Institute of Science and Technology, Cluj, Romania
{catalin.stoean,ruxandra.stoean}@inf.ucv.ro
[2] University of Craiova, Craiova, Romania
[3] Singidunum University, Belgrade, Serbia
nbacanin@singidunum.ac.rs
[4] ORT Braude College, Karmiel, Israel
vlvolkov@braude.ac.il
[5] Restoration and Conservation Lab, Oltenia Museum, Craiova, Romania

Abstract. This paper proposes the use of pre-trained semantic inpainting deep learning architectures to reach a high-fidelity, visually plausible filling content suggestion for the restoration of museum textile objects with traditional motifs. Two state-of-the-art models are selected and their reconstructions are additionally given to an autoencoder trained on a specific collection of textiles. The results show some potential of the tandem and the viability of an automatic support for artefact restoration.

Keywords: semantic inpainting · deep learning · diffusion model · Fourier convolutions · artefacts

1 Introduction

Semantic inpainting is a technique used in image processing that involves filling in the missing parts of an image by predicting what should be present, based on the surrounding context. Deep learning (DL) models for this task are often pre-trained on regular images of common objects and scenes, in order to already have the learning of basic geometrical features achieved and avoid overfitting with insufficient data. While promising results are obtained for natural images that belong to similar categories to the ones that form the pre-training data, the same may not hold true for those with very specific content, e.g. with traditional motifs. The traditional motifs may differ significantly from the objects in regular pictures, as they have specific visual characteristics and patterns. Moreover, their disposal may not conform to the standard distribution of features of standard images. Their symmetry is also distinct from that of faces or buildings.

Therefore, it is necessary to check the suitability of DL semantic inpainting models pre-trained on natural data for images with traditional motifs. In this way, the applicability and effectiveness of semantic inpainting techniques can be

I. Rojas et al. (Eds.): IWANN 2023, LNCS 14135, pp. 479–490, 2023.
https://doi.org/10.1007/978-3-031-43078-7_39

increased to a wider range of practical scenarios and domains. The current paper thus tests the performance of two state-of-the-art DL models on a collection of museum artefacts with traditional content. The objects are first photographed before reaching the decision regarding their restoration procedure. Then the regular reconstruction is performed based on the expertise of the human specialist. A first automatic attempt to fill the degraded parts by a DL architecture was made in [9], however, the visual match was not satisfactory, due to the small number of available samples for training. Hence, the current solution appoints pre-trained DL models to support a higher-fidelity inpainting. An autoencoder is additionally subsequently appointed to refine the images obtained from the pre-trained models on the given collection of traditional textiles.

The rest of the paper is organized as follows. Section 2 introduces the data set and outlines the pre-trained models, explaining the reasons for their choice among those in the state of the art, as well as the role of the autoencoder. The selection of specific image masks and metrics is given in Sect. 3. The experiments are unfolded in Sect. 4 and the conclusions reached in Sect. 5.

2   Data and Models

Traditional Romanian costumes (blouses, shirts, vests, skirts) were selected from the web in order to further accommodate the models pre-trained on general images to the current task. The tests were carried on images of the collection from Oltenia Museum in Craiova, Romania. They were pictures of table cloths, shirts, vests, as it can be seen in the later figures with the visualization of the results. The current paper appoints two pre-trained inpainting methods for the semantic completion of textile degraded artefacts. These are high-performing architectures among the recent models in the state of the art.

2.1   State of the Art

A recent review devoted to semantic image inpainting can be found in [12]. Various objectives and types of inputs suggest different DL approaches to this task admitting, as usual, several possible solutions, of which some were deemed unsatisfactory, especially for images with significantly large damage. In the study [4], an improved GAN is put forward for natural images inpainting, through dynamic partial convolutions, a point-wise normalization and a hybrid weighted feature map merge. The paper [5] proposes a progressive GAN, also in application to faces and buildings images. The image is considered from the edge to the centre, recursively using the repaired features as conditions, and thus making the content of the centre of the image the most important. An adaptive attention module is additionally included in order to merge the results of the recursions. The importance of the auxiliary guidelines in helping the inpainting model is investigated in [3] by introducing a multi-expansion loss and showing the augmentation on data sets with building images. Closer to the current work, the paper [2] compares three state-of-the-art inpainting models applied from the repairing perspective of an art painting, namely *CoModGAN*, *LaMa* and *GLIDE*. As it was noted here, when large areas of the piece are degraded, little surrounding information remains to conduct the inpainting model to fill in the correct content choice. The *CoModGAN* relies on the GAN architecture common for semantic inpainting tasks, but co-modulating the stochastic with the conditional approaches in the generator. *LaMa* replaces the GAN with ResNet with fast Fourier convolutions. *GLIDE* is a diffusion model of multimodal application that also embeds guidance from text.

2.2 Architectures

The image restoration of traditional artefacts is therefore an endeavour less explored in the field of semantic inpainting. Symmetry and structure will be helpful traits for training the models. However, contrary to their closest match, i.e. paintings, these historical pieces have a very diverse content, all laid on a small area. Moreover, the pattern variation within the same type of object is very high, depending on the skills and imagination of the artisan. Out of the current existing semantic inpainting approaches, we have chosen two architectures to adapt to the current scenario of traditional artefacts: the LaMa and the latent diffusion model (LDM).

The LaMa model is a deep convolutional neural network (CNN) able to handle large mask inpainting and generate high-quality output [11]. It consists of an encoder-decoder architecture, where the encoder is responsible for extracting features from the input image, and the decoder is responsible for generating the inpainted image. The key innovation of the LaMa model lies in the use of Fourier convolutions [1]. Fourier convolutions operate in the frequency domain, which allows for the propagation of information across the image, making the model more efficient in capturing global context information.

A second recent and well-performing architecture is a diffusion model, trained in the learned latent space of an autoencoder [8]. Additionally, cross-attention is introduced into the architecture for multi-modal inputs and convolution is employed for realistic inpainting. The strong point of the LDM in comparison to the traditional diffusion architectures is that is successfully achieves a high-fidelity image but under reasonable computational resources.

2.3 Autoencoder

An experiment of the current study presumes the application of a simple autoencoder that is trained to reach the initial images from the outputs of LaMa, on the one hand, and of LDM, on the other, in an attempt to further refine the results. Afterwards, the autoencoder is applied to the outputs of these two superior models on the test set. The training data consists of samples of a similar type, i.e. with traditional motifs. The experiment aims to evaluate whether a simple autoencoder can gain extra information from data that has such particularity. It is true that both LaMa and LDM would lead to better results if trained directly on such data, but their requirements are very demanding both with respect to the processing power and the quantity of data.

3   Input and Performance Evaluation

Before proceeding to experiments, the mask types and performance metrics are decided.

3.1   Masks

We will use two types of image masks: squared masks and irregular masks. The square one has in all cases the top left corner as the quarter of the image height and width, and the dimension varies from 10% to 50% of the image size in steps of 10%. Each case is separately applied for all test images in order to verify how the models behave when different mask sizes are considered.

Irregular masks are generated randomly with lines, circles, and ellipses and the size of the mask is controlled by the thickness of these shapes. The number of these objects is randomly generated and, for each shape type, there can be up to 20 pieces. Accordingly, irregular masks can have different shapes and sizes. To make the inpainting process more challenging, we added three thickness levels for the objects that form the masks: small, medium, and large. The small objects cover about 20% of the image, the medium objects cover about 50%, and the large objects cover about 70%. Figure 1 shows examples of masks of both types having different coverage levels.

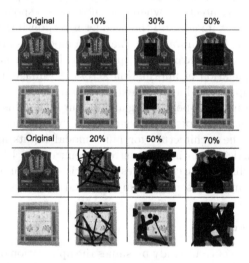

Fig. 1. Examples of squared and irregular masks with different sizes. The first two rows show samples with squared masks and the last two illustrate irregular ones.

Besides these two types of masks, we considered some special cases where we deliberately covered some parts from the materials that possess particular problems. Figure 2 illustrates four such samples. For these scenarios, it is interesting to evaluate whether the models *fix* the problem through their applications, which is the actual restoration challenge.

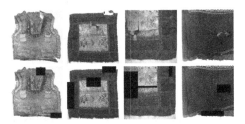

Fig. 2. Examples of photos for which dedicated masks are generated. First row illustrates the original images and the problematic regions are covered in the second one.

3.2   Metrics

To evaluate the performance of the inpainting algorithms, the following measurement metrics are appointed: Mean Squared Error (MSE), Peak Signal-to-Noise Ratio (PSNR) and Structural Similarity Index (SSIM). As shown in the survey [7], these metrics are often used for image inpainting. They are briefly presented next.

MSE is a commonly used metric for evaluating the quality of an inpainted image. It measures the average of the squared differences between the original and the inpainting. The lower the MSE, the better the quality of the inpainted image. The MSE can be calculated using the Eq. (1), where m and n are the dimensions of the image, $I_{i,j}$ is the original image pixel value at location (i,j), and $\hat{I}_{i,j}$ is the corresponding inpainted image pixel value.

$$MSE = \frac{1}{mn} \sum_{i=1}^{m} \sum_{j=1}^{n} (I_{i,j} - \hat{I}_{i,j})^2 \tag{1}$$

PSNR is another commonly used metric for evaluating the quality of an inpainting result. It measures the ratio between the maximum possible pixel value and the MSE. The higher the PSNR, the better the quality of the inpainted image. The PSNR can be calculated using the Eq. (2), where $max(I)$ is the maximum pixel value of the original image.

$$PSNR = 10 \log_{10} \left(\frac{max(I)^2}{MSE} \right) \tag{2}$$

SSIM is a metric that takes into account the structural information of the image and is commonly used for evaluating the perceptual quality of an inpainting outcome. It measures the similarity between the original and inpainted image in terms of luminance, contrast, and structure. The SSIM is calculated using Eq. (3), where μ_I and $\mu_{\hat{I}}$ are the average pixel values of the original and inpainted images, σ_I^2 and $\sigma_{\hat{I}}^2$ are their variances, and $\sigma_{I\hat{I}}$ is their covariance. The constants c_1 and c_2 are small positive numbers used to avoid division by zero. The value of SSIM ranges from -1 to 1, with a value of 1 indicating perfect similarity between the original and inpainted images.

$$SSIM = \frac{(2\mu_I\mu_{\hat{I}} + c_1)(2\sigma_{I\hat{I}} + c_2)}{(\mu_I^2 + \mu_{\hat{I}}^2 + c_1)(\sigma_I^2 + \sigma_{\hat{I}}^2 + c_2)} \qquad (3)$$

4   Experiments

The setup of the experiments is next outlined and obtained results are discussed.

4.1   Experimental Setup

Several details that allow the replication of the experiments are provided in the current subsection.

Smaller vs Larger Images Both LaMa and LDM perform very well on large images, i.e. having a minimum of 512×512 pixels. This is very practical, however, it is very hard to train such models locally due to the limitation of equipment of the usual users. In the current work, two scenarios are used, one in which the images have a resolution of 512×512 pixels and one with 256×256 pixels. In order to provide the smaller images to LaMa and LDM, they are resized. The outputs from the 2 models in this scenario with smaller images are also resized to 256×256, in order to compare them directly with the original inputs.

In the scenario with the larger images, no resizing is necessary. The aim of this experiment is to evaluate which of the 2 models is more sensitive to resizing.

Autoencoder Setup. There are 2 autoencoder models tried, a smaller one (further named AS) and a larger one (AL), depending on the size of the expected input images. While AS takes images of 128×128 pixels, AL expects double the size. The first convolutional layer has 256 (AL) or 128 (AS) filters, followed by 2 having half (128 and 64 for AL, and 64 and 32, respectively, for AS) layers of filters. The max pooling layer reduces the spatial dimensions of the feature maps by a factor of 2, which helps to capture the most salient features of the input image.

After the downsampling stage, the model uses a series of convolutional layers with smaller filter sizes to reconstruct the input image. The upsampling layer doubles the spatial dimensions of the feature maps and is followed by a convolutional layer with 32 for AS or 64 filters for AL, followed by two more layers that double the number of filters, respectively. The final layer is a convolutional layer with 3 filters, corresponding to the three color channels of the output image.

Each model is trained using the mean squared error loss function and the Adam optimizer with a learning rate of 1e-3. During training, the model is fed with batches of 4 for AL and 16 for AS input images and their corresponding ground-truth images. The training process continues for 100 epochs, during which the model updates its weights to minimize the training loss. The data set considered for training the autoencoder contains 310 images with a size of 512×512 pixels or larger with traditional motifs, downloaded from the Internet.

The limited number of samples is explained by the fact that, to the best of our knowledge, there is no dedicated data set for such a task. The test set comprises 25 images of samples from Oltenia Museum in Craiova, Romania.

4.2   Results and Visualization

Figures 3 and 4 present the results obtained by applying the LDM and LaMa directly on the test images for the squared masks in the first figure, and for the irregular masks in the second one, respectively.

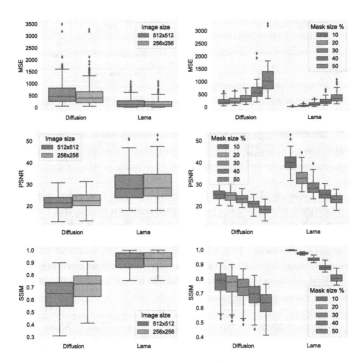

Fig. 3. Box plots showing MSE, PSNR and SSIM for the test cases with the squared masks. While the left column illustrates the comparison between results from larger vs smaller images, the right column shows them for each mask size in turn. While for MSE smaller values correspond to better results, it is reverse for the other two measures.

Table 1 shows the average numerical results obtained for the cases when the irregular masks that cover 50% of the images are used. A comparison between all tried models is provided.

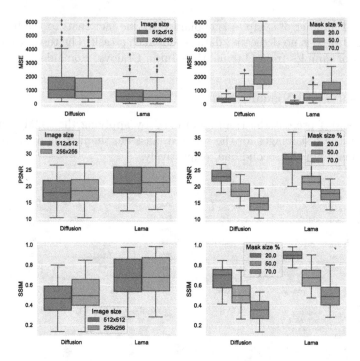

Fig. 4. Box plots showing MSE, PSNR and SSIM for the test cases with the irregular masks. The left column illustrates the comparison between results from larger vs smaller images, and the right column shows results for the different mask sizes.

Table 1. Comparison between the original pre-trained models and the variants with additional training using autoencoders for the irregular masks covering around 50% of the images. AS and AL denote the smaller and the larger variants of the autoencoders.

Model	MSE			PSNR			SSIM		
	Mean	Median	Std	Mean	Median	Std	Mean	Median	Std
Lama	610.069	496.125	361.175	20.966	21.175	2.527	0.679	0.678	0.103
AS-Lama	676.489	581.163	330.492	20.270	20.488	1.966	0.605	0.605	0.106
AL-Lama	1005.753	968.039	504.433	18.681	18.272	2.390	0.601	0.609	0.103
Diffusion	1124.451	1036.686	585.853	18.205	17.979	2.240	0.449	0.451	0.128
AS-Diffusion	1103.117	987.222	556.445	18.186	18.187	2.066	0.459	0.473	0.122
AL-Diffusion	1054.953	904.862	590.780	18.449	18.565	2.163	0.458	0.469	0.123

Figure 5 illustrates some samples from the images that are used for producing the outputs in Table 1. Each image is presented as original, masked, and as it was inpainted by each model. The different metrics are computed for each particular case (with regard to the original version) and are shown above the images.

Figure 6 illustrates how the two models are able to actually improve the initial version of the images by eliminating their defects.

4.3   Discussion

The first column in Figs. 3 and 4 illustrate the differences between the smaller and the larger images. Overall, the differences for LaMa are rather absent between the 2 cases, while the LDM is affected by the changes. We calculated a Mann-Whitney U test to verify the differences between the results obtained by each model in turn for the large and small images. LaMa model showed no significant differences between results for any measure. This occurred for both types of masks, as well. On the other hand, LDM had significant differences between the images of different sizes for each measure, for the cases when the masks were squares, and only for SSIM when irregular.

The difference in results between LaMa and LDM can be observed overall in the box plots from the left columns or, for each mask size considered in turn, in the box plots from the right columns. LaMa does not only lead to better average results but it can be observed that at least for MSE and SSIM the upper and lower quartiles are closer to the median than in the case of LDM, hence the results for the different samples are closer to each other.

Table 1 details the results of the models when applied for images containing irregular masks that cover approximately 50% of the original ones. They are applied only for larger images of 512 by 512 pixels. As detailed in the previous subsection, we included another experiment that employed 2 versions of autoencoder that would ideally learn to convert the images produced either by LaMa or LDM into the original ones. The results from the table represent the mean results over the 25 images from the test set. As it can be seen, the best results are obtained by LaMa in all cases, while occasionally the autoencoders lead to better results than the ones of Diffusion. Figure 5 illustrates several particular cases and, in some of them, the outputs of the autoencoder models even exceed the results of LaMa by a small margin. However, despite the better values in metrics, the images produced by the autoencoders are smoother and with a smaller contrast. Such cases can be seen in rows 1–3, 6 and 8. It is actually the smaller model that reaches better results than the larger one, especially as there are some cases, like row 4 or row 7, when AL-Lama leads to very poor results, especially as concerns MSE and PSNR. The opposite happens for LDM, where the larger autoencoder model leads to better results than the original model itself. From Fig. 5, it can be observed that the LDM samples have a better contrast than Lama and even than the original.

The comparison between the Lama and LDM models can be observed also in the special cases in Fig. 6. While both models *repair* the issues hidden by the masks, it can be seen in these cases as well that LaMa produces more blurred solutions (see especially row 3), while the LDM model is generally more creative (e.g., in the same third row it introduces a flower). The creativity of LDM goes

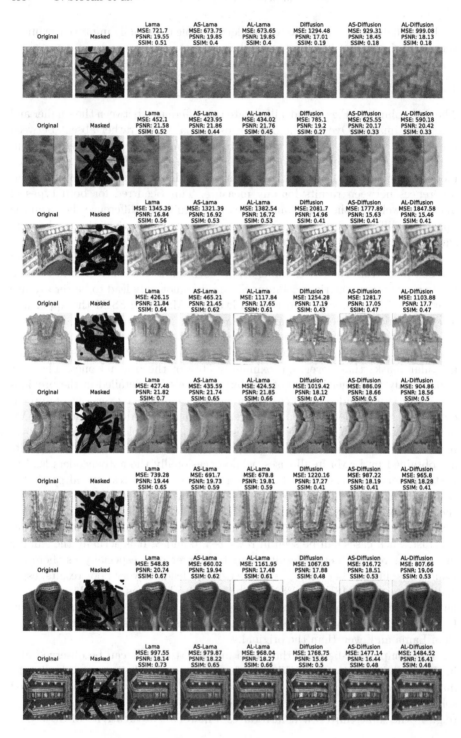

Fig. 5. Results comparison for 8 test samples.

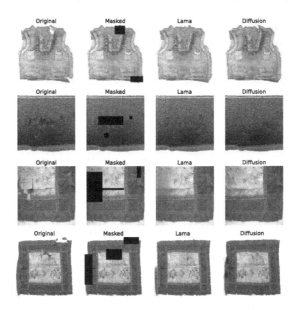

Fig. 6. Corrections added to some samples.

sometimes beyond what is expected, as it even suggests drawing a face, for instance, in certain empty regions where that would not suit well, especially in artefacts with traditional motifs.

5   Conclusions

Two successful pre-trained models for image inpainting, LaMa and Latent Diffusion, are compared based on the results they obtain on several images of cloths with traditional motifs. Squared masks of different sizes are used, covering from 10% up to 50%, in steps of 10%. Irregular masks covering approximately 20%, 50% and 70% are also used. One experiment evaluates the robustness of the 2 models to image resizing and overall the results obtained by LaMa are independent, while the ones from LDM are not. Another experiment uses two versions of autoencoder models to learn to transform the outputs obtained by either LaMa or LDM into the original samples. They are trained on a relatively small data set of 310 images containing garments with traditional motifs. Generally, the produced results are more blurred than the inputs, although the measures show encouraging results, especially for LDM. In future work, we intend to gather a larger data set of training samples and invest more in tuning the hyperparameters [6] of the autoencoder to boost its performance. Also, several computed features [10] for the geometrical shapes of the context could be added an auxiliary information to the models.

Acknowledgement. This work was supported by a grant of the Romanian Ministry of Research and Innovation, CCCDI - UEFISCDI, project number 178PCE/2021, PN-III-P4-ID-PCE-2020-0788, *Object Perception and Reconstruction with deep neural Architectures (OPERA)*, within PNCDI III.

References

1. Chi, L., Jiang, B., Mu, Y.: Fast Fourier convolution. In: Larochelle, H., Ranzato, M., Hadsell, R., Balcan, M., Lin, H. (eds.) Advances in Neural Information Processing Systems. vol. 33, pp. 4479–4488. Curran Associates, Inc. (2020)
2. Cipolina-Kun, L., Caenazzo, S., Mazzei, G.: Comparison of comodgans, lama and glide for art inpainting completing m.c escher's print gallery. In: 2022 IEEE/CVF Conference on Computer Vision and Pattern Recognition Workshops (CVPRW), pp. 715–723. IEEE Computer Society, Los Alamitos, CA, USA (2022)
3. He, J., Zhang, X., Lei, S., Wang, S., Lu, C.T., Xiao, B.: Semantic inpainting on segmentation map via multi-expansion loss. Neurocomputing **501**, 306–317 (2022)
4. Li, G., Li, L., Pu, Y., Wang, N., Zhang, X.: Semantic image inpainting with multi-stage feature reasoning generative adversarial network. Sensors **22**(8), 2854 (2022)
5. Li, H.A., Hu, L., Zhang, J.: Irregular mask image inpainting based on progressive generative adversarial networks. Imag. Sci. J. **0**(0), 1–14 (2023)
6. Postavaru, S., Stoean, R., Stoean, C., Caparros, G.J.: Adaptation of deep convolutional neural networks for cancer grading from histopathological images. In: Rojas, I., Joya, G., Catala, A. (eds.) IWANN 2017. LNCS, vol. 10306, pp. 38–49. Springer, Cham (2017). https://doi.org/10.1007/978-3-319-59147-6_4
7. Rojas, D.J.B., Fernandes, B.J.T., Fernandes, S.M.M.: A review on image inpainting techniques and datasets. In: 2020 33rd SIBGRAPI Conference on Graphics, Patterns and Images (SIBGRAPI), pp. 240–247 (2020)
8. Rombach, R., Blattmann, A., Lorenz, D., Esser, P., Ommer, B.: High-resolution image synthesis with latent diffusion models. In: IEEE/CVF Conference on Computer Vision and Pattern Recognition (CVPR), pp. 10684–10695 (2022)
9. Stoean, C., et al.:On using perceptual loss within the u-net architecture for the semantic inpainting of textile artefacts with traditional motifs. In: 24th Intl Symposium on Symbolic and Numeric Algorithms for Scientific Computing (2022)
10. Stoean, C., Stoean, R., Sandita, A., Ciobanu, D., Mesina, C., Gruia, C.L.: SVM-based cancer grading from histopathological images using morphological and topological features of glands and nuclei. In: De Pietro, G., Gallo, L., Howlett, R.J., Jain, L.C. (eds.) Intelligent Interactive Multimedia Systems and Services 2016. SIST, vol. 55, pp. 145–155. Springer, Cham (2016). https://doi.org/10.1007/978-3-319-39345-2_13
11. Suvorov, R., et al.: Resolution-robust large mask inpainting with Fourier convolutions. CoRR abs/2109.07161 (2021)
12. Xiang, H., Zou, Q., Nawaz, M.A., Huang, X., Zhang, F., Yu, H.: Deep learning for image inpainting: a survey. Pattern Recogn. **134**, 109046 (2023)

A Novel Approach to Jominy Profile Prediction Based on 1D Convolutional Neural Networks and Autoencoders that Supports Transfer Learning

Marco Vannucci$^{(\boxtimes)}$ ⓘ and Valentina Colla ⓘ

TeCIP Institute, Scuola Superiore Sant'Anna, Pisa, Italy
{marco.vannucci,valentina.colla}@santannapisa.it

Abstract. This paper introduces a novel method for the estimation the Jominy profile of steel based on its composition, by combining autoencoders and 1-D Convolutional Neural Networks. The approach has two goals: firstly, to enhance the accuracy of hardenability prediction by exploiting the capability of the 1-D CNN to learn how the chemical composition of steel affects the shape of the Jominy profile; secondly, to use transfer learning to apply the knowledge gained from training on a specific dataset to new types of production with less available data or data with different characteristics as it often occurs in the industrial context. The proposed approach was tested on two industrial datasets aiming to assess the effectiveness of the methods on the two goals achieving satisfactory results.

Keywords: steel hardenability · 1-D convolutional neural networks · transfer learning

1 Introduction

The Jominy test is important in steelmaking as it allows assessing steel hardenability, which is its ability to be hardened by quenching. This information is crucial in selecting and designing materials for specific applications where hardness, strength, and wear resistance are important factors.

The test involves the heating of one end of a cylindrical steel specimen up to the austenatizing temperature and its subsequent quenching through a water jet. This creates a gradient of cooling rates along the length of the sample, which produces a hardness profile when the sample is tested at fixed distances from the quenched end. The resulting Jominy profile is used to predict the ability of a steel alloy to be hardened by quenching, and is an important factor in determining the suitability of a steel for specific applications. By understanding the steel hardenability, steel producers can optimize the alloying and heat treatment processes to provide the product with the desired properties. This can save time

I. Rojas et al. (Eds.): IWANN 2023, LNCS 14135, pp. 491–502, 2023.
https://doi.org/10.1007/978-3-031-43078-7_40

and resources in the production process by avoiding costly trial-and-error methods. Moreover, the Jominy test is widely used as a quality control tool to ensure that the produced steel meets specific requirements. It is also an essential test for research and development of new steel alloys, as it allows evaluation and comparison of different materials.

The ability of steel to harden is primarily determined by its chemical composition, as the content of certain elements such as C, B, Cr, N, Si, Mo, Nb, V, Ti has a significant impact. While the influence of these elements has been extensively studied, both individually and in combination, there is still some uncertainty regarding the exact relationship between chemical composition and the Jominy profile. This is due to the complex interactions between the chemical elements that affect the cooling behavior and ultimately determine the microstructure of the steel, which in turn affects its hardenability at different distances.

The Jominy end-quench test is very costly and time-consuming, therefore there is a strong economical interest in developing models to estimate hardenability from chemical composition. In this context, the cost of the Jominy test limits the availability of experimental data to set–up and tune such models when new types of steel are designed or produced.

In this paper a novel approach for estimating the Jominy profile from steel composition based on a combination between Autoencoders and Convolutional Neural Networks (CNN) is proposed. This method has a twofold purpose: the first is to improve the accuracy in hardenability prediction by learning how the steel chemical composition affects the *shape* of the Jominy profile through the CNN, and the second is to generalize and embed such knowledge obtained from an arbitrary training dataset to be able to reuse it, via transfer learning, while training the model for new types of productions with lower data availability.

The paper is organized as follows: in Sect. 2 the current state of art about the Jominy profile prediction is outlined with a special focus on approaches that employ Artificial Intelligence (AI) and, in particular, Artificial Neural Networks (ANNs) in Sect. 2.2. The proposed approach is described in detail in Sect. 3. The experimental set–up for the assessment of the method in the light of previously introduced objectives is depicted in Sect. 4, while the achieved results are reported and discussed in Sect. 5. Finally, Sect. 6 is devoted to drawing conclusions and outlining future lines of development of the approach.

2   Related Works

The strategical importance of the information gained from the Jominy profile together with its cost led through the years to development of a multitude of models aiming to the prediction of the Jominy profile from the chemical composition. These methods can be grouped according to the exploited approach.

2.1   Numerical Approaches

Phase field models simulate the evolution of the microstructure during the heat treatment process [2], and can predict Jominy profiles. These models consider the thermodynamics and kinetics of the phase transformations, and can provide information on the distribution of the different phases in the material which can be correlated to steel hardenability at the Jominy distances [5,9]. Similarly, Finite Elements Modelling (FEM) can simulate the heat transfer and phase transformations during the Jominy test, and can provide a detailed picture of the temperature and microstructure evolution [8]. These models require significant computational resources, but can provide a high level of accuracy. Quench Factor Analysis (QFA) involves analyzing the cooling curves and correlating them to the metallurgical response. This technique is used to estimate hardness from simulated cooling curves and has been shown to have a good correlation between the predicted and measured hardness. This approach developed in the 70-ies was adopted in [11] achieving an acceptable accuracy only for high hardness values.

2.2 Data Driven Approaches

Over time, the availability of information from numerous quenched–end Jominy tests has led to the formation of datasets that include the chemical composition of steels and their respective Jominy profile. These datasets have made the development of data-driven models possible.

Initially, traditional statistical techniques were used to make the first attempts in this area. One example of this is the approach was based on *multiplicators* and was improved and made more general in the following decades by various researchers, such as in [4] where the author proposed multiple statistical and empirical methods to estimate hardenability, while also discussing their advantages and limitations. In addition to these methods, regression analysis has also been employed with favorable results to predict hardenability, focusing not only on the method feasibility and optimization of models accuracy, but also investigating the key variables that impact each hardness value of the Jominy profile [6] paving the road to numerous successive researches. These methods, mainly due to their low complexity and linearity, demonstrated their accuracy only within limited ranges of steel grades. Such models often fail to generalize due to the fact that the relationship between the model parameters and the steel chemical composition is mostly based on empirical analysis and is challenging to learn. Furthermore, these models only offer satisfactory accuracy for a small number of points on the Jominy curve. The individual effect of each alloy element is typically examined, while interactions between them are ignored, resulting in reduced model accuracy. To address these issues, ANNs have been used as a tool to predict the Jominy profile since the 1990s.

Many studies used Multi-Layer Perceptron (MLP) ANNs to estimate the Jominy curve based solely on chemical composition. Some of these studies focused on microalloyed steel grades, while others considered steel for specific applications and included not only chemical composition, but also mechanical properties like yield strength and ultimate tensile strength as input of the ANN. However, all these approaches neglected the correlations between neighboring hardness values in the Jominy profile. To address this issue, in [3] a parametric approach is proposed in which the Jominy profile is represented by a parametric mathematical function of the distance from the quenched end, such as a *quasi–sigmoidal* monotonic decreasing function. Wavelet neural networks are then applied to correlate the steel chemistry with the function parameters. In [1] a sequential predictor was developed, in which each point of the Jominy profile is predicted by a dedicated ANN that exploits the content of specific chemical elements selected according to theoretical knowledge of the phenomenon and the predicted hardness at other distances. In this latter work a measure of the reliability of the predicted hardness is associated to each point of the profile. Finally, despite the dizzying development of applications based on the use of deep learning, there are still very few works that employ such technology for Jominy profile prediction. Among these, it is worth mentioning [7], in which an optimized Deep Convolutional Neural Network (DCNN) is used and in which the convolutional layer is applied directly to steel chemistry.

3 Proposed Approach

In this paper, a novel approach to the design of a Jominy profile predictor using the steel chemical composition is proposed. The model aims at improving the accuracy of existing predictors not through a punctual estimation of hardness at different distances from the specimen quenched-end, but through the determination of the profile shape and by linking it to the steel chemical composition. This kind of approach aims at overcoming some typical deformations of the profile, which are usually fixed during post-processing and result in an accuracy degradation. Furthermore, relating the chemistry to the profile shape instead of single points improves the robustness to outliers within the data.

In addition, the proposed model aims at *storing* the relationship between chemical composition and profile shape to make it available in a transfer-learning context. Such an approach may be useful for training models for new types of steel for which the available datasets are small, or for transferring the model to different steel plants, which are characterized by different ways of collecting experimental data or performing the experimental tests (i.e. different instrumentation) that may alter the aforementioned relationship.

The proposed Jominy profile predictor consists of two main models depicted in Figs. 1 and 2:

1. an *autoencoder* named *JAutoencoder* (Fig. 1) that has the task of mapping an arbitrary Jominy profile measured at the 15 standard points (such domain is subsequently referred as D_{J15}) into a latent space L characterized by a significantly smaller dimension;
2. a DNN named *JNetwork* (Fig. 2) that performs the actual prediction of the profile, by mapping the steel chemistry into the 15 points of profile, exploiting the JAutoencoder.

JAutoencoder is a quasi typical autoencoder whose goal is to learn a compressed representation of the input dataset composed by Jominy profiles, such

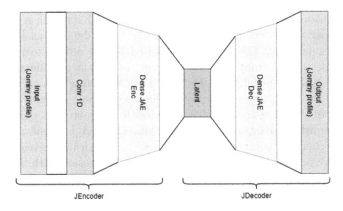

Fig. 1. Schematic architecture of the JAutoencoder.

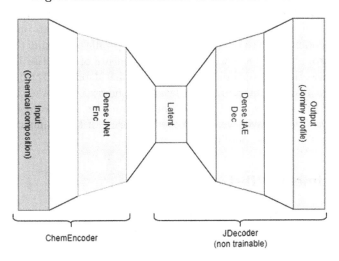

Fig. 2. Schematic architecture of the JNetwork putting into evidence the *ChemEncoder* and the *JDecoder*, non–trainable within this network.

that the original data can be accurately reconstructed from the compressed representation. Its main peculiarity is that the first layer performs a 1-Dimensional convolution (coupled to a max-padding operator) aiming at extracting knowledge about the shape of the fed Jominy profiles. After the convolutional layer, the autoencoder is composed of different fully connected layers, symmetrical with respect to the latent space L that can represent compactly the original D_{J15} domain, that accounts the shape of the handled Jominy profiles because of the 1D convolutional layer. The fully connected layers of JAutoencoder are activated through the *LeakyRelu* function, which is widely used for autoencoders implementation. In addition, both $L1$ and $L2$ type regularisation are employed to reduce the risk of overfitting.

Once JAutoencoder is trained using experimental data, it is decomposed into an encoder and a decoder, as shown in Fig. 1 and denoted as JEncoder and JDecoder, respectively. The latter is exploited for its ability to reconstruct a Jominy profile from a compact representation of it that considers the original shape of the profile. Specifically, JDecoder is embedded within a DNN denoted as *JNetwork* in Fig. 2, downstream of a chemical encoder (*ChemEncoder*) that has the task of mapping the chemical composition of a steel to the corresponding L representation. JNetwork, the DNN thus obtained, is connected and takes as input the chemical composition of the steel to return the corresponding predicted Jominy profile. In this context, the JDecoder present within JNetwork is **not trainable** and uses its own internal parameter values as got from JAutoencoder training. JNetwork is then trained with an experimental dataset resulting from Jominy testing that thus includes the chemical compositions of steel and the respective Jominy hardness profiles. This training procedure only affects the *trainable* parameters within the chemical encoder layers, supporting the mapping of chemical compositions into a latent representation that, once fed to the (frozen) JDecoder, leads to the predicted Jominy profile exploiting the knowledge it embeds through transfer learning. The architecture of both the JAutoencoder and the JNetwork were optimized through empirical tests described in Sect. 4. Optimization involves the number of layers of the two networks, including the dimension of the latent space L, as well as the number of neurons in each layer. The 1D convolutional layer was fine tuned in terms of the number of employed filters, filters dimensions, pooling dimension.

4   Experimental Tests

The performance of the proposed approach was evaluated through a test campaign that involves two datasets. The pursued tests aim at evaluating both the model accuracy and its capability of transferring knowledge regarding the Jominy profile shape through the JDecoder sub-system as discussed in Sect. 3.

4.1   Available Datasets

The datasets come from two distinct industrial plants and include a different number of samples: the bigger one, referred as *Dataset A*, is formed by 1500 observations; the other, *Dataset B*, by 250. Observations include steel chemical composition and the result of the Jominy quench-end test.

The two datasets correspond to different products in terms of both chemical composition and mechanical properties. Figure 3 puts into evidence such difference comparing the average Jominy profile throughout the datasets, reported in the figure together with the punctual standard deviation of hardness.

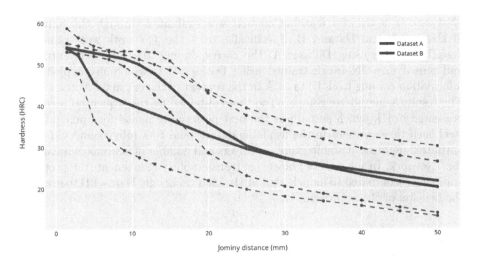

Fig. 3. Comparison between the Jominy profiles present in Dataset A and Dataset B. Average profile is shown together with point–wise standard deviation.

4.2   Tests Description

The accuracy in the prediction of Jominy profile from the steel chemical composition of the proposed approach was evaluated mainly based on Dataset A, which contains a number of observations large enough to support the training of the models and a reliable performance assessment on validation and test data. To this aim, Dataset A was preliminary divided into two parts: the first one, composed of 80% of the data, was used for training and validation purposes, the remaining 20% for test. Training and validation data are used for selecting the hyper–parameters values of JAutoencoder and JNetwork. The considered parameters include:

- number of layers and neurons per layer within the JEncoder and JDecoder (symmetric in this work) of the JAutoencoder
- number of layers and neurons per layer within the ChemEncoder of JNetwork
- latent space (L) dimension within JAutoencoder and JNetwork
- number of filters used within the 1D convolutional layer of the JAutoencoder
- dimension of 1D convolutional filters within JAutoencoder
- pooling dimension of the MaxPooling layer following the 1D convolution of the JAutoencoder.

The tested values for the listed hyper–parameters are summarized in Table 1. For each combination of hyper–parameters a 10–fold cross validation is set up and the models are tuned by using training data and evaluated by using the validation data. The combination resulting in a lower average prediction error throughout the points of the Jominy profile is then selected for comparison with other approaches. The ability to use the approach used for the development of

JNetwork in a transfer learning context was evaluated through the joint use of Dataset A and Dataset B. Specifically, once the JNetwork was trained as described above using Dataset A, the corresponding JDecoder was extracted and reused in a JNetwork trained using Dataset B that actually exploits the information coming from Dataset A in the context of the original JAutoencoder. The results achieved are compared to those obtained by the sequential predictor presented in [1], which actually is the best performing model for micro–alloyed steel such those present in the available dataset and to a fully–connected ANN optimized in terms of architecture (i.e. layers and number of neurons) similarly to the JNetwork. In the case of knowledge transferability assessment, the proposed approach is compared to models that are trained using only Dataset B to evaluate the benefits of the approach.

Table 1. Values of hyper–parameters tested during the models tuning phase. Layers are described through the number of neurons for each layers in brackets.

Hyper–parameter	Values
1D Conv. filters	2,3,5,10
Dimension 1D Conv. filters	2,3,4
Pooling dimension	2,3,4
Latent space dimension	2,3,4,5,6
JEncoder–JDecoder layers	(10,10), (10,10,10), (20,10), (20,20), (20,20,20), (30,30)
ChemEncoder layers	(10,10), (10,10,10), (20,10), (20,20), (20,20,20), (30,30)

5   Results and Discussion

5.1   Base Model Evaluation

The best performing hyper–parameters combinations resulting from the grid search on the values reported in Table 1 are shown in Table 2 in terms of mean absolute error (MAE) achieved by JAutoencoder and by JNetwork on the validation sets (within the CV framework) and by JNetwork on the test dataset.

In addition to those shown in the table, most of the combinations tested achieve satisfactory results in terms of MAE, in line with the industrial requirements of the application. Among the best ones, it is worth highlighting the low number of convolutional filters used and their size (2), which is sufficient to extract the necessary features from the Jominy profile likely due to the low number of points the profile itself is formed by. Furthermore, the optimal dimension of the latent space L resulting from the test is limited (4 in best performing cases), which, considering the achieved low JAutoencoder error, highlights that the main characteristics of the Jominy profile shape can be efficiently compacted.

Table 2. Best performing combinations of the hyper–parameters reported oin table 1 achieved while training the models on Dataset A.

1D–Conv filters	Filters dim.	Pool. dim.	L dim.	JAut layers	JNet layers	JAut. CV MAE	JNet. CV MAE	JNet. TS MAE
3	2	2	4	(20, 20)	(20, 20)	0.22	0.85	0.96
2	2	2	4	(30, 30)	(30, 30)	0,20	0,95	1,04
4	2	3	4	(10, 10)	(20, 20)	0,38	0,98	1,07
2	2	2	5	(20, 20)	(20, 20)	0,22	1,00	1,07
2	2	2	3	(20, 20)	(20, 20)	0,22	1,05	1,13

The best performing model was selected according to the performance of the JNetwork in terms of MAE within the CV (first row of Table 2). The average prediction error (MAE) of this model on the test data within the Dataset A is 0.96 HRC. The above introduced *sequential* model achieves a MAE of 1.10 HRC (+16% with respect to JNetwork) on the same data after being trained with the remaining observations of Dataset A.

A fully connected DNN was tested as well by using the same data. The DNN architecture was optimized testing different number of layers and hidden neurons similarly to what was done for the hyper–parameters optimization of the proposed approach. The best performing set–up in this case, a three layers network with (20,20,20) neurons in the hidden layers respectively and *ReLU* activation function, achieves a MAE of 1.12 HRC (+17% with respect to the proposed approach). The punctual behaviour of the three compared methods are shown in Fig. 4 in terms of prediction error. The barplot shows that the sequential model outperforms the other ones in the initial part of the profile, which are strongly related to a few chemical elements. On the other hand, the JNetwork is much more accurate in the central part of the profile (which is crucial from the industrial point of view) and obtains a more balanced performance throughout the whole profile, demonstrating the correct *learning* of how chemical composition affects the overall shape of the profile. The JNetwork also performs excellently qualitatively as shown in Fig. 5, in which two predicted Jominy profiles are compared to the corresponding measured ones are shown as examples.

5.2 Transfer Learning Evaluation

The effectiveness of the proposed transfer learning approach based on the reuse of the JDecoder component was evaluated according to the procedure described in Sect. 4. The straightforward approach to feed Dataset B input samples to the JNetwork trained by using only Dataset A led to the achievement of a MAE higher than 4 HRC throughout the profile, which is unacceptable from the industrial point of view. This failure is probably due to the difference, both in terms of chemistry and Jominy profile shape, between the two datasets. This result encourages the application of the proposed approach. The results obtained were compared to those of the sequential model and a fully connected ANN trained

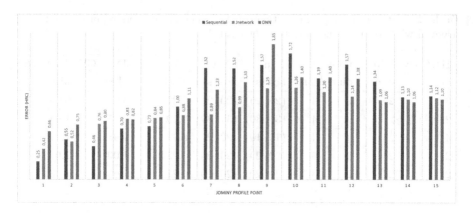

Fig. 4. Point–wise error of JNetwork, sequential model and DNN model on test data after training with Dataset A.

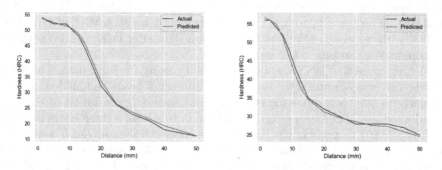

Fig. 5. Examples of two predictions performed by the JNetwork on test data from the Dataset A.

using only the data in Dataset B, partitioned into training, validation and testing with proportions 70%, 15%, 15% respectively. The ANN model was optimized in terms of architecture in a similar manner to what was done for Dataset A, obtaining a network of smaller size with respect to the previous one: two layers both holding 10 neurons activated via ReLU. The average error on test data of the JNetwork is 1.09 HRC, while the one of the sequential mode and of the fully connected ANN are 1.34 HRC (23% greater) and 1.40 HRC (29% greater), respectively. These error values are sensibly higher than the one obtained by the proposed approach. The punctual errors of the three approaches are depicted in Fig. 6 that confirms the goodness of the proposed method and puts into evidence that the prediction of the other approaches shows low accuracy in the central region of the profile, where the uncertainty of the shape is higher. The poor performance of these methods is likely due to the low number of samples available for training and validation of the employed ANNs. This latter issue is overcome by the JNetwork approach by transferring useful knowledge from the pre–trained JDecoder.

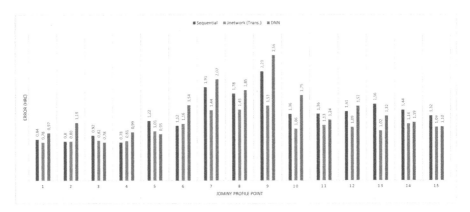

Fig. 6. Point–wise error of JNetwork trained via a transfer learning approach on Dataset B after the training of the JDecoder with Dataset A, sequential model and DNN model on test data after training with Dataset B.

6   Conclusions and Future Work

The paper presented a new approach to Jominy hardness profile prediction based on the use of an autoencoder using a 1D convolutional layer to learn a compressed encoding of the profile shape. Subsequently, the steel chemistry is mapped into that encoding using a fully connected ANN. The method exploits the idea of *learning* the Jominy profile shape rather than the point value of hardness at standard distances, one of the weaknesses of many existing approaches. In addition, the method is suitable to transfer learning by using the pre-trained JDecoder component and exploiting its ability to reconstruct the complete profile from its compressed encoding. The proposed approach was tested using two different industrial datasets showing very good results both in terms of accuracy of predictions and in the possibility of being used for transfer learning purposes in the rather common cases of experimental data scarcity (i.e. new product types, different plants). These results encourage further developments of this technology that will involve testing in a material design context as in [10], the evaluation of different and types of autoencoders for profile encoding, and development of a *chemical encoder* that exploits the theoretical knowledge regarding the influence of various chemical elements in different regions of the profile.

References

1. Cateni, S., Colla, V., Vannucci, M., Vannocci, M.: Prediction of steel hardenability and related reliability through neural networks. In: IASTED Multiconferences-Proceedings of the IASTED International Conference on Artificial Intelligence and Applications, AIA, pp. 169–174 (2013)
2. Colla, V., Desanctis, M., Dimatteo, A., Lovicu, G., Valentini, R.: Prediction of continuous cooling transformation diagrams for dual-phase steels from the intercritical region. Neural Comput. Appl. **33**(23), 16451–16470 (2021)

3. Colla, V., Reyneri, L.M., Sgarbi, M.: Neuro-wavelet parametric characterization of Jominy profiles of steels. Integr. Comput.-Aid. Eng. **7**(3), 217–228 (2000)
4. Doane, D.V.: A critical review of hardenability predictors. Hardenability Concepts with Applications to Steel, pp. 351–396 (1977)
5. Kirkaldy, J.: Prediction of microstructure and hardenability in low alloy steels. In: Proceedings of the International Conference on Phase Transformation in Ferrous Alloys, 1983. AIME (1983)
6. Komenda, J., Sandström, R., Tukiainen, M.: Multiple regression analysis of Jominy hardenability data for boron treated steels. Steel Res. **68**(3), 132–137 (1997)
7. Li, C., Yin, C., Xu, X.: Hybrid optimization assisted deep convolutional neural network for hardening prediction in steel. J. King Saud Univ.-Sci. **33**(6), 101453 (2021)
8. Li, M.V., Niebuhr, D.V., Meekisho, L.L., Atteridge, D.G.: A computational model for the prediction of steel hardenability. Metall. and Mater. Trans. B. **29**, 661–672 (1998)
9. Saunders, N., Guo, U., Li, X., Miodownik, A., Schillé, J.P.: Using JMatPro to model materials properties and behavior. JOM **55**(12), 60–65 (2003)
10. Vannucci, M., Colla, V.: Automatic steel grades design for Jominy profile achievement through neural networks and genetic algorithms. Neural Comput. Appl. **33**(23), 16451–16470 (2021)
11. Yazdi, A.Z., Sajjadi, S.A., Zebarjad, S.M., Nezhad, S.M.: Prediction of hardness at different points of Jominy specimen using quench factor analysis method. J. Mater. Process. Technol. **199**(1–3), 124–129 (2008)

Ordinal Classification

Gramian Angular and Markov Transition Fields Applied to Time Series Ordinal Classification

Víctor Manuel Vargas[1] , Rafael Ayllón-Gavilán[1(✉)] ,
Antonio Manuel Durán-Rosal[2] , Pedro Antonio Gutiérrez[1] ,
César Hervás-Martínez[1] , and David Guijo-Rubio[1,3]

[1] Department of Computer Sciences, Universidad de Córdoba, 14014 Córdoba, Spain
i72aygar@uco.es
[2] Universidad Loyola Andalucía, Escritor Castilla Aguayo, 4, 14004 Córdoba, Spain

[3] School of Computing Sciences, University of East Anglia, NR4 7TQ Norwich, UK

Abstract. This work presents a novel ordinal Deep Learning (DL) approach to Time Series Ordinal Classification (TSOC) field. TSOC consists in classifying time series with labels showing a natural order between them. This particular property of the output variable should be exploited to boost the performance for a given problem. This paper presents a novel DL approach in which time series are encoded as 3-channels images using Gramian Angular Field and Markov Transition Field. A soft labelling approach, which considers the probabilities generated by a unimodal distribution for obtaining soft labels that replace crisp labels in the loss function, is applied to a ResNet18 model. Specifically, beta and triangular distributions have been applied. They have been compared against three state-of-the-art deep learners in the Time Series Classification (TSC) field using 13 univariate and multivariate time series datasets. The approach considering the triangular distribution (O-GAMTF$_T$) outperforms all the techniques benchmarked.

Keywords: Gramian Angular Fields · Markov Transition Fields · Time Series Ordinal Classification · Soft Labelling

1 Introduction

Time Series Ordinal Classification (TSOC) is a yet unexplored field with great projection. It consists in applying a specific task of the Machine Learning (ML) field, known as ordinal classification, to a specific type of data, time series. To begin with, a time series is a series of data points collected chronologically, i.e. there is an even temporal order relationship between features. Examples of time series can be found in many fields, such as atmospheric events prediction [8] or seizure detection [4]. Furthermore, there is a wide variety of tasks that could be applied to this temporal data. One of the most popular tasks is Time Series

I. Rojas et al. (Eds.): IWANN 2023, LNCS 14135, pp. 505–516, 2023.
https://doi.org/10.1007/978-3-031-43078-7_41

Classification (TSC), which has an extensive literature, in which several different techniques have been published. TSC is the task of predicting a discrete target variable from a time series. TSC has been applied to a range of fields, such as arrhythmia classification [16] or cryptocurrencies prediction [1], among others.

On the other hand, in this work we focus on ordinal classification, a special type of classification dealing with ordinal categorical variables. In other words, the output variable of a time series takes values in an ordinal scale, and hence, these values follow an order relationship between them which is determined by the real-world task. Ordinal classification problems are present in all areas of science, including the assessment of neurological damage in Parkinson's disease [3] or the development of decision support systems for aesthetic quality control [17], among others.

Algorithms belonging to TSC are organised in several categories, depending on the core data representation used [2]. Up to seven categories can be identified: 1) distance-based approaches are based on computing the distance between time series; 2) interval-based methodologies rely on obtaining temporal features present in intervals of time series; 3) shapelet-based techniques consist in finding phase independent subsequences of the time series able to discriminate between them; 4) dictionary-based approaches compute the frequency of recurring patterns, which are obtained by extracting sequences of symbolic words; 5) convolutional-based methods focus on producing a large number of summary stats using random convolutional kernels and then selecting the most useful employing a linear classifier; 6) Deep Learning (DL)-based techniques are based on adapting the well-known residual and convolutional networks, as well as the attention and inception modules [7]; and 7) ensemble-based methods consider the combination of several algorithms belonging to the previous categories, in such a way that each of the members of the ensemble is built on a different representation.

As aforementioned, TSOC is a yet unexplored subarea of TSC. Up to now, only the shapelet-based approaches have covered this type of problem. [9] proposed an ordinal version of the Shapelet Transform (ST), in which ordinal information is exploited in two phases of the algorithm. First of all, the candidate shapelets are assessed using an ordinal shapelet quality measure. The goal of using an ordinal metric is to reduce the misclassification errors involving more jumps in the ordinal scale. In this sense, the Pearson's correlation coefficient is the one achieving the best performance. And secondly, an adaptation to the ordinal paradigm of the Support Vector Classifier, known as SVORIM, is applied to the transform.

In this paper we focus on the DL-based approaches, where several methodologies have been presented. [12] presented a review of the most popular DL architectures and their adaptation to time series. Recently [7] presented a current survey of deep learners to tackle both classification and extrinsic regression tasks. From these reviews, it can be observed that the best two standard DL approaches adapted to time series are the Fully Convolutional Network (FCN) [12] and the Residual Networks (ResNet) [12]. However, the best performing approach is InceptionTime [13], an ensemble combining residual networks with inception modules.

In this work, we present a novel approach consisting on imaging time series by means of three well-known representations for encoding time series as images: Gramian Angular Summation Field (GASF), Gramian Angular Difference Field (GADF), and the Markov Transition Field (MTF) [21,22]. These three representations are capable of extracting significant features hidden in time series. Specifically, GASF and GADF images are Gramian matrices, in which each element represents the trigonometric sum or difference, respectively, between different time intervals. Regarding MTF, images are the first order Markov Transition probability along one dimension and temporal dependency along the other one.

These three representation are introduced in different channels of a single image. In this way, each time series is encoded into an image with three channels, enabling the use of DL techniques from computer vision. Considering the ordinal characterisation of the time series considered in this work, the deep learners have to take into consideration these especial arrangements. Hence, a Convolutional Neural Network (CNN) along with a soft labelling encoding are used to improve the competitive results of other state-of-the-art DL-based approaches in TSOC. This soft labelling approach considers the probabilities generated by an unimodal distribution for obtaining a new labelling encoding. Two different types of unimodal distributions are employed: beta [20] and triangular distributions [19], being the latter the one achieving the best results in terms of the ordinal performance metrics.

The remainder of this paper is organised as follows: Sect. 2 details the methodology proposed in this work. Section 3 presents the experimental settings used for benchmarking the DL approaches. Section 4 shows and describes the results obtained, as well as presents a statistical test carried out to demonstrate superiority of the proposal presented. Finally, Sect. 5 closes this work.

2   Methodology

This section describes the methodology proposed to build images from time series, which can be then classified using an ordinal classifier.

2.1   Imaging Time Series

Two different time series imaging methodologies are applied to extract an image representation from time series: 1) Gramian Angular Field (GAF), that represents time series in a polar coordinate system instead of the standard Cartesian coordinates; and 2) MTF, which builds a Markov matrix of quantile bins and encodes the dynamic transition probability in a quasi-Gramian matrix.

Gramian Angular Field (GAF). The GAF [21] is a way to represent a time series as a matrix that can be interpreted as an image. The GAF takes a time series $\mathbf{x} = (x_1, x_2, \ldots, x_n)$ of n real-valued observations and rescales $\mathbf{x}$ so that all values are in the interval $[-1, 1]$ or $[0, 1]$. Then, the rescaled time series $\tilde{\mathbf{x}} = (\tilde{x}_1, \tilde{x}_2, \ldots, \tilde{x}_n)$ is converted to polar coordinates by encoding the value as the angular cosine and the time stamp as the radius. Therefore, for each point, the polar coordinates are computed as follows:

$$
\begin{cases}
\phi_i = \arccos(\tilde{x}_i), -1 \leq \tilde{x}_i \leq 1, \tilde{x}_i \in \tilde{\mathbf{x}}, \\
r_i = \frac{t_i}{N}, t_i \in \mathbb{N},
\end{cases}
\tag{1}
$$

where t_i is the i-th timestamp and N is a regularisation factor. Then, using the time series polar coordinates representation, the GASF and GADF are defined as follows:

$$
\text{GASF} = \begin{bmatrix}
\cos(\phi_1 + \phi_1) & \cdots & \cos(\phi_1 + \phi_n) \\
\cos(\phi_2 + \phi_1) & \cdots & \cos(\phi_2 + \phi_n) \\
\vdots & \vdots & \vdots \\
\cos(\phi_n + \phi_1) & \cdots & \cos(\phi_n + \phi_n)
\end{bmatrix},
\tag{2}
$$

$$
\text{GADF} = \begin{bmatrix}
\sin(\phi_1 - \phi_1) & \cdots & \sin(\phi_1 - \phi_n) \\
\sin(\phi_2 - \phi_1) & \cdots & \sin(\phi_2 - \phi_n) \\
\vdots & \vdots & \vdots \\
\sin(\phi_n - \phi_1) & \cdots & \sin(\phi_n - \phi_n)
\end{bmatrix}.
\tag{3}
$$

Then, following the trigonometric relations $\sin^2(a) + \cos^2(a) = 1$, $\sin(a - b) = \sin(a)\cos(b) - \sin(b)\cos(a)$ and $\cos(a + b) = \cos(a)\cos(b) - \sin(a)\sin(b)$, and the aforementioned polar definition of time series:

$$
\text{GASF} = \tilde{\mathbf{x}}^{\mathrm{T}} \cdot \tilde{\mathbf{x}} - \sqrt{I - \tilde{\mathbf{x}}^2}^{\mathrm{T}} \cdot \sqrt{I - \tilde{\mathbf{x}}^2},
\tag{4}
$$

$$
\text{GADF} = \sqrt{I - \tilde{\mathbf{x}}^2}^{\mathrm{T}} \cdot \tilde{\mathbf{x}} - \tilde{\mathbf{x}}^{\mathrm{T}} \cdot \sqrt{I - \tilde{\mathbf{x}}^2},
\tag{5}
$$

where I is the unit row vector $[1, 1, \ldots, 1]$ and the square roots are computed element-wise. Given that the size of the matrices obtained using these methods are based on the length of the time series, with the aim to obtain images of different sizes (k being the size), Piecewise Aggregation Approximation [14], which is a dimensionality reduction technique, is performed over the time series before computing the GAF.

Markov Transition Field (MTF). The MTF is another matrix representation of a time series $\mathbf{x}$. To create it, we first identify the quantile bins k, and each point of time series (x_i) is mapped to the corresponding bins $q_j (j \in [1, k])$. Then, the MTF matrix (W) is constructed in such a way that the element $w_{i,j}$ represents the transition probability between quantile bins q_i and q_j. In this way, a matrix of $k \times k$ is obtained. Note that, by adjusting the number of quantile bins, matrices (or images) of different sizes can be constructed.

To sum up, the images used in this work have three separate channels, which are constructed using the GADF, the GASF and the MTF, for the red, green and blue channels, respectively, adopting the same idea described in [21]. The whole process is graphically summarised in Fig. 1.

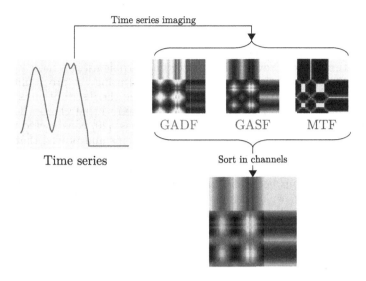

Fig. 1. Time series imaging process, from the original time series to the final images with three channels.

2.2 Ordinal Classifier

This section shows the ordinal classifier used to classify the images constructed following the process described in the previous subsection. However, before describing the classifier, a brief definition of ordinal problems and ordinal classification is provided.

In the context of DL, solving a classification problem with J classes is defined as the task of, given an input data $\mathbf{x} \in \mathcal{X} \subseteq \mathbb{R}^d$, obtaining its label $y \in \mathcal{Y}$ where $\mathcal{Y} \in \{\mathcal{C}_1, \mathcal{C}_2, ..., \mathcal{C}_J\}$. When these labels are naturally ordered, the problem becomes an ordinal classification task. In this way, in an ordinal classification problem, the aforementioned order between labels can be defined as $\mathcal{C}_1 \prec \mathcal{C}_2 \prec ... \prec \mathcal{C}_J$. In these terms, $\mathcal{O}(\mathcal{C}_j) = j$ defines the rank of class j as an integer. Therefore, in this kind of problem, the cost of classifying a sample in an adjacent class should be smaller than the cost of classifying it in the furthest class. Following this reasoning, the rest of the Section describes the type of deep learner employed and the ordinal methodology used to train this model accounting for the aforementioned constraint.

Images constructed following the process previously described are classified using a CNN model. To define the network, the well-known and proven ResNet18 architecture has been used. The most important advantage of using an existing CNN architecture is that we can use the pre-trained ImageNet weights, reducing the number of epochs needed to optimise the model.

The aforementioned model is optimised using the well-known Adam [15] optimiser. This algorithm uses a loss function that must be minimised to guide the optimisation process. Commonly, the Categorical Cross-Entropy (CCE) loss function is employed for classification problems. However, this function encodes the labels of each sample as a binary vector in which the target class is set to 1 and the other classes are 0 (i.e. crisp labels). For a nominal classification problem, this is the optimal encoding since none of the classes are related or follow a natural order. Nevertheless, for ordinal problems, it is more appropriate to use a soft label encoding that can be achieved using the probabilities generated by a unimodal distribution. In this way, when the distribution is centred in the target class, most of the probability is assigned to that class, whereas some probability is still assigned to the adjacent classes. [20] constructed soft labels using beta distributions for several ordinal problems and proved that this approach obtained better results compared to state-of-the-art nominal and ordinal approaches. Then, in [19], the authors proposed to use triangular distributions, instead of the previous beta, also improving the state-of-the-art results. For this reason, in this work, we are employing these two approaches, that can be formally expressed as below. The expression of the standard CCE can be defined as:

$$\mathscr{L}(\mathbf{x}, k) = \sum_{j=1}^{J} q(j, k)[-\log P(y = C_j | \mathbf{x})], \tag{6}$$

where $k = \mathcal{O}(\mathcal{C}_k)$, and $q(j, k) = \delta_{j,k}$ defines the aforementioned crisp labels, i.e. $\delta_{j,k}$ is 1 if $j = k$. However, following the process described in [19], these crisp labels can be converted to soft labels using the probabilities sampled from beta or triangular distributions. In this way, the $q(j, k)$ term can be replaced with the soft term $q'(j, k)$:

$$q'(j, k) = (1 - \eta) \cdot q(j, k) + \eta \cdot P_j(k), \tag{7}$$

where $\eta \in [0, 1]$ is a hyperparameter which is set to the values used by the authors of the original works, and $P_j(k)$ is the probability for $\mathcal{C}_j$ when the target is $\mathcal{C}_k$, sampled from the probability distribution. This probability can be obtained using the probability density function of the beta or triangular distribution associated with class k. Moreover, in both works, the authors proposed an approach to compute the parameters of the distributions for each class:

1. The parameters of the beta distributions are computed using the analytical process described in [20].
2. The parameters a, b and c of the triangular distributions are computed using a similar process based on a free parameter $\alpha_\mathcal{T}$ adjusting the probability in the adjacent classes. This parameter should be conveniently adjusted for each dataset using a crossvalidation procedure.

To sum up, in this work, two variants of our methodology are proposed, changing the ordinal loss function employed to train the model: 1) O-GAMTF$_\beta$ uses beta distributions to construct the soft labels, and 2) O-GAMTF$_T$ employs triangular distributions to sample the probabilities.

3   Experimental Settings

This Section describes the experiments carried out to compare the proposed methodology with other state-of-the-art alternatives. First, the compared approaches are presented. Next, the ordinal time series datasets used to compare these methodologies are presented. Finally, the procedure followed to train and evaluate the model is explained.

3.1 Compared Methodologies

In this Section, several methodologies from the TSC literature are proposed for comparison with the novel TSOC approach proposed in this paper. Thus, the experimental comparison includes the following DL-based methodologies:

1. O-GAMTF$_T$ and O-GAMTF$_\beta$, which are the proposed methodologies.
2. Fully Convolutional Network (FCN) [23]. The authors proposed a FCN consisting of three stacked convolutional blocks. Each block is a convolutional layer followed by a batch normalisation and a Rectified Linear Unit (ReLU). The convolution operation is performed by three 1-D kernels of size $\{8, 5, 3\}$ without striding. Pooling operations are excluded to avoid overfitting.
3. Residual Network (ResNet) [23]. ResNet models have demonstrated excellent results for different vision-related tasks like object detection or image classification [11]. [23] proposed a ResNet model for TSC. This model contains three residual blocks of $\{64, 128, 128\}$ filters. Note that this ResNet model is directly applied to the time series (1-D convolutions), whereas the ResNet18 model included with the proposed ordinal classifier uses 2-D convolutions and takes images as input.
4. InceptionTime [13] is currently one of the most advanced and best performing methods for TSC. The methodology consists of an ensemble of five randomly initialised inception networks.

3.2 Datasets

In order to conduct a comprehensive benchmarking, we have considered a collection of 13 TSOC problems from a variety of fields. This Section presents the sources of the datasets. Thee complete list of datasets and their characteristics are available online[1]. There are the three sources of data:

1. The UEA/UCR TSC archive, from which we have chosen 9 ordinal tasks [10].
2. The Monash/UEA/UCR time series extrinsic regression archive, from which we have chosen datasets with equal length and no missing values. Thus, we have included two more datasets (*AppliancesEnergy* and *Covid3Month*) in our experiments. These datasets originally had continuous output variables that have been discretised into five intervals of equal width.
3. Buoy data from the national data buoy center, which includes two problems: *USASouthwestEnergyFlux* and *USASouthwestSWH*. The former comprises 468 time series, each of which is constructed using 112 energy fluctuation measurements collected over four weeks (four measurements per day). The objective is to estimate the level of energy fluctuation during that period

---

[1] http://www.uco.es/grupos/ayrna/tsoc-gamtf-iwann.

of time, with values ranging from 0 (minimum level) to 3 (maximum level). The latter includes $1,872$ time series of length 28, representing the variation in sea wave height over a week (four measurements per day). The goal is to estimate the wave height level during that period of time, with values ranging from 0 (lowest height) to 3 (highest height).

3.3   Performance Metrics

In order to analyse the performance of the different methodologies considered in this work, the following metrics are calculated for each of the experiments:

1. Correct Classification Rate (CCR) (also known as Accuracy), which determines the ratio of correctly classified samples. It is the most commonly used metric in nominal classification problems.
2. Quadratic Weighted Kappa (QWK), which is based on the Kappa index (κ), and was defined and used in [18]. It is an ordinal metric given that it applies different weights to the errors based on the distance to the target class and a penalisation matrix.
3. The Mean Absolute Error (MAE) measures the average deviation from the target classes measured as the number of categories between the predicted one and the target. This metric has been commonly used for measuring the performance of ordinal classification problems [5].
4. The 1-off accuracy metric measures the proportion of samples whose prediction is no more than one class away, on the ordinal scale, from the target.

3.4   Experimental Procedure

In this Section, we describe the process followed to train and evaluate the models. While all datasets used in this study come with a pre-defined train and test split, we conducted multiple resamples with different seeds to increase the robustness of our results and verify the independence of the results from the chosen splits. In particular, we generated four additional train and test partitions for each dataset using different seeds. This approach allowed us to obtain more reliable and statistically significant results. Furthermore, the same seeds used to create the different samples were used to generate different initialisations of the model. Note that for the convolutional part, the pre-trained ImageNet weights were used. Therefore, only the weights of the final layers were randomly initialised.

   To fit the hyperparameters of the proposed methodology, a validation set was set apart from the training set by taking a stratified split containing 20% of the data. Then, the MAE computed over this validation set was minimised during the process. More concretely, O-GAMTF$_T$ has two hyperparameters that has to be adjusted: 1) α_I from the triangular regularised loss, and 2) k for the time series imaging methodology. The former has been selected from $\{0.04, 0.06, 0.08, 0.1, 0.12, 0.14, 0.16\}$ for all the datasets. However, the latter has been adjusted depending on the length of the time series of each dataset. The candidate list for each dataset contains six values, which are evenly distributed between a minimum value, the minimum size of the images generated, and a maximum, the maximum image size. The minimum size was set to 4 for all datasets, since using images smaller than 4×4 pixels does not make sense, even if the dataset is simple. The maximum size is calculated by the expression $\frac{L}{4}$, where L is the length of the time series, so that images are not too big. The list of k values evaluated for each dataset is available online(See footnote 1).

4   Results and Comparisons

This Section presents the results of the experiments described in Sect. 3. Table 1 shows the mean MAE for each method in all the datasets. In addition, it displays the average ranking of each method and the number of datasets in which each method has been the best or second best. The results for the other metrics are available online(See footnote 1). As can be observed from Table 1, the proposed method O-GAMTF$_T$ obtains the best result in 5 datasets, whereas the other proposed approach, O-GAMTF$_\beta$, achieved the best mean value in 3 datasets. Also, note that the O-GAMTF$_T$ obtained the best ranking value (2.0).

Table 1. Average results for each dataset and method regarding the MAE metric. Note that OAG stands for OutlineAgeGroup.

	InceptionTime	ResNet	FCN	O-GAMTF$_T$	O-GAMTF$_\beta$
AppliancesEnergy	**0.752**$_{0.076}$	*0.800*$_{0.161}$	0.903$_{0.295}$	1.390$_{0.419}$	1.452$_{0.281}$
AtrialFibrillation	0.826$_{0.101}$	0.907$_{0.126}$	0.893$_{0.118}$	*0.787*$_{0.110}$	**0.760**$_{0.146}$
Covid3Month	1.177$_{0.422}$	1.384$_{0.917}$	1.341$_{0.203}$	*1.062*$_{0.193}$	**0.964**$_{0.104}$
DistalPhalanxOAG	**0.232**$_{0.020}$	0.279$_{0.088}$	*0.242*$_{0.047}$	0.259$_{0.056}$	0.250$_{0.082}$
DistalPhalanxTW	0.455$_{0.114}$	*0.414*$_{0.047}$	0.460$_{0.091}$	**0.394**$_{0.065}$	0.565$_{0.143}$
EthanolConcentration	0.837$_{0.153}$	1.182$_{0.110}$	1.086$_{0.087}$	*0.418*$_{0.047}$	**0.343**$_{0.078}$
EthanolLevel	**0.174**$_{0.074}$	0.880$_{0.489}$	1.228$_{0.203}$	*0.220*$_{0.055}$	0.224$_{0.047}$
MiddlePhalanxOAG	0.451$_{0.067}$	0.486$_{0.046}$	0.483$_{0.088}$	**0.343**$_{0.065}$	*0.386*$_{0.098}$
MiddlePhalanxTW	0.755$_{0.118}$	0.778$_{0.063}$	0.757$_{0.100}$	**0.588**$_{0.066}$	*0.709*$_{0.061}$
ProximalPhalanxOAG	0.202$_{0.059}$	0.263$_{0.084}$	0.247$_{0.117}$	**0.180**$_{0.026}$	*0.191*$_{0.033}$
ProximalPhalanxTW	**0.212**$_{0.014}$	0.497$_{0.330}$	0.830$_{0.984}$	*0.271*$_{0.020}$	0.668$_{0.070}$
USASouthwestEnergyFlux	*0.401*$_{0.162}$	0.438$_{0.544}$	0.482$_{0.608}$	**0.313**$_{0.041}$	0.450$_{0.147}$
USASouthwestSWH	0.888$_{0.578}$	*0.545*$_{0.249}$	**0.483**$_{0.148}$	0.671$_{0.047}$	0.748$_{0.057}$
Rankings	*2.462*	3.923	3.769	**2.000**	2.846
Best (second best)	*4 (1)*	0 (3)	1 (1)	**5 (5)**	3 (3)

4.1   Statistical Analysis

Besides the analysis performed on the average results for the MAE metric, this Section analyses the results by means of rank tests considering all metrics. Thus, Fig. 2 shows the Critical Difference Diagram (CDD) [6] for each of the metrics. The significance value α is set to 0.1. The critical difference value is computed

pairwise and is equal to 0.456. As shown in the CDD, the proposed methodology O-GAMTF$_T$ obtained significantly better results than all the DL-based
approaches considered in terms of MAE metric. Also, it performs better than
O-GAMTF$_\beta$ for all the metrics except for the QWK. Finally, it is worth of
mentioning that the proposed O-GAMTF$_T$ methodology performs significantly
better than ResNet and FCN for all metrics.

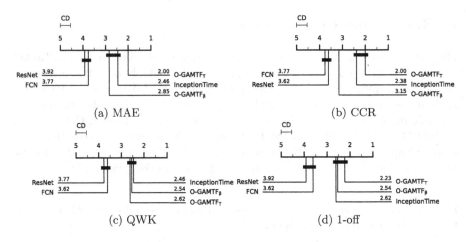

Fig. 2. Critical difference diagram for each metric.

5   Conclusions

This work proposed a novel approach for imaging ordinal time series and classifying the resulting images using a CNN model bearing in mind the order information of the problem. The process followed to create the images from time
series as well as the classifier employed to solve the ordinal classification problem associated with those images was explained. The construction of the images
was done through a channel by channel composition, aggregating the matrices
obtained from three different time series imaging methods from the literature:
Gramian Angular Summation Field (GASF), Gramian Angular Difference Field
(GADF) and Markov Transition Field (MTF). On the other hand, the proposed
ordinal classifier consists in a ResNet18 model, trained using a unimodal regularised loss function that encourages the model to respect the order between
labels defined in the problem. Such unimodal regularisation is achieved by using
soft labels instead of the standard crisp labels when computing the loss value. To
obtain the aforementioned soft labels, two different types of distributions have
been proposed: beta (O-GAMTF$_\beta$) and triangular (O-GAMTF$_T$) distributions.
    The proposed methodology has been tested using 13 different time series
ordinal datasets obtained from different sources, which were presented in the
paper. It has been compared against three TSC methodologies from the literature (ResNet, FCN, and InceptionTime) taking into account one nominal

and three ordinal metrics. The experimental results show that the proposed O-GAMTF$_T$ outperformed the rest of the methods from the literature, including InceptionTime. Furthermore, statistical tests showed that the proposed approach performed significantly better than all the other state-of-the-art methodologies for the MAE metric and significantly better than ResNet and FCN for all metrics.

Acknowledgements. This work has been partially subsidised by "Agencia Española de Investigación (España)" (grant ref.: PID2020-115454GB-C22 / AEI / 10.13039 / 501100011033). Víctor Manuel Vargas's research has been subsidised by the FPU Predoctoral Program of the Spanish Ministry of Science, Innovation and Universities (MCIU), grant reference FPU18/00358. David Guijo-Rubio's research has been subsidised by the University of Córdoba through grants to Public Universities for the requalification of the Spanish university system of the Ministry of Universities, financed by the European Union - NextGenerationEU (grant reference: UCOR01MS).

References

1. Ayllón-Gavilán, R., Guijo-Rubio, D., Gutiérrez, P.A., Hervás-Martínez, C.: Assessing the efficient market hypothesis for cryptocurrencies with high-frequency data using time series classification. In: García Bringas, P., et al (eds.) 17th International Conference on Soft Computing Models in Industrial and Environmental Applications (SOCO 2022). SOCO 2022. LNCS, vol. 531. Springer, Cham (2022). https:// doi.org/10.1007/978-3-031-18050-7_14
2. Bagnall, A., Lines, J., Bostrom, A., Large, J., Keogh, E.: The great time series classification bake off: a review and experimental evaluation of recent algorithmic advances. Data Min. Knowl. Disc. **31**(3), 606–660 (2017)
3. Barbero-Gómez, J., Gutiérrez, P.A., Vargas, V.M., Vallejo-Casas, J.A., Hervás-Martínez, C.: An ordinal CNN approach for the assessment of neurological damage in Parkinson's disease patients. Expert Syst. Appl. **182**, 115271 (2021)
4. Boonyakitanont, P., Lek-Uthai, A., Chomtho, K., Songsiri, J.: A review of feature extraction and performance evaluation in epileptic seizure detection using EEG. Biomed. Signal Process. Control **57**, 101702 (2020)
5. Cruz-Ramírez, M., Hervás-Martínez, C., Sánchez-Monedero, J., Gutiérrez, P.A.: Metrics to guide a multi-objective evolutionary algorithm for ordinal classification. Neurocomputing **135**, 21–31 (2014)
6. Demšar, J.: Statistical comparisons of classifiers over multiple data sets. J. Mach. Learn. Res. **7**, 1–30 (2006)
7. Foumani, N.M., Miller, L., Tan, C.W., Webb, G.I., Forestier, G., Salehi, M.: Deep learning for time series classification and extrinsic regression: a current survey. arXiv preprint arXiv:2302.02515 (2023)
8. Guijo-Rubio, D., Gutiérrez, P., Casanova-Mateo, C., Sanz-Justo, J., Salcedo-Sanz, S., Hervás-Martínez, C.: Prediction of low-visibility events due to fog using ordinal classification. Atmos. Res. **214**, 64–73 (2018)
9. Guijo-Rubio, D., Gutiérrez, P.A., Bagnall, A., Hervás-Martínez, C.: Time series ordinal classification via shapelets. In: 2020 International Joint Conference on Neural Networks (IJCNN), pp. 1–8. IEEE (2020)

10. Guijo-Rubio, D., Gutiérrez, P.A., Bagnall, A., Hervás-Martínez, C.: Ordinal versus nominal time series classification. In: Lemaire, V., Malinowski, S., Bagnall, A., Guyet, T., Tavenard, R., Ifrim, G. (eds.) AALTD 2020. LNCS (LNAI), vol. 12588, pp. 19–29. Springer, Cham (2020). https://doi.org/10.1007/978-3-030-65742-0_2

11. He, K., Zhang, X., Ren, S., Sun, J.: Deep residual learning for image recognition. In: Proceedings of the IEEE Conference on Computer Vision and Pattern Recognition, pp. 770–778 (2016)

12. Ismail Fawaz, H., Forestier, G., Weber, J., Idoumghar, L., Muller, P.-A.: Deep learning for time series classification: a review. Data Min. Knowl. Disc. **33**(4), 917–963 (2019). https://doi.org/10.1007/s10618-019-00619-1

13. Ismail Fawaz, H., et al.: InceptionTime: finding AlexNet for time series classification. Data Min. Knowl. Disc. **34**(6), 1936–1962 (2020)

14. Keogh, E., Chakrabarti, K., Pazzani, M., Mehrotra, S.: Dimensionality reduction for fast similarity search in large time series databases. Knowl. Inf. Syst. **3**, 263–286 (2001)

15. Kingma, D.P., Ba, J.: Adam: a method for stochastic optimization. arXiv preprint arXiv:1412.6980 (2014)

16. Liu, P., Sun, X., Han, Y., He, Z., Zhang, W., Wu, C.: Arrhythmia classification of LSTM autoencoder based on time series anomaly detection. Biomed. Signal Process. Control **71**, 103228 (2022)

17. Rosati, R., et al.: A novel deep ordinal classification approach for aesthetic quality control classification. Neural Comput. Appl. **34**(14), 11625–11639 (2022)

18. de la Torre, J., Puig, D., Valls, A.: Weighted kappa loss function for multi-class classification of ordinal data in deep learning. Pattern Recogn. Lett. **105**, 144–154 (2018)

19. Vargas, V.M., Gutiérrez, P.A., Barbero-Gómez, J., Hervás-Martínez, C.: Soft labelling based on triangular distributions for ordinal classification. Inf. Fusion **93**, 258–267 (2023)

20. Vargas, V.M., Gutiérrez, P.A., Hervás-Martínez, C.: Unimodal regularisation based on beta distribution for deep ordinal regression. Pattern Recogn. **122**, 108310 (2022)

21. Wang, Z., Oates, T.: Imaging time-series to improve classification and imputation. In: 24th International Joint Conference on Artificial Intelligence (IJCAI), pp. 3939–3945 (2015)

22. Wang, Z., Oates, T., et al.: Encoding time series as images for visual inspection and classification using tiled convolutional neural networks. In: Workshops at the Twenty-ninth AAAI Conference On Artificial Intelligence, vol. 1. AAAI Menlo Park, CA, USA (2015)

23. Wang, Z., Yan, W., Oates, T.: Time series classification from scratch with deep neural networks: a strong baseline. In: 2017 International Joint Conference on Neural Networks (IJCNN), pp. 1578–1585. IEEE (2017)

Ordinal Classification Approach for Donor-Recipient Matching in Liver Transplantation with Circulatory Death Donors

Marcos Rivera-Gavilán[1] , Víctor Manuel Vargas[1(✉)] ,
Pedro Antonio Gutiérrez[1] , Javier Briceño[2] , César Hervás-Martínez[1] ,
and David Guijo-Rubio[1,3]

[1] Department of Computer Sciences, Universidad de Córdoba, 14014 Córdoba, Spain
{i92rigam,vvargas,pagutierrez,chervas,dguijo}@uco.es
[2] Hospital Universitario Reina Sofía, Córdoba, Spain
[3] School of Computing Sciences, University of East Anglia, Norwich NR4 7TQ, UK

Abstract. This paper tackles the Donor-Recipient (D-R) matching for Liver Transplantation (LT). Typically, D-R matching is performed following the knowledge of a team of experts guided by the use of a prioritisation system. One of the most extended, the Model for End-stage Liver Disease (MELD), aims to decrease the mortality in the waiting list. However, it does not take into account the result of the transplant. In this sense, with the aim of developing a system able to bear in mind the survival benefit, we propose to treat the problem as an ordinal classification one. The organ survival will be predicted at four different thresholds. The results achieved demonstrate that ordinal classifiers are capable of outperforming nominal approaches in the state-of-the-art. Finally, this methodology can help experts make more informed decisions about the appropriateness of assigning a recipient for a specific donor, maximising the probability of post-transplant survival in LT.

Keywords: Donor-recipient matching · Liver transplantation · Ordinal classification · Ordinal Binary Decomposition

1 Introduction

Liver Transplantation (LT) is a highly effective treatment for patients with end-stage liver disease. However, the limited availability of suitable donors makes it

This work has been partially subsidised by "Agencia Española de Investigación (España)" (grant ref.: PID2020-115454GB-C22/AEI/10.13039/501100011033). Víctor Manuel Vargas's research has been subsidised by the FPU Predoctoral Program of the Spanish Ministry of Science, Innovation and Universities (MCIU) (grant reference: FPU18/00358). David Guijo-Rubio's research has been subsidised by the University of Córdoba through grants to Public Universities for the requalification of the Spanish university system of the Ministry of Universities, financed by the European Union - NextGenerationEU (grant reference: UCOR01MS).

I. Rojas et al. (Eds.): IWANN 2023, LNCS 14135, pp. 517–528, 2023.
https://doi.org/10.1007/978-3-031-43078-7_42

imperative to identify the optimal Donor-Recipient (D-R) pairing to maximise the benefits of the procedure. Despite the importance of this task, it remains a matter of controversy, and medical experts do not usually rely on Decision Support System (DSS) to aid in their selection process [26].

Several scoring systems have been developed for this purpose, each with different goals. For instance, BAlance of Risk (BAR) [8], Survival Output Following liver Transplantation (SOFT) [27] or Model for End-stage Liver Disease (MELD) are the most known. Specifically, MELD [15], one of the earliest systems, is primarily focused on reducing mortality in the waiting list, without taking into account the outcome of the transplant. Despite this, it remains the most widely used system. However, it is worth noting that reducing waiting list mortality is not necessarily associated with longer survival times after the LT procedure [3].

In the last years, there have been an explosion of interest in developing organ allocation systems improving the performance of MELD or other scores used in the field [10]. For instance, [6] proposes a novel method employing multi-objective evolutionary artificial neural networks. The two goals of the multi-objective methodology are optimising the probability of graft survival and, at the same time, minimising the probability of graft failure. [1] also validates the use of artificial neural networks as the basis of a rule-based system for facilitating the decision about the most appropriate D-R matching. [5] describes a multi-step, consensus-based approach with the goal of developing a "blended principle model" that incorporates both objective and subjective factors. This model aims to improve the fairness and efficiency of organ allocation, while also addressing concerns around issues such as geographic disparities and the prioritisation of sickest patients. [24] presents a system that takes into account the imbalanced nature of the data by using semi-supervised learning and considers two sources of unsupervised data to improve the model: recent transplants and virtual D-R pairs. Finally, [18] develops a score to predict the risk of dropout in LT candidates. The score is found to be effective in identifying high-risk candidates for organ allocation.

The objective of this study is to develop a methodology able to predict the graft survival time in LT given a D-R pair. This task involves classification and can be approached using either binary (two classes) or multi-class (more than two classes) classification techniques. To achieve this goal, we analyse a dataset of LT patients obtained from various hospitals in Spain. The dataset contains variables related to the donor, the recipient and the surgical procedure. Notably, the dataset is different with respect to those reported in the literature in the sense that it includes donors who suffered circulatory death. Controlled Donations after Circulatory Deaths (cDCDs) have increased significantly in recent years and now represent nearly 25% of all donors in Spain. CDCD donors carry additional risks compared to other donor types, such as extended criteria donors or donors after brain death. In particular, two different procedures are used for cDCDs: the Abdominal Normothermic Regional Perfusion (A-NRP) and the Super-Rapid Recovery (SRR) technique. However, the former is the more common procedure in Spain. Our study employs a dataset of 255 patients to develop a DSS for optimal D-R matching in LT [13].

The dataset used in this study involves four classes: 1) graft failure within the first month, 2) failure between one month and 12 months, 3) failure between one year and 3 years, and 4) no failure presented. These classes have been decided by medical experts. Notably, this problem can be addressed as a multi-class classification problem. However, the classes are ordered according to graft survival time, from shortest to longest. If standard multi-class techniques are employed, the ordinality of the classes is not taken into account, leading to suboptimal results. Instead, ordinal classifiers are well-suited for this type of data, as they can utilise the order information to improve classification accuracy.

Ordinal classification [12] have gained popularity in recent years. Examples of ordinal classification can be found in several areas, such as atmospheric event detection [11], age estimation [16], bank failure prediction [20] or shotgun stocks quality assessment [28]. It is designed to handle multi-class classification problems where the classes are ordered. Ordinal classifiers consider the order information of the classes, rather than treating them as independent categories, which can improve the accuracy of classification. They can also handle imbalanced datasets and are robust to noisy data, as is the case, given that the classes are not balanced (most of the grafts do not present failure). An example of this paradigm can be found in the age estimation field, where a person can be classified as *baby, child, teenager, adult* or *elderly* according to the age. Thus, misclassifying an adult as baby should be far more penalised than misclassifying as elderly. Note that standard multi-class approaches treats all misclassification equally.

The ordinal classification paradigm has been previously validated for standard D-R allocation in LT. For instance, [23] employs a hierarchical methodology that combines binary decomposition methods with ordinal support vector machines, demonstrating competitiveness in both ordinal and nominal metrics. However, in contrast to previous studies, the dataset used in our work introduces an additional level of complexity, as the donors are cDCD, making it difficult to accurately estimate the optimal D-R pair. To the best of our knowledge, our proposed methodology is the first to address cDCD D-R allocation in LT.

The remainder of the paper is organised as follows: Sect. 2 describes the LT dataset considered. Section 3 presents the ordinal methodology proposed for tackling the D-R matching in LT. Section 4 introduces the experimental settings used, as well as the methodologies used for validating the effectiveness of the ordinal classification. Section 5 shows the results achieved. Finally, Sect. 6 closes the paper with some final remarks, and suggests directions for future work.

2   Dataset Description

The dataset used in this study is derived from a cohort study that analysed all cDCD LTs performed in Spain from 2012 to 2019, with outcomes evaluated up to at least December 31, 2020. It is worth noting that cDCD was initiated in Spain in 2009 and was formally regulated in 2012. Therefore, this is the first dataset comprising cDCD-R matching in LT. From the two types of procedures used for cDCD, only those performed through the A-NRP, which is the most commonly used in Spain [13], were included in the analysis.

As previously stated in the Introduction Section, a total of 255 D-R pairs were considered in this study. Note that all censored graft survival has been excluded. Specifically, the class distribution (where the classes are specified by experts in the field) is as follows: 21 grafts failed within the first month (8.24%), 23 grafts failed between the first month and 12 months (9.02%), 25 grafts failed between the first year and 3 years (9.80%), and finally, 186 grafts presented no failure (72.94%). As can be observed, the dataset is very imbalanced, the majority of the patterns belonging to the last class. This fact highlights the effectiveness of this treatment for patients with end-stage liver disease.

The study collected several donor and recipient variables, including age, sex, body mass, height, and MELD score. In addition, variables related to the surgical procedure, such as cold ischaemia time and A-NRP duration, were also recorded. Table 1 presents the main variables included in the study.

3 Methodology

This section outlines the methodology used to address the problem presented in Sect. 2. As the problem is an ordinal classification problem, an ordinal methodology is employed to achieve accurate classification based on ordinal metrics. Firstly, a definition for an ordinal classification problem is provided. Subsequently, several Ordinal Binary Decomposition (OBD) strategies that have been previously proposed in the literature [12] are considered.

In machine learning, solving a classification problem with J classes is defined as the task of, given an input data $\mathbf{x} \in \mathcal{X} \subseteq \mathbb{R}^d$, predicting its label $y \in \mathcal{Y}$ where $\mathcal{Y} = \{\mathcal{C}_1, \mathcal{C}_2, ..., \mathcal{C}_J\}$. The input data can have an arbitrary number of dimensions d. The main distinguishing feature of ordinal problems from nominal problems is the fact that the classes in $\mathcal{Y}$ follow a natural order that is determined by the actual problem associated with the data set. In these terms, the order of class $\mathcal{C}_j$, which is the j-th class following the aforementioned order, can be defined as $\mathcal{O}(\mathcal{C}_j) = j$. Therefore, the order constraint can be expressed in two different ways: 1) $\mathcal{C}_1 \prec \mathcal{C}_2 \prec ... \prec \mathcal{C}_J$, where the $\prec$ operator determines that there is a precedence order between the elements but the distance between them is not known; or 2) $\mathcal{O}(\mathcal{C}_1) < \mathcal{O}(\mathcal{C}_2) < ... < \mathcal{O}(\mathcal{C}_J)$, using the order defined for each class. Note that, unlike regression problems, ordinal problems do not define a distance between categories and may have different distances between them. Taking into account the definition of an ordinal problem, an ordinal classifier should guarantee that the errors, if any, are located in the adjacent classes rather than committing errors in distant classes.

The authors of [12] categorised various OBD approaches that were previously proposed in the literature. These approaches aim to break down a multi-class problem into multiple binary problems that can be solved independently. However, the method of decomposition can differ from one approach to another. This methodology is not necessarily designed only for ordinal problems. However, by

Table 1: Set of variables included in this study. Variables collected during the surgical procedure are listed following the donor variables. Mean [Range] is provided for continuous variables, whereas % is used for binary variables.

Donor	
Age (y)	56 [48–66]
Sex male (%)	169 (66.27)
Body Mass Index	26.5 [24.2–28.4]
Intubation prior to WLST (days)	9 [4–10]
Antemortem canulation (%)	233 (91.37)
Withdrawal of life-sustaining Intensive Care Unit (%)	91 (35.68)
Total warm ischemia time (min)	20 [14–24]
Functional warm ischemia (min)	13 [9–16]
Asystolic warm ischemia time (min)	7 [5–7]
A-NRP duration (min)	111 [84–130]
Cold ischemia time (min)	331 [274–376]
Recipient	
Age (y)	57 [53–62]
Sex male (%)	197 (77.25)
Body Mass Index	27.3 [24.2–30.1]
Laboratory MELD score	13 [9–16]
Re-transplantation (%)	9 (3.52)
Outcome (Graft survival)	
Class 1 (≤ 1 month) (%)	21 (8.24)
Class 2 (1 month $<$ X $\leq$ 1 year) (%)	23 (9.02)
Class 3 (1 year $<$ X $\leq$ 3 years) (%)	25 (9.80)
Class 4 ($>$3 years) (%)	186 (72.94)

using a suitable decomposition matrix, the order constraint can be encoded in the labels.

In general terms, OBD methods can be divided into two wide groups: 1) multiple model approaches, which train a different model for each subproblem, and 2) multiple-output single models, which learn a single model to address all the subproblems using different outputs. In addition to this taxonomy based on the number of models trained, OBD algorithms can also be categorised based on the type of decomposition matrix used. Two naïve binary decomposition approaches, namely *OneVsAll* and *OneVsOne*, can be considered. However, since these approaches do not encode the order of the classes in the encoding of the subproblems, they are not suitable for ordinal classification and hence not considered in this work. There are four other approaches that are well-suited for ordinal classification, as they account for the order between categories when

Table 2: OBD matrices for an ordinal problem with four classes. Each column represents one subproblem and each row is associated with one class. The value of each cell represents the role of each class for each subproblem (+: positive class, −: negative class, ·: not considered in that problem).

Nominal decompositions

$OneVsAll$	$OneVsOne$

$$\begin{bmatrix} + & - & - & - \\ - & + & - & - \\ - & - & + & - \\ - & - & - & + \end{bmatrix} \qquad \begin{bmatrix} - & - & - & \cdot & \cdot & \cdot \\ + & \cdot & \cdot & - & - & \cdot \\ \cdot & + & \cdot & + & \cdot & - \\ \cdot & \cdot & + & \cdot & + & + \end{bmatrix}$$

Ordinal decompositions

$OrderedPartitions$	$OneVsNext$	$OneVsFollowers$	$OneVsPrevious$

$$\begin{bmatrix} - & - & - \\ + & - & - \\ + & + & - \\ + & + & + \end{bmatrix} \; \begin{bmatrix} - & \cdot & \cdot \\ + & - & \cdot \\ \cdot & + & - \\ \cdot & \cdot & + \end{bmatrix} \; \begin{bmatrix} - & \cdot & \cdot \\ + & - & \cdot \\ + & + & - \\ + & + & + \end{bmatrix} \; \begin{bmatrix} + & + & + \\ + & + & - \\ + & - & \cdot \\ - & \cdot & \cdot \end{bmatrix}$$

dividing the subtasks. These include *OrderedPartitions*, *OneVsNext*, *OneVsFollowers*, and *OneVsPrevious*. Table 2 provides a summary of the different types of binary decomposition matrices that can be employed for both nominal and ordinal scenarios.

In this work, only multiple models with single output approaches have been considered, as the linear classifier used to solve the problem is not appropriate for producing multiple outputs. Therefore, the following multiple models OBD methodologies have been employed:

1. The *OrderedPartitions* [9] approach solves the problem of determining whether the label of a given sample $\mathbf{x}$ is greater than or equal to $\mathcal{C}_j$, where j is a class label. Each subproblem can be considered independently, and then the predictions can be combined to obtain the predictions of the original problem. The authors of [9] defined the probabilities that a sample belongs to the true class in the different subproblems as $p_j = P(y \succ \mathcal{C}_j), j = 1, ..., J - 1$. Thus, the probability for each class can be obtained as follows:

$$P(y = \mathcal{C}_1|\mathbf{x}) \approx 1 - p_1, \quad P(y = \mathcal{C}_J|\mathbf{x}) \approx p_{J-1},$$
$$P(y = \mathcal{C}_j|\mathbf{x}) \approx p_{j-1} - p_j, \quad 2 \le j \le J - 1. \tag{1}$$

Since there are no constraints that guarantee that $p_{j-1} < p_j$, the probabilities obtained with this method can be negative. However, in cases where it is not necessary to obtain adequate probability estimates, the prediction can be obtained by simply taking the maximum [29].

2. *OneVsPrevious* [30] involves creating several subproblems with the goal of distinguishing between each class and the previous one. To achieve this, the

approach considers $J - 1$ subproblems in such a way that subproblem j separates classes $\mathcal{C}_1 \cup \ldots \cup \mathcal{C}_{J-j-1}$ from class $\mathcal{C}_{J-j}$.

3. *OneVsNext* [17] considers that each binary problem j distinguishes class $\mathcal{C}_j$ from class $\mathcal{C}_{j+1}$.

4. *OneVsFollowers* [17] proposes several subproblems where the goal is to separate class $\mathcal{C}_j$ from classes $\mathcal{C}_{j+1} \cup \ldots \cup \mathcal{C}_J$. In this case, the final predictions are obtained in the same way described for the *OneVsNext*.

The ultimate model employed to address the initial ordinal problem is a LogisticRegression ensemble model that resolves subtasks using one of the decomposition techniques discussed in this section. All models in the ensemble share similar characteristics and use the same parameters.

4   Experimental Settings

This section presents the experimental setup conducted to compare the proposed ordinal methodologies with other state-of-the-art machine learning approaches. The section begins by describing the compared methodologies. It then proceeds to discuss the performance metrics used to evaluate these approaches. Finally, the experimental procedure employed to train and evaluate the models is presented.

4.1   Compared Methodologies

Four different variants of the OBD methodology described in Sect. 3 are considered: *OrderedPartitions* [9], *OneVsPrevious* [30], *OneVsNext* [17], and *OneVs-Followers* [17].

To compare the performance of the proposed ordinal methodologies against nominal methodologies for solving the proposed ordinal problem, several nominal machine learning approaches from the literature are considered. Thus, the following algorithms are included in the experiments in addition to the OBD methodologies: Logistic Regression (LR), which is a simple linear model for nominal classification, Decision Tree (DT) [19], Multilayer Perceptron (MLP) [14], ordinal Logistic regression with the All-Threshold variant (LAT) [22] Ridge Classifier (RC) [21], Random Forest (RF) [2], and Support Vector Classification (SVC) [4].

4.2   Performance Metrics

In order to evaluate the performance of each of the ordinal and nominal models considered in this paper, different metrics have been considered, including two nominal classification metrics and one metric that aims to evaluate the ordinal performance, known as Mean Absolute Error (MAE). These metrics are defined as:

1. The Mean Absolute Error (MAE) is a metric that quantifies the average deviation between the predicted and target classes, expressed as the number of categories separating them. It is a widely adopted evaluation measure for ordinal classification tasks [7].

$$\text{MAE} = \frac{1}{N} \sum_{i=1}^{N} |\mathcal{O}(y_i) - \mathcal{O}(\hat{y_i})|, \tag{2}$$

where N is the total number of samples, $y_i \in \{\mathcal{C}_1, \mathcal{C}_2, ..., \mathcal{C}_J\}$ is the target label of i-th sample and $\hat{y_i} \in \{\mathcal{C}_1, \mathcal{C}_2, ..., \mathcal{C}_J\}$ is its predicted label.

2. Correct Classification Rate (CCR), also known as Accuracy, represents the proportion of correctly classified samples out of the total number of samples. It is the most widely used evaluation metric in classification tasks.
3. The Area Under the Curve (AUC) is a performance metric that assesses the capability of a model to differentiate between positive and negative samples. It measures the area under the Receiver Operating Characteristic (ROC) curve, which plots the True Positive Rate (TPR) versus the False Positive Rate (FPR) for varying classification thresholds. The values of the AUC range between 0 and 1, with higher values for superior performance.

4.3 Experimental Procedure

In this section, we explain the process used to train and evaluate our models. Since the dataset we used did not include pre-determined train and test partitions, we employed a 3-fold strategy to evaluate our models using different exhaustive partitions. Furthermore, for models that required random initialisation, we ran 10 different seeds, resulting in a total of 30 executions for non-deterministic models and 3 executions for deterministic models. This approach helped us obtain more reliable results.

In order to obtain the best performance for some methodologies, parameter tuning is required. To do this, we employ a grid search using a 3-fold strategy to create three different validation sets. For each partition, we train the model with the training set and compute the MAE on the validation set to evaluate its performance. The values specified in Table 3 are used during hyperparameter tuning. The entire experimental procedure described in this section is summarised in Fig. 1.

5   Results

This Section presents the results of the experiments described in Sect. 4. As shown in Table 4, the OBD *OneVsFollowers* methodology achieved the best results for two out of the three considered metrics (CCR, and AUC). Additionally, it obtained the second-best performance for the MAE metric. Meanwhile, the OBD *OrderedPartitions* approach had the best performance regarding the MAE metric and the second-best for the CCR and AUC metrics. Taking all of this into account, we can conclude that the OBD *OneVsFollowers* methodology is the best alternative. Furthermore, it is worth noting that all of the ordinal methodologies outperformed the nominal alternatives considered in this study.

Table 3: Crossvalidated parameters for each of the methodologies.

Method	Parameters
LR	$C \in \{0.001, 0.01, 0.1, 1, 10, 100, 1000\}$
	$max\_iter \in \{100, 1000, 1500, 2000, 2500\}$
RF	$n\_estimators \in \{10, 50, 100, 150\}$
	$max\_depth \in \{3, 4, 5, 6, None\}$
	$min\_samples\_split \in \{2, 3, 4\}$
MLP	$hidden\_layer\_sizes \in \{10, 25, 50, 100\}$
	$\alpha \in [10^{-4}, 10^4]$
	$max\_iter \in \{100, 1000, 1500, 2000, 2500\}$
	$learning\_rate\_init \in \{10^{-5}, 10^{-4}, 10^{-3}\}$
RC	$\alpha \in [10^{-4}, 10^4]$
	$max\_iter \in \{100, 1000, 1500, 2000, 2500\}$
SVC	$C \in \{10^{-3}, 10^{-2}, 10^{-1}, 1, 10, 100, 1000\}$
DT	$\alpha \in [10^{-4}, 10^4]$
LAT	$\alpha \in [10^{-4}, 10^4]$
	$max\_iter \in \{100, 1000, 1500, 2000, 2500\}$

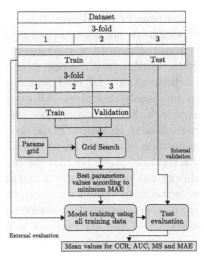

Fig. 1: Experimental procedure followed to train the model, crossvalidate the parameters and evaluate the performance on the test set.

Table 4: Mean results and standard deviation on the test set. Best and second-best results for each metric are highlighted in bold and italics, respectively.

Methodology	CCR ($\uparrow$)	AUC ($\uparrow$)	MAE ($\downarrow$)
LR	$0.439_{0.087}$	$0.623_{0.018}$	$1.082_{0.150}$
DT	$0.618_{0.063}$	$0.569_{0.044}$	$0.762_{0.118}$
MLP	$0.700_{0.116}$	$0.615_{0.065}$	$0.622_{0.352}$
LAT	$0.702_{0.048}$	$0.638_{0.051}$	$0.576_{0.101}$
RC	$0.392_{0.116}$	$0.643_{0.013}$	$1.192_{0.263}$
RF	$0.646_{0.073}$	$0.550_{0.054}$	$0.688_{0.132}$
SVC	$0.380_{0.279}$	$0.451_{0.041}$	$1.192_{0.427}$
OBD *OrderedPartitions*	$0.725_{0.030}$	$0.656_{0.052}$	$\mathbf{0.533_{0.095}}$
OBD *OneVsNext*	$0.608_{0.065}$	$0.594_{0.011}$	$0.761_{0.114}$
OBD *OneVsFollowers*	$\mathbf{0.733_{0.036}}$	$\mathbf{0.662_{0.043}}$	$0.545_{0.083}$
OBD *OneVsPrevious*	$0.643_{0.024}$	$0.62_{0.041}$	$0.71_{0.078}$

6   Conclusions

This paper addresses the prediction of graft survival in the context of LT for D-R matching with cDCD donors. The aim of this strategy is to reduce the waiting time in the list by mitigating the imbalance between donors and recipients. In particular, this study focuses on a dataset comprising all transplants performed using grafts from cDCD donors. The objective is to estimate the time of graft survival, which has been categorised into four classes based on the survival time, ranging from the shortest to the longest. In this way, the use of ordinal classifiers is straightforward, as it is the best-performing methodology for this type of data.

Four different ordinal methodologies have been considered to address the problem previously described. All of these methodologies are variants of the OBD approach, which decomposes the original ordinal classification problem into several binary problems. These problems can be independently solved using a decomposition matrix. The results of these subproblems are then combined to obtain the final prediction for the original problem. The performance of this methodology has been evaluated using different matrices. To compare with other machine learning methodologies from the literature, seven nominal methodologies were used conducting the same experiments.

The results of the experiments were analysed using three different nominal and ordinal classification metrics. On average, the OBD methodology with the *OneVsFollowers* decomposition strategy outperformed all other alternatives considered in this study with respect to all metrics. This indicates that binary decomposition and ordinal methods, in general, perform better on ordinal and imbalanced problems. Therefore, the proposed methodology can be integrated into a DSS to assist field experts in making decisions about D-R matching. In future work, to further improve the results obtained in this paper, ordinal oversampling techniques [25] can be applied to increase the overall number of samples and reduce the imbalance degree, given that the number of samples in the available dataset is quite limited.

References

1. Ayllón, M.D., et al.: Validation of artificial neural networks as a methodology for donor-recipient matching for liver transplantation. Liver Transpl. **24**(2), 192–203 (2018)
2. Breiman, L.: Random forests. Mach. Learn. **45**, 5–32 (2001)
3. Briceño, J., Cruz-Ramírez, M., Prieto, M., et al.: Use of artificial intelligence as an innovative donor-recipient matching model for liver transplantation: results from a multicenter Spanish study. J. Hepatol. **61**(5), 1020–1028 (2014)
4. Chang, C.C., Lin, C.J.: LibSVM: a library for support vector machines. ACM Trans. Intell. Syst. Technol. (TIST) **2**(3), 1–27 (2011)
5. Cillo, U., et al.: A multistep, consensus-based approach to organ allocation in liver transplantation: toward a "blended principle model". Am. J. Transplant. **15**(10), 2552–2561 (2015)
6. Cruz-Ramirez, M., Hervas-Martinez, C., Fernandez, J.C., Briceno, J., De La Mata, M.: Predicting patient survival after liver transplantation using evolutionary multi-objective artificial neural networks. Artif. Intell. Med. **58**(1), 37–49 (2013)
7. Cruz-Ramírez, M., Hervás-Martínez, C., Sánchez-Monedero, J., Gutiérrez, P.A.: Metrics to guide a multi-objective evolutionary algorithm for ordinal classification. Neurocomputing **135**, 21–31 (2014)
8. Dutkowski, P., et al.: Are there better guidelines for allocation in liver transplantation?: A novel score targeting justice and utility in the model for end-stage liver disease era. Ann. Surg. **254**(5), 745–754 (2011)

9. Frank, E., Hall, M.: A simple approach to ordinal classification. In: De Raedt, L., Flach, P. (eds.) ECML 2001. LNCS (LNAI), vol. 2167, pp. 145–156. Springer, Heidelberg (2001). https://doi.org/10.1007/3-540-44795-4_13
10. Ge, J., Kim, W.R., Lai, J.C., Kwong, A.J.: "beyond meld"–emerging strategies and technologies for improving mortality prediction, organ allocation and outcomes in liver transplantation. J. Hepatol. **76**(6), 1318–1329 (2022)
11. Guijo-Rubio, D., et al.: Ordinal regression algorithms for the analysis of convective situations over Madrid-Barajas airport. Atmos. Res. **236**, 104798 (2020)
12. Gutiérrez, P.A., Perez-Ortiz, M., Sanchez-Monedero, J., Fernandez-Navarro, F., Hervas-Martinez, C.: Ordinal regression methods: survey and experimental study. IEEE Trans. Knowl. Data Eng. **28**(1), 127–146 (2015)
13. Hessheimer, A.J., et al.: Abdominal normothermic regional perfusion in controlled donation after circulatory determination of death liver transplantation: outcomes and risk factors for graft loss. Am. J. Transplant. **22**(4), 1169–1181 (2022)
14. Hinton, G.E.: Connectionist learning procedures. In: Machine Learning, pp. 555–610. Elsevier, Amsterdam (1990)
15. Kamath, P.S., Kim, W.R.: The model for end-stage liver disease (MELD). Hepatology **45**(3), 797–805 (2007)
16. Kong, C., Wang, H., Luo, Q., Mao, R., Chen, G.: Deep multi-input multi-stream ordinal model for age estimation: Based on spatial attention learning. Futur. Gener. Comput. Syst. **140**, 173–184 (2023)
17. Kwon, Y.S., Han, I., Lee, K.C.: Ordinal pairwise partitioning (OPP) approach to neural networks training in bond rating. Intell. Syst. Account. Finance Manag. **6**(1), 23–40 (1997)
18. Lai, Q., et al.: Sarco-model: a score to predict the dropout risk in the perspective of organ allocation in patients awaiting liver transplantation. Liver Int. **41**(7), 1629–1640 (2021)
19. Loh, W.Y.: Classification and regression trees. Wiley Interdisc. Rev. Data Min. Knowl. Discov. **1**(1), 14–23 (2011)
20. Manthoulis, G., Doumpos, M., Zopounidis, C., Galariotis, E.: An ordinal classification framework for bank failure prediction: methodology and empirical evidence for us banks. Eur. J. Oper. Res. **282**(2), 786–801 (2020)
21. McDonald, G.C.: Ridge regression. Wiley Interdisc. Rev. Comput. Stat. **1**(1), 93–100 (2009)
22. Pedregosa, F., Bach, F., Gramfort, A.: On the consistency of ordinal regression methods. J. Mach. Learn. Res. **18**(1), 1769–1803 (2017)
23. Pérez-Ortiz, M., Cruz-Ramírez, M., Ayllón-Terán, M.D., Heaton, N., Ciria, R., Hervás-Martínez, C.: An organ allocation system for liver transplantation based on ordinal regression. Appl. Soft Comput. **14**, 88–98 (2014)
24. Pérez-Ortiz, M., et al.: Synthetic semi-supervised learning in imbalanced domains: constructing a model for donor-recipient matching in liver transplantation. Knowl.-Based Syst. **123**, 75–87 (2017)
25. Pérez-Ortiz, M., Gutiérrez, P.A., Carbonero-Ruz, M., Hervás-Martínez, C.: Semi-supervised learning for ordinal kernel discriminant analysis. Neural Netw. **84**, 57–66 (2016)
26. Polyak, A., Kuo, A., Sundaram, V.: Evolution of liver transplant organ allocation policy: current limitations and future directions. World J. Hepatol. **13**(8), 830–839 (2021)
27. Rana, A., et al.: Survival outcomes following liver transplantation (soft) score: a novel method to predict patient survival following liver transplantation. Am. J. Transplant. **8**(12), 2537–2546 (2008)

28. Vargas, V.M., et al.: Exponential loss regularisation for encouraging ordinal constraint to shotgun stocks quality assessment. Appl. Soft Comput. 110191 (2023)
29. Waegeman, W., Boullart, L., et al.: An ensemble of weighted support vector machines for ordinal regression. Int. J. Comput. Syst. Sci. Eng. **3**(1), 47–51 (2009)
30. Wu, H., Lu, H., Ma, S.: A practical SVM-based algorithm for ordinal regression in image retrieval. In: Proceedings of the Eleventh ACM International Conference on Multimedia, pp. 612–621 (2003)

Evaluating the Performance of Explanation Methods on Ordinal Regression CNN Models

Javier Barbero-Gómez[1,3(✉)] [ID], Ricardo Cruz[2,3] [ID], Jaime S. Cardoso[2,3] [ID],
Pedro A. Gutiérrez[1,3] [ID], and César Hervás-Martínez[1,3] [ID]

[1] Department of Computer Science and Numerical Analysis,
University of Córdoba, Córdoba, Spain
{jbarbero,pagutierrez,chervas}@uco.es
[2] INESC TEC, Porto, Portugal
[3] Faculty of Engineering, University of Porto, Porto, Portugal
{rpcruz,jaime.cardoso}@fe.up.pt

Abstract. This paper introduces an evaluation procedure to validate the efficacy of explanation methods for Convolutional Neural Network (CNN) models in ordinal regression tasks. Two ordinal methods are contrasted against a baseline using cross-entropy, across four datasets. A statistical analysis demonstrates that attribution methods, such as Grad-CAM and IBA, perform significantly better when used with ordinal regression CNN models compared to a baseline approach in most ordinal and nominal metrics. The study suggests that incorporating ordinal information into the attribution map construction process may improve the explanations further.

Keywords: Convolutional Neural Networks · Interpretability · Ordinal Regression

1 Introduction

CNN models have become increasingly prevalent in a wide range of applications in image classification, from object detection to face recognition to medical diagnosis. However, one major challenge with these models is their lack of interpretability: it can be difficult to understand how a model arrives at its predictions, making it challenging to debug, validate, or audit its behaviour. To address this problem, researchers have developed a variety of explanation methods that attempt to provide insight into the inner workings of CNN models by providing a visualization of the relevant regions of the input image.

This work has been partially subsidised by "Agencia Española de Investigación" (Spain) (grant ref.: PID2020-115454GB-C22/AEI/10.13039/501100011033). Javier Barbero-Gómez's research has been subsidised by the FPI Predoctoral Program of the Spanish Ministry of Science, Innovation and Universities (MCIU) [grant reference PRE2018-085659].

I. Rojas et al. (Eds.): IWANN 2023, LNCS 14135, pp. 529–540, 2023.
https://doi.org/10.1007/978-3-031-43078-7_43

This line of work started with simple occlusion analysis [22] but has evolved into more sophisticated techniques that use backpropagation of gradients to obtain saliency maps [1,13,18,20]. Those gradients can also be combined with layer activations at different depths of the network to produce Class Activation Maps (CAM) [17,23].

However, existing work in this field fails to address the validity of these methods when applied to ordinal regression tasks. These tasks are a specific case of classification where an order relationship exists between the class labels. In this context, regular performance metrics are no longer as relevant and a new set of metrics is needed to address the ordinality of the classes.

In this work, we propose an evaluation procedure to assess the validity of explanation methods when applied to ordinal regression CNN models. Furthermore, we use this procedure to compare the ordinal performance of evaluation methods when used on models that exploit ordinal information as opposed to models that do not.

This work is structured as follows: in Sect. 2 the framework of ordinal regression is described and two different ordinal regression approaches are presented. In Sect. 3 two different CNN attribution methods are shown and in Sect. 4 a new scheme for the evaluation of their performance is discussed. Then, in Sect. 5 a set of experiments is proposed to test our hypothesis and, in Sect. 6, the results of these experiments are presented. Finally, in Sect. 7 conclusions are drawn from the results and we propose some future work.

2   Ordinal Regression

A classification task consists on assigning a label y to an input vector $\mathbf{x}$, where $\mathbf{x} \in \mathcal{X} \subseteq \mathbb{R}^K$ and $y \in \mathcal{Y} = \{\mathcal{C}_1, \mathcal{C}_2, \ldots, \mathcal{C}_Q\}$, i.e., $\mathbf{x}$ is a vector in K dimensions and y is a class label in a finite set of Q categories. A classification model is a mapping $r \colon \mathcal{X} \to \mathcal{Y}$ that tries to predict the labels of new patterns given a training dataset $D = \{(\mathbf{x}_i, y_i) \mid \mathbf{x}_i \in \mathcal{X},\, y_i \in \mathcal{Y},\, i \in \{1, \ldots, N\}\}$.

Additionally, in the ordinal regression framework, a specific ordering relation $\prec$ exists between the class labels: $\mathcal{C}_1 \prec \mathcal{C}_2 \prec \cdots \prec \mathcal{C}_Q$. This is similar to traditional regression tasks, where $y \in \mathbb{R}$, and real values can be ordered by the $<$ operator, but, in this case, the labels are discrete [8].

2.1   Ordinal Binary Decomposition (OBD)

One approach to adapt CNNs to ordinal outputs was presented in [2] and consists on decomposing the original Q-class ordinal problem into $Q - 1$ binary decision problems. This technique is known as Ordinal Binary Decomposition (OBD). Each q problem consists of deciding if label $y \succ \mathcal{C}_q$ conditioned to sample $\mathbf{x}$ $(1 \leq q < Q)$ (this is referred to as the "Ordered partitions" scheme in [8]).

Following this idea, the output of the model is set to $Q - 1$ neurons with sigmoid activation. Each of the $Q - 1$ outputs of the model o_q is trying to predict the probability $\mathrm{P}(y \succ \mathcal{C}_q \mid \mathbf{x})$.

The OBD model outputs cumulative probabilities and requires combining several outputs for the decision rule. Ideal output vectors for each class are considered based on the Error-Correcting Output Codes (ECOC) framework to circumvent the problem of inconsistent probabilities. The decision rule is based on determining the ideal vector that minimizes the distance to the obtained output vector $\mathbf{o}$ using the L_2 norm as the distance metric:

$$\hat{y}_i = \underset{\mathcal{C}_1 \preceq \mathcal{C}_q \preceq \mathcal{C}_Q}{\arg\min} \; \|\mathbf{o} - \mathbf{v}(\mathcal{C}_q)\|_2 , \tag{1}$$

where $\mathbf{v}(\mathcal{C}_q)$ is the ideal output vector for class q.

To align with this, categorical cross-entropy is substituted by the Squared Error loss because it copes better with the distance function used for the ECOC decision:

$$\mathcal{L}(\mathbf{x}_i) = \sum_{k=1}^{Q-1} (1\{y_i \succ \mathcal{C}_k\} - \mathrm{P}(y_i \succ \mathcal{C}_k \mid \mathbf{x}_i))^2. \tag{2}$$

2.2 Unimodal Distribution Output

For a class $\mathcal{C}_\ell$ for which the model outputs the highest posterior probability, $\ell = \arg_k \max \mathrm{P}(y = \mathcal{C}_k \mid \mathbf{x})$, it would be expected for the second highest posterior probability to be either $\mathrm{P}(y = \mathcal{C}_{\ell-1} \mid \mathbf{x})$ or $\mathrm{P}(y = \mathcal{C}_{\ell+1} \mid \mathbf{x})$, given the previously defined class order, $\mathcal{C}_{\ell-1} \prec \mathcal{C}_\ell \prec \mathcal{C}_{\ell+1}$. In fact, it is common for ordinal regression metrics to more strongly penalize errors when the distance between the predicted and the true class is farther apart than when it is closer.

The previous reasoning also applies to the ranking order from all the subsequent posterior probabilities. Motivated by that, the authors of [15] propose that the probability distribution produced by the model must necessarily be unimodal. A parametric approach to ensure this is to restrict the output of the model to a discrete probability distribution, such as the Binomial or Poisson distribution.

For the Binomial distribution, $\mathcal{B}(n, p)$, the support of the distribution is known, $n = Q-1$, and the only parameter left to be estimated is the shape of the success probability distribution, p. The model outputs a single output, the shape of the distribution, which is then converted into posterior probabilities using the Binomial probability mass function, $\mathrm{P}(y = \mathcal{C}_q \mid \mathbf{x}) = \binom{n}{q-1} p^{q-1}(1-p)^{n-q-1}$, for each $q \in \{1, \ldots, Q\}$, where p is the output produced by the model. This may be seen as an activation function. To train the model, a common loss such as cross-entropy may then be used.

3 Visualizing the Decisions of a CNN

The goal of CNN explanation methods is to provide the user with a map of which regions of the image are relevant in the decision taken by the model. For a given image of height H and width W, the output of these methods is a matrix

$E \in \mathbb{R}^{H \times W}$ with values between 0 and 1 representing the relevance of each pixel of the original image in the classification decision (0 meaning no relevance and 1 maximum relevance).

Numerous methods have already been proposed. The authors of [7] present a general taxonomy based on the principles they use to generate their explanation: perturbation-based methods, propagation-based methods and activation-based methods. For our experiments, we choose to examine a perturbation-based method as well as an activation-based method. Propagation-based methods have been excluded due to their noisiness and poor performance [10].

3.1 IBA

The Information Bottleneck for Attribution (IBA) method, as proposed by [16], is a perturbation-based method. However, unlike other perturbation methods that alter the information at the input of the model, it consists on injecting a perturbation amidst its information flow, creating a bottleneck in the network. This bottleneck helps evaluate the impact in the output of the regions from the input image.

To achieve this, it introduces a new random variable Z by maximizing the amount of information it shares with the output y while minimizing the information it shares with the model input $\mathbf{x}$:

$$\max\ I[y; Z] - \beta I[\mathbf{x}; Z], \tag{3}$$

where I denotes the mutual information and β controls the trade-off between predicting the labels well and using little information of the input. $Z \in \mathbb{R}^{h \times w}$ acts as a substitute for the output of one of the intermediate layers $R \in \mathbb{R}^{h \times w}$ adding a certain noise $\epsilon \in \mathbb{R}^{h \times w}$:

$$Z = \lambda(\mathbf{x})R + (1 - \lambda(\mathbf{x}))\epsilon, \tag{4}$$

where $\lambda(\mathbf{x}) \in [0, 1]^{h \times w}$ adjusts how much of the original signal is passed along (see Fig. 1). A loss function $\mathcal{L}$ aligned with the objective posed in Eq. (3) is designed, and $\lambda(\mathbf{x})$, parametrised as $\lambda(\mathbf{x}) = \sigma(\alpha(\mathbf{x}))$ (where $\alpha \in \mathbb{R}^{h \times w}$ and σ is the sigmoid function), is optimized for $\mathcal{L}$ using the Adam backpropagation method [11]. Regions of the image with relevant information will present a λ value close to 1 and, conversely irrelevant parts will present a value close to 0. For this reason, the output explanation map E is just λ upsampled to the original input size.

3.2 Grad-CAM

Grad-CAM is an activation-based method proposed by [17]. It works by using the information flowing into any of the convolutional layers in a CNN to assign an importance level to each region of the original image, based on the principle that convolutional layers preserve location information before it is lost in fully-connected layers.

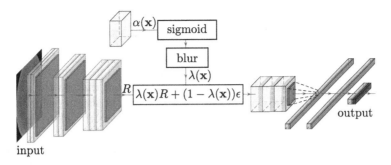

Fig. 1. IBA adds an information bottleneck in the middle of the network, where λ adjusts how much information is allowed to pass

First, it computes the gradient of the score for the target class c, y^c, with respect to the activations of each feature map k of a convolutional layer, A^k, namely $\frac{\partial y^c}{\partial A^k}$. This is then averaged over the whole width W_A and height H_A of the activation to obtain the weights α_k^c:

$$\alpha_k^c = \frac{1}{W_A \times H_A} \sum_i^{H_A} \sum_j^{W_A} \frac{\partial y^c}{\partial A_{ij}^k}. \tag{5}$$

These represent the importance of each feature map k on the final decision of the network. Thus, the explanation map can be computed as a linear combination of the activation feature maps A^k based on their importance α_k^c followed by the application of the Rectified Linear Unit ($ReLU$) to eliminate negative interactions:

$$E = ReLU \left(\sum_k \alpha_k^c A^k \right). \tag{6}$$

4 Evaluating the Performance of Explanations

A popular approach to evaluate the real impact of a region of an image in the classification decision is performing a perturbation analysis, that is, occluding parts of the input images and observing the change in the model outputs. It is expected that, if the occluded parts of the input $\mathbf{x}_i$ are relevant according to an explanation E_i, the classification performance should drop in a specific way.

One way to implement this idea presented in [3] is simply multiplying the input $\mathbf{x}_i$ by the explanation E_i in order to obtain an occluded image $\tilde{\mathbf{x}}_i$:

$$\tilde{\mathbf{x}}_i = \mathbf{x}_i \circ E_i, \tag{7}$$

where $\circ$ represents the element-wise multiplication. We can then look at the average effect on the score of the target class C, $f_C(\mathbf{x})$, for each example where the score drops. This is called the average drop:

$$\text{Average drop} = \frac{1}{N} \sum_{i}^{N} \frac{\max(0, f_C(\mathbf{x}_i) - f_C(\tilde{\mathbf{x}}_i))}{f_C(\mathbf{x}_i)}. \tag{8}$$

It is expected that if the explanation is good, the drop in score should be as small as possible. Thus, this is a minimizing feature. However, this metric ignores the ordering information in the class labels. That is, if confidence drops for the target class, it is preferable that it in turn increases for the nearby classes instead of for distant classes, but this is not evaluated by the average drop.

A different approach is taken by [16]: the explanation map is divided into tiles (e.g. 8×8) and the tiles are ranked according to the total sum of relevance within them. The input image is then occluded tile by tile, starting from the most relevant to the least relevant, and the target class score is plotted against the degradation level of the image (the number of occluded tiles) to obtain the Most Relevant Features (MoRF) curve. For a relevant explanation map, it is expected that the score should drop sharply at the beginning, when the most relevant parts of the image are being occluded. The same procedure is then repeated in reverse order of relevance to obtain the Least Relevant Features (LeRF) curve. The inverse is expected of this curve: the score should not drop dramatically until the most relevant parts are occluded. The extremes of these curves can then be normalized between 0 and 1 and the signed area between them can be computed[1]. A large area between the MoRF and LeRF curves is expected for a relevant explanation map. An example is shown in Fig. 2.

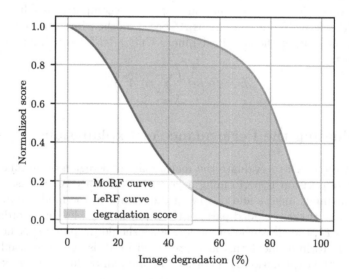

Fig. 2. Example of the MoRF and LeRF curves and the area between them or "degradation score"

---

[1] Note that the area contribution can be negative for those cases where the performance of the LeRF curve is higher than that of the MoRF curve.

This procedure presents an advantage: the behaviour of any metric can be studied in this fashion, not only the target score. Thus, we propose studying the degradation of ordinal metrics such as Mean Absolute Error (MAE), Quadratic Weighted Kappa (κ) and Spearman's rank correlation coefficient (ρ) [4].

5   Experiment Design

5.1   Model and Training

A base ResNet34 CNN model pre-trained on ImageNet-1K [5] is trained on 4 different ordinal regression datasets. The base model is used as-is as a baseline categorical classification procedure. Two other adaptations of the same model are tested: an OBD model using ECOC as described in Sect. 2.1 and a unimodal binomial output activation function paired with the cross-entropy loss function as described in Sect. 2.2.

Each model is trained using a different random initialization and a different 80/10/10 training/validation/testing split 60 different times using the Adam backpropagation method [11]. Training is done in batches of 64 samples for a maximum of 200 epochs, stopping early when the loss on the validation set does not improve for 20 epochs. Two different explanation methods are then applied: GradCAM [17] and IBA [16]. Degradation of the following metrics are obtained as outlined in Sect. 4: Correct Classification Rate (CCR), Average Area under the ROC curve ($AvAUC$), MAE, κ and ρ.

5.2   Data

The following 4 image datasets are selected to test the performance of the explanation methods. They were chosen because they all present an ordinal prediction label and cover a large spectrum of different classification tasks. In all cases, the original images were resized to a resolution of 224×224 and normalized to ImageNet's original mean and standard deviation per input channel.

Herlev Pap Smear Dataset. 917 images of single Pap smear cells classified by doctors and technicians into 7 different classes, 3 of them normal from different parts of the cervix (242 images in total) and 4 of them abnormal in different stages of dysplasia (675 images in total) [9]. These are condensed into 4 ordinal classes, following the Bethesda System standard [14].

Diabetic Retinopathy Dataset.[2] A collection of 53 569 high-resolution retina images, rated by a clinician on the presence of Diabetic Retinopathy (DR), an eye disease present in a large proportion of diabetes patients, on a scale from 0 (no DR) to 4 (proliferative DR) for a total of 5 classes.

---

[2] https://www.kaggle.com/c/diabetic-retinopathy-detection/data.

Adience Dataset. A set of 17 702 photos of people scraped from the web and pre-aligned to fit their face, categorized into 8 different age groups [6] of increasing value: 0 to 2 years, 4 to 6 years, 8 to 13 years, 15 to 20 years, 25 to 32 years, 38 to 43 years, 48 to 53 years, and 60 years and up.

Curated Breast Imaging Subset of Digital Database for Screening Mammography (CBIS-DDSM). A database of 2620 scanned film mammography studies curated from the larger DDSM and each one assigned a Breast Imaging Reporting and Data System (BI-RADS) assessment [12] by a trained mammographer. The assessment is done on a scale of 0 to 5 according to the standard for a total of 6 classes (there are no cases of class 6 as there is no biopsy information). For this dataset, before being resized, all images were cropped into a square centered around the Region of Interest of the lesion.

6   Results

The averages over the 60 runs of all the experiments for each dataset are shown in Tables 1, 2, 3 and 4. Some examples of the generated attributions can be seen in Fig. 3.

Table 1. Mean metric degradation score for the Herlev dataset. The best result for each explanation method is highlighted in bold

		CCR	$AvAUC$	MAE	κ	ρ
Grad-CAM	Baseline	0.1972	0.2048	0.1621	**0.1518**	**0.1597**
	Binomial	0.1555	0.1561	0.1735	0.1309	0.1383
	OBD	**0.2090**	**0.2543**	**0.1805**	0.1372	0.1464
IBA	Baseline	0.1365	0.1265	0.1328	0.1351	0.1234
	Binomial	**0.1463**	**0.1401**	**0.1577**	**0.1573**	**0.1448**
	OBD	0.1315	0.1067	0.1292	0.1141	0.1036

Table 2. Mean metric degradation score for the Retinopathy dataset. The best result for each explanation method is highlighted in bold

		CCR	$AvAUC$	MAE	κ	ρ
Grad-CAM	Baseline	0.7022	0.2057	0.3820	0.1759	0.2024
	Binomial	**1.0447**	**0.2365**	0.4343	0.2039	0.2719
	OBD	1.0240	0.2095	**0.5392**	**0.2166**	**0.2840**
IBA	Baseline	−0.0001	**0.2951**	0.0101	0.1991	0.2397
	Binomial	−0.0689	0.2663	0.0914	0.2166	0.2420
	OBD	**0.5082**	0.2915	**0.2829**	**0.2191**	**0.2668**

Table 3. Mean metric degradation score for the Adience dataset. The best result for each explanation method is highlighted in bold

		CCR	$AvAUC$	MAE	κ	ρ
Grad-CAM	Baseline	**0.0644**	**0.0706**	**0.0346**	**0.0121**	**0.01934**
	Binomial	0.0267	0.0027	0.0129	−0.0039	0.0008
	OBD	0.0312	0.0242	0.0109	−0.0005	0.0052
IBA	Baseline	0.1198	0.1235	0.1394	0.1419	0.1429
	Binomial	0.1018	**0.1730**	**0.1710**	**0.1993**	**0.1751**
	OBD	**0.1251**	0.1474	0.1614	0.1767	0.1722

Table 4. Mean metric degradation score for the CBIS-DDSM dataset. The best result for each explanation method is highlighted in bold

		CCR	$AvAUC$	MAE	κ	ρ
Grad-CAM	Baseline	0.4919	0.2008	**0.6553**	0.1969	0.1892
	Binomial	**0.9410**	**0.3177**	0.1218	**0.3034**	**0.3056**
	OBD	0.4179	0.1660	0.4180	0.1789	0.1634
IBA	Baseline	0.9131	0.3606	**2.7559**	0.4650	0.4783
	Binomial	1.2288	**0.4996**	0.7670	0.5112	0.5416
	OBD	**1.7439**	0.4713	2.2170	**0.5499**	**0.5577**

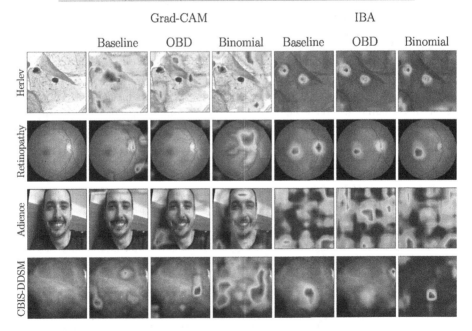

Fig. 3. Example input images from the four datasets (first column) and some example attributions for each combination of classifier type and explanation method

The results seem to be highly dependent on the combination of the dataset and attribution method used, so more sophisticated statistical techniques are required to draw any conclusion on the performance comparison between the classification methodologies.

6.1 Statistical Analysis

In this section, a statistical analysis is performed to determine whether the differences observed between the ordinal models and the baseline are statistically significant. To do this, we first check if the distributions are normal or not using a Kolmogorov-Smirnov test [19], for each experiment with 60 runs for each database (Adience, CBIS-DDSM, Retinopathy, Herlev), explanation method (Grad-CAM, IBA) and classification methodology (nominal, OBD, unimodal). The results obtained show that the normality hypothesis is not verified for most of these tests, since the significance level $\alpha = 0.05$ is greater than the p-value for each test.

Therefore, we have considered non-parametric Wilcoxon comparison tests of pairs of dependent variables [21], where the results achieved for the different datasets and explanation methods have been concatenated. In these terms, we have carried out pairwise Wilcoxon tests for the three methodologies: baseline, OBD and unimodal, using the values achieved by these approaches for the five-evaluation metrics considered: CCR, $AvAUC$, MAE, κ and ρ.

A significance level of $\alpha = 0.05$ is considered, and the corresponding Bonferroni correction for the number of comparisons is included. Thus, considering that three algorithms are being compared, the total number of comparisons for each metric is 3, being the corrected level of significance $\alpha^* = \frac{0.05}{3} = 0.017$.

Table 5 shows the results obtained for all the pairwise Wilcoxon tests performed. It is noteworthy that the OBD approach improves in CCR, MAE, κ and ρ over the baseline, whereas the ordinal binomial unimodal approach obtains significantly better results with respect to the baseline for the $AvAUC$, κ and ρ metrics. Between the two ordinal approaches, the unimodal one is outperformed in CCR but improves on $AvAUC$ and κ. Interestingly, the baseline (cross-entropy) has shown better explanation power for the only non-ordinal metric (CCR), while the ordinal methods have shown better explainability for all ordinal metrics.

Table 5. Results for the pairwise Wilcoxon test. Significant findings are highlighted in bold. For significant findings ($\alpha^* = 0.017$), "(1)" means the first methodology significantly improves over the second one and "(2)" the inverse

Methodology	p-values				
	CCR	$AvAUC$	MAE	κ	ρ
OBD vs. baseline	**<0.001** (1)	0.208	**<0.001** (1)	**0.002** (1)	**0.002** (1)
Unimodal vs. baseline	**<0.001** (2)	**<0.001** (1)	0.041	**<0.001** (1)	**<0.001** (1)
Unimodal vs. OBD	**<0.001** (2)	**<0.001** (1)	0.022	**0.013** (1)	0.078

7 Conclusions and Future Work

Statistical analysis shows significant improvements for the ordinal regression methods across all ordinal metrics. The ordinal unimodal model is only outperformed by the baseline approach with respect to Accuracy but performs equivalently or better in all other cases. In the case of the OBD model, it is never outperformed by the baseline while improving in most metrics, including all the ordinal ones. Thus, the effectiveness of Grad-CAM and IBA has been validated to improve when using it on a model better suited to an ordinal task domain.

Nonetheless, this is only the first step in the development of attribution methods for ordinal regression models. All proposals to date ignore ordinal information when constructing their attribution map, so there exists potential in this regard to improve the explanations further by incorporating this information into the process.

References

1. Bach, S., Binder, A., Montavon, G., Klauschen, F., Müller, K.R., Samek, W.: On pixel-wise explanations for non-linear classifier decisions by layer-wise relevance propagation. PLoS ONE **10**(7), e0130140 (2015). https://doi.org/10.1371/journal. pone.0130140
2. Barbero-Gómez, J., Gutiérrez, P.A., Hervás-Martínez, C.: Error-correcting output codes in the framework of deep ordinal classification. Neural Process. Lett. (2022). https://doi.org/10.1007/s11063-022-10824-7
3. Chattopadhyay, A., Sarkar, A., Howlader, P., Balasubramanian, V.N.: Grad-CAM++: improved visual explanations for deep convolutional networks. In: 2018 IEEE Winter Conference on Applications of Computer Vision (WACV), pp. 839–847 (2018). https://doi.org/10.1109/WACV.2018.00097
4. Cruz-Ramírez, M., Hervás-Martínez, C., Sánchez-Monedero, J., Gutiérrez, P.A.: Metrics to guide a multi-objective evolutionary algorithm for ordinal classification. Neurocomputing **135**, 21–31 (2014)
5. Deng, J., Dong, W., Socher, R., Li, L.J., Li, K., Fei-Fei, L.: ImageNet: a large-scale hierarchical image database. In: 2009 IEEE Conference on Computer Vision and Pattern Recognition, pp. 248–255, June 2009. https://doi.org/10.1109/CVPR. 2009.5206848
6. Eidinger, E., Enbar, R., Hassner, T.: Age and gender estimation of unfiltered faces. IEEE Trans. Inf. Forensics Secur. **9**(12), 2170–2179 (2014). https://doi.org/10. 1109/TIFS.2014.2359646
7. Fu, R., Hu, Q., Dong, X., Guo, Y., Gao, Y., Li, B.: Axiom-based Grad-CAM: towards accurate visualization and explanation of CNNs. Technical report. arXiv:2008.02312, arXiv, August 2020
8. Gutiérrez, P.A., Pérez-Ortiz, M., Sánchez-Monedero, J., Fernández-Navarro, F., Hervás-Martínez, C.: Ordinal regression methods: survey and experimental study. IEEE Trans. Knowl. Data Eng. **28**(1), 127–146 (2016). https://doi.org/10.1109/ TKDE.2015.2457911
9. Jantzen, J., Norup, J., Dounias, G., Bjerregaard, B.: Pap-smear benchmark data for pattern classification. In: Nature Inspired Smart Information Systems (NiSIS), January 2005

10. Khakzar, A., Khorsandi, P., Nobahari, R., Navab, N.: Do explanations explain? Model knows best. In: 2022 IEEE/CVF Conference on Computer Vision and Pattern Recognition (CVPR), pp. 10234–10243. IEEE, New Orleans, LA, USA, June 2022. https://doi.org/10.1109/CVPR52688.2022.01000

11. Kingma, D.P., Ba, J.: Adam: a method for stochastic optimization. arXiv:1412.6980 [cs], January 2017

12. Lee, R.S., Gimenez, F., Hoogi, A., Miyake, K.K., Gorovoy, M., Rubin, D.L.: A curated mammography data set for use in computer-aided detection and diagnosis research. Sci. Data 4(1), 170177 (2017). https://doi.org/10.1038/sdata.2017.177

13. Montavon, G., Lapuschkin, S., Binder, A., Samek, W., Müller, K.R.: Explaining nonlinear classification decisions with deep Taylor decomposition. Pattern Recogn. **65**, 211–222 (2017). https://doi.org/10.1016/j.patcog.2016.11.008

14. Nayar, R., Wilbur, D.C.: The Bethesda System for Reporting Cervical Cytology: Definitions, Criteria, and Explanatory Notes. Springer, Cham (2015). https://doi.org/10.1007/978-3-319-11074-5

15. Pinto da Costa, J.F., Alonso, H., Cardoso, J.S.: The unimodal model for the classification of ordinal data. Neural Netw. **21**(1), 78–91 (2008). https://doi.org/10.1016/j.neunet.2007.10.003

16. Schulz, K., Sixt, L., Tombari, F., Landgraf, T.: Restricting the Flow: Information Bottlenecks for Attribution, May 2020

17. Selvaraju, R.R., Cogswell, M., Das, A., Vedantam, R., Parikh, D., Batra, D.: Grad-CAM: visual explanations from deep networks via gradient-based localization. Int. J. Comput. Vision **128**(2), 336–359 (2019). https://doi.org/10.1007/s11263-019-01228-7

18. Simonyan, K., Vedaldi, A., Zisserman, A.: Deep Inside Convolutional Networks: Visualising Image Classification Models and Saliency Maps, April 2014

19. Smirnov, N.: Table for estimating the goodness of fit of empirical distributions. Ann. Math. Stat. **19**(2), 279–281 (1948). https://doi.org/10.1214/aoms/1177730256

20. Springenberg, J.T., Dosovitskiy, A., Brox, T., Riedmiller, M.: Striving for simplicity: the all convolutional net. Technical report. arXiv:1412.6806, arXiv, April 2015

21. Wilcoxon, F.: Individual comparisons by ranking methods. In: Kotz, S., Johnson, N.L. (eds.) Breakthroughs in Statistics: Methodology and Distribution. Springer Series in Statistics, pp. 196–202. Springer, New York (1992). https://doi.org/10.1007/978-1-4612-4380-9_16

22. Zeiler, M.D., Fergus, R.: Visualizing and Understanding Convolutional Networks, November 2013. https://doi.org/10.48550/arXiv.1311.2901

23. Zhou, B., Khosla, A., Lapedriza, A., Oliva, A., Torralba, A.: Learning deep features for discriminative localization. Technical report. arXiv:1512.04150, arXiv, December 2015

A Dictionary-Based Approach to Time Series Ordinal Classification

Rafael Ayllón-Gavilán[1], David Guijo-Rubio[1,2]([✉]),
Pedro Antonio Gutiérrez[1], and César Hervás-Martínez[1]

[1] Department of Computer Sciences, Universidad de Córdoba, 14014 Córdoba, Spain
{i72aygar,dguijo,pagutierrez,chervas}@uco.es
[2] School of Computing Sciences, University of East Anglia,
Norwich NR4 7TQ, UK

Abstract. Time Series Classification (TSC) is an extensively researched field from which a broad range of real-world problems can be addressed obtaining excellent results. One sort of the approaches performing well are the so-called dictionary-based techniques. The Temporal Dictionary Ensemble (TDE) is the current state-of-the-art dictionary-based TSC approach. In many TSC problems we find a natural ordering in the labels associated with the time series. This characteristic is referred to as ordinality, and can be exploited to improve the methods performance. The area dealing with ordinal time series is the Time Series Ordinal Classification (TSOC) field, which is yet unexplored. In this work, we present an ordinal adaptation of the TDE algorithm, known as ordinal TDE (O-TDE). For this, a comprehensive comparison using a set of 18 TSOC problems is performed. Experiments conducted show the improvement achieved by the ordinal dictionary-based approach in comparison to four other existing nominal dictionary-based techniques.

Keywords: time series · dictionary-based approaches · ordinal classification

1 Introduction

Machine Learning (ML) focuses on developing computer algorithms able to learn from previous experience, in such a way that they could be applied to solve real-world problems or, at least, provide support for human activities. Inside the ML paradigm, different sub-domains can be found, which emerge according to the sort of data used. Specifically, this work deals with the classification of time series. A time series is a set of values collected chronologically. This type of data can be found in a wide range of fields. For instance, the prices of a market asset over a certain period of time, or the monthly sales of a shop.

In this study, we focus on Time Series Classification (TSC), a task in which a discrete label is associated with each time series specifying some property of

This work has been partially subsidised by "Agencia Española de Investigación (España)" (grant ref.: PID2020-115454GB-C22/AEI/10.13039/501100011033). David Guijo-Rubio's research has been subsidised by the University of Córdoba through grants to Public Universities for the requalification of the Spanish university system of the Ministry of Universities, financed by the European Union - NextGenerationEU (grant reference: UCOR01MS).

I. Rojas et al. (Eds.): IWANN 2023, LNCS 14135, pp. 541–552, 2023.
https://doi.org/10.1007/978-3-031-43078-7_44

interest about it. The main goal is finding a model that learns the correspondence between labels and time series, so that it is capable of labelling new, unknown patterns accurately. Examples of applications can be found in medical research [3], psychology [14] and industry [21], among others. Due to its versatility, the TSC paradigm has been greatly enhanced over the last decades. The main reason is the establishment of the UEA/UCR archive, a set of benchmark problems, that has made easier the validation of novel techniques.

TSC approaches are divided into different groups according to the methodology adopted. A first detailed taxonomy of the state of the art was presented in [1], where six main categories were distinguished: *whole series, intervals, shapelets, dictionary-based, combinations* and *model-based* techniques. In subsequent years, three additional groups emerged in the literature: the *convolutional-based* models, introduced with the Random Convolutional Kernel Transform (ROCKET) method [5]; *deep learning-based* techniques, which mainly raised from the adaptation of residual and convolutional networks to the TSC case [30]; and ensemble-based methods, in which the Hierarchical Vote Collective of Transformation-based Ensembles (HIVE-COTE) [19] particularly stands out due to its superiority in terms of accuracy in comparison to the rest of the state-of-the-art methodologies. Later on, an improved version of this last technique, named as HIVE-COTE 2.0 (HC2), was introduced in [24]. The HC2 approach combines four methods from different categories: Arsenal, an ensemble of the ROCKET algorithm; Shapelet Transform Classifier (STC) [13], a standard classifier applied to a transformation built from the distances between the phase independent subsequences, known as shapelets, and the original time series; the interval-based Diverse representation Canonical Interval Forest (DrCIF) [24], a random forest-based technique applied to statistical features extracted from dependent subsequences of the original time series; and the Temporal Dictionary Ensemble (TDE) [23], an approach using bag of words representations of time series. TDE is the basis for the methodology proposed in this work.

More specifically, this work deals with the classification of ordinal time series, a special type of time series in which the associated discrete target values present a natural order relationship between them. This vaguely explored subdomain of TSC is known as Time Series Ordinal Classification (TSOC) and was firstly presented in [11]. One example of this type of series was introduced in [16], in which the task is to associate a spectrograph of 1751 observations (i.e. time series) with a label that can take four different values, *E35, E38, E40* and *E45*, ordered by the ethanol level of the sample. With this setting, during model training, misclassifying an *E45* sample as *E35* should be far more penalized than misclassifying it as *E40*. This property is known as *ordinality*, and can be exploited in a wide variety of domains including industry [29], image classification [20], atmospheric events detection [9], finance [7], and medicine [31], among others.

Finally, the goal of this work is to develop a new dictionary-based approach for the TSOC paradigm. For this, the TDE, the state-of-the-art approach in this category of TSC, is considered as the basis. For this, a TDE methodology capable of exploiting the ordinal information of the output variable is proposed. Specifically, more appropriate strategies in the ensemble member selection and in the computation of the time series symbolic representation are employed.

The remainder of this paper is organized as follows: related works are described in Sect. 2; Sect. 3 describes the methodology developed, i.e. the Ordinal Temporal Dictionary Ensemble (O-TDE); Sect. 4 presents the datasets and experimental settings; Sect. 5 shows the obtained results; and finally, Sect. 6 provides the conclusions and future research of our work.

2   Related Works

The first dictionary-based method for time series classification was the Bag Of Patterns (BOP) presented in [18]. The BOP algorithm is divided into four phases: 1) a sliding window is applied to the time series; 2) a dimensionality reduction method called Symbolic Aggregate approXimation (SAX) [17] is used to transform each window to a symbolic representation. This representation is known as word; 3) the frequency of occurrence of each word is counted; and finally 4) histograms of words counts are computed for the time series of the training set. The prediction of new patterns is obtained through a k-Nearest Neighbours (kNN) classifier measuring the similarity between their histograms and those of the training instances.

Most of the state-of-the-art methods follow the structure of the BOP algorithm. This is the case of Bag of Symbolic Fourier approximation Symbols (BOSS) [26]. BOSS also transforms the input time series into symbolic representations (words). For this purpose, instead of SAX, it uses the Discrete Fourier Transform (DFT) [12] method. DFT avoids issues related with noisy time series, achieving a more representative transformation.

Another distinguishing feature of BOSS is that it conforms an ensemble of BOSS approaches trained with different window sizes. Only those BOSS members achieving an over-threshold accuracy are included in the ensemble. BOSS significantly outperformed BOP. Given the performance of this approach, several BOSS-based methods were proposed in the literature. In this sense, we have the Word ExtrAction for time SEries cLassification (WEASEL) [27] method. WEASEL applies an ANOVA test to obtain a subset of the most significant DFT coefficients for each class. From this subset it builds the bag of words for each time series. Then a chi-square test is performed to select the most significant words to compute the histograms. This feature selection methodology makes WEASEL more scalable and faster than previous proposals.

On the same line, contractable BOSS (cBOSS) [25] performs a random selection on the parameter space making the BOSS ensemble lighter. cBOSS is significantly more scalable than BOSS but performs equally. Spatial Pyramids (SP) BOSS [15] incorporates the SP method, widely used in computer vision problems, to the BOSS technique. SP recursively segments the input time series and computes histograms for these segments. This allows the combination of temporal and phase independent features in the symbolic transformation process, slightly improving the robustness of the algorithm.

Finally, the latest and most successful dictionary-based technique is the Temporal Dictionary Ensemble (TDE) [23]. TDE implements the same structure than BOSS, but makes use of a Gaussian process of the parameter space to do the ensemble member selection. Its superiority over competing dictionary-based methods led it to replace BOSS in the second version of the HIVE-COTE technique, HIVE-COTE2.0 (HC2) [24].

Focusing now on TSOC, only one type of approaches have been developed. This is the Ordinal Shapelet Transform Classifier (O-STC) [11]. O-STC extracts phase independent features from the time series keeping those that satisfy a minimum shapelet quality (measured through a specific ordinal metric). The resulting set of shapelets are fed to an ordinal classifier such as a Proportional Odds Model (POM) [22] or an ordinal support vector machine technique [4].

In this work, we focus on implementing the ordinal version of the TDE approach, given its superiority over the existing dictionary-based approaches in TSC. This technique is known as Ordinal Temporal Dictionary Ensemble (O-TDE).

3   Ordinal Temporal Dictionary Ensemble (O-TDE)

First of all, a time series can be categorised according to the number of dimensions d as univariate ($d = 1$) or multivariate ($d > 1$). A univariate time series $\mathbf{x}$ of length l is an ordered set of l real values, $\mathbf{x} = (x_1, \ldots, x_l)$. Conversely, a multivariate time series with d dimensions (or channels) and length l is a collection of d ordered sets, each containing l real values denoted as $\mathbf{x} = \{(x_{1,1}, \ldots, x_{1,l}), \ldots, (x_{d,1}, \ldots x_{d,l})\}$. A time series dataset is then defined as $D = \{(\mathbf{x}_1, y_1), (\mathbf{x}_2, y_2), \ldots, (\mathbf{x}_N, y_N)\}$, where N is the number of available time series, $\mathbf{x}_i$ is a time series (either univariate or multivariate), and y_i is the output label associated with the respective time series. Both in this paper and in the wider TSC literature, our analyses rely on datasets comprising time series that are uniformly spaced, meaning that the observations within each time series are collected at equally-spaced time intervals. Additionally, all of the time series in the datasets are of equal length.

Focusing now on the proposal, as BOSS and TDE, O-TDE also consists of several individual techniques which, to prevent ambiguity, will be referred to as individual O-TDE. In the O-TDE algorithm, a guided parameter selection is performed to build the ensemble members. This parameter selection is guided by a Gaussian process [28] intended to predict the Mean Absolute Error (MAE) values for specific O-TDE configurations, basing its prediction on previous parameters-MAE pairs [23]. This helps to reduce the computational complexity of the ensemble construction. This process is similar to that followed in the original TDE algorithm, but considering the MAE metric instead of the accuracy. Note that MAE quantifies the error committed in the ordinal scale. Hence, it helps to boost the performance achieved for ordinal problems.

Regarding the individual O-TDE, i.e. the method considered in the ensemble, it consists of a sequence of steps, summarised in the following lines. Firstly, a given input time series of size l is processed by sliding windows of length w, in such a way that $w \ll l$. Then, a Discrete Fourier Transform (DFT) [12] is applied to each window, decomposing it into a set of w orthogonal basis functions using sinusoidal waves. The set of waves obtained through Fourier analysis is commonly referred to as Fourier coefficients. In practice, only the first c coefficients are typically retained, while the remaining coefficients, which contribute to higher frequencies, are discarded ($c \ll w$). This selection process serves two purposes: 1) since the first Fourier coefficients are related to the smoothest sections of the time series, potentially noisy parts can be eliminated. And 2) the dimensionality of the representation can be substantially reduced from w coefficients to just c. This reduction can provide computational benefits, particularly for large or complex datasets.

At this point, from the initial time series, c Fourier coefficients are kept. The j-th Fourier coefficient extracted from the i-th time series is represented by a complex number $F_{i,j} = (\text{real}_{i,j}, \text{imag}_{i,j})$. With this setting, the following matrix A is built:

$$A = \begin{bmatrix} \text{real}_{1,1} & \text{imag}_{1,1} & \cdots & \text{real}_{1,c} & \text{imag}_{1,c} \\ \text{real}_{2,1} & \text{imag}_{2,1} & \cdots & \text{real}_{2,c} & \text{imag}_{2,c} \\ \vdots & \vdots & \cdots & \vdots & \vdots \\ \text{real}_{N,1} & \text{imag}_{N,1} & \cdots & \text{real}_{N,c} & \text{imag}_{N,c} \end{bmatrix}, \tag{1}$$

where N is the number of time series of the training dataset. For each column of A, $C_m = (C_{1,m}, C_{2,m}, \ldots, C_{N,m})$, with $m \in \{1, 2, \ldots, 2c\}$, a set of thresholds $\beta_m = (\beta_{m,0}, \beta_{m,1}, \ldots, \beta_{m,T})$ is extracted through a process called Information Gain Binning (IGB) that will covered below. The $\beta_{m,0}$ and $\beta_{m,T}$ thresholds are set to $-\infty$ and $+\infty$ respectively. Note that as coefficients are represented by complex numbers (with real and imaginary parts), m takes values up to $2c$. With this setting, the C_m real-valued elements are discretised according to β_m and a finite alphabet $\Sigma = \{\alpha_1, \alpha_2, \ldots, \alpha_T\}$, where T is the size of the dictionary. An element C_{im} of A is mapped to a symbol α_t of Σ if $\beta_{m,t-1} \leq C_{i,m} \leq \beta_{m,t}$, with $t \in \{1, 2, \ldots, T\}$.

The resulting symbolic representation of each column is what is called a *word*. The IGB process finds the optimal set β for each column by fitting a Decision Tree Regressor (DTR). Each $\beta_{m,i}$ corresponds to a threshold value used in a given splitting node of the tree. The impurity criterion i used in the DTR is the Mean Squared Error (MSE) with an improvement score proposed in [8]:

$$i = \frac{w_l \cdot w_r}{w_l + w_r} (\bar{y}_l - \bar{y}_r), \tag{2}$$

where $\bar{y}_l$, $\bar{y}_r$ are the left and right child nodes response means, and w_l, w_r are the corresponding sums of the weights. The utilisation of this criterion instead of the accuracy (considered in the original TDE proposal) greatly enhances the performance in ordinal problems. This criteria is usually known in the literature as *friedman-MSE*.

In base of all the above, an individual O-TDE transforms an input time series into a set of words (one word for each sliding window). Then, a histogram of words counts is built from this set. The label for a testing time series is obtained by computing the distances between its histogram and those of the training time series and returning the label of the closest one.

4   Experimental Settings

The experiments are performed on an extended version of the TSOC archive. To avoid possible randomisation biases, 30 runs have been performed. To measure the performance of the techniques, both nominal and ordinal metrics have been considered to get a better analysis on how the proposed ordinal methodology performs.

4.1   Datasets Considered

With the aim of performing a robust experimentation, a set of 18 TSOC problems from a wide variety of domains has been considered. In this section, we present these datasets and the source from which they have been collected. Table 1 provides a summary of the complete set of problems. We can distinguish four different data sources: 1) The UEA/UCR TSC archive[1], where a subset of 9 ordinal problems has been identified [10]. 2) The Monash/UEA/UCR Time Series Extrinsic Regression (TSER) archive[2]. From this repository, we limited our selection to equal-length problems without missing values, adding two more datasets to our experiments. The originally continuous output variable of these datasets has been discretised into five equally wide bins. 3) Historical price data from 5 of the most important companies in the stock market. We have taken this data from Yahoo Finance[3] website, extracting weekly price data from the earliest available date to March 2023. Each time series is built with the returns over 53 weeks (the number of weeks of a year) prior to a given date t, and the output label corresponds to the price return in t (r_t). This value is discretised according to a set of predefined symmetrical thresholds $(-\infty, -0.05, -0.02, 0.02, 0.05, \infty)$. In this way, our experimentation is extended with 5 more problems. 4) Buoy data from the National Data Buoy Center (NDBC)[4]. Two problems from this source has been considered, which are *USASouthwestEnergyFlux* and *USASouthwestSWH*. The first comprises a set of 468 time series. Each time series is built on 112 energy fluctuation measurements collected during 4 weeks (4 measures per day). The objective is to estimate the level of energy fluctuation during that period of time, being 0 the minimum level, and 3 the highest energy level. The second problem consists on 1872 time series of length 28 representing sea waves height variation along a week (4 measures per day). The purpose is to estimate

---

[1] https://www.timeseriesclassification.com/dataset.php.

[2] http://tseregression.org/.

[3] https://es.finance.yahoo.com/.

[4] https://www.ndbc.noaa.gov/.

the wave height level during that period of time, ranging from 0 (the lowest height) to 3 (the highest height).

Table 1. Information about the datasets considered. OAG stands for OutlineAge-Group.

Dataset name	# Train	# Test	# Classes	Length	# Dimensions
AAPL	1720	431	5	53	1
AMZN	1035	259	5	53	1
AppliancesEnergy	95	42	5	144	24
AtrialFibrillation	15	15	3	640	2
Covid3Month	140	61	5	84	1
DistalPhalanxOAG	400	139	3	80	1
DistalPhalanxTW	400	139	6	80	1
EthanolConcentration	261	263	4	1751	3
EthanolLevel	504	500	4	1751	1
GOOG	732	183	5	53	1
META	408	103	5	53	1
MSFT	1501	376	5	53	1
MiddlePhalanxOAG	400	154	3	80	1
MiddlePhalanxTW	399	154	6	80	1
ProximalPhalanxOAG	400	205	3	80	1
ProximalPhalanxTW	400	205	6	80	1
USASouthwestEnergyFlux	327	141	4	112	7
USASouthwestSWH	1310	562	4	28	7

4.2 Experimental Setup

With the goal of demonstrating that ordinal approaches can outperform nominal techniques when dealing with ordinal datasets, the proposed methodology O-TDE is compared against 4 state-of-the-art approaches in dictionary-based techniques: BOSS, cBOSS, WEASEL, and TDE.

The performance of these approaches is measured in terms of four metrics (1 nominal and 3 ordinal). The Correct Classification Rate (CCR), also known as accuracy, is the most spread measure when dealing with nominal time series. It measures the percentage of correctly classified instances.

The first ordinal measure is the Mean Absolute Error (MAE), that quantifies the error committed in the ordinal scale:

$$\text{MAE} = \frac{1}{N} \sum_{i=1}^{N} |\hat{y}_i - y_i|, \tag{3}$$

where N represents the number of patterns, and $\hat{y}_i$ and y_i are the predicted and real labels, respectively.

The second ordinal measure is the Quadratic Weighted Kappa (QWK). QWK establishes different weights depending on the different disagreement levels between real and predicted values. As MAE, it penalises to a greater extent errors made in farther classes in the ordinal scale:

$$\text{QWK} = 1 - \frac{\sum_{i,j}^{N} \omega_{i,j} O_{i,j}}{\sum_{i,j}^{N} \omega_{i,j} E_{i,j}}, \tag{4}$$

where ω is the penalization matrix with quadratic weights, O is the confusion matrix, $E_{ij} = \frac{O_{i\bullet} O_{\bullet j}}{N}$, with $O_{i\bullet}$ and $O_{\bullet j}$ being the accumulated sum of all the elements of the i-th row and the j-th column, respectively.

The remaining ordinal metric considered is the 1-OFF accuracy (1-OFF) which is the same as the CCR but also considering as correct the predictions one category away from the actual class on the ordinal scale.

Furthermore, given that the employed methodologies have a stochastic behaviour, the experiments have been performed using 30 different resamples. The first run is with the default data and subsequent runs are carried out with data resampled using the same train/test proportion as the original.

Finally, the code of the nominal approach is open source and is available in the **aeon** toolkit[5], a scikit-learn compatible implementation of the time series approaches. The ordinal version of the TDE will be included in **aeon**. Detailed results can be found in the supporting website of this work[6].

5   Results

Table 2 shows the results achieved in terms of MAE. Results are shown as the as the mean and standard deviation of the 30 runs carried out. As can be seen, O-TDE is the approach achieving the best results, as is the best and second best in 10 and 4 of the 18 ordinal datasets, respectively. The second best approach is the nominal version of TDE, which obtained the best results for 3 datasets (tied with WEASEL) but is the second-best in other 6 datasets, whereas WEASEL only is the second-best in 4.

Furthermore, to compare the results obtained for multiple classifiers over multiple datasets, Critical Difference Diagrams (CDDs) are used [6]. The post-hoc Nemenyi test is replaced by a comparison of all classifiers using pairwise Wilcoxon signed-rank tests. Finally, cliques are formed using the Holm correction [2]. Figure 1 shows the CDDs for the four measures detailed in Sect. 4.2.

---

[5] https://github.com/aeon-toolkit/aeon.
[6] http://www.uco.es/grupos/ayrna/tsoc-dictionaries-iwann.

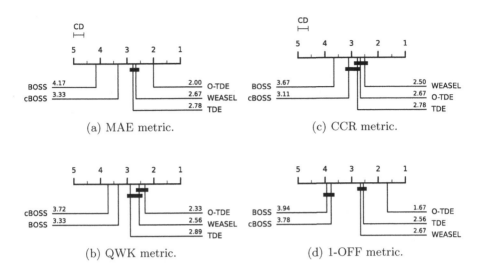

(a) MAE metric.

(c) CCR metric.

(b) QWK metric.

(d) 1-OFF metric.

Fig. 1. CDDs in terms of MAE (a), QWK (b), CCR (c) and 1-OFF (d). The significance value α is set to 0.1. The critical difference (CD) value is computed pairwise and is equal to 0.456.

Table 2. Results achieved in terms of MAE for the 5 dictionary-based approaches considered in this work. Results are exposed as the Mean and Standard Deviation (SD) of the 30 runs: Mean$_{SD}$.

Dataset	BOSS	cBOSS	WEASEL	TDE	O-TDE
AAPL	$1.383_{0.042}$	$1.376_{0.045}$	$\mathbf{1.292_{0.040}}$	$1.380_{0.051}$	$\mathit{1.364_{0.048}}$
AMZN	$1.390_{0.059}$	$1.379_{0.070}$	$\mathbf{1.293_{0.063}}$	$1.368_{0.068}$	$\mathit{1.365_{0.062}}$
AppliancesEnergy	$0.572_{0.010}$	$0.571_{0.000}$	$0.561_{0.028}$	$\mathit{0.544_{0.044}}$	$\mathbf{0.508_{0.058}}$
AtrialFibrillation	$0.958_{0.129}$	$\mathbf{0.802_{0.146}}$	$0.953_{0.125}$	$0.951_{0.178}$	$\mathit{0.813_{0.130}}$
Covid3Month	$0.755_{0.063}$	$0.767_{0.050}$	$0.777_{0.053}$	$\mathbf{0.737_{0.043}}$	$\mathit{0.747_{0.036}}$
DistalPhalanxOAG	$\mathbf{0.180_{0.029}}$	$\mathit{0.204_{0.025}}$	$0.213_{0.025}$	$0.207_{0.032}$	$0.205_{0.025}$
DistalPhalanxTW	$\mathit{0.386_{0.033}}$	$0.388_{0.037}$	$\mathbf{0.365_{0.026}}$	$0.406_{0.030}$	$0.404_{0.041}$
EthanolConcentration	$0.725_{0.057}$	$0.790_{0.050}$	$\mathbf{0.513_{0.043}}$	$0.552_{0.086}$	$\mathit{0.539_{0.061}}$
EthanolLevel	$0.561_{0.040}$	$0.585_{0.036}$	$\mathit{0.466_{0.061}}$	$0.478_{0.097}$	$\mathbf{0.425_{0.057}}$
GOOG	$1.082_{0.051}$	$1.098_{0.055}$	$1.012_{0.055}$	$\mathit{0.966_{0.059}}$	$\mathbf{0.952_{0.063}}$
META	$1.193_{0.082}$	$1.185_{0.085}$	$\mathit{1.127_{0.098}}$	$1.150_{0.082}$	$\mathbf{1.106_{0.072}}$
MSFT	$1.101_{0.044}$	$1.106_{0.038}$	$\mathit{1.017_{0.042}}$	$1.051_{0.040}$	$\mathbf{1.009_{0.042}}$
MiddlePhalanxOAG	$0.361_{0.034}$	$0.335_{0.036}$	$0.388_{0.039}$	$\mathit{0.315_{0.038}}$	$\mathbf{0.314_{0.039}}$
MiddlePhalanxTW	$0.657_{0.042}$	$0.604_{0.046}$	$0.622_{0.046}$	$\mathbf{0.579_{0.047}}$	$\mathit{0.593_{0.040}}$
ProximalPhalanxOAG	$0.176_{0.018}$	$\mathbf{0.145_{0.020}}$	$0.159_{0.018}$	$\mathit{0.145_{0.020}}$	$0.147_{0.020}$
ProximalPhalanxTW	$0.249_{0.027}$	$\mathit{0.216_{0.020}}$	$0.218_{0.022}$	$\mathbf{0.212_{0.020}}$	$0.220_{0.017}$
USASouthwestEnergy	$0.221_{0.015}$	$0.217_{0.012}$	$\mathbf{0.189_{0.022}}$	$0.223_{0.025}$	$\mathit{0.205_{0.023}}$
USASouthwestSWH	$0.677_{0.048}$	$0.391_{0.021}$	$\mathbf{0.383_{0.013}}$	$0.392_{0.015}$	$\mathit{0.385_{0.014}}$
Best (second best)	1 (1)	2 (2)	*6* (3)	*3* (4)	**6** (8)
Rank	4.167	3.333	*2.667*	2.778	**2.000**

The best results are highlighted in bold, whereas the second-best are in italics.

From these results, it can be said that a solid superiority of the O-TDE method is observed against the nominal methodologies. O-TDE outperforms all the nominal techniques not only in terms of ordinal performance measures (MAE and QWK) but also in terms of CCR, a nominal measure. Even though improving the results in CCR is not the final goal of the ordinal approaches, this superiority demonstrates the potential of the ordinal techniques over nominal ones. Finally, indicate that this difference becomes statistically significant for the MAE and 1-OFF metrics, indicating an excellent performance of the O-TDE proposed approach.

6   Conclusion and Future Scope

Time Series Ordinal Classification is still an unexplored paradigm in the time series literature, being a subset of the popular nominal Time Series Classification (TSC) task. However, it has a wealth of real-world applications in a wide range of fields such as finances, medicine or energy, among others. In this work, it has been shown that when this sort of problems are approximated through ordinal methods, such as the presented Ordinal Temporal Dictionary Ensemble (O-TDE), a significant boost in performance is obtained. This superiority is mainly achieved by penalising more severely those predictions that fall far away from the real class in the ordinal scale.

From the original set of 7 datasets previously identified, this work provides another 11 datasets, taking the ordinal archive to 18 ordinal datasets, including 13 univariate and 5 multivariate, making the obtained results more robust. The performance of the 5 approaches has been measured in terms of accuracy, the most used one in nominal TSC, and three ordinal metrics, Mean Average Error (MAE), Quadratic Weighted Kappa (QWK) and 1-OFF accuracy (1-OFF). These three measures help to properly quantify the capacity of the approaches to model the ordinal scale. Consequently, the biggest differences in performance between nominal and ordinal methodologies are obtained in terms of these last three metrics, being the difference in terms of MAE and 1-OFF statistically significant.

For future works, the TSOC archive is sought to be expanded. In addition, multiple well-known TSC methods such as kernel-based, ensemble-based or interval-based techniques will be explored for the ordinal paradigm.

References

1. Bagnall, A., Lines, J., Bostrom, A., Large, J., Keogh, E.: The great time series classification bake off: a review and experimental evaluation of recent algorithmic advances. Data Min. Knowl. Discov. **31**(3), 606–660 (2017). https://doi.org/10.1007/s10618-016-0483-9
2. Benavoli, A., Corani, G., Mangili, F.: Should we really use post-hoc tests based on mean-ranks? J. Mach. Learn. Res. **17**(1), 152–161 (2016). http://jmlr.org/papers/v17/benavoli16a.html

3. Buza, K., Koller, J., Marussy, K.: PROCESS: projection-based classification of electroencephalograph signals. In: Rutkowski, L., Korytkowski, M., Scherer, R., Tadeusiewicz, R., Zadeh, L.A., Zurada, J.M. (eds.) ICAISC 2015. LNCS (LNAI), vol. 9120, pp. 91–100. Springer, Cham (2015). https://doi.org/10.1007/978-3-319-19369-4_9

4. Chu, W., Keerthi, S.S.: New approaches to support vector ordinal regression. In: Proceedings of the 22nd International Conference on Machine Learning, pp. 145–152 (2005). https://doi.org/10.1145/1102351.1102370

5. Dempster, A., Petitjean, F., Webb, G.I.: Rocket: exceptionally fast and accurate time series classification using random convolutional kernels. Data Min. Knowl. Discov. **34**, 1454–1495 (2020). https://doi.org/10.1007/s10618-020-00701-z

6. Demšar, J.: Statistical comparisons of classifiers over multiple data sets. J. Mach. Learn. Res. **7**, 1–30 (2006). http://jmlr.org/papers/v7/demsar06a.html

7. Fernandez-Navarro, F., Campoy-Munoz, P., de la Paz-Marin, M., Hervas-Martinez, C., Yao, X.: Addressing the EU sovereign ratings using an ordinal regression approach. IEEE Trans. Cybern. **43**(6), 2228–2240 (2013). https://doi.org/10.1109/TSMCC.2013.2247595

8. Friedman, J.H.: Greedy function approximation: a gradient boosting machine. Ann. Stat. **29**(5), 1189–1232 (2001). https://doi.org/10.1214/aos/1013203451

9. Guijo-Rubio, D., et al.: Ordinal regression algorithms for the analysis of convective situations over Madrid-Barajas airport. Atmos. Res. **236**, 104798 (2020). https://doi.org/10.1016/j.atmosres.2019.104798

10. Guijo-Rubio, D., Gutiérrez, P.A., Bagnall, A., Hervás-Martínez, C.: Ordinal versus nominal time series classification. In: Lemaire, V., Malinowski, S., Bagnall, A., Guyet, T., Tavenard, R., Ifrim, G. (eds.) AALTD 2020. LNCS (LNAI), vol. 12588, pp. 19–29. Springer, Cham (2020). https://doi.org/10.1007/978-3-030-65742-0_2

11. Guijo-Rubio, D., Gutiérrez, P.A., Bagnall, A., Hervás-Martínez, C.: Time series ordinal classification via shapelets. In: 2020 International Joint Conference on Neural Networks (IJCNN), pp. 1–8 (2020). https://doi.org/10.1109/IJCNN48605.2020.9207200

12. Harris, F.J.: On the use of windows for harmonic analysis with the discrete Fourier transform. Proc. IEEE **66**(1), 51–83 (1978). https://doi.org/10.1109/PROC.1978.10837

13. Hills, J., Lines, J., Baranauskas, E., Mapp, J., Bagnall, A.: Classification of time series by shapelet transformation. Data Min. Knowl. Discov. **28**(4), 851–881 (2014). https://doi.org/10.1007/s10618-013-0322-1

14. Kurbalija, V., von Bernstorff, C., Burkhard, H.D., Nachtwei, J., Ivanović, M., Fodor, L.: Time-series mining in a psychological domain. In: Proceedings of the Fifth Balkan Conference in Informatics, pp. 58–63 (2012). https://doi.org/10.1145/2371316.2371328

15. Large, J., Bagnall, A., Malinowski, S., Tavenard, R.: On time series classification with dictionary-based classifiers. Intell. Data Anal. **23**(5), 1073–1089 (2019). https://doi.org/10.3233/IDA-184333

16. Large, J., Kemsley, E.K., Wellner, N., Goodall, I., Bagnall, A.: Detecting forged alcohol non-invasively through vibrational spectroscopy and machine learning. In: Phung, D., Tseng, V.S., Webb, G.I., Ho, B., Ganji, M., Rashidi, L. (eds.) PAKDD 2018. LNCS (LNAI), vol. 10937, pp. 298–309. Springer, Cham (2018). https://doi.org/10.1007/978-3-319-93034-3_24

17. Lin, J., Keogh, E., Wei, L., Lonardi, S.: Experiencing sax: a novel symbolic representation of time series. Data Min. Knowl. Discov. **15**, 107–144 (2007). https://doi.org/10.1007/s10618-007-0064-z

18. Lin, J., Khade, R., Li, Y.: Rotation-invariant similarity in time series using bag-of-patterns representation. J. Intell. Inf. Syst. **39**, 287–315 (2012). https://doi.org/10.1007/s10844-012-0196-5

19. Lines, J., Taylor, S., Bagnall, A.: Time series classification with hive-cote: the hierarchical vote collective of transformation-based ensembles. ACM Trans. Knowl. Discov. Data **12**(5) (2018). https://doi.org/10.1145/3182382

20. Liu, Y., Wang, Y., Kong, A.W.K.: Pixel-wise ordinal classification for salient object grading. Image Vision Comput. **106** (2021). https://doi.org/10.1016/j.imavis.2020.104086

21. Malhotra, P., Vig, L., Shroff, G., Agarwal, P., et al.: Long short term memory networks for anomaly detection in time series. In: ESANN, vol. 2015, p. 89 (2015). https://api.semanticscholar.org/CorpusID:43680425

22. McCullagh, P.: Regression models for ordinal data. J. Roy. Stat. Soc. Ser. B (Methodol.) **42**(2), 109–127 (1980). https://www.jstor.org/stable/2984952

23. Middlehurst, M., Large, J., Cawley, G., Bagnall, A.: The temporal dictionary ensemble (TDE) classifier for time series classification. In: Hutter, F., Kersting, K., Lijffijt, J., Valera, I. (eds.) ECML PKDD 2020. LNCS (LNAI), vol. 12457, pp. 660–676. Springer, Cham (2021). https://doi.org/10.1007/978-3-030-67658-2_38

24. Middlehurst, M., Large, J., Flynn, M., Lines, J., Bostrom, A., Bagnall, A.: Hive-cote 2.0: a new meta ensemble for time series classification. Mach. Learn. **110**(11–12), 3211–3243 (2021). https://doi.org/10.1007/s10994-021-06057-9

25. Middlehurst, M., Vickers, W., Bagnall, A.: Scalable dictionary classifiers for time series classification. In: Yin, H., Camacho, D., Tino, P., Tallón-Ballesteros, A.J., Menezes, R., Allmendinger, R. (eds.) IDEAL 2019. LNCS, vol. 11871, pp. 11–19. Springer, Cham (2019). https://doi.org/10.1007/978-3-030-33607-3_2

26. Schäfer, P.: The boss is concerned with time series classification in the presence of noise. Data Min. Knowl. Discov. **29**, 1505–1530 (2015). https://doi.org/10.1007/s10618-014-0377-7

27. Schäfer, P., Leser, U.: Fast and accurate time series classification with weasel. In: Proceedings of the 2017 ACM on Conference on Information and Knowledge Management, pp. 637–646 (2017). https://doi.org/10.1145/3132847.3132980

28. Schulz, E., Speekenbrink, M., Krause, A.: A tutorial on Gaussian process regression: modelling, exploring, and exploiting functions. J. Math. Psychol. **85**, 1–16 (2018). https://doi.org/10.1016/j.jmp.2018.03.001

29. Vargas, V.M., Gutiérrez, P.A., Rosati, R., Romeo, L., Frontoni, E., Hervás-Martínez, C.: Deep learning based hierarchical classifier for weapon stock aesthetic quality control assessment. Comput. Ind. **144**, 103786 (2023). https://doi.org/10.1016/j.compind.2022.103786

30. Wang, Z., Yan, W., Oates, T.: Time series classification from scratch with deep neural networks: a strong baseline. In: 2017 International Joint Conference on Neural Networks (IJCNN), pp. 1578–1585. IEEE (2017). https://doi.org/10.48550/arXiv.1611.06455

31. Zhou, Z., et al.: Methods to recognize depth of hard inclusions in soft tissue using ordinal classification for robotic palpation. IEEE Trans. Instrum. Meas. **71**, 1–12 (2022). https://doi.org/10.1109/TIM.2022.3198765

Real World Applications of BCI Systems

Effects of Stimulus Sequences on Brain-Computer Interfaces Using Code-Modulated Visual Evoked Potentials: An Offline Simulation

Jordy Thielen[✉][ⒾD]

Donders Institute for Brain, Cognition and Behaviour, Radboud University,
Thomas van Aquinostraat 4, 6525 GD Nijmegen, The Netherlands
jordy.thielen@donders.ru.nl
https://neurotechlab.socsci.ru.nl/

Abstract. Brain-computer interfaces (BCIs) translate brain activity into computer commands opening a novel non-muscular channel for communication and control. Before BCIs can be used effectively, they require extensive calibration of their machine learning algorithms using labeled training data. This tedious calibration process is not only time-consuming and costly but also restricts the exploration and optimization of stimulus parameters that could greatly enhance BCI performance. To overcome the challenge of acquiring large training datasets, a simulation framework was developed to eliminate the need for recording calibration data. Unlike previous studies, this simulation framework incorporates a biologically plausible forward model of the code-modulated visual evoked potential (c-VEP). By utilizing synthetic data generated by this improved simulation framework, an offline study was conducted to systematically compare five different stimulus conditions: almost perfect autocorrelation (APA) sequence, de Bruijn sequence, Golay sequence, Gold code, and m-sequence. The results of the study revealed that the Golay sequence achieved the highest grand average performance, followed by the APA sequence, m-sequence, Gold code, and finally the de Bruijn sequence. Furthermore, when the stimulus sequence was optimized for individual participants, typically the Golay and APA sequences exhibited the highest classification accuracy, leading to a significant improvement in overall accuracy as compared to using the Golay sequence for all participants. This research represents an important initial step towards optimizing stimulus parameters for BCIs using simulated data. This approach has the potential to accelerate the development and optimization of BCIs, resulting in more effective applications that can be tailored to individual users.

Keywords: Brain-computer interfacing · Code-modulated visual evoked potentials · Electroencephalography · Simulation

I. Rojas et al. (Eds.): IWANN 2023, LNCS 14135, pp. 555–568, 2023.
https://doi.org/10.1007/978-3-031-43078-7_45

1 Introduction

A brain-computer interface (BCI) is emerging neurotechnology that translates brain activity into digitized commands. As such, a BCI provides a novel non-muscular output pathway. A BCI can be particularly useful for people with motor disabilities, such as patients suffering from amyotrophic lateral sclerosis (ALS) [13]. The brain activity is typically recorded using electroencephalography (EEG), because it is affordable, practical, and non-invasive.

A major challenge in the BCI field is that prior to their usage, BCIs typically require a substantial amount of labeled training data to calibrate the machine learning algorithms to the user's brain activity. This requirement not only results in extensive training times that prevent immediate use, but also largely impede the exploration of novel stimulus protocols.

One of the fastest BCIs for communication uses the code-modulated visual evoked potential (c-VEP) as recorded by EEG [5]. The c-VEP is observed in the EEG when the user attends to visually presented pseudo-random sequences of flashes. As each of the presented symbols in a BCI speller concurrently flickers with a random but unique sequence of flashes, specific brain activity is evoked when the user attends to one of the symbols. Subsequently, machine learning algorithms infer the attended symbol from the user's evoked brain activity. Such visual BCI spellers allow users to select symbols or commands one-by-one, opening a novel non-muscular channel for communication and control [7].

The pseudo-random stimulation sequences that evoke the c-VEP are commonly borrowed from telecommunication. These sequences exhibit excellent correlation properties that may make the classification of their responses easier [1]. Indeed, most c-VEP BCI studies use such predefined sequences such as an m-sequence, Gold codes, Kasami codes, a Barker sequence, a Golay sequence, almost-perfect auto-correlation (APA) sequences, or a de Bruijn sequence [5]. Several recent studies have shown that other sequences such as random sequences and hand-crafted sequences may lead to faster c-VEP BCIs [5].

The continuous growth of studies in this field adds to the expanding repertoire of potential stimulus sequences that could enhance c-VEP BCI performance. However, several significant challenges arise from this situation. Firstly, the exploration, validation, and comparison of these sequences are greatly hindered by the limited availability of recorded EEG data. Acquiring large datasets for such evaluations is a time-consuming and expensive endeavor. Consequently, the optimization of stimulus protocols is often deemed impractical due to the scarcity of calibration data and the vast space of potential stimulus sequences.

Secondly, many studies that compare or validate new stimulus sequences typically focus on the population level, seeking to determine the optimal stimulus for the "average" user. However, it is essential to recognize that individual users exhibit subject-to-subject variability. By acknowledging this variance and undertaking the optimization of stimulus parameters tailored to each individual user, significant improvements in c-VEP BCI performance could be achieved.

To overcome these challenges, an effective approach could involve the utilization of synthetic (i.e., simulated) data [3]. This strategy has shown promise in a recent study conducted by Torres and Daly, in which they compared various stimulus sequences such as the APA sequence, Golay sequence, de Bruijn sequence, m-sequence, and Gold code [12]. Their findings revealed that the APA, Golay, and de Bruijn sequence exhibited superior performance compared to the more traditional m-sequence and Gold codes.

However, the existing simulation framework employed by Torres and Daly generated the c-VEP as a simple binary square-wave response by combining the pure binary stimulus sequence with additive EEG noise [12]. It is unlikely that the brain's response to a sequence of flashes follows a binary square-wave pattern. To enhance the biological plausibility of their forward model, it would be beneficial to incorporate the brain's response to a single flash, known as the flash-VEP. Additionally, by incorporating the flash-VEP, their model could be improved by considering the unique characteristics of individual participants, such as the amplitude and latency of each component (i.e., peaks) within the flash-VEP. Consequently, this approach would also allow for a more comprehensive modeling of inter-individual differences.

By integrating the flash-VEP into the simulation framework, researchers can achieve a more realistic representation of the brain's response and better tailor the stimulus parameters to individual users. This advancement would contribute to the overall understanding of the neural mechanisms underlying the c-VEP as well as an enhanced accuracy and effectiveness of the BCI system.

In this work, two aspects of the c-VEP BCI are investigated using an improved simulation framework for c-VEP data: (1) which of several commonly used stimulus sequences leads to the highest BCI performance on the population level, and (2) whether different stimulus sequences lead to the highest performance for individual participants. By addressing these aspects, this work aims to improve the efficiency and performance of BCIs for communication and control, by making use of an improved simulation framework for c-VEPs.

2   Methods

The improved forward model incorporates empirically derived flash-VEPs, which were obtained through the recording of flash-VEPs across various flash duration conditions for each individual participant. The methodology employed to record and estimate these empirical flash-VEPs is detailed in Sect. 2.1. Subsequently, in Sect. 2.2, the simulation framework utilizing these empirically derived flash-VEPs to generate synthetic c-VEPs is explained. Finally, Sect. 2.3 outlines the classification methods employed to analyze the generated synthetic datasets for each stimulus condition. For a visual representation of the entire processing pipeline, please refer to Fig. 1.

2.1   Empirical Data

Twenty-six participants (17 female) between the ages of 18 and 31 years (average 23 years) took part in the EEG experiment. All participants reported normal

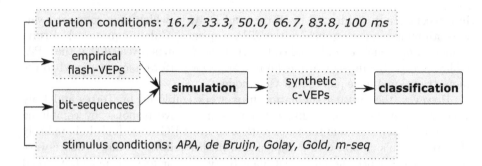

Fig. 1. Processing pipeline. This study employed a well-defined pipeline that encompassed several key components. Firstly, empirical participant-specific flash-VEPs were estimated, capturing individual variations in response to flashes of six different duration of flashes. These empirically derived flash-VEPs, along with specific bit-sequences, were then input into a carefully designed simulation framework. The simulation framework generated datasets containing synthetic c-VEPs tailored to each participant and corresponding to different stimulus conditions, including the almost perfect autocorrelation (APA) sequence, de Bruijn sequence, Golay sequence, Gold code, and m-sequence. These datasets served as the basis for subsequent classification tasks, allowing for the evaluation of the effectiveness of different stimulus conditions for individual participants using simulated EEG data.

or corrected-to-normal vision, no central nervous system abnormalities, and no history of epilepsy. Prior to the experiment, participants gave written informed consent and after the experiment they received payment for their participation. The experimental procedure and methods were approved by and performed in accordance with the guidelines of the local ethical committee of the Faculty of Social Sciences of the Radboud University.

The EEG data were recorded using 64 sintered Ag/AgCl active electrodes placed according to the 10-10 system and amplified by a Biosemi ActiveTwo amplifier, with a sampling frequency of 2 kHz. Raw continuous data were bandpass filtered with a 6th order Butterworth zero-phase forward and reverse digital IIR filter, with cut-off frequencies at 2 and 30 Hz. The data were then downsampled to 600 Hz. To match the methodology of Torres and Daly [12], only 9 channels were used for further analysis (P3, Pz, P4, PO7, POz, PO8, O1, Oz, O2).

During the experiment, participants were seated 60 cm away from a 24 inch Benq XL2420Z monitor with a resolution of 1920 × 1080 pixels and a refresh rate of 120 Hz. Despite the 120 Hz refresh rate, stimuli were presented at a presentation rate of 60 Hz by presenting every bit in a sequence for a duration of two frames.

The experiment contained several tasks, although only one specific task is relevant to the current study. In this particular task, participants were presented with a square box at the center of the screen, covering an angular subtense of 3°. The box appeared against a gray mean-luminance background. During this task, participants were instructed to maintain fixation on a cross inside the box while completing 90 trials. These trials were evenly divided into two runs, with a self-paced break provided between the runs. Each trial consisted of a pseudo-random sequence of flashes, with durations varying across six distinct conditions. These duration conditions corresponded to different lengths of the flashes, ranging from 2 to 12 frames in 2-frame increments. In terms of time, this equated to durations of 16.7 ms to 100 ms in steps of 16.7 ms, considering a refresh rate of 120 Hz. Notably, these duration conditions align with the basic components of the stimulus sequences that were examined in this study, forming the fundamental units to be able to generate synthetic data.

In each trial, all six duration conditions were presented twice, followed by a final flash of the shortest duration, resulting in a total of 13 flashes per trial. The order of flashes was pseudo-randomized within and across trials using a balanced Latin square design to mitigate any order effects. To decrease the overlap between subsequent flash-VEPs, a random inter-stimulus interval between 250 and 500 ms was included between each flash. Additionally, an inter-trial interval of 2 s was included between trials to allow for rest and eye blinks.

In summary, each of the 90 trials took between 4.5 and 5.5 s and contained 13 flashes. For each participant, responses to a total of 180 flashes were collected for each duration condition, and responses to 270 flashes for the shortest duration condition.

These empirical data were analyzed per participant using Python (version 3.9.12) and the MNE library (version 1.2.2). The continuous preprocessed EEG data were sliced into epochs $\mathbf{X}_i \in \mathbb{R}^{c \times m}$ of $c = 9$ channels and $m = 300$ samples starting 200 ms before until 300 ms after flash onset. Epochs were baseline corrected using the 200 ms before flash onset, after which the baseline was removed ($m = 180$). Epochs that contained amplitudes exceeding 150 μV were rejected from the datasets. Subsequently, multi-channel flash-VEPs $\mathbf{R}_j \in \mathbb{R}^{c \times m}$ were computed for each duration condition j by averaging over all k epochs that belong to the same jth flash duration:

$$\mathbf{R}_j = \frac{1}{k} \sum_i^k \mathbf{X}_i \tag{1}$$

Finally, the multi-channel flash-VEPs $\mathbf{R}_j$ were spatially filtered following:

$$\mathbf{r}_j = \mathbf{w}^\top \mathbf{R}_j \tag{2}$$

with $\mathbf{r}_j \in \mathbb{R}^m$ is the spatially filtered flash-VEP for the jth duration condition and $\mathbf{w} \in \mathbb{R}^c$ is the spatial filter as estimated with canonical correlation analysis (CCA). CCA was applied to the concatenated epochs $\tilde{\mathbf{X}} = [\mathbf{X}_1, \ldots, \mathbf{X}_k]$ and concatenated flash-VEPs $\tilde{\mathbf{R}} = [\mathbf{R}_1, \ldots, \mathbf{R}_k]$ as follows [8]:

$$\underset{\mathbf{w}, \mathbf{v}}{\arg \max} \, \rho(\mathbf{w}^\top \tilde{\mathbf{X}}, \mathbf{v}^\top \tilde{\mathbf{R}}) \tag{3}$$

where also $\mathbf{v} \in \mathbb{R}^c$ is a spatial filter, but in general only $\mathbf{w}$ is used [8]. Note, only the first component of the CCA is used.

2.2   Synthetic Data

Synthetic datasets were generated for each participant for five stimulus conditions, including (1) APA sequence, (2) de Bruijn sequence, (3) Golay sequence, (4) Gold code, and (5) m-sequence. Figure 2 shows the time-series of these bit-sequences for one full cycle. The APA sequence and Golay sequence were taken from Wei and colleagues [14]. The de Bruijn sequence was generated with an alphabet of size two and a sequence order of 6. The Gold code was generated with the preferred pair of primitive polynomials $f(x) = x^6 + x$ and $f(x) = x^6 + x^5 + x^2 + x^1$. Gold codes come in sets, from which the first balanced code with a run-length distribution limited to the six duration conditions was selected. The m-sequence was generated with the primitive polynomial $f(x) = x^6 + x$. The APA, de Bruijn, and Golay sequences each have a length of 64 bits, while the Gold code and m-sequence have a length of 63 bits. At a presentation rate of 60 Hz this corresponds to a duration of $64/60 = 1.07$ and $63/60 = 1.05$ seconds, respectively.

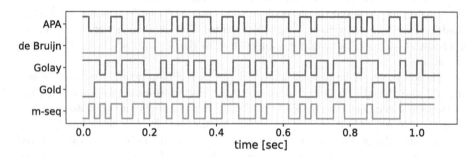

Fig. 2. Stimulus conditions. The study examined five distinct stimulus sequences, which were the almost perfect autocorrelation (APA) sequence, de Bruijn sequence, Golay sequence, Gold code, and m-sequence. Each of these are binary sequences with a "1" representing a white background (i.e., a flash) and a "0" representing a black background, for instance of the symbols in a visual matrix speller. For the purpose of creating a 32-class speller BCI with one of these stimulus conditions, the sequence was circularly shifted to generate the stimulus sequences for the remaining 31 classes. This circular shifting ensured the utilization of the same underlying sequence structure while varying the starting position, thereby enabling the distinction between different classes within the speller. In this figure, the solid and dashed gray vertical lines denote the frame rate of the presentation screen, which was 120 Hz. The solid gray vertical lines indicate the individual bits of the sequence as they were presented at a presentation rate of 60 Hz (i.e., each bit was duplicated to match the 120 Hz screen refresh rate).

For each of the five stimulus conditions, a 32-class c-VEP BCI was simulated using the specific bit-sequence that was circularly shifted with a lag of 2 bits to form 32 classes, which is a standard procedure for c-VEP BCIs [5]. Five trials were generated for each of the 32 classes for each stimulus condition and participant, resulting in 160 synthetic EEG trials per stimulus condition and participant. Each trial lasted one full code cycle, which was $64/60 = 1.07\,\mathrm{s}$ for the APA, de Bruijn and Golay sequences, and $63/60 = 1.05\,\mathrm{s}$ for the Gold code and m-sequence. The synthetic EEG data were sampled at $600\,\mathrm{Hz}$.

The simBCI toolbox [4] was used to generate synthetic EEG data using Matlab (version 9.11.0 (R2021b), The MathWorks Inc.). Each trial was modeled as the linear superposition of the spatially filtered flash-VEPs $\mathbf{r}_j$ according to the trial-specific bit-sequence as follows [9–11]:

$$y(t) = \sum_{j}^{6} \sum_{\tau}^{e} I_j(t) r_j(t - \tau) \tag{4}$$

Here, $y(t)$ is the single-source c-VEP at time t, I_j is an indicator function specifying the onset of the jth duration condition within a stimulus sequence, and $r_j(t)$ is the flash-VEP of the jth duration condition at time t, which has a length of $e = 180$ samples (i.e., 300 ms at $600\,\mathrm{Hz}$). In other words, assuming the linear superposition hypothesis [2], the c-VEP is assumed to be the result of the (discrete) convolution where the bit-sequence acts as a sequence of Dirac delta pulses and the flash-VEP as the impulse response function [9–11].

The resulting single-source c-VEP signal was propagated to the $c = 9$ EEG channels using a linear superposition model that employed a leadfield matrix to encode the electrical propagation of the head model, assuming one occipital source. The head model was taken from Torres and Daly [12], which is "a physiologically realistic nonspecific brain model, which projects uniform brain volume data to the surface electrodes, constrained by the cortical surface normal. The conductive parameters of the model consist of scalp, skull, and brain mesh layers, with normalized conductivities of 1, 1/15, and 1, respectively." Included in this head model are also the dipole sources at frontal areas that represent the participant's eyes, used to model any eye artefacts.

Lastly, additive noise was added to the signal component of the modeled EEG. Firstly, pink noise was added which represents the $1/f$ characteristic of EEG. Secondly, white noise was added representing the measurement noise at the cortical level. Thirdly, occasional eye blinks were added with a probability of occurrence of 0.3. Finally, occasional eye movements were added with a probability of occurrence of 0.1 and a random maximum duration of $1\,\mathrm{s}$.

Similar to the empirical data, the synthetic data were band-pass filtered using a sixth order Butterworth filter with cut-off frequencies at 2 and $30\,\mathrm{Hz}$. The single-trial signal-to-noise ratio (SNR), computed as follows:

$$\mathrm{SNR} = \frac{\sqrt{\frac{1}{k} \sum_i^k \mathbf{T}_{\mathbf{y}_i}^2}}{\sqrt{\frac{1}{k} \sum_i^k (\mathbf{X}_i - \mathbf{T}_{\mathbf{y}_i})^2}} \tag{5}$$

was tuned to be on average 0.259 with a standard deviation of 0.054, which is similar to empirically observed SNR [11].

2.3   Classification

The synthetic datasets were evaluated using Python (version 3.9.12) and the PyNT library (version 0.0.2)[1]. The data were classified using the so-called 'reference pipeline' as outlined by Martinez-Cagigal and colleagues [5], which defines a template matching classifier as follows:

$$\hat{y} = \arg\max_{i} \rho(\mathbf{w}^\top \mathbf{X}, \mathbf{w}^\top \mathbf{T}_i) \tag{6}$$

where $\hat{y}$ is the predicted class-label, $\rho(.)$ is the Pearson's correlation, $\mathbf{w} \in \mathbb{R}^c$ is a spatial filter for $c = 9$ channels, $\mathbf{X} \in \mathbb{R}^{c \times m}$ is the single-trial EEG for $m = 64/60 * 600 = 640$ samples for the APA, de Bruijn, and Golay sequences, and $m = 63/60 * 600 = 630$ samples for the Gold code and m-sequence, and $\mathbf{T}_i \in \mathbb{R}^{c \times m}$ is the multi-channel template of expected responses for the ith class.

Given training data $D = \{(\mathbf{X}_i, y_i), \ldots, (\mathbf{X}_k, y_k)\}$ with k labeled training trials, the multi-channel templates can be computed by averaging the synchronized trials. Specifically, because the different classes are responses to the same but circularly shifted stimulus sequence, the EEG can be synchronized by undoing the latency shift. Subsequently, the template of zero latency is found by:

$$\mathbf{T}_0 = \frac{1}{k} \sum_{i}^{k} \mathbf{X}_i \tag{7}$$

Then, each of the other templates $\mathbf{T}_i$ are created by adding the latency shifts.

The final step is to find the spatial filter $\mathbf{w}$, which is done with CCA using concatenated single-trials $\tilde{\mathbf{X}} = [\mathbf{X}_0, \ldots, \mathbf{X}_k]$ and concatenated templates $\tilde{\mathbf{T}} = [\mathbf{T}_{y_0}, \ldots, \mathbf{T}_{y_k}]$ as follows [8]:

$$\arg\max_{\mathbf{w}, \mathbf{v}} \rho(\mathbf{w}^\top \tilde{\mathbf{X}}, \mathbf{v}^\top \tilde{\mathbf{T}}) \tag{8}$$

Note, both $\mathbf{w}$ and $\mathbf{v}$ are spatial filters, but typically only $\mathbf{w}$ and only the first CCA component is used.

3   Results

For each of the 26 participants six spatially-filtered flash-VEPs were estimated from empirical data, one for each of six duration conditions, see Fig. 3. Each of these filtered flash-VEPs showed a dominant negative peak at 72 ms and a positive peak at 117 ms after flash onset, which is typical for flash-VEPs. Between participants there was variability in the flash-VEPs, specifically in its latency and amplitude, see the participant-specific time-series in Fig. 3.

Using the participant-specific spatially-filtered flash-VEPs, synthetic c-VEP datasets were generated for each of the five different stimulus conditions. The

---

[1] PyNT can be installed from source: https://gitlab.socsci.ru.nl/jthielen/pynt.

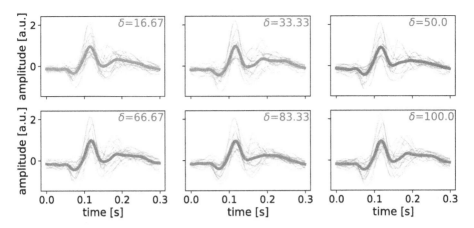

Fig. 3. Empirical visual evoked potentials. This study estimated an empirical visual evoked potential (VEP) for each of six flash duration conditions. This figures shows the grand average flash-VEPs in color, and the participant-specific flash-VEPs in gray. The duration condition is marked with δ and denotes the length of the flash in milliseconds. Note, these flash-VEPs are spatially filtered with canonical correlation analysis (CCA) which is agnostic to amplitude scaling.

autocorrelation function of the stimulus sequences and estimated c-VEP templates for sub-01 are shown in Fig. 4. The autocorrelation functions of the bit-sequences substantially deviate from those of the evoked responses, which is likely caused by the convolution with the flash-VEP, which can be considered as a filtering of the bit-sequence. For example, from the stimulus domain (top row of Fig. 4) one would conclude that the m-sequence is favorable. However, when considering the response domain (bottom row of Fig. 4), in which the classification will be performed, the correlation properties seem less ideal, and seem more similar to those of the other sequences.

The effectiveness of each stimulus condition was offline evaluated by measuring the classification accuracy of a simulated 32-class c-VEP BCI using a 5-fold cross-validation within each stimulus condition and participant. In decreasing order, the grand-average accuracy and standard error for the Golay sequence, APA sequence, m-sequence, Gold code, and de Bruijn sequence were 0.926 ± 0.018, 0.912 ± 0.024, 0.904 ± 0.024, 0.872 ± 0.029, and 0.868 ± 0.030, respectively, see Fig. 5.

Pair-wise one-sided Wilcoxon signed-rank tests showed that the Golay sequence achieved a significantly higher accuracy than the de Bruijn sequence ($p < .001$), the Gold code ($p < .001$) and m-sequence ($p = .010$), but not the APA sequence ($p = .180$). The APA sequence obtained a higher accuracy than the de Bruijn sequence ($p < .001$) and Gold code ($p = 0.001$), but not the m-sequence ($p = .176$). The m-sequence performed significantly better than the Gold code ($p = .002$) and the de Bruijn sequence ($p < .001$). The Gold code did not perform significantly higher than the de Bruijn sequence ($p = .314$).

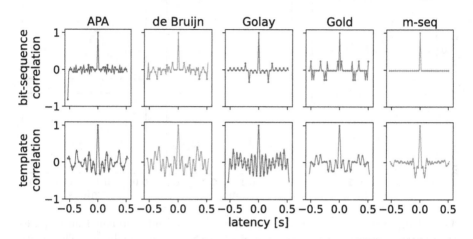

Fig. 4. Autocorrelation function of bit-sequences and template responses.
This study estimated the autocorrelation function of each of the stimulus conditions of
both their bit-sequences as well as their template responses. The template responses
used for this figure were generated from synthetic c-VEP data from participant sub-01.
The autocorrelation is given for all latency shifts, where the dots represent the latencies
as used by the 32-class c-VEP BCI.

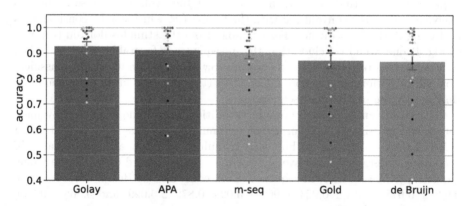

Fig. 5. Classification accuracy of simulated 32-class c-VEP BCI. This study
offline evaluated the performance of several stimulus sequences using synthetic c-VEP
data. This figure shows the grand average classification accuracy for each of the stimulus conditions in descending order: Golay sequence, almost perfect autocorrelation
(APA) sequence, m-sequence, Gold code and de Bruijn sequence. The bars represent
the average accuracy and the error bars represent the standard error. The gray dots
depict the performance as obtained by individual participants. The theoretical chance-
level classification accuracy is 1/32.

Instead of comparing the average accuracy at the population level across different stimulus conditions, an alternative approach is to optimize the stimulus sequence for each individual participant. In this study, the highest accuracy was observed with the Golay sequence for 7 participants, the APA sequence for 7 participants, the m-sequence for 2 participants, and the Gold code for 2 participants. For the remaining 8 participants, two to four different sequences yielded equal performances. By selecting the best performing stimulus condition for each participant individually, a significantly higher grand-average accuracy of 0.937 ± 0.017 ($p < .001$) was achieved compared to using the Golay sequence for all participants, which was the best-performing sequence at the population level with an accuracy of 0.926 ± 0.018. This individual optimization approach highlights the importance of tailoring the stimulus condition to the specific characteristics of each participant, resulting in an improved overall performance compared to a one-size-fits-all approach.

4   Discussion

The primary objective of this study was to assess the performance of c-VEP BCIs under five distinct stimulus conditions, utilizing simulated EEG datasets. The results revealed that the Golay sequence achieved the highest classification accuracy, surpassing the APA sequence, m-sequence, Gold code, and the de Bruijn sequence in that order. Although the Golay sequence achieved the highest accuracy, it was not significantly different than the accuracy obtained with the APA sequence.

Notably, when considering the performance at the population level, there was a difference of 0.058 between the accuracy of the best-performing sequence and the worst-performing sequence. This discrepancy highlights the significance of evaluating and discerning the disparities among various stimulus sequences.

It is particularly intriguing that despite the m-sequence possessing an ideal autocorrelation function, both the Golay sequence and the APA sequence outperformed it in terms of classification accuracy. This finding underscores the importance of considering factors beyond just autocorrelation in the stimulus domain when selecting stimulus sequences for c-VEP BCIs. Instead, one should consider the characteristics of the sequences in their response domain.

Moreover, upon examining individual participants, it was observed that the majority achieved their highest performance levels with either the Golay or the APA sequence. Conversely, only a subset of participants attained their peak performance with the m-sequence or Gold code. Notably, none of the participants achieved their highest performance with the de Bruijn sequence.

When the best performing stimulus condition was determined individually for each participant, an average increase of 0.011 in classification accuracy was observed. This finding strongly suggests that tailoring stimulus sequences to the specific characteristics of individual participants leads to a statistically significant improvement in decoding performance compared to using an average stimulus sequence for all participants.

This study provides further support for the findings of Torres and Daly [12] regarding the effectiveness of the Golay and APA sequences as stimulus sequences in c-VEP BCIs. Interestingly, these sequences demonstrate promising potential, outperforming the current standards for state-of-the-art c-VEP BCIs, namely the m-sequence and Gold code [5]. By confirming the superiority of the Golay and APA sequences over these established sequences, this study highlights the potential for further advancements in c-VEP BCI technology.

These findings contrast with the previous findings of Torres and Daly [12], where a clear distinction in performance was observed among the APA, de Bruijn, and Golay sequences, all achieving high levels of accuracy, while the Gold code and m-sequence demonstrated comparatively lower performance. Notably, in the present study, the de Bruijn sequence did not emerge as one of the top-performing sequences and in fact was the least performing stimulus.

Furthermore, the effect sizes observed in the current study were relatively small compared to the work of Torres and Daly. This discrepancy could potentially be attributed to the utilization of different formulations of the forward model for generating the c-VEP. In the study by Torres and Daly, the forward model incorporated the actual bit-sequence as the signal with additive noise. In contrast, in the current study, the bit-sequences were convolved with participant-specific flash-VEPs, aiming to enhance the biological plausibility of the model. By incorporating such a more biologically plausible approach, the present study introduces a novel perspective that could potentially influence the outcomes and understanding of c-VEP BCIs.

It is crucial to acknowledge certain limitations in this study. Firstly, the analyses were conducted offline using synthetic data, and it is essential to validate these findings in an online BCI using real EEG data. The reliance on synthetic data generated from an empirical simulation framework might restrict the generalizability of the results to different participant populations.

Furthermore, the incorporation of empirically estimated flash-VEPs in the simulation framework introduces another limitation. While using real EEG data in the simulation enhances realism, it also implies a dependency on the specific characteristics of the participant population from which the flash-VEPs were derived. However, this limitation could potentially be addressed by modeling the flash-VEP as well, thereby allowing for more flexibility and broader applicability.

Additionally, it is important to note that the forward model assumes perfect synchronization between the stimulus presentation and EEG recording, which may not always hold true in real-world scenarios (e.g., vertical blanking on presentation screens). This lack of synchronization has been demonstrated to have a negative impact on performance if not properly corrected for [6]. Also, the assumption that the c-VEP is a linear addition of flash-VEPs may oversimplify the complexity of the flash-VEP characteristics, which could potentially depend on the temporal proximity of previous flashes. Thus, a more sophisticated model of the flash-VEP might be necessary.

Lastly, the forward model assumes that the duration of a flash defines an event for which a unique flash-VEP is learned. However, alternative event definitions, such as edge responses, have been proposed in the literature [11,15], which might provide a more plausible representation.

Taking these limitations into account, future research should aim to address these aspects and validate the findings in online BCI settings with real EEG data, consider alternative models for flash-VEP characteristics, and explore different event definitions to further enhance the understanding and application of c-VEP BCIs.

To summarize, this study introduced an enhanced simulation framework for c-VEP responses, enabling the optimization of stimulus parameters tailored to individual participants in c-VEP BCIs. The findings demonstrate a noteworthy improvement in classification accuracy by using Golay or APA sequences, or by customizing the BCI for each individual, surpassing the performance achieved with average stimulus conditions. The future trajectory of this research aims to further refine the c-VEP simulation framework, ultimately elucidating the neural mechanisms underlying the c-VEP as well as facilitating the development of high-speed c-VEP BCIs that are specifically tailored to individual users, eliminating the need for extensive training data.

Still, it is essential to approach these results with caution, considering the limitations associated with the simulation framework and the underlying assumptions of the forward model. Careful consideration of these factors is necessary to ensure the accurate interpretation and application of the findings in real-world scenarios. Further investigations are required to address these limitations and enhance the robustness and reliability of the c-VEP BCI technology.

References

1. Bin, G., Gao, X., Wang, Y., Hong, B., Gao, S.: VEP-based brain-computer interfaces: time, frequency, and code modulations. IEEE Comput. Intell. Mag. **4**(4), 22–26 (2009)
2. Capilla, A., Pazo-Alvarez, P., Darriba, A., Campo, P., Gross, J.: Steady-state visual evoked potentials can be explained by temporal superposition of transient event-related responses. PLoS ONE **6**(1), e14543 (2011)
3. Eilts, H., Putze, F.: Is that real? A multifaceted evaluation of the quality of simulated EEG signals for passive BCI. In: 2022 IEEE International Conference on Systems, Man, and Cybernetics (SMC), pp. 2639–2644. IEEE (2022)
4. Lindgren, J.T., Merlini, A., Lecuyer, A., Andriulli, F.P.: simBCI-a framework for studying BCI methods by simulated EEG. IEEE Trans. Neural Syst. Rehabil. Eng. **26**(11), 2096–2105 (2018)
5. Martínez-Cagigal, V., Thielen, J., Santamaría-Vázquez, E., Pérez-Velasco, S., Desain, P., Hornero, R.: Brain-computer interfaces based on code-modulated visual evoked potentials (c-VEP): a literature review. J. Neural Eng. **18**, 061002 (2021)
6. Nagel, S., Dreher, W., Rosenstiel, W., Spüler, M.: The effect of monitor raster latency on VEPs, ERPs and brain-computer interface performance. J. Neurosci. Methods **295**, 45–50 (2018)
7. Rezeika, A., Benda, M., Stawicki, P., Gembler, F., Saboor, A., Volosyak, I.: Brain-computer interface spellers: a review. Brain Sci. **8**(4), 57 (2018)

8. Spüler, M., Walter, A., Rosenstiel, W., Bogdan, M.: Spatial filtering based on canonical correlation analysis for classification of evoked or event-related potentials in EEG data. IEEE Trans. Neural Syst. Rehabil. Eng. **22**(6), 1097–1103 (2013)
9. Thielen, J., Marsman, P., Farquhar, J., Desain, P.: Re(con)volution: accurate response prediction for broad-band evoked potentials-based brain computer interfaces. In: Guger, C., Allison, B., Lebedev, M. (eds.) Brain-Computer Interface Research. SECE, pp. 35–42. Springer, Cham (2017). https://doi.org/10.1007/978-3-319-64373-1_4
10. Thielen, J., van den Broek, P., Farquhar, J., Desain, P.: Broad-band visually evoked potentials: re(con)volution in brain-computer interfacing. PLoS ONE **10**(7), e0133797 (2015)
11. Thielen, J., Marsman, P., Farquhar, J., Desain, P.: From full calibration to zero training for a code-modulated visual evoked potentials for brain-computer interface. J. Neural Eng. **18**(5), 056007 (2021)
12. Torres, J.A.R., Daly, I.: How to build a fast and accurate code-modulated brain-computer interface. J. Neural Eng. **18**, 046052 (2021)
13. Verbaarschot, C., et al.: A visual brain-computer interface as communication aid for patients with amyotrophic lateral sclerosis. Clin. Neurophysiol. **132**(10), 2404–2415 (2021)
14. Wei, Q., et al.: A novel c-VEP BCI paradigm for increasing the number of stimulus targets based on grouping modulation with different codes. IEEE Trans. Neural Syst. Rehabil. Eng. **26**(6), 1178–1187 (2018)
15. Yasinzai, M.N., Ider, Y.Z.: New approach for designing cVEP BCI stimuli based on superposition of edge responses. Biomed. Phys. Eng. Express **6**(4), 045018 (2020)

Evaluation of Visual Parameters to Control a Visual ERP-BCI Under Single-Trial Classification

Álvaro Fernández-Rodríguez[1] , Ricardo Ron-Angevin[1(✉)] ,
Francisco Velasco-Álvarez[1] , Jaime Diaz-Pineda[2], Théodore Letouzé[3] ,
and Jean-Marc André[3]

[1] Departamento de Tecnología Electrónica, Instituto Universitario de Investigación en
Telecomunicación de la Universidad de Málaga (TELMA), Universidad de Málaga,
29071 Malaga, Spain
{afernandezrguez,rron}@uma.es, fvelasco@dte.uma.es
[2] Thales Avionics France, Bordeaux, France
jaime.diazpineda@fr.thalesgroup.com
[3] Bordeaux University, INP Bordeaux-ENSC, Laboratoire IMS - UMR CNRS 5218,
33400 Talence, France
{tletouze,jean-marc.andre}@ensc.fr

Abstract. A brain-computer interface (BCIs) based on event-related potentials
(ERPs) is a technology that provides a communication channel between a device
and a user through their brain activity. These systems could be used to assist
and facilitate decision making in applications such as an air traffic controller
(ATC). Thus, this work attempts to be an approximation to determine whether
it is possible to detect the stimulus through a single presentation of a stimulus
(single-trial classification) and furthermore, to evaluate the effects of the type of
stimulus to be detected, or not knowing the position of the stimulus appearance in
an ERP-BCI. This experiment has involved six participants in four experimental
conditions. Two conditions varied only in the type of stimulus used, faces (a type
of stimulus that has shown high performance in previous ERP-BCI proposals)
versus radar planes; and two conditions varied in the prior knowledge of where
the stimulus would appear on the screen (knowing vs. not knowing). The results
suggest that the use of single-trial classification could be adequate to correctly
detect the desired stimulus using and ERP-BCI. In addition, the results reveal no
significant effect on either of the two factors. Therefore, it seems that radar planes
may be as suitable stimuli as faces and that not knowing the location of the target
stimulus is not a significant problem, at least in a standard BCI scenario without
distracting stimuli. Therefore, future studies should consider these findings for the
design of an ATC using an ERP-BCI for stimulus detection.

Keywords: Brain-Computer Interface (BCI) · Event-Related Potential (ERP) ·
Single-Trial Classification · Air Traffic Controller (ATC)

I. Rojas et al. (Eds.): IWANN 2023, LNCS 14135, pp. 569–579, 2023.
https://doi.org/10.1007/978-3-031-43078-7_46

1 Introduction

Brain-computer interfaces (BCIs) use brain activity to create a communication pathway between a device and a user [1]. The most common method to measure brain activity in a BCI is electroencephalography (EEG) [2]. EEG has several advantages such as its low cost, non-invasive nature, and good temporal resolution [3]. BCIs have been employed in several areas, including clinical and recreational applications [4]. Recent research suggests that BCIs could also be useful in decision-making and monitoring user states during surveillance tasks in situational awareness contexts [5, 6]. Situational awareness refers to the comprehension of environmental conditions and events, considering their temporal and spatial context, as well as predicting their potential future states. A hierarchical framework, proposed by [7], identifies three levels to approach SA: (i) perception of current situation elements, (ii) comprehension of the current situation, and (iii) prediction of future situations.

Air traffic control (ATC) is a scenario where a trained operator guides planes on the ground and through a specific area of regulated airspace. The primary objectives of ATC are to prevent collisions, organize air traffic flow, and provide pilots with relevant information and support. Therefore, ATC could be a suitable scenario for the use of brain-computer interfaces (BCIs) to aid decision-making, where a user needs to be aware of different cues and respond accordingly [8–10]. This paper focuses on the applications of BCIs for ATC, with the aim of enhancing the safety and precision of the controlled system. Two types of BCI systems can be distinguished to achieve these objectives: passive and active. A passive BCI aims to recognize the user's state during task execution, such as their level of tiredness or mental workload [10]. This information could be valuable for the system to detect potential errors in detecting critical cues for preventing incidents [11, 12]. On the other hand, an active BCI would assist with decision-making, such as detecting the appearance of new relevant elements on the map. To our knowledge, there is no previous work that has employed an active BCI for detecting new elements in the ATC scenario. Hence, this study focuses on active BCIs and the first level of the situational awareness framework, i.e., perceiving elements in the current situation. This involves detecting the appearance of new key elements—such as new planes on the map—using the user's EEG signal to control the system.

ATC operators are required to attend to planes as visual stimuli on a virtual map, so this study uses visual event-related potentials (ERPs) recorded through EEG as the input signal for detection. Visual ERPs refer to potential changes in brain activity that occur in response to the presentation of visual stimuli. ERPs are influenced by factors such as the type [13], size [14, 15], and luminosity [16] of the stimuli. When designing a visual ERP-BCI for an ATC scenario, it's important to consider these factors based on previous research. There are some key differences between visual ERP-BCI applications like wheelchair [17] or virtual keyboard [18] control and ATC. For instance, the number of times the target stimulus is presented (only one) and the location of its appearance (unknown) are especially relevant in the case of an ATC application. In most visual ERP-BCI applications, the target stimuli are displayed multiple times to increase the likelihood of accurate selection. However, in applications such as ATC where alert messages are presented, it is crucial that the target stimulus can be recognized just after one presentation. This requires the visual ERP-BCI to operate with single-trial classification, where the detection of a target stimulus is identified from a single presentation of the stimulus. However, this presents a challenge as ERP-BCIs typically require multiple stimulus presentations to effectively distinguish the relevant components of the EEG signal from the noise, such as muscle artifacts. The noise level decreases as more presentations are made, allowing better observation of ERP components associated with the presentation of a target stimulus. However, previous ERP-BCI proposals that focus on using single-trial classification have shown acceptable performance (~80% accuracy [19–21]). However, these previous works employed a different scenario than the one used in an ATC, i.e., they did not address the characteristics that could constrain the performance of an ATC, such as the type of visual stimuli to be attended, the use of a stimulus-rich map as background, moving planes, or small target stimuli like the planes to be detected. Therefore, exploring the use of single-trial classification under some specific characteristics presented in an ATC scenario could be worthwhile. In visual ERP-BCIs, the best performing stimuli to date are the red faces on a white background [22], and they are presented in a specific location that the user knows beforehand; however, in an ATC, the used stimulus are planes that appear in an unknown location. Therefore, it would be interesting to assess whether the type of stimulus to be attended and not knowing the position of stimulus appearance affects performance.

The objective of this study was to explore the use of single-trial classification and the impact of two visual factors on the accuracy of a visual ERP-BCI system in detecting new planes in a situational awareness scenario by an ATC. The utilization of an active BCI to aid an ATC is a unique approach; hence, two experiments were carried out to explore this approach. The initial experiment aimed to test the single-trial classification and BCI single-character paradigm (SCP) [23] to analyze the effects of different variables. It involved the presentation of two types of stimuli (faces and radar planes) and determining the impact of knowing or not knowing the location where the target stimulus would appear.

2 Method

2.1 Participants

The study has involved six participants (22.6 $\pm$ 1.52 years old, one woman, named P01-P06). Only P01 and P02 had previous experience in the control of an ERP-BCI. All subjects gave their written informed consent on the anonymous use of their EEG data. They declared having normal or corrected-to-normal vision. The study was approved by the Ethics Committee of the University of Malaga and met the ethical standards of the Declaration of Helsinki.

2.2 Data Acquisition and Signal Processing

Signals were recorded through eight active electrodes, namely Fz, Cz, Pz, Oz, P3, P4, PO7, and PO8 (10/10 international system). A reference electrode was placed on the left mastoid, and a ground electrode was placed at AFz. An acti-CHamp amplifier (Brain

Products GmbH, Gilching, Germany) was used, with a sample rate of 250 Hz, a band-pass filter of 0.1–30 Hz, a notch filter of 50 Hz, and an epoch length of 800 ms. The data were collected by BCI2000 [24]. When offline tasks were over, the weights of a classifier were calculated from the data of the condition tested through a stepwise linear discriminant analysis (SWLDA), using the P300Classifier, a BCI2000 tool. These weights were later used to carry out online tasks and to offer feedback to participants.

An HP Envy 15-j100 laptop was used (2.20 GHz, 16 GB, Windows 10), but the display was an Acer P224W screen of 46.47 × 31.08 cm (16:10 ratio), connected through HDMI, at a resolution of 1680 × 1050 pixels. The refresh rate of the screen was 60.014 Hz.

2.3 Experimental Conditions

This experiment aimed to investigate the impact of the type of stimulus used and whether the participant was aware of where the stimulus would appear on the performance of a visual ERP-BCI. The experiment used the BCI2000 software [24] and employed the SCP [23] with single-trial classification. The SCP involves presenting each stimulus sequentially at a different position on the display, with nine possible locations arranged in a 3x3 matrix. The stimuli used in the experiment varied based on the experimental condition, but they all measured 3.4 × 3.4 cm and were displayed on a black background. The user's distance from the screen was approximately 60 cm. The goal was to validate the use of an active BCI for detecting a stimulus presented only once in a specific position on the screen, which is similar to the case of plane detection for an ATC. The following experimental conditions were used:

C1-faces. The stimuli used were red celebrity faces with a white square background, a type of stimuli that has been suggested by recent work as one of the most appropriate to obtain high accuracy in the control of a visual ERP-BCI [22]. Both target and non-target stimuli were presented, and the user knew in advance the exact position of appearance of the target stimulus.

C2-planes. It was the same as C1-faces—the presence of target and non-target stimuli and the user knew the specific location of the target stimulus—but employed symbols similar to those used for planes on radars.

C3-known. The stimuli were also radar planes and the user knew in advance the exact position of the target stimulus. However, the non-target stimuli were not presented, i.e., only the target stimulus to be attended by the user appeared on the screen.

C4-unknown. It was similar to C3-known as it also employed radar plans, and non-target stimuli were not presented; however, in this condition, the user did not know in advance where the target stimulus would appear (Fig. 1).

The present study has had a progressive approach in order to evaluate relevant factors in the use of an ERP-BCI for the detection of new elements in an ATC scenario. For this purpose, different conditions have been evaluated until reaching C4-unknown, in which the stimuli were radar planes appearing in an unknown position of the interface, as it would happen in an ATC scenario. Thanks to this progressive approach, in addition to the

C1-faces   **C2-planes, C3-known, C4-unknown**

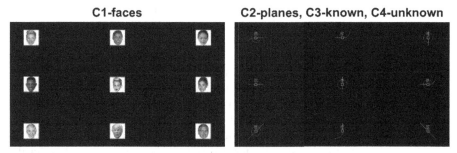

Fig. 1. Stimuli and locations used to present them on the screen. The C1-faces condition used celebrity faces, while the C2-planes, C3-known, and C4-unknown conditions used stimuli that simulated those used on flight radar. Images of celebrity faces have been pixelated for copyright reasons. The celebrity faces were (from left to right and from top to bottom): Scarlett Johansson, Cristiano Ronaldo, Rihanna, Will Smith, Miley Cyrus, Ariana Grande, Ellen DeGeneres, Donald Trump, and George Clooney.

use of single-trial classification, two factors have been evaluated during the experiment across the different conditions. Specifically, the aim of these conditions was to study the effect of two factors on system performance when detecting the presence of specific target stimuli in the interface based on the user's EEG signal. On the one hand, comparison between C1-faces and C2-planes allowed evaluating the effect of the type of stimulus. On the other hand, comparison between C3-known and C4-unknown allowed evaluating the effect of knowing in advance the exact location of appearance of the target stimulus.

2.4 Procedure

The participant arrived at the laboratory and received an explanation of the experimental procedure. They provided informed consent, the EEG electrodes and cap were placed, and the tasks could begin. The testing involved a design where each participant completed all conditions, which included a calibration task to adjust the system and an online task where the system aimed to detect specific stimuli. During the online task, the user received feedback on their performance based on specific parameters (i.e., the weights for the P300Classifier) already calculated after the calibration task. The terms used to detail the procedure of the experiments included the following. A run is the process to detect a single target stimulus. To complete a run, all the stimuli that compose the interface must be presented. A block is the interval from when the interface is started until it stops automatically; it is composed of the different runs made by the user.

The experiment was divided into two consecutive sessions: a first session with conditions C1-faces and C2-planes, and a second session with conditions C3-known and C4-unknown. The order of the conditions of each session was counterbalanced among the subjects. The approximate duration of the experiment was 80 min from the time the participant arrived at the laboratory until the end of the tasks. The four conditions used in this experiment had similar timing. Before the start of each block there was a waiting time of 1920 ms, after which the different runs began. Moreover, at the beginning of each run (except for C4-unknown), a message was presented in Spanish ("Atiende a:"

[Focus on:]) for 960 ms, after which the stimulus to be attended to was presented for another 960 ms. For C4-unknown, this information was replaced by a black background for 1920 ms. Before the first stimulus of the run was presented, all conditions included a pause time of 1920 ms. The stimulus duration was 384 ms, and the inter-stimulus interval (ISI) was 96 ms, resulting in a stimulus onset asynchrony (SOA) of 480 ms. Likewise, in the online task in all conditions, a message was presented at the end of each run ("Resultado:" [Result:]) for 960 ms, after which the stimulus selected by the system was presented for 960 ms. The attention and result messages were accompanied by an auditory cue to facilitate the user's attention to the task. For both the calibration and online tasks, a pause time of 1920 ms was added. The specific procedure for the C1-faces and C2-planes conditions was identical, as was the specific procedure for C3-known and C4-unknown, so the particularities of each condition in this experiment are detailed below.

C1-faces and C2-planes. The calibration task consisted of three blocks of six runs of 55 s each (Fig. 2). In each block, the following stimuli were selected from left to right: for the first block, the three stimuli in rows 1 and 2; for the second block, the stimuli in rows 2 and 3; and for the third block, the stimuli in rows 1 and 3. Each block of the calibration task had a duration of 55 s. The online task consisted of presenting as target stimuli all stimuli of the interface in row-major order, that is, nine runs in one block, which had a duration of 111 s. (E01 and E02 performed 18 runs instead of 9).

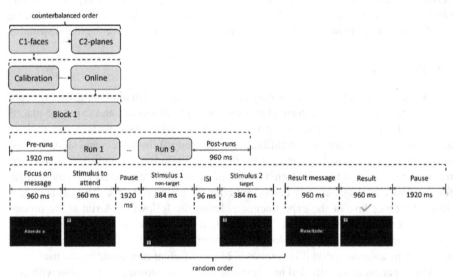

Fig. 2. Procedure and timing used in conditions C1-faces and C2-planes. Specifically, the figure shows the execution of the first run of the C1-faces condition during the online task. ISI stands for inter-stimulus interval.

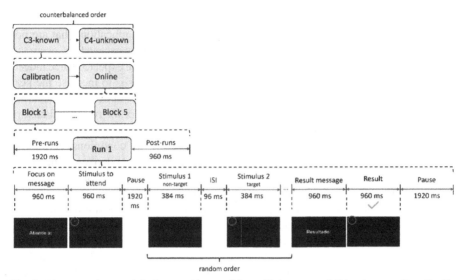

Fig. 3. The procedure and timing used in conditions C3-known and C4-unknown. Specifically, the figure shows the execution of the first selection of the C3-known condition during the online task. Due to the small size of the stimulus in the figure, compared with when it was presented on the screen during the experiment, the stimulus has been marked with a red circle here. ISI stands for inter-stimulus interval.

C3-known and C4-unknown. The calibration task consisted of 16 blocks of one run, resulting in a duration of 11 s per block (Fig. 3). The online task used five blocks of one selection, with a duration of 14 s per block (E01 and E02 performed 10 blocks of one selection). For both tasks, the target stimulus order to be attended to was randomly selected with replacement.

2.5 Evaluation

In all conditions, the classifier had to select a target stimulus from nine possible stimuli (including E1-know, C4-unknown, in which the non-target stimuli were invisible to the user). The accuracy (%) corresponds to the percentage of correct selections divided by the total number of selections made. The accuracy was calculated for the online task of each condition. The Wilcoxon signed-rank test, a non-parametric test for the comparison of two related samples, was used to compare between the conditions. All these analyses were carried out using SPSS software [25].

3   Results and Discussion

In this experiment, in addition to single-trial classification, two factors were evaluated: (i) the stimulus type (faces versus radar planes), using visible non-targets; and (ii) the knowledge of the location of the stimulus to attend to before it appears (known versus unknown), using the radar plane stimulus type and invisible non-target stimuli (Table 1).

In general, the results obtained (between 60% and 80% accuracy depending on the condition) are below those usually employed by other ERP-BCI applications that are not based on the single-trial classification approach (which can easily exceed 90% accuracy in applications such as virtual keyboards [26]). These results were expected since indeed the reason for using several presentations of the target stimulus is to increase the performance. Therefore, the results obtained highlight the challenge of detecting the target stimulus after a single presentation of it. The following results for the two visual factors studied—the type of stimulus and the knowledge of the place of appearance of the target stimulus—are detailed next. First, the C1-faces and C2-planes conditions were compared (64.81 ± 34.73%, and 69.45 ± 30.98%, respectively). The Wilcoxon signed-rank test showed that there was no significant difference between the conditions ($Z = 0.406$; $p = 0.684$). Therefore, it seems that type of stimulus does not have a significant impact on performance. Second, to test the effect of prior knowledge of the stimulus location, the C3-known and C4-unknown accuracies were compared (75 ± 25.1%, and 76.67 ± 15.06%, respectively). The Wilcoxon signed-rank test showed that knowing the location of the stimulus beforehand did not affect accuracy ($Z = 0.378$; $p = 0.705$). Therefore, these results showed that knowing where to attend to the incoming target did not affect performance.

Table 1. Mean ± standard deviation accuracy (%) for each user in the online task.

User	C1-faces	C2-planes	C3-known	C4-unknown
P01	100	88.89	90	80
P02	100	94.44	100	80
P03	88.89	88.89	60	80
P04	33.33	66.67	100	100
P05	33.33	66.67	60	60
P06	33.33	11.11	40	60
Mean	64.81 ± 34.72	69.45 ± 30.98	75 ± 25.1	76.67 ± 15.06

There are two important aspects related to performance that can be discussed: (i) the impact of the type of stimulus used and (ii) the effect of the size of the appearance surface of the target stimulus. Regarding the type of stimulus used, there was no significant effect on the performance of the system when using an ERP-BCI under the SCP (faces vs. radar planes), which is consistent with previous research that did not find that face stimuli offered significantly better performance than alternative stimuli [27, 28]. Therefore, using radar planes as visual stimuli could be appropriate in the use of an ATC system managed through an ERP-BCI. On the other hand, not knowing the exact place of appearance of the target stimulus has not led to a decrease in the performance of the ERP-BCI when detecting these stimuli. This evidence could indicate that in applications such as an ATC it should not be, initially, a problem to lack knowledge of the place of appearance of the target stimulus. However, it should be considered that in the current experiment the interface where the stimulus appeared had no distracting elements, which could be interesting to study in future studies and would be closer to a real use of these applications. Some examples of factors that could make the task more difficult in a real ATC could be the presence of multiple moving planes on the screen, a smaller size of the target stimuli or a map on the background with additional information.

The accuracy results have been very heterogeneous, from participants who have even obtained 100% to others with a lower accuracy than 50%. It is worth emphasizing that most of the participants in the study had no previous experience in the control of an ERP-BCI. Indeed, users P01 and P02, the only participants with previous experience in the use of ERP-BCI systems, were the only ones who presented an accuracy of at least 80% in all the conditions, even reaching 100% in C1-faces. Therefore, we cannot exclude the possibility that through extended training in the use of the system, the performance may be better, which would allow the use of single-trial classification to accurately detect the target stimuli.

4 Conclusions

The present work has been a preliminary study on the use of an ERP-BCI under the single-trial classification approach and its future application to an air traffic controller. Specifically, it has been shown that (i) it is possible to achieve an adequate performance under the single-trial classification approach, (ii) radar plane stimuli may be suitable for use as visual stimuli in an ERP-BCI visual, and (iii) not knowing the location of occurrence may not have a significant effect on their performance. As we said, these results can be applied to the use of an ERP-BCI in the control of an ATC. However, the accuracy shown confirms that the use single-trial classification is a challenge in the BCI domain and the user experience could be an important factor. As the combination of an ATC and a BCI is a relatively novel area, there is considerable scope for future proposals. For instance, the results are promising to be implemented in a real ATC scenario; it would be interesting to test these findings in a real ATC, where, for example, there are other distractor stimuli or the size of the area in which the target stimulus could appear is specifically studied. Also, future studies should focus on improving the performance of the visual ERP-BCI systems by considering what has been previously studied in other types of BCI devices, such as spellers which are the most studied ERP-BCI applications [26]. Possible areas of improvement include those related to human factors [29] and different signal processing and classification techniques [30]. While BCI systems have been used previously in the field of ATC to assess the cognitive state of users (assessment of mental workload [11] or the presence of microsleep states [12]), it would be interesting to use them with the dual purpose of measuring the cognitive state of the user (passive BCIs) and supporting the correct perception of stimuli at the interface (active BCIs). Overall, the use of an ERP-BCI for stimulus detection in an ATC is an interesting area that could be further explored, as the present work has shown that the presentation of a radar plane under a black background produces an ERP waveform that can be discriminated by a BCI system, even when the location of the stimulus is previously unknown to the user.

Acknowledgements. This work was partially supported by the project PID2021-127261OB-I00 (SICODIS), funded by MCIN (Ministerio de Ciencia e Innovación) /AEI (Agencia Estatal de Investigación) /https://doi.org/10.13039/501100011033/ FEDER, UE (Fondo Europeo de Desarrollo Regional). The work was also partially supported by the University of Málaga (Universidad de Málaga) and by THALES AVS in the context of a GIS Albatros project. The authors would also like to thank all participants for their cooperation.

References

1. Wolpaw, J.R., Birbaumer, N., McFarland, D.J., Pfurtscheller, G., Vaughan, T.M.: Brain-computer interfaces for communication and control. Clin. Neurophysiol. **113**, 767–791 (2002). https://doi.org/10.1016/S1388-2457(02)00057-3
2. Xu, L., Xu, M., Jung, T.P., Ming, D.: Review of brain encoding and decoding mechanisms for EEG-based brain–computer interface (2021). https://doi.org/10.1007/s11571-021-09676-z
3. Nicolas-Alonso, L.F., Gomez-Gil, J.: Brain computer interfaces, a review. Sensors **12**, 1211–1279 (2012). https://doi.org/10.3390/s120201211
4. Bonci, A., Fiori, S., Higashi, H., Tanaka, T., Verdini, F.: An introductory tutorial on brain–computer interfaces and their applications. Electron **10**, 1–43 (2021). https://doi.org/10.3390/electronics10050560
5. Gaume, A., Dreyfus, G., Vialatte, F.B.: A cognitive brain–computer interface monitoring sustained attentional variations during a continuous task. Cogn. Neurodyn. **13**, 257–269 (2019). https://doi.org/10.1007/s11571-019-09521-4
6. Bhattacharyya, S., Valeriani, D., Cinel, C., Citi, L., Poli, R.: Anytime collaborative brain–computer interfaces for enhancing perceptual group decision-making. Sci. Rep. **11**, 1–16 (2021). https://doi.org/10.1038/s41598-021-96434-0
7. Endsley, M.R.: Toward a theory of situation awareness in dynamic systems. Hum. Factors. **37**, 32–64 (1995). https://doi.org/10.1518/001872095779049543
8. Aricò, P., et al.: Adaptive automation triggered by EEG-based mental workload index: a passive brain-computer interface application in realistic air traffic control environment. Front. Hum. Neurosci. **10**, 1–13 (2016). https://doi.org/10.3389/fnhum.2016.00539
9. Di Flumeri, G., et al.: Brain–computer interface-based adaptive automation to prevent out-of-the-loop phenomenon in air traffic controllers dealing with highly automated systems. Front. Hum. Neurosci. **13** (2019). https://doi.org/10.3389/fnhum.2019.00296
10. Aricò, P., Borghini, G., Di Flumeri, G., Colosimo, A., Pozzi, S., Babiloni, F.: A passive brain–computer interface application for the mental workload assessment on professional air traffic controllers during realistic air traffic control tasks. Prog. Brain Res. **228**, 295–328 (2016). https://doi.org/10.1016/bs.pbr.2016.04.021
11. Li, W., Li, R., Xie, X., Chang, Y.: Evaluating mental workload during multitasking in simulated flight. Brain Behav. **12**, 1–11 (2022). https://doi.org/10.1002/brb3.2489
12. Boyle, L.N., Tippin, J., Paul, A., Rizzo, M.: Driver performance in the moments surrounding a microsleep. Transp. Res. Part F Traffic Psychol. Behav. **11**, 126–136 (2008). https://doi.org/10.1016/j.trf.2007.08.001
13. Kaufmann, T., Schulz, S.M., Grünzinger, C., Kübler, A.: Flashing characters with famous faces improves ERP-based brain-computer interface performance. J. Neural Eng. **8**, 056016 (2011). https://doi.org/10.1088/1741-2560/8/5/056016
14. Pfabigan, D.M., Sailer, U., Lamm, C.: Size does matter! Perceptual stimulus properties affect event-related potentials during feedback processing. Psychophysiology **52**, 1238–1247 (2015). https://doi.org/10.1111/psyp.12458

15. Fernández-Rodríguez, Á., Darves-Bornoz, A., Velasco-Álvarez, F., Ron-Angevin, R.: Effect of Stimulus Size in a Visual ERP-Based BCI under RSVP. Sensors. 22, (2022). https://doi.org/10.3390/s22239505

16. Li, Y., Bahn, S., Nam, C.S., Lee, J.: effects of luminosity contrast and stimulus duration on user performance and preference in a P300-based brain-computer interface. Int. J. Hum. Comput. Interact. **30**, 151–163 (2014). https://doi.org/10.1080/10447318.2013.839903

17. Fernández-Rodríguez, A., Velasco-Álvarez, F., Ron-Angevin, R.: Review of real brain-controlled wheelchairs. J. Neural Eng. **13** (2016). https://doi.org/10.1088/1741-2560/13/6/061001

18. Alrumiah, S.S., Alhajjaj1, L.A., Alshobaili, J.F., Ibrahim, D.M.: A review on brain-computer interface spellers: P300 speller. Biomed. Commun. **13**, 1191–1199 (2020). https://doi.org/10.1016/s0022-4804(03)00693-0

19. Cecotti, H., Ries, A.J.: Best practice for single-trial detection of event-related potentials: application to brain-computer interfaces. Int. J. Psychophysiol. **111**, 156–169 (2017). https://doi.org/10.1016/j.ijpsycho.2016.07.500

20. Tian, Y., Zhang, H., Pang, Y., Lin, J.: Classification for single-trial N170 during responding to facial picture with emotion. Front. Comput. Neurosci. **12** (2018). https://doi.org/10.3389/fncom.2018.00068

21. Goljahani, A., D'Avanzo, C., Silvoni, S., Tonin, P., Piccione, F., Sparacino, G.: Preprocessing by a Bayesian single-trial event-related potential estimation technique allows feasibility of an assistive single-channel P300-based brain-computer interface. Comput. Math. Methods Med. **2014** (2014). https://doi.org/10.1155/2014/731046

22. Zhang, X., Jin, J., Li, S., Wang, X., Cichocki, A.: Evaluation of color modulation in visual P300-speller using new stimulus patterns. Cogn. Neurodyn. 0123456789, (2021). https://doi.org/10.1007/s11571-021-09669-y

23. Pires, G., Nunes, U., Castelo-Branco, M.: Comparison of a row-column speller vs. a novel lateral single-character speller: assessment of BCI for severe motor disabled patients. Clin. Neurophysiol. **123**, 1168–1181 (2012). https://doi.org/10.1016/j.clinph.2011.10.040

24. Schalk, G., McFarland, D.J., Hinterberger, T., Birbaumer, N., Wolpaw, J.R.: BCI2000: a general-purpose brain-computer interface (BCI) system (2004). https://doi.org/10.1109/TBME.2004.827072

25. IBM Corp.: IBM SPSS Statistics for Windows, Version 24.0 (2016)

26. Rezeika, A., Benda, M., Stawicki, P., Gembler, F., Saboor, A., Volosyak, I.: Brain–computer interface spellers: a review. Brain Sci. **8** (2018). https://doi.org/10.3390/brainsci8040057

27. Kübler, A., et al.: The user-centered design as novel perspective for evaluating the usability of BCI-controlled applications. PLoS ONE **9**, 1–22 (2014). https://doi.org/10.1371/journal.pone.0112392

28. Lotte, F., et al.: A review of classification algorithms for EEG-based brain-computer interfaces: a 10 year update. J. Neural Eng. **15** (2018). https://doi.org/10.1088/1741-2552/aab2f2

29. Kellicut-Jones, M.R., Sellers, E.W.: P300 brain-computer interface: comparing faces to size matched non-face stimuli. Brain-Comput. Interfaces **5**, 30–39 (2018). https://doi.org/10.1080/2326263X.2018.1433776

30. Ron-Angevin, R., et al.: Performance analysis with different types of visual stimuli in a BCI-Based speller under an RSVP paradigm. Front. Comput. Neurosci. **14** (2021). https://doi.org/10.3389/fncom.2020.587702

Toward Early Stopping Detection for Non-binary c-VEP-Based BCIs: A Pilot Study

Víctor Martínez-Cagigal[1,2](✉) , Eduardo Santamaría-Vázquez[1,2] ,
and Roberto Hornero[1,2]

[1] Biomedical Engineering Group (GIB), E.T.S. Ingenieros de Telecomunicación,
University of Valladolid, Paseo de Belén, 15, 47011 Valladolid, Spain
`victor.martinez.cagigal@uva.es`
[2] Centro de Investigación Biomédica en Red en Bioingeniería,
Biomateriales y Nanomedicina (CIBER-BBN), Valladolid, Spain

Abstract. Code-modulated visual evoked potentials (c-VEPs) have
potential as a reliable and non-invasive control signal for brain-computer
interfaces (BCIs). However, these systems need to become more user-
friendly. Non-binary codes have been proposed to reduce visual fatigue,
but there is still a lack of adaptive methods to shorten trial durations.
To address this, we propose a nonparametric early stopping algorithm
for the non-binary circular shifting paradigm. The algorithm analyzes
the distribution of unattended commands' correlations and stops stimulation when the most probable correlation is considered an outlier.
This proposal was evaluated offline with 15 healthy participants using
p-ary maximal length sequences encoded with shades of gray. Results
showed that the algorithm could stop stimulation in under two seconds
for all sequences, achieving mean accuracies over 95%. The highest performances were achieved by bases $p = 2$ and $p = 5$, attaining 98.3%
accuracy with ITRs of 164.8 bpm and 121.7 bpm, respectively. The proposed algorithm reduces required cycles without compromising accuracy
for c-VEP-based BCI systems.

Keywords: Early stopping · non-binary codes · code-modulated
visual evoked potential (c-VEP) · brain–computer interface (BCI) ·
electroencephalography (EEG)

1 Introduction

Non-invasive brain-computer interface (BCI) systems have the capability of
interpreting users' intentions directly from their electroencephalographic (EEG)
signals and converting them into commands for controlling external devices or
applications [13]. However, decoding such intentions is challenging and requires
the use of control signals that generate measurable responses in the EEG. These

© The Author(s), under exclusive license to Springer Nature Switzerland AG 2023
I. Rojas et al. (Eds.): IWANN 2023, LNCS 14135, pp. 580–590, 2023.
https://doi.org/10.1007/978-3-031-43078-7_47

control signals can be generated either by processing external stimuli (exogenous approach) or by performing cognitive tasks (endogenous approach) [13]. Among other exogenous signals, code-modulated visual evoked potentials (c-VEPs) stand out as a promising strategy to develop non-invasive BCIs with high accuracy and speed [7].

In the most common paradigm, known as circular shifting, selectable commands flicker following uncorrelated shifted versions of a binary pseudorandom sequence [7]. In real-time, the identification of the desired command is determined by analyzing the correlation between the EEG response and these shifted templates [7]. Despite the excellent performances, several studies have reported that the high-contrast changes produced by binary codes, which use black and white flashes to encode commands, may cause visual fatigue for some users [4,5,12]. In a previous study, we proposed the use of non-binary sequences encoded with different shades of gray to improve user friendliness [6]. The results indicated that these non-binary codes are suitable for achieving high speed and accuracy while reducing visual fatigue.

Although c-VEP-based BCIs have great potential, they need to be further adapted to become more user-friendly technologies. Apart from addressing the visual fatigue, the adoption of adaptive methods to reduce as much as possible the trial decoding duration has been also identified as a current challenge in the literature [7]. In this sense, early stopping techniques that adaptively stop visual stimulation whenever the BCI is ready to deliver a command selection are still limited. Many of the previous approaches are incompatible with the circular shifting paradigm [9,11], require parameter optimization, or are dependent on the classifier stage [2,3]. Furthermore, none of these methods have been applied to non-binary stimulation.

The aim of this pilot study is to present a new nonparametric early stopping technique that is applicable to non-binary c-VEP-based BCIs. The method was offline tested with 5 different sequences of bases 2, 3, 5, 7 and 11; displayed at a rate of 120 Hz. The base indicates the number of distinct events that are encoded within the m-sequence, represented as a shades of gray. For instance, a base of 2 (i.e., binary) is encoded using solely black and white; whereas a base of 11 uses black, white, and an additional nine intermediate shades of gray [6]. Our algorithm is noteworthy due to its classifier-independence (filter-based), lack of need for parameter training (nonparametric), and ability to be implemented in real-time without being trained with additional EEG recordings. Moreover, to the best of our knowledge, this is the first early stopping method for non-binary visual stimulation based on the circular shifting paradigm.

2 Signals

We conducted our study using an offline database consisting of 15 healthy participants (mean age: 28.80 ± 5.02 years, 10 males, 5 females), who performed BCI spelling tasks using the "P-ary c-VEP Speller" application of MEDUSA©, which is publicly available at www.medusabci.com [10]. Prior to their participation, all users provided informed consent. EEG data was collected using a

Table 1. Details regarding the generation of the p-ary m-sequences.

	Base	Order	Length (bits)	Polynomial	Duration* (s/cycle)
GF(2^6)	2	6	63	$x^6 + x^5 + 1$	0.525
GF(3^4)	3	4	80	$x^4 + 2x^3 + 1$	0.667
GF(5^3)	5	3	124	$3x^3 + 2x^2 + 1$	1.033
GF(7^2)	7	2	48	$4x^2 + 1$	0.400
GF(11^2)	11	2	120	$3x^2 + x + 1$	1.000

* Computed using a monitor refresh rate of 120 Hz

g. USBamp device (g.Tec, *Guger Technologies*, Austria) and recorded from 16 active channels: F3, Fz, F4, C3, Cz, C4, CPz, P3, Pz, P4, PO7, POz, PO8, Oz, I1 and I2. The EEG device was grounded at AFz and referenced to the right ear-lobe. Visual stimuli were presented on a LED FullHD @ 144 Hz monitor (KEEP OUT XGM24F+ 23.8"), with a refresh rate of 120 Hz. A computer with an Intel Core i9-11900KF 3.5 GHz processor and 64 GB of RAM (Windows 10 OS) was used to display the visual stimuli. For additional details, refer to [6].

3   Methods

3.1   Paradigm

The circular shifting paradigm relies on the use of shifted versions of a pseudo-random sequence to encode individual commands. Therefore, it is crucial that the sequence exhibits low autocorrelation to facilitate subsequent decoding [7]. Maximal length sequences (i.e., m-sequences) are pseudorandom time series that demonstrate almost optimal autocorrelation properties, and can be generated by linear-feedback shift registers (LFSR). M-sequences are determined by: (1) the base p, i.e. the number of levels (e.g., $p = 2$ for binary m-sequences); (2) the order r, i.e. the number of LFSR taps; and (3) the generator polynomial expressed as a Galois Field of p elements, GF(p^r), i.e. the arrangement of the LFSR taps [1]. Apart from other mathematical constraints, the length of p-ary m-sequences is exactly $N = p^r - 1$ bits, and it is repeated cyclically [6]. Since commands are encoded with shifted versions of the p-ary m-sequences, the length of the sequence is directly related to the number of commands that can be encoded with the same code.

In this study, we utilized five distinct p-ary m-sequences with different bases, including binary GF(2^6) with base 2, GF(3^5) with base 3, GF(5^3) with base 5, GF(7^2) with base 7, and GF(11^2) with base 11. The detailed characteristics of each code, such as their length and duration when presented at a 120 Hz rate, are presented in Table 1. The paradigm consisted of a 16-command speller with adequately spaced lags to prevent any misclassifications. It is worth noting that a deterministic algorithm was utilized to avoid spurious correlations, as non-binary ($p > 2$) m-sequences exhibit periodic phase shifts that lead to high

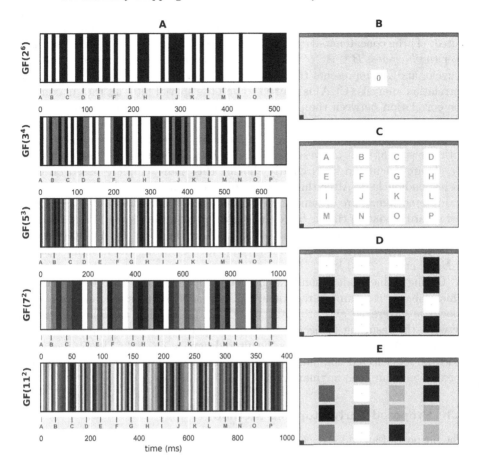

Fig. 1. (A) Gray encoding of each p-ary m-sequence and associated lags for each command. From top to bottom: base 2, base 3, base 5, base 7, and base 11. (B) Calibration stage, where a single command flashes according to the original p-ary m-sequence. (C) Online stage, showing the alphabetical arrangement of the 16 commands. (D) Snapshot of the binary m-sequence, $GF(2^6)$. (E) Snapshot of the $GF(11^2)$ m-sequence.

anti-/correlations. Additional information about this procedure can be found in [6]. Figure 1 depicts the associated lags for each command, the arrangement of commands, and the gray encoding of each p-ary m-sequence, as well as several snapshots of the application [10].

3.2 Signal Processing

In the calibration stage, the participant is instructed to focus on a single command encoded by the original p-ary m-sequence without lag, for a duration of k cycles. First, the EEG signal is preprocessed by a filter bank of bandpass filters (1–60 Hz, 12–60 Hz, and 30–60 Hz) and a notch filter at 50 Hz, generating three filtered EEG signals. For each signal, two versions of the EEG response are computed: (1) the concatenated epochs, $A \in \mathbb{R}^{[kN_s \times N_c]}$; and (2) the epochs averaged over the k cycles, $B \in \mathbb{R}^{[N_s \times N_c]}$. Here, N_s represents the number of samples of a cycle, and N_c represents the number of channels. Subsequently, a canonical correlation analysis (CCA) is trained to find the spatial filter ω_b that maximizes the correlation between the projected versions of A and B. In this procedure, B is replicated k times to match the dimensions of A. The main template (i.e., for the command without lag) is computed by projecting the averaged signal with the spatial filter ω_b, resulting in $x_0 = B\omega_b$. Templates for the other commands are calculated by circularly shifting this main template based on their corresponding lags. After this process, 16×3 templates, each for a command and filtered signal, are obtained. It is worth noting that calibration epochs with a standard deviation that is three times greater than the average standard deviation of all epochs were discarded before training the CCA [6]. A raster latency correction was applied to the trained templates, following the recommendation of Nagel et al. (2018) [8].

During online mode, a similar approach is applied to identify the command at which the user is looking in real-time. EEG signal is preprocessed, and individual epochs are averaged and projected using the spatial filter ω_b. The correlation between the resulting projection and all templates is then computed, yielding $\hat{\rho} \in \mathbb{R}^{16 \times 3}$. After averaging across the filtered signals, $\rho \in \mathbb{R}^{16 \times 1}$ is obtained. The index of the selected command corresponds to the one that yields the highest correlation value, i.e., $\arg\max_i(\rho)$ [7].

3.3 Proposed Early Stopping Method

The purpose of an effective early stopping method is to select a command before a fixed number of cycles have elapsed, enabling real-time adaptation of the signal processing pipeline to the characteristics of the EEG signal. In a hypothetical scenario, the selection of a command is expected to take more time when the user is slightly distracted or when the EEG is contaminated with artifacts, and less time under ideal conditions. Importantly, a trial is no longer composed of a fixed number of cycles, but rather a variable number of cycles. Thus, the early stopping algorithm must make a binary decision each time a cycle is fully displayed: (1) select the most probable command; or (2) continue the visual stimulation for one additional cycle.

As detailed in Sect. 3.2, the online signal processing pipeline calculates a comparison between the EEG response from the start of the trial to the end of the current cycle and the command templates, resulting in a correlation vector $\rho \in \mathbb{R}^{16 \times 1}$. After arranging this vector in descending order, ρ_1 corresponds to the most likely command as it represents the highest correlation. The remaining correlations, $\rho_2, \rho_3, \ldots, \rho_{16}$, can be considered as spurious correlations associated with non-attended commands. Additionally, we can widen the number of observations of spurious correlations by computing the correlation of the EEG response with all possible shifted versions of the template, not just with those lags associated to the selectable commands. We end up with a correlation vector of length N, where N is the length of the p-ary m-sequence. Thus, ρ_1 corresponds to the selected command, and $\boldsymbol{\rho}_{spu} = [\rho_2, \ldots, \rho_N]$ constitutes the distribution of spurious correlations. A reliable approach to determining whether ρ_1 indeed corresponds to the attended command would be to verify whether it is an outlier from the distribution $\boldsymbol{\rho}_{spu}$ (i.e., its correlation is statistically higher than that of the presumable non-attended commands).

Various techniques can be used to detect outliers from distributions, including those based on hypothesis testing or on the interquartile range. In this study, we suggest employing z-scores due to their simplicity. Assuming that the spurious distribution is normal, i.e., $\boldsymbol{\rho}_{spu} \sim \mathcal{N}(\mu, \sigma)$, we can identify ρ_1 as an outlier if $\rho_1 - \mu > h\sigma$, where $h = 3$. Thus, we can ascertain that ρ_1 is an outlier if it exceeds the 99.87% percentile of the spurious distribution. Consequently, if ρ_1 is an outlier, the command selection is delivered; otherwise, the visual stimulation continues with the next cycle.

3.4   Evaluation Protocol

This pilot study entailed an exploratory analysis of offline data gathered from 15 healthy participants who completed spelling tasks utilizing a 16-command c-VEP speller, which was encoded with the p-ary m-sequences $GF(2^6)$, $GF(3^5)$, $GF(5^3)$, $GF(7^2)$, $GF(11^2)$. Specifically, a total of 300 calibration cycles (6 runs $\times$ 5 trials $\times$ 10 cycles) and 320 test cycles (2 runs $\times$ 16 trials $\times$ 10 cycles) per participant were acquired for each p-ary m-sequence. During the test cycles, participants selected all commands in alphabetical order twice [6].

4   Results and Discussion

4.1   Correlation Distributions

Figure 2 depicts the correlations for the selected commands ρ_1, as well as for the non-attended ones $\boldsymbol{\rho}_{spu}$. Results obtained from the Kolmogorov-Smirnov test reveal that all distributions (ρ_1 and $\boldsymbol{\rho}_{spu}$ for both calibration and test data) are normal (p-value < 0.01). Given that the normality assumption is satisfied for the z-score, the estimated value of 99.87% can be considered accurate. Moreover, a significant similarity in the distributions between calibration and test data (p-value < 0.01, Wilcoxon-signed rank test) indicates the potential for optimizing h using only calibration data without acquiring additional recordings. It is worth noting that an increase in the number of cycles results in a greater separation between ρ_1 and $\boldsymbol{\rho}_{spu}$ distributions. This phenomenon highlights the tradeoff between speed and accuracy. For instance, stopping in the very early cycles poses a higher risk of misclassification, but it allows for a faster selection speed.

4.2   Performance Analysis

Table 2 displays the performance results, including accuracy and number of cycles, of each participant and p-ary m-sequence. The theoretical maximum accuracy, i.e. the minimum number of cycles required to achieve the highest accuracy

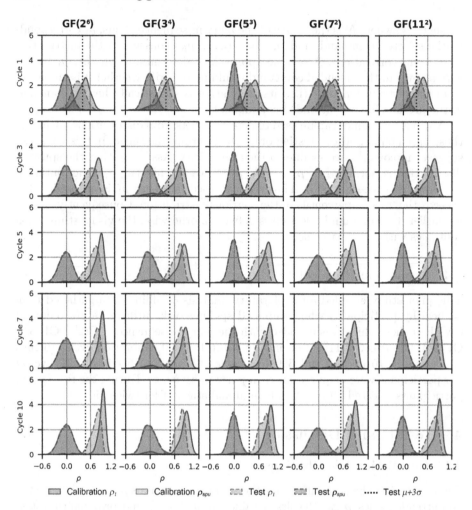

Fig. 2. Distribution of the correlations of the selected commands, ρ_1; and the spurious distributions ρ_{spu} for all p-ary m-sequences. Both calibration (blue and orange) and test (green and red) distributions are shown, including the estimated 99.87% percentile in test. Only cycles 1, 3, 5, 7 and 10 are depicted for visualization purposes. (Color figure online)

(equivalent to 10 cycles in this database), is also included for comparison purposes. As shown, the mean visual stimulation duration of all p-ary m-sequences is below 2 s, with all sequences achieving accuracies exceeding 95%. The top performances sorted by accuracy are as follows: 98.3% with 1.8 cycles for GF(5^3), 98.3% with 2.6 cycles for GF(2^6), 98.1% with 2.4 cycles for GF(3^4), 98.1% with 1.8 cycles for GF(11^2), and 95.6% with 4.1 cycles for GF(7^2). Additionally, the mean information transfer rates (ITR) range from 121.7 to 164.8 bpm. These findings suggest that the proposed early stopping algorithm can deliver fluent command selection while achieving high accuracy.

Table 2. Offline results applying early stopping with all p-ary m-sequences.

| | GF(2^6) | | | | GF(3^4) | | | | GF(5^3) | | | | GF(7^2) | | | | GF(11^2) | | | |
| | E.S. | | T.M. | | E.S. | | T.M. | | E.S. | | T.M. | | E.S. | | T.M. | | E.S. | | T.M. | |
	%	N_c	%	N_c	%	N_c	%	N_c	%	N_c	%	N_c	%	N_c	%	N_c	%	N_c	%	N_c
U01	100	1.7	100	1.2	100	1.8	100	1.2	100	1.3	100	1.2	96.9	3.1	100	1.7	100	1.4	100	1.1
U02	100	2.2	100	1.3	100	1.9	100	1.1	100	1.2	100	1.0	96.9	3.6	100	1.8	100	1.5	100	1.2
U03	100	3.1	100	1.2	96.9	1.3	100	1.1	100	1.6	100	1.0	100	2.5	100	1.2	100	1.2	100	1.1
U04	100	2.0	100	1.1	100	1.5	100	1.0	100	1.1	100	1.0	100	8.2	100	1.4	100	1.1	100	1.0
U05	96.9	2.8	100	1.8	100	3.7	100	1.6	100	1.8	100	1.3	93.8	5.0	96.9	2.0	100	1.5	100	1.1
U06	100	2.5	100	1.6	100	2.7	100	1.4	96.9	1.3	100	1.2	100	3.0	100	1.8	100	1.7	100	1.1
U07	100	2.6	100	1.4	100	1.5	100	1.1	100	1.7	100	1.2	93.8	2.4	100	1.6	100	1.7	100	1.1
U08	100	2.2	100	1.2	100	1.3	100	1.1	100	1.8	100	1.1	96.9	3.0	100	1.7	100	1.3	100	1.0
U09	100	1.8	100	1.2	100	2.0	100	1.2	100	1.6	100	1.3	100	2.6	100	1.3	78.1	3.9	84.4	2.5
U10	100	2.2	100	1.3	100	1.6	100	1.1	100	1.4	100	1.0	100	2.7	100	1.7	100	1.1	100	1.0
U11	96.9	4.1	96.9	2.1	84.4	4.3	90.6	2.7	81.2	4.2	93.8	2.2	75.0	8.4	90.6	3.4	100	3.0	100	1.3
U12	96.9	4.2	96.9	2.0	93.8	3.0	100	1.7	96.9	2.8	100	1.9	96.9	6.2	100	2.4	96.9	1.9	100	1.3
U13	93.8	1.9	100	1.3	100	2.0	100	1.0	100	1.1	100	1.1	96.9	3.0	100	1.4	100	1.3	100	1.1
U14	90.6	4.2	100	2.2	96.9	4.8	100	2.2	100	2.5	100	1.6	90.6	4.6	100	2.4	96.9	2.9	100	1.6
U15	100	2.3	100	1.3	100	2.3	100	1.3	100	2.0	100	1.2	96.9	2.8	100	1.5	100	1.4	100	1.1
avg.	**98.3**	**2.6**	99.6	1.5	**98.1**	**2.4**	99.4	1.4	**98.3**	1.8	99.6	1.3	**95.6**	**4.1**	99.2	1.8	**98.1**	**1.8**	99.0	1.2
std	2.8	0.8	1.1	0.3	4.1	1.1	2.3	0.5	4.7	0.8	1.6	0.3	6.1	1.9	2.4	0.5	5.4	0.8	3.9	0.4
ITR	**164.8**		170.5		**143.3**		148.2		**121.7**		125.9		**131.6**		143.7		**126.9**		129.6	
dur.	**1.4 s**		0.8 s		**1.6 s**		0.9 s		**1.9 s**		1.3 s		**1.6 s**		0.7 s		**1.8 s**		1.2 s	

E.S.: early stopping, T.M.: theoretical maximum (minimum number of cycles to attain the accuracy that would have been obtained using 10 cycles), %: accuracy, N_c: number of cycles, avg.: average, std.: standard deviation, ITR: mean information transfer rate in bit per minute (bpm), dur.: mean duration of the visual stimulation in seconds

As could be expected, the shortest m-sequence, GF(7^2), yielded the lowest accuracy (95.6%). Figure 1 indicates that the ρ_1 and $\boldsymbol{\rho}_{spu}$ distributions overlapped more in this m-sequence than in the others, especially in the first cycle. While the average number of cycles for GF(7^2) is high relative to the others, which does not necessarily imply a longer trial duration, it would have been expected to be even higher to cope with this uncertainty. Therefore, individual optimization of h for each p-ary m-sequence could potentially benefit the system's performance, though further analyses are necessary to gain insight into this phenomenon.

The theoretical maximum accuracy suggests that trial duration could have been reduced to between 0.7–1.3 s, with accuracy exceeding 99% achievable through an ideal early stopping algorithm. This is equivalent to stopping between the first and second cycle. However, Fig. 1 shows that the ρ_1 and $\boldsymbol{\rho}_{spu}$ distributions are difficult to separate in the first cycle, resulting in unreliable selections using our method. It remains an open question whether other approaches can reach this theoretical maximum. Interestingly, the increase in accuracy by the theoretical maximum is not significant by all p-ary m-sequences (p-value > 0.05, Wilcoxon signed-rank test), except for the GF(7^2) m-sequence (p-value = 0.0036). Although we assert that our results demonstrate the usefulness of our algorithm, this observation indicates that there is still room for improvement.

4.3   Limitations and Future Lines of Research

Despite the success of the proposed early stopping algorithm, there are still opportunities for enhancing its reliability. To begin with, it is crucial to conduct an online proof of concept and increase the sample size to improve the statistical power of the results. Additionally, it would be desirable to evaluate the algorithm's efficacy with motor-disabled participants. A promising research direction would be also to complement the algorithm with an asynchronous stage to monitor users' attention. Currently, the cumulative correlation across cycles presents a challenge in detecting a change from non-control to control cycles, as previous non-control epochs could negatively impact the correlation analysis. Therefore, an asynchronous algorithm could be focused on detecting attention in single cycles. Another possible avenue for investigation would be to explore whether an optimization of h between users or p-ary m-sequences could enhance the final classification.

5   Conclusions

This study introduces a novel early stopping algorithm for non-binary c-VEP-based BCIs, which presents a promising alternative for minimizing visual fatigue for end-users. The main strengths of this method are its classifier-independence, the lack of need for parameter training, and its real-time application without requiring additional EEG recordings. In an offline analysis, the algorithm was found to reduce trial duration to less than 2 s while achieving over 95% accuracy for five different p-ary m-sequences, namely GF(2^6), GF(3^4), GF(5^3), GF(7^2), and GF(11^2). Although the algorithm's efficacy was demonstrated for all the tested m-sequences, the highest accuracy was achieved with GF(2^6) and GF(5^3), which attained 98.3% accuracy with 1.4 s and 1.9 s of stimulation, equivalent to ITRs of 164.8 bpm and 121.7 bpm, respectively. In conclusion, the proposed early stopping algorithm represents a valuable metric for significantly reducing the required number of cycles without compromising the system's accuracy in the circular shifting paradigm.

Acknowledgements. This research was supported by projects TED2021-12991 5B-I00, RTC2019-007350-1 and PID2020-115468RB-I00 funded by MCIN/AEI/ 10.13039/501100011033 and 'European Union NextGenerationEU/PRTR'; and by 'Centro de Investigación Biomédica en Red en Bioingeniería, Biomateriales y Nanomedicina (CIBER-BBN)' through 'Instituto de Salud Carlos III' co-funded with European Regional Development Fund (ERDF) funds. E. Santamaría-Vázquez was in receipt of a PIF grant by the 'Consejería de Educación de la Junta de Castilla y León'.

References

1. Buračas, G.T., Boynton, G.M.: Efficient design of event-related fMRI experiments using m-sequences. NeuroImage **16**(3 I), 801–813 (2002). https://doi.org/10.1006/nimg.2002.1116
2. Gembler, F., et al.: A dictionary driven mental typewriter based on code-modulated visual evoked potentials (cVEP). In: Proceedings - 2018 IEEE International Conference on Systems, Man, and Cybernetics, SMC 2018, pp. 619–624. IEEE (2018). https://doi.org/10.1109/SMC.2018.00114
3. Gembler, F., Volosyak, I.: A novel dictionary-driven mental spelling application based on code-modulated visual evoked potentials. Computers **8**(2) (2019). https://doi.org/10.3390/computers8020033
4. Gembler, F.W., Rezeika, A., Benda, M., Volosyak, I.: Five shades of grey: exploring quintary m -sequences for more user-friendly c-VEP-based BCIs. Comput. Intell. Neurosci. **2020** (2020). https://doi.org/10.1155/2020/7985010
5. Ladouce, S., Darmet, L., Torre Tresols, J.J., Velut, S., Ferraro, G., Dehais, F.: Improving user experience of SSVEP BCI through low amplitude depth and high frequency stimuli design. Sci. Rep. **12**(1), 1–12 (2022). https://doi.org/10.1038/s41598-022-12733-0
6. Martínez-Cagigal, V., Santamaría-Vázquez, E., Pérez-Velasco, S., Marcos-Martínez, D., Moreno-Calderón, S., Hornero, R.: Non-binary m-sequences for more comfortable brain-computer interfaces based on c-VEPs. Expert Syst. Appl. (2023). https://doi.org/10.1016/j.eswa.2023.120815
7. Martínez-Cagigal, V., Thielen, J., Santamaría-Vázquez, E., Pérez-Velasco, S., Desain, P., Hornero, R.: Brain-computer interfaces based on code-modulated visual evoked potentials (c-VEP): a literature review. J. Neural Eng. **18**(6), 061002 (2021). https://doi.org/10.1088/1741-2552/ac38cf
8. Nagel, S., Dreher, W., Rosenstiel, W., Spüler, M.: The effect of monitor raster latency on VEPs, ERPs and brain-computer interface performance. J. Neurosci. Methods **295**, 45–50 (2018). https://doi.org/10.1016/j.jneumeth.2017.11.018

9. Nagel, S., Spüler, M.: World's fastest brain-computer interface: combining EEG2Code with deep learning. PLoS ONE **14**(9), 1–15 (2019). https://doi.org/10.1371/journal.pone.0221909

10. Santamaría-Vázquez, E., et al.: MEDUSA: a novel Python-based software ecosystem to accelerate brain-computer interface and cognitive neuroscience research. Comput. Methods Programs Biomed. **230**(107357) (2023). https://doi.org/10.1016/j.cmpb.2023.107357

11. Thielen, J., Marsman, P., Farquhar, J., Desain, P.: From full calibration to zero training for a code-modulated visual evoked potentials brain computer interface. J. Neural Eng. **18**(5), 56007 (2021). https://doi.org/10.1088/1741-2552/abecef

12. Wandell, B.A., Dumoulin, S.O., Brewer, A.A.: Visual field maps in human cortex. Neuron **56**(2), 366–383 (2007). https://doi.org/10.1016/j.neuron.2007.10.012

13. Wolpaw, J., Wolpaw, E.W.: Brain-Computer Interfaces: Principles and Practice. OUP, New York (2012)

Gender Influence on cVEP-Based BCI Performance

Ivan Volosyak$^{(\boxtimes)}$, Foluke Adepoju , Piotr Stawicki , Paul Rulffs ,
Atilla Cantürk , and Lisa Henke

Faculty of Technology and Bionics, Rhine-Waal University of Applied Sciences,
47533 Kleve, Germany
ivan.volosyak@hochschule-rhein-waal.de
https://bci-lab.hochschule-rhein-waal.de

Abstract. A Brain-Computer Interface (BCI) is a technical system that creates a direct communication pathway between the human brain and an external device, such as a computer, without generally necessitating any physical movement. BCIs use various methods to detect and interpret brain activity, the most common being electroencephalography (EEG).

BCIs can be beneficial for people with disabilities since they provide an alternative communication and control method that can extend and replace traditional means, such as speech or muscular movements.

This paper investigates performance differences in code-modulated visual evoked potentials (cVEP) based BCI system between subjects of different gender identities. In this regard, the cVEP-based spelling interface with four targets was tested between two gender groups - 18 females and 18 males each, with ages ranging from 20 to 39 years. Three different spelling tasks were performed - writing of two command-balanced words and a pangram sentence, and cVEP stimuli were rendered on a monitor with a vertical refresh rate of 240 Hz.

Both groups (female and male) successfully completed all spelling tasks, achieving for the pangram task, a mean information transfer rate (ITR) of 29.38 bits per minute (bpm) and 28.09 bpm, respectively. Although the difference was not statistically significant for the pangram task, some recognizable differences were observed for the command-balanced tasks. Consequently, a trend (rather than a substantial difference) was realized between the male and female groups' performance in the pangram task. Regarding the level of annoyance, subjects from both groups rated similar results on the visual stimulation setup.

Keywords: Brain-Computer Interface (BCI) · Human-Computer
Interaction (HCI) · Code-modulated Visual Evoked Potentials
(cVEP) · BCI Speller · Gender Differences

1 Introduction

BCI is a technical system for real-time communication and control that creates a direct pathway between the human brain and other external equipment [5,6]. The

I. Rojas et al. (Eds.): IWANN 2023, LNCS 14135, pp. 591–602, 2023.
https://doi.org/10.1007/978-3-031-43078-7_48

brain signals from a BCI user are translated into a desired output by the BCI system, such as for computer-based communication purposes or to control an external device. One of the most common techniques for evaluating brain waves is the electroencephalogram (EEG) [1]. The ability to record brain activities with high temporal and spatial precision has been made feasible by developments in EEG technology, thus giving researchers more thorough knowledge of the brain's electrical activity and enabling the creation of cutting-edge BCI applications.

EEG activities collected at the scalp may be used to derive the Visual Evoked Potentials (VEPs), or in general, evoked electrophysiological potentials. As a result, important diagnostic data on the visual system's functional integrity can be provided by the VEPs. These are electrical signals which are produced in reaction to visual stimuli used to interpret the user's intent and convert it into a control signal for a computer or other device.

In the past, the male gender made up the majority of participants in medical research. Studies have demonstrated, for instance, that symptoms, treatment reactions, and side effects may differ between males and females. However, if research studies solely contain male individuals, these changes can be overlooked and the female gender could receive less effective care leading to negative results.

Moreover, gender has been a major topic in the European Commission (EC) for several years and it continues to be a top priority today. The Commission is therefore dedicated to promoting gender equality and strengthening the visibility of the female gender at all levels, through several initiatives such as Gender Equality Strategy 2020–2025. Similarly, several Universities now offer *Gender and Diversity* as a course of study as well as the promotion of *Girls' Day* in many establishments. Thus, this study would contribute to the optimization of BCI design and the reduction of gender bias in research, while ensuring that the development of assistive technologies is inclusive and equitable.

A variation of the standard VEP known as code-modulated Visual Evoked Potentials (cVEPs), a pseudorandom code employed to regulate various visual stimuli, has gained increased popularity in recent years [4]. A cVEP is induced when someone responds to one of those stimuli and may thus be utilized to regulate the BCI. cVEP uses the potential to control a "three-step-speller" which requires three selections to choose a letter. Three-step spellers are particularly dependable and robust, only requiring four independent visual stimuli [7].

A user is provided with a collection of flickering targets in a cVEP application, each of which is connected to a unique binary code pattern, the so-called m-sequence, that controls whether the stimulus is displayed or not during the actual frame. The user's brain signals are evaluated in real-time using pre-recorded target-specific EEG templates from a training session for categorization.

The information transfer rate (ITR), measured in bits per minute (bpm), is typically employed in the evaluation of the BCI performance. It is the speed at which information is transferred, defining the main property of every informa-tion channel. ITR is influenced by several factors such as classification speed, accuracy, and the number of targets, for instance see this interactive BCI ITR Calculator https://bci-lab.hochschule-rhein-waal.de/en/itr.html.

In earlier SSVEP-based BCI studies, a tendency for female participants to perform better with a higher Information Transfer Rate (ITR) and lower number of errors than their male counterparts was observed [2,9], with similar results during our recent study comparing different BCI paradigms [8]. This study was therefore aimed at investigating this inclination further by performing the three-step-speller cVEP-based BCI experiment on an *equal number of male and female participants*. With a total number of 36 subjects - 18 male and 18 female participants - the accuracies and the ITRs were evaluated, while comparing the data for both genders. Furthermore, information regarding the degree of fatigue of participants before and after the experiment was collected. In addition, the indication of how the flickering stimuli affected the research participants after the experiment was also gathered in the form of questionnaire responses.

2   Methods and Materials

This section describes the hardware setup, stimulus presentation, experimental design, and presents details about the classification methods and spelling interface.

2.1   Participants

For the experiment, 36 healthy participants (18 males and 18 females) were recruited and divided into two groups based on their gender; one group referred to as the female (mean age 24.4 years, SD 4.8, median 22 years, range 20 to 39 years, all identifying as female gender in the questionnaire), the other referred to as the male group (mean age 24.6 years, SD 2.0, median 24 years, range 21 to 28 years, all identifying as male).

According to the demographic distribution of the participants depicted in Table 1, both groups were quite diverse with 8 different native languages in the female group and 9 in the male group, adding up to 12 different native languages in total, which supports the diversification of research. All subjects had normal or corrected to normal vision, but not all with corrected vision utilized their vision aid during the experiment as shown in Table 1.

BCI studies are non-invasive and do not inflict pain on the subjects. The Medical Faculty's Ethical Committee at the University of Duisburg-Essen gave its approval to the study. The probable risks and the experimental process were explained to the volunteers prior to commencing the experiment. The participants completed a consent form in line with the Helsinki declaration after giving consent to participate. Information about the subjects was saved anonymously. By taking part in the study, each volunteer was compensated with 10 €. The test subjects had the option to withdraw at any point throughout the experiment, without having to give any reasons.

Table 1. Participants demographic data, with subjects grouped into cohorts according to their gender. Yes* indicates the cases of subjects which were not wearing their prescripted vision aid.

Gender	Subject Nr	Age	Native language	Need for vision correction
f	6	22	German	no
f	8	21	Russian	no
f	10	21	German	no
f	12	29	German	no
f	14	30	Arabic	no
f	17	20	English/German	no
f	18	21	English	yes
f	19	26	German	no
f	21	26	Nepali	no
f	23	22	Russian	yes*
f	24	28	German	yes
f	26	23	German	no
f	28	25	Georgian	yes*
f	29	22	Spanish	no
f	32	22	Arabic	yes
f	34	22	Dutch	no
f	35	20	German	yes
f	36	39	English	yes*
m	1	27	German	no
m	2	23	Hindi/English	yes
m	3	23	Urdu	no
m	4	23	Arabic	yes
m	5	24	Portuguese	no
m	7	24	Urdu	no
m	9	25	Puntabi	no
m	11	24	German	no
m	13	27	Spanish	no
m	15	21	German	no
m	16	28	Spanish	yes
m	20	25	Urdu	no
m	22	26	Russian/German	yes*
m	25	25	German	no
m	27	24	Arabic	no
m	30	22	English	yes
m	31	28	Spanish	no
m	33	23	English	yes

2.2   Experimental Protocol

An identical experimental design was employed for both gender groups, and the following protocols were adopted to ensure that all participants were prepared in a standardized way, minimizing variation and potential sources of error in the data collection process.

The experiment started by taking the participant's head size measurement and selecting a cap with an ideal fit. Afterwards, the participant is given the necessary documents including an information sheet, a consent form, and an experiment question sheet, to fill out. Once the documents are completed, the cap is placed on the participant's head and the electrodes are connected to the EEG amplifier. Subsequently, a program is used to check the impedance, while the conductive electrolyte gel is applied between the electrodes and scalp to reduce impedance levels. The training phase begins once the impedance levels for all electrodes are below 5 kΩ.

2.3   Graphical User Interface (GUI)

The graphical user interface presents four selection options as depicted in Fig. 1. The GUI spelling software was organised in a three-step speller as utilised in previous research [7,8], with 26 letters and one underscore/space character divided into three boxes (nine characters each, 'A-I', 'J-R', 'S-Z' + space character). The copy-spelling task and the user output were both presented in the centre of the screen. Every selection was accompanied by audio and visual feedback (the size of the selected box is increased for a short time).

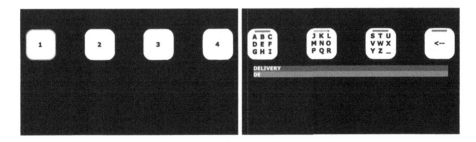

Fig. 1. *Left* presents the GUI screen shortly before the start of the training phase. Here, the 1st target is marked with the green frame, this mark changes to the consecutive targets during the training phase. *Right* is the GUI speller screen during the spelling of the DELIVERY task. Letter selection required three steps: In order to type e.g. the letter D, participant first has to select 'A-I' box (1st target) followed by 'D-F' box (2nd target) and finally the D box (1st target of the group of letters 'D', 'E', 'F'). (Color figure online)

Stimulus Presentation: Four boxes (230×230 pixels) arranged as 1×4 stimulus matrices were utilized (see Fig. 1 for more details), which corresponded to $K = 4$ stimulus classes/targets.

A distinct flashing pattern was assigned to each of these targets using the cVEP paradigm. When flickering, the color of the target stimuli corresponding to the codes alternated between the background color 'black' (represented by '0') and 'white' (represented by '1'). The m-sequences used in this experiment, c_i, $i = 1, \ldots, K$, were set recursively. The initial code, c_1 was set to

$$c_1 = 101011001101110110100100111000101111001010001100001000001111110.$$

The remaining $K - 1$ sequences were determined by employing a circular shift of 16 bits (c_1 had no shift, c_2 was shifted by 16 bits to the left, c_3 was shifted by 32 bits to the left, etc.). The duration of the stimulation cycle is calculated by dividing the code length (here 63 bits) by the monitor refresh rate r in Hz. Further details about used cVEP signal processing methods can be found in [8].

Training Phase: During the training phase, four stimuli were looked at sequentially from 1 to 4 by the participant, see *Fig.* 1. The recording was grouped into six blocks of training, $n_b = 6$. In each block, each stimulus was attended once, resulting in a total of $6 * 4 = 24$ trials. Each of these trials lasted 2.1 s, i.e. the code pattern was repeated for 2 cycles. A green frame indicated which box the participant needed to gaze at. After each trial, the next box the user needed to focus on was highlighted, and the flickering paused for one second. After each block of four trials (all four targets), the user was allowed to rest. In order to continue, the user needed to press the space bar, thus the duration of this break was self-initiated and mostly kept short.

Copy-Spelling Phase: In the copy-spelling exercise, four boxes were displayed on the screen. As indicated in Fig. 1 *Right*, the first 3 boxes from left to right contain letters, and the fourth is the "UNDO" function which is used as either a backspace or delete button (when going back is not an option). The box turns green when the correct box is selected, and red when the wrong one is selected. Participants spelt the word BCI as a brief familiarization run. The automatically generated classification parameters like **threshold** and **gaze shift** were slightly modified when necessary. Subsequently, the words CONTRARY, DELIVERY, and the pangram (THE_FIVE_BOXING_WIZARDS_JUMP_QUICKLY) were written out. The order of CONTRARY and DELIVERY was switched depending on the subject number being even or odd. The underscore stands in for the *space* character. Errors are corrected using the UNDO feature of the spelling interface. Following the completion of the spelling task, the participants were permitted to type any word or sentence they desired to demonstrate they are in control of the letter selection. Upon the conclusion of the experiment, the participants were surveyed concerning their degree of irritation with the flickering light, level of exhaustion, and their perspectives on BCI control methods. The total duration of the experiment was about an hour.

2.4   Hardware

The used computer (Dell Precision, RTX3070 graphic card) operated on Microsoft Windows 10 (21H2) Education, running on an Intel processor (Intel Core i9-10900K, 3.70 GHz). A liquid crystal display screen (Asus ROG Swift PG258Q, 1920 × 1080 pixel, 240 Hz maximal refresh rate) was utilised for stimulus presentation.

An EEG amplifier (g.USBamp, Guger Technologies, Graz, Austria) was used, utilising all 16 signal channels which were placed according to the international system of EEG electrode placement: P_7, P_3, P_Z, P_4, P_8, PO_7, PO_3, PO_Z, PO_4, PO_8, O_1, O_Z, O_2, O_9, I_Z, and O_{10}. Additionally, the reference electrode was placed at C_Z and the ground electrode at AF_Z. Standard abrasive electrolytic electrode gel was applied between the electrodes and the scalp to bring impedances below 5 kΩ during the preparation phase.

3   Results

The results section presents both the objective measures in terms of Accuracies and ITRs, as well as the subjective measures which entail the degree of exhaustion and the disturbing effect of the flickering stimuli. Each participant's BCI performance was assessed by computing the typical ITR in bpm. The spelling exercise was successfully completed by every subject.

3.1   Evaluation of the BCI Performance (Accuracies and ITRs)

The results of the spelling tasks are presented in Table 2. It shows the objective measures of each task consisting of the ITRs and accuracies for both female and male gender groups. Also depicted are the accuracy standard deviation (SD) and ITR SD for each set of tasks, for each gender group. Task 1 had an accuracy SD of 7.06 and an ITR SD of 17.98 for females while the accuracy SD for the male group was estimated to be 7.66 with an ITR SD of 14.73. For Task 2, the accuracy SD of 6.25 and ITR SD of 16.97 were computed for the females, at the same time, the males were estimated to have an accuracy SD of 8.82 and an ITR SD of 14.73. Finally, the accuracy SD for the pangram task was calculated to be 3.68 with an ITR SD of 9.57 for the female gender, and for the male gender 6.98 and 10.69, respectively.

Asides from these, Table 2 equally shows the mean accuracies and the mean ITRs. However, they are better visualized in Fig. 2. For the CONTRARY spelling task in Fig. 2, the mean ITR for the female gender group was 43.4 bpm which is higher compared to an ITR of 35.4 bpm for the male group. Likewise for the DELIVERY task, the female gender had a larger ITR of 47.1 bpm as against 38.2 bpm for the male gender. Although the ITRs were generally low for the pangram task for both gender groups, there was still a difference of 1.3 bpm between the mean ITRs, with the female group obtaining the higher value of 29.4 bpm as against 28.1 bpm for the male group.

Table 2. Results for three spelling tasks (Task 1: CONTRARY, Task 2: DELIVERY, and Task 3: THE_FIVE_BOXING_WIZARDS_JUMP_QUICKLY). Shown are the subject numbers, accuracies, and ITRs for tasks 1-3.

#	Female						#	Male					
	Task1		Task 2		Task 3			Task1		Task 2		Task 3	
	Acc [%]	ITR [bpm]	Acc [%]	ITR [bpm]	Acc [%]	ITR [bpm]		Acc [%]	ITR [bpm]	Acc [%]	ITR [bpm]	Acc [%]	ITR [bpm]
6	96.2	60.69	100.0	78.90	91.3	24.09	1	84.2	17.45	100.0	46.41	92.9	28.50
8	100.0	75.49	96.2	66.12	85.4	40.18	2	88.2	16.61	96.2	23.05	85.2	16.37
10	92.6	38.86	100.0	56.75	90.7	29.13	3	93.3	35.86	92.9	35.91	97.4	35.08
12	93.8	41.27	100.0	68.90	85.9	13.55	4	90.6	31.64	88.2	34.43	82.2	18.42
14	96.2	41.34	96.2	38.36	90.1	22.11	5	100.0	50.22	100.0	54.60	97.4	33.10
17	96.2	60.41	83.3	42.92	86.6	26.83	7	93.3	40.85	85.3	23.06	76.4	21.26
18	96.2	48.42	96.0	54.71	90.4	40.75	9	100.0	30.59	81.6	14.30	94.3	21.46
19	100.0	45.21	96.4	43.16	95.2	27.46	11	96.2	51.91	96.4	49.77	90.4	41.15
21	81.3	13.13	88.6	17.37	88.9	18.39	13	88.0	33.67	76.0	18.82	76.0	13.99
23	92.9	41.40	96.2	48.86	84.5	31.47	15	100.0	49.83	100.0	56.36	96.0	27.49
24	75.4	21.04	90.0	37.75	85.0	22.56	16	93.8	14.57	100.0	47.56	92.0	21.70
26	100.0	65.31	96.4	55.13	92.9	37.13	20	100.0	69.07	100.0	63.02	96.6	51.70
28	90.0	60.85	78.6	40.51	87.0	45.35	22	77.1	36.14	75.9	31.34	86.1	24.50
29	86.8	27.00	86.8	30.23	95.9	24.90	25	93.3	38.82	96.2	51.61	97.4	45.05
32	78.3	23.60	85.3	23.45	85.8	26.96	27	100.0	52.17	100.0	51.85	93.6	30.95
34	90.6	57.66	96.2	72.33	86.5	40.43	30	96.2	22.93	86.7	17.59	91.9	27.59
35	96.4	48.31	96.4	48.31	95.9	43.40	31	100.0	29.88	100.0	47.06	80.7	10.54
36	93.3	10.97	100.0	24.51	88.8	14.23	33	74.1	15.01	78.3	21.57	94.4	36.74
Mean	92.0	43.39	93.5	47.13	89.3	29.38	Mean	92.7	35.40	91.9	35.40	90.0	28.09
SD	7.06	17.98	6.25	16.97	3.68	9.57	SD	7.66	14.73	8.82	14.73	6.98	10.69

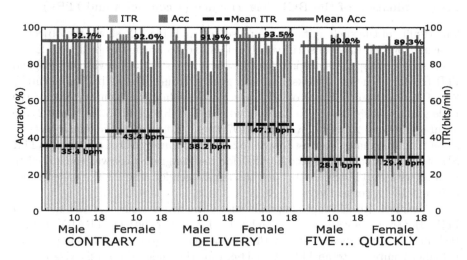

Fig. 2. Results of the spelling tasks for the male and female participants. Horizontal bars indicate the mean values for accuracy and ITR.

With respect to the mean accuracy for the tasks, the male gender group obtained slightly higher values of 92.7% and 90% for the CONTRARY and Pangram tasks respectively, while that of the female gender was higher (93.5%) for the DELIVERY task. Overall, the mean accuracy levels for both genders are nearly identical, with females and males attaining an average accuracy of 91.6% and 91.53%, respectively for all three tasks. Notably, the mean ITR value for all three tasks, for the female group is 39.96 bpm, exceeding that of the male group, which stands at 33.96 bpm.

The two sample t-test showed no statistically significant differences in the ITR of Task 1, Task 2, and Task 3 between the female and male groups: $t(34) = 1.42$, $p = 0.17$, $t(34) = 1.61$, $p = 0.12$, and $t(34) = 0.37$, $p = 0.71$, respectively.

A *Mann-Whitney U Test* was conducted to compare the accuracy of Task 1, Task 2, and Task 3 between the female and male groups for which the p-values were approximately 0.68, 0.92, and 0.27, respectively.

3.2   Evaluation of the Questionnaire

The subjective measures presented in the form of the degree of exhaustion of the participants before and after the experiment are depicted in Fig. 3.

Of the 18 participants in each group, 2 females and 4 males respectively stated that they consumed alcohol in the last 24 h prior to the experiment, 3 females and 2 males had smoked in the last 2 h and 3 females and 7 males stated that they had coffee during the last 2 h before the experiment. This might have an impact on their performance and could be investigated in further studies. Owing to the limited number of subjects, a meaningful analysis cannot be performed. 11 females and 12 males did not have a vision prescription, while the others had. 3 of the female participants and 1 of the male participants with vision prescriptions did not wear their reading aid.

To investigate possible differences in the level of tiredness of the BCI speller usage depending on gender, all subjects were asked to rate how tired they felt before and after the experiment on a scale of 1 (not at all) to 5 (very tired). Furthermore, they were to state whether they feel more, less, or equally as tired afterwards. Considering that some participants stated no change but chose a different number than before the experiment or vice versa, this makes the absolute reliability of the results limited. The overall subjects' responses to the questionnaire are depicted in Fig. 3 for both groups. Before the experiment started, the female subjects stated to be slightly less tired (median 2, mean 2.17, SD 1.12) than the male subjects (median 2, mean 2.11, SD 0.94), which is not a significant difference, especially when considering the small sample size and the subjectivity of the chosen values. Both female and male subjects stated to be more tired after the experiment, as can also be seen in Fig. 3. Afterwards, the tiredness in case of the female group led to the same median of 2, a mean of 2.61 with a slightly decreased SD of 1.01. In case of the male group, the mean and the SD increased to 2.67 and 1.05 respectively. The tiredness decreased by one unit on the scale of roughly 6% for the female subjects and 11% for the male subjects.

It remained the same in most cases (56% of females, 54% of males), increased by one in case of 38% females and 8% males. An increase by two was stated by 23% male subjects, but not by female subjects.

Even though the participants were asked about their hours of sleep during the previous night, the resulting subgroups of males and females with less than 7 h of sleep were too small (3 females, 4 males with consistent answers) to investigate its influence on the level of tiredness. If this should be investigated further, the individuality of the ideal amount of sleep and possible sleep deprivation in previous nights should be taken into account as well.

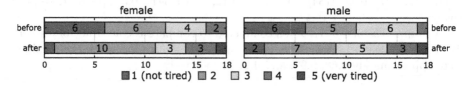

Fig. 3. Subjective level of tiredness before and after the experiment of both groups, rated on a scale of 1 (not at all) to 5 (very tired) without inconsistent data as described above.

Concerning the disturbing effect of the flickering, the participants were asked to rate their perception again on a scale of 1 to 5. As the meaning of 1 to 5 respectively was subjectively interpreted, there might have been different impressions. Assuming they followed the more (1), less (5) mentioned above on the questionnaire, the female group described the flickering as slightly less annoying (mean value of 3.3, SD of 1.3) than the male group (mean of 2.8, SD of 1.3). Regardless of their interpretation of the scale, both groups had a median of 3, suggesting that the flickering did have some disturbing effect on the subjects. Therefore, the subjects' perception of how long the system could be used without breaks is interesting. The female group estimated this duration to be around 1 h (median, mean of 1.3 h with SD 0.7), which is twice the duration the male group stated (median 0.5 h, mean 1.8 h, SD 1.3). Especially in case of the male group, the median is to be preferred as single outliers like a maximum duration of 2 min have a smaller impact. When asked whether they could use the system daily, 4 females and 5 males said yes. Most females (12) chose the option "maybe", while 9 males chose the option "no". Similarly, the responses provided to the question as to whether they consider BCI a reliable control method (female group: 9 yes, 9 maybe; male group: 10 yes, 6 maybe, 2 no) varied.

4   Discussion

The main objective of the presented study is to evaluate the difference in cVEP-based BCI performance between the female and male gender groups. The data shown in Fig. 2 supports this notion through the information transfer rate.

For Task 1 (CONTRARY), the level of cluster of the accuracy data around the mean is almost identical for both groups (7.06 for females and 7.66 for males). However, the accuracy data for Task 2 (DELIVERY) and Task 3 (Pangram) are closer to the mean for the female group compared to that of the male gender. Regarding the ITR data, the female group had more data spread out around the mean values for Tasks 1 and 2 compared to the male group, while Task 3 values are slightly more clustered for the female gender than the male. The accuracy SDs (6.25 to 8.82) are generally lower than that of the ITR SDs (9.57 to 17.98), and this indicates that the accuracy data are more clustered around their mean values than the ITR data. The reason for the far ITR values could be due to the adjusted threshold and gaze shift for some participants during the experiment.

While the mean accuracies for the CONTRARY and pangram tasks are slightly higher for the male gender compared to the female group, the accuracy of the DELIVERY task is higher for the females than the males. It is evident from the results presented in Fig. 2 that the mean ITRs achieved for the balanced tasks, namely CONTRARY and DELIVERY, were significantly higher for the female group than the male group. For the BCI performance, higher ITR values were generally observed for the female group. However, the difference is more pronounced in the spelling tasks for CONTRARY and DELIVERY when juxtaposed with that of the pangram task.

Based on the above observations, a considerable difference was identified in BCI performance of the female and male genders for the balanced tasks, with the female group obtaining higher ITR values. This aligns with earlier research utilizing the SSVEP paradigm, which demonstrated that the ITR performance of females surpassed that of males [3]. However, there was a discernible trend in the performance of the female and male groups for the pangram task, although the difference may not be significant. The results might have been influenced by the adjustment of threshold and gaze shift for some participants.

The data concerning the flickering effects may be fairly limited, as the interpretation of the values assigned to the 5-point Likert scale was subjective.

Apart from the performance, the flickering speed has a high impact on the user comfort of the BCI. To investigate the user-friendliness of the system, the participants were queried regarding the level of annoyance caused by the flickering. This was done by comparing their subjective fatigue levels (tiredness) before and after the experiment. Thus, the use of a BCI speller seems to cause a slight increase in tiredness (fatigue) in most cases, regardless of the subject's gender. In the case of the male group, the level of tiredness was more widespread and included greater changes.

Consequently, based on the various inferences from the objective and subjective data, the difference in the BCI performance of pangram tasks between the female and male groups requires further investigation.

5   Conclusion

A study to confirm the trend that female participants performed with better ITR than male participants was performed. The experiment was carried out on a considerable number of healthy female and male subjects (18 females and 18 males). The study corroborates that female subjects have higher BCI performance for balanced tasks but only an inclination (and not a significant difference) was observed for the pangram task. Further research might be considered in the area of the BCI performance between the female gender and the male gender, for pangram tasks.

Acknowledgment. This research was supported by the German Federal Ministry of Education and Research funding program Forschung an Fachhochschulen under grant number 13FH033EX0.

The authors gratefully acknowledge the financial support by the association "The Friends of the University Rhine-Waal - Campus Cleve".

We also appreciate each and every one of the research study's participants as well as our student assistants.

References

1. Abiri, R., Borhani, S., Sellers, E.W., Jiang, Y., Zhao, X.: A comprehensive review of EEG-based brain-computer interface paradigms. J. Neural Eng. **16**(1), 011001 (2019). https://doi.org/10.1088/1741-2552/aaf12e

2. Allison, B., Luth, T., Valbuena, D., Teymourian, A., Volosyak, I., Graser, A.: BCI demographics: how many (and what kinds of) people can use an SSVEP BCI? IEEE Trans. Neural Syst. Rehabil. Eng. **18**(2), 107–116 (2010). https://doi.org/10.1109/TNSRE.2009.2039495

3. Gembler, F., Stawicki, P., Volosyak, I.: Autonomous parameter adjustment for SSVEP-based BCIs with a novel BCI wizard. Front. Neurosci. **9**, 474 (2015). https://doi.org/10.3389/fnins.2015.00474

4. Martínez-Cagigal, V., Thielen, J., Santamaria-Vazquez, E., Pérez-Velasco, S., Desain, P., Hornero, R.: Brain-computer interfaces based on code-modulated visual evoked potentials (c-VEP): a literature review. J. Neural Eng. **18**(6), 061002 (2021). https://doi.org/10.1088/1741-2552/ac38cf

5. Rezeika, A., Benda, M., Stawicki, P., Gembler, F., Saboor, A., Volosyak, I.: Brain-computer interface spellers: a review. Brain Sci. **8**(4), 57 (2018). https://doi.org/10.3390/brainsci8040057

6. Stegman, P., Crawford, C.S., Andujar, M., Nijholt, A., Gilbert, J.E.: Brain-computer interface software: a review and discussion. IEEE Trans. Hum.-Mach. Syst. **50**(2), 101–115 (2020). https://doi.org/10.1109/THMS.2020.2968411

7. Volosyak, I., Gembler, F., Stawicki, P.: Age-related differences in SSVEP-based BCI performance. Neurocomputing **250**, 57–64 (2017). https://doi.org/10.1016/j.neucom.2016.08.121

8. Volosyak, I., Rezeika, A., Benda, M., Gembler, F., Stawicki, P.: Towards solving of the Illiteracy phenomenon for VEP-based brain-computer interfaces. Biomed. Phys. Eng. Express **6**(3), 035034 (2020). https://doi.org/10.1088/2057-1976/ab87e6

9. Volosyak, I., Valbuena, D., Luth, T., Malechka, T., Graser, A.: BCI demographics II: how many (and what kinds of) people can use a high-frequency SSVEP BCI? IEEE Trans. Neural Syst. Rehabil. Eng. **19**(3), 232–239 (2011). https://doi.org/10.1109/TNSRE.2011.2121919

Bit-Wise Reconstruction of Non-binary Visual Stimulation Patterns from EEG Using Deep Learning: A Promising Alternative for User-Friendly High-Speed c-VEP-Based BCIs

Eduardo Santamaría-Vázquez[1,2](✉) , Víctor Martínez-Cagigal[1,2] ,
and Roberto Hornero[1,2]

[1] Biomedical Engineering Group (GIB), E.T.S Ingenieros de Telecomunicación,
University of Valladolid, Paseo de Belén 15, 47011 Valladolid, Spain
`eduardo.santamaria.vazquez@uva.es`
[2] Centro de Investigación Biomédica en Red en Bioingeniería,
Biomateriales y Nanomedicina, (CIBER-BBN), Madrid, Spain

Abstract. Brain-computer interfaces (BCI) based on code-modulated visual evoked potentials (c-VEP) have shown great potential for communication and device control. These systems encode each command using different sequences of visual stimuli. Normally, the stimulation pattern is binary (i.e., black and white), but non-binary stimuli sequences with different shades of gray could reduce eyestrain and improve user-friendliness. This study introduces a novel approach to decode non-binary visual stimuli patterns from electroencephalography (EEG) signals using deep learning. The proposed method uses, for the first time, a bit-wise reconstruction strategy for stimulation patterns encoded with 2, 3, 5, 7 and 11 levels of gray. The performance of the proposed approach was evaluated on a dataset of 16 subjects, reaching an average command decoding accuracy over 95% for all stimulation sequences. The high accuracy and speed of the proposed method make it a promising alternative for user-friendly, high-speed c-VEP-based BCIs.

Keywords: Brain-computer interfaces · c-VEP · EEG · Deep learning

1 Introduction

Brain-Computer Interfaces (BCI) enable direct communication between the brain and an external device, bypassing traditional motor pathways such as muscles and nerves [14]. BCIs have the potential to transform the way we interact with technology and the world around us, providing new possibilities for people with disabilities to control their environment and enhancing the performance of able-bodied individuals. In this regard, the field of non-invasive BCI research is rapidly advancing, with new technologies and applications being developed and tested in both clinical and non-clinical settings [14].

I. Rojas et al. (Eds.): IWANN 2023, LNCS 14135, pp. 603–614, 2023.
https://doi.org/10.1007/978-3-031-43078-7_49

The electroencephalography (EEG) is the most used technique to measure the brain activity in BCIs due to its non-invasiveness and high temporal resolution, providing a cost-effective option for real-time control of external devices in a variety of settings [14]. On the other hand, this signal has low-spatial resolution and is greatly affected by noisy artifacts, such as power line frequency, muscular activity or eye blinks, among others. As a consequence of its low signal-to-noise ratio (SNR), EEG cannot be used to decode fine neural activity, such as inner speech or activation of individual muscles [14].

In order to increase the SNR of EEG, BCIs use different paradigms and control signals to decode user's intentions from EEG. Non-invasive BCIs are typically categorized as exogenous and endogenous [1]. Exogenous BCIs rely on external stimuli, such as flashing lights, sounds, or tactile sensations, to evoke neural responses that encode user's intentions in the EEG. These BCIs allow to select commands among a predefined set of options, being suitable for communication and control applications due to their greater accuracy and robustness [1]. In contrast, endogenous BCIs rely on self-generated brain activity to control the BCI system, being more adequate for clinical applications, such as neurorehabilitation [1].

The most widespread exogenous paradigms are P300 evoked potentials and steady-state visual evoked potentials (SSVEP). The P300 paradigm detects differences in the neural response to target and non-target stimuli, showing to be effective in a variety of BCI applications [11]. However, this paradigm has important limitations in terms of accuracy and speed, needing long calibration sessions to achieve peak performance [10]. The SSVEP paradigm solved some of these drawbacks by encoding each command with visual stimuli that flicker at different frequencies, increasing the overall performance [5]. However, its susceptibility to visual fatigue, and the low number of selectable commands that can be displayed at the same time without affecting systems's accuracy, limit the use of SSVEP BCIs for practical applications [5].

Code-modulated visual evoked potentials (c-VEPs) represent a promising alternative to these paradigms, as they offer several advantages, including higher accuracy, selection speed and robustness to environmental factors [7]. C-VEP-based BCIs present specific stimulation patterns for each command following a predefined sequence. The most widely used technique is the circular shifting paradigm [7]. This method uses shifted versions of maximal length sequences (m-sequences), which present very low correlation between them, to encode the commands [7]. Therefore, the different options are encoded with the same stimulation sequence, which is circularly shifted a certain number of bits. In the calibration phase, the EEG response to this sequence is recorded, calculating a subject-specific template. Then, the target command is decoded by correlating the EEG signal of each trial with the different circularly-shifted versions of the template. The main advantages of this paradigm are: (1) it reduces the calibration time, as it only needs to record the EEG response to 1 stimulation sequence; and (2) it ensures that the shifted versions of the m-sequence have very low correlation, maximizing the performance of the system. However, as only 1 stimulation sequence is used, the number of commands that can be encoded with the circular shifting paradigm is limited. To solve this limitation, Nagel et al. [8] introduced EEG2Code, a method that models the brain response to individual stimulation events. Its main advantage is that it allows to decode commands by predicting, bit by bit, the stimulation sequence. Thus, it allows the use of random stimulation patterns to encode more commands than m-sequences. EEG2Code showed high accuracy, which was later improved in more recent studies using deep learning [9]. Despite these advances, which show the potential of c-VEPs to significantly enhance the usability and effectiveness of exogenous BCIs, there is still room for improvement. For instance, as SSVEPs, c-VEPs also provoke visual fatigue and eyestrain to the user, which reduces the usability of the system for practical applications. To alleviate this effect, Gembler et al. [4] recently proposed low-contrast stimuli based on different shades of gray as a mean to reduce user discomfort. They applied the m-sequences with 5 levels of gray and canonical correlation analysis (CCA) to classify between 8 different commands with accuracies above 95% [4]. Despite of the novelty of Gembler's study, their method can only decode a limited number of commands due to the use of the circular shifting paradigm. On the other hand, bit-wise reconstruction approaches, such as EEG2Code, could cope with an unlimited number of commands [9]. However, these techniques have not been applied to non-binary stimulation patterns yet.

In this study, we propose, for the first time, a novel classification paradigm based on bit-wise code reconstruction for non-binary stimulation patterns encoded with 2, 3, 5, 7 and 11 shades of gray. In our approach, the stimulation code is reconstructed bit by bit, rather than use the whole trial to calculate correlations, as in the circular shifting paradigm. Our method uses a flavored version of EEG-Inception, a convolutional neural network (CNN) for EEG classification tasks, which has been specifically tailored to better fit the needs of the bit-wise reconstruction approach [10].

2 Materials and Methods

2.1 Subjects and Signals

In this study, 16 healthy participants (11 males, 5 females; 28.8 ± 5.0 years old) took part in the experiments [6]. EEG signals were recorded using a g.USBamp amplifier with 16 channels at F3, Fz, F4, C3, Cz, C4, CPz, P3, Pz, P4, PO7, PO7 and Oz, according to the International System 10-10. The reference was placed at the right earlobe, whereas ground was placed at FPz. The sampling rate was 256 Hz. MEDUSA©, a novel Python-based BCI platform, was used to present and process stimuli in real-time [12]. Concretely, we used the app "P-ary c-VEP Speller", which is publicly available at www.medusabci.com [6]. The experiment was conducted on a computer with an Intel Core i7-7700 processor, 32 GB RAM, and a LED FullHD screen with a refresh rate at 144 Hz [6].

Table 1. P-ary m-sequences used to generate the stimulation sequences.

	Base	Order	Length	Cycle duration
$GF(2^6)$	2	6	63 bits	0.525 s
$GF(3^4)$	3	4	80 bits	0.667 s
$GF(5^3)$	5	3	124 bits	1.033 s
$GF(7^2)$	7	2	48 bits	0.408 s
$GF(11^2)$	11	2	120 bits	1.000 s

GF: Galois Field; Cycle duration: time of each stimulation cycle with an stimulus presentation rate of 120 Hz

2.2 System Design

Stimulation Patterns. The stimulation patterns of each command were designed using m-sequences, which are pseudorandom periodic time series that exhibit almost orthogonal behavior to circularly shifted versions of themselves [7]. These sequences, which are widely used in c-VEP-based BCIs, are generated through linear-feedback shift registers (LFSRs) that utilize a linear function of the immediate previous state to compute new values. The base p, order r, and arrangement of taps determine the LFSR, which is expressed as a polynomial whose coefficients are bounded on a Galois Field (GF). More information about the generation of m-sequences is available in the study of Buračas et al. [2]. In our study, 5 different p-ary m-sequences were used, as displayed in Table 1. Importantly, an m-sequence cannot be of arbitrary length and instead consists of exactly $N = p^r - 1$ bits. The 16 commands of our BCI system are encoded with shifted versions of the each m-sequence, which were selected to ensure near-to-zero correlation to maximize the performance of the c-VEP-based BCI. See the study of Martínez-Cagigal et al. [6] for more information about the design of the m-sequences. The stimuli were presented with a refresh rate of 120 Hz.

Pre-processing. The preprocessing stage of the algorithm involves two main steps: the application of a bandpass Infinite Impulse Response (IIR) Butterworth filter between 1 and 60 Hz, and a notch IIR Butterworth filter between 49 and 51 Hz. Both filters had order 7. The purpose of the band-pass filter is to remove waveforms outside of the desired range, which can include noise and artifacts that can interfere with the analysis of the signal. The notch filter is designed to eliminate the powerline interference at 50 Hz.

Feature Extraction. After pre-processing, the feature extraction stage decimates the EEG signals to a sampling frequency of 200 Hz. Then, the system extracts an EEG epoch for each stimulus, with a temporal window of 0 to 500 ms after the stimulus onset. This temporal window is designed to capture the entire VEP response of the brain to each stimulus. Z-score baseline normalization is applied by taking the 250 ms of signal before the stimulus onset. After this process, each observation has a feature vector of 100 samples × 16 channels.

Classification. The classification of EEG epochs is performed with a modified version of EEG-Inception [10]. This CNN was specifically designed for P300 detection. Therefore, it has been adapted for its use in c-VEP-based BCIs. However, most of the architectural advantages, which proved to be effective for P300 detection, can be applied in this context as well. These include [10]: (1) efficient integration of depthwise convolutions, dropout regularization, batch normalization, and average pooling; (2) inclusion of Inception modules specifically designed for EEG processing to allow a multi-scale analysis in the temporal domain; (3) special design to avoid overfitting, including an output block that synthesizes the information extracted by Inception modules in very few, high-level features; and (4) optimized hyperparameters (i.e., dropout rate, activation functions and learning rate) for EEG classification tasks. For this study, we the input layer has been modified to allow the new shape of input feature vectors, the temporal scales of the 3 Inception branches, the number of filters of each branch and the dropout rate. Additionally, it must be noticed that the number of classes for each m-sequence is different, being equal to p, in accordance to the different shades of gray. Table 2 provides a detailed overview of the architecture and hyperparameters of the model used in this study. The dropout rate is set to 0.15. The model is trained using the calibration trials from the evaluation experiment (see Sect. 2.3). In this study, we use subject-specific models, i.e., models that are trained and tested with data from the same subject. Specifically, the training dataset consisted of $30 \times L_s \times 10$ observations, where L_s is the length of each m-sequence with base p and order r, as described in Table 1. Then, the test trials are used to evaluate the model. To decode the test trials, the epochs corresponding to individual stimulation events are fed to each instance of EEG-Inception. Finally, the model predicts the class (i.e., shade of gray) of each stimulus based on the EEG data to reconstruct the m-sequence bit by bit.

Command Decoding. Once the stimulation pattern has been reconstructed with EEG-Inception, the system correlates the predicted stimulation sequence with each one of the sequences of the 16 commands. The command with maximum Pearson correlation coefficient is selected.

2.3 Experimental Procedure

Participants performed a single evaluation session with the c-VEP speller. The session had 5 blocks, one for each m-sequence, with calibration and test recordings. The graphical user interface of the speller, with the different modes, is displayed in Fig. 1. The stimulus presentation rate was 120 Hz. For each m-sequence, participants performed 30 calibration trials focusing on the unshifted version of the corresponding m-sequence using the calibration matrix. Then, the test task consisted of selecting the 16 commands, 2 times each. Therefore, we recorded 32 test trials with the test matrix for each m-sequence and participant. In summary, the experiment had 5 blocks of 30 calibration trials and 32 test trials per participant. Importantly, each trial consisted of 10 stimulation cycles

Table 2. Details of the modified version of EEG-Inception for c-VEP detection.

Block	Type	Filt.	Depth	Kernel	Padding	Output	Conn. to
IN	Input	–	–	–	–	$100 \times 16 \times 1$	C1, C2, C3
C1	Conv2D	12	–	50×1	Same	$100 \times 16 \times 12$	CO1
C2	Conv2D	12	–	25×1	Same	$100 \times 16 \times 12$	CO1
C3	Conv2D	12	–	12×1	Same	$100 \times 16 \times 12$	CO1
CO1	Concatenate	–	–	–	–	$100 \times 1 \times 36$	A1
A1	AveragePooling2D	–	–	2×1	–	$50 \times 1 \times 36$	D1
D1	DepthwiseConv2D	–	2	1×16	Valid	$50 \times 1 \times 72$	A1
A2	AveragePooling2D	–	–	2×1	–	$25 \times 1 \times 72$	C4, C5, C6
C4	Conv2D	12	–	12×1	Same	$25 \times 1 \times 12$	CO2
C5	Conv2D	12	–	6×1	Same	$25 \times 1 \times 12$	CO2
C6	Conv2D	12	–	3×1	Same	$25 \times 1 \times 12$	CO2
CO2	Concatenate	–	–	–	–	$25 \times 1 \times 36$	A3
A3	AveragePooling2D	–	–	2×1	–	$12 \times 1 \times 36$	C7
C7	Conv2D	18	–	6×1	Same	$12 \times 1 \times 18$	A4
A4	AveragePooling2D	–	–	2×1	–	$6 \times 1 \times 18$	C8
C8	Conv2D	9	–	3×1	Same	$6 \times 1 \times 9$	A5
A5	AveragePooling2D	–	–	2×1	–	$3 \times 1 \times 9$	OUT
OUT	Dense	–	–	–	–	p	–

The type specifies the class of each block in Keras framework [3]. All convolutional blocks (i.e., Conv2D and DethpwiseConv2D) include batch normalization, activation and dropout regularization (dropout rate of 0.15). The model has 26948 parameters, of which 26606 are fitted during training. The base of the m-sequence p is the number of output classes of the model, one for each shade of gray

(i.e., repetitions of the stimulation sequence). This allows to analyze the performance of the system as a function of this parameter. More stimulation cycles will increase the decoding accuracy of the system at the expense of reducing the selection speed, as more stimulation time is needed. The order of the blocks was randomized across participants to avoid bias. Users were not aware of which specific p-ary m-sequence was used to avoid unintentional biases.

3 Results

The results of the experiments are presented in Tables 3, 4, 5, 6 and 7. Each of these tables show the command decoding accuracy per participant and number of cycles considered in the analysis for each m-sequence. The mean accuracy across cycles for each subject is detailed at the left side of the tables, whereas the mean accuracy across subjects for each number of cycles is reported at the bottom. Moreover, the grand-average accuracy provided in the bottom-right corner gives an overall picture of the system's performance in a single metric.

The results show that the proposed approach achieved high command decoding accuracy for all m-sequences. Unsurprisingly, the number of cycles has a positive impact in the systems' performance: more cycles, which imply longer stimulation times, increase the command decoding accuracy. The results indicate that

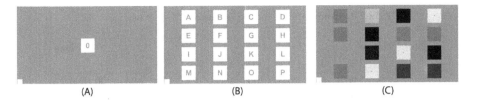

Fig. 1. Screenshots of the "P-ary c-VEP Speller" app in MEDUSA$^{©}$. (A) Calibration matrix with 1 command that is highlighted following the original p-ary m-sequence. (B) Test matrix with 16 commands that are highlighted following shifted versions of the p-ary m-sequence. (C) Screenshot during the stimulation period for $GF(11^2)$.

Table 3. Command decoding accuracy (%) for base $p = 2$.

Subj.	Cycles										Mean
	1	2	3	4	5	6	7	8	9	10	
1	87.5	96.9	100	100	100	100	100	100	100	100	98.4
2	96.9	100	100	100	100	100	100	100	100	100	99.7
3	93.8	100	100	100	100	100	100	100	100	100	99.4
4	96.9	100	100	100	100	100	100	100	100	100	99.7
5	75	93.8	100	100	100	100	100	100	100	100	96.9
6	100	100	100	100	100	100	100	100	100	100	100
7	96.9	100	100	100	100	100	100	100	100	100	99.7
8	87.5	96.9	100	100	100	100	100	100	100	100	98.4
9	93.8	100	100	100	100	100	100	100	100	100	99.4
10	93.8	100	100	100	100	100	100	100	100	100	99.4
11	75	87.5	90.6	93.8	93.8	93.8	93.8	93.8	96.9	100	91.9
12	53.1	84.4	93.8	96.9	96.9	96.9	96.9	96.9	96.9	96.9	90.9
13	90.6	96.9	96.9	100	100	100	100	100	100	100	98.4
14	50	71.9	84.4	84.4	96.9	93.8	96.9	100	100	100	87.8
15	93.8	100	100	100	100	100	100	100	100	100	99.4
16	81.2	96.9	100	100	100	100	100	100	100	100	97.8
Mean	85.4	95.3	97.9	98.4	99.2	99.0	99.2	99.4	99.6	99.8	97.3

binary m-sequences with high-contrast black and white stimulus achieved the highest overall performance, reaching a grand-average accuracy of 97.3%. Non-binary m-sequences achieved slightly lower performances in the range between 90.9% for base 3 and 93.7% for base 5.

4 Discussion

In this study, the feasibility of bit-wise reconstruction of non-binary stimulation patterns from EEG data using deep learning has been tested. To this end, we

Table 4. Command decoding accuracy (%) for base $p = 3$.

Subj.	Cycles										Mean
	1	2	3	4	5	6	7	8	9	10	
1	75	96.9	96.9	100	100	100	100	100	100	100	96.9
2	90.6	93.8	93.8	93.8	93.8	93.8	96.9	100	100	100	95.6
3	96.9	96.9	100	100	96.9	96.9	100	100	100	100	98.8
4	87.5	96.9	100	100	100	100	100	100	100	100	98.4
5	37.5	78.1	96.9	93.8	96.9	100	100	100	100	100	90.3
6	87.5	90.6	93.8	93.8	93.8	93.8	93.8	96.9	96.9	96.9	93.8
7	96.9	100	100	100	100	100	100	100	100	100	99.7
8	96.9	100	100	100	100	100	100	100	100	100	99.7
9	59.4	71.9	81.2	84.4	90.6	90.6	93.8	93.8	90.6	90.6	84.7
10	96.9	100	100	100	100	100	100	100	100	100	99.7
11	46.9	53.1	68.8	71.9	71.9	68.8	78.1	78.1	81.2	81.2	70
12	40.6	59.4	78.1	90.6	87.5	93.8	96.9	96.9	96.9	96.9	83.8
13	96.9	100	100	100	96.9	100	96.9	100	100	100	99.1
14	15.6	28.1	31.2	43.8	50	50	50	59.4	59.4	75	46.2
15	87.5	93.8	100	100	100	100	100	100	100	100	98.1
16	96.9	100	100	100	100	100	100	100	100	100	99.7
Mean	**75.6**	**85.0**	**90.0**	**92.0**	**92.4**	**93.0**	**94.1**	**95.3**	**95.3**	**96.3**	**90.9**

Table 5. Command decoding accuracy (%) for base $p = 5$.

Subj.	Cycles										Mean
	1	2	3	4	5	6	7	8	9	10	
1	87.5	100	100	100	100	100	100	100	100	100	98.8
2	90.6	96.9	96.9	100	100	100	100	100	100	100	98.4
3	100	100	100	100	100	100	100	100	100	100	100
4	90.6	93.8	93.8	93.8	93.8	93.8	93.8	96.9	96.9	96.9	94.4
5	31.2	75	90.6	84.4	93.8	96.9	96.9	100	100	100	86.9
6	100	96.9	100	100	100	100	100	100	100	100	99.7
7	93.8	93.8	100	100	100	100	100	100	100	100	98.8
8	93.8	96.9	100	100	100	100	100	100	100	100	99.1
9	87.5	96.9	96.9	96.9	96.9	100	100	100	100	100	97.5
10	84.4	100	100	100	100	100	100	100	100	100	98.4
11	56.2	68.8	75	78.1	78.1	75	75	75	78.1	78.1	73.8
12	34.4	59.4	68.8	81.2	84.4	90.6	96.9	100	100	100	81.6
13	87.5	100	100	100	100	100	100	100	100	100	98.8
14	53.1	65.6	71.9	75	81.2	81.2	87.5	87.5	87.5	90.6	78.1
15	78.1	96.9	100	100	100	100	100	100	100	100	97.5
16	90.6	96.9	96.9	96.9	100	100	100	100	100	100	98.1
Mean	**78.7**	**89.8**	**93.2**	**94.1**	**95.5**	**96.1**	**96.9**	**97.5**	**97.7**	**97.9**	**93.7**

Table 6. Command decoding accuracy (%) for base $p = 7$.

Subj.	Cycles										Mean
	1	2	3	4	5	6	7	8	9	10	
1	71.9	90.6	96.9	100	100	100	100	100	100	100	95.9
2	65.6	87.5	100	100	100	100	100	100	100	100	95.3
3	96.9	100	100	100	100	100	100	100	100	100	99.7
4	71.9	90.6	100	100	100	100	100	100	100	100	96.2
5	68.8	84.4	93.8	100	100	96.9	96.9	100	100	100	94.1
6	78.1	87.5	87.5	90.6	90.6	90.6	90.6	93.8	93.8	96.9	90
7	96.9	90.6	100	100	100	100	100	100	100	100	98.8
8	75	90.6	96.9	96.9	96.9	100	100	100	100	100	95.6
9	84.4	90.6	96.9	96.9	96.9	96.9	96.9	96.9	96.9	96.9	95
10	81.2	100	100	100	100	100	100	100	100	100	98.1
11	43.8	46.9	56.2	68.8	75	71.9	71.9	75	75	75	65.9
12	59.4	62.5	93.8	96.9	100	100	100	100	100	100	91.2
13	81.2	90.6	87.5	93.8	96.9	100	100	100	100	100	95
14	46.9	59.4	71.9	78.1	75	81.2	84.4	81.2	87.5	84.4	75
15	96.9	100	96.9	100	100	100	100	100	100	100	99.4
16	90.6	100	100	100	100	100	100	100	100	100	99.1
Mean	**75.6**	**85.7**	**92.4**	**95.1**	**95.7**	**96.1**	**96.3**	**96.7**	**97.1**	**97.1**	**92.8**

Table 7. Command decoding accuracy (%) for base $p = 11$.

Subj.	Cycles										Mean
	1	2	3	4	5	6	7	8	9	10	
1	96.9	100	100	100	100	100	100	100	100	100	99.7
2	87.5	93.8	100	100	100	100	100	100	100	100	98.1
3	96.9	100	100	100	100	100	100	100	100	100	99.7
4	100	100	100	100	100	100	100	100	100	100	100
5	75	96.9	96.9	96.9	96.9	100	100	100	100	100	96.2
6	100	100	100	100	100	100	100	100	100	100	100
7	96.9	100	100	100	100	100	100	100	100	100	99.7
8	96.9	100	100	100	100	100	100	100	100	100	99.7
9	46.9	53.1	50	53.1	56.2	56.2	53.1	53.1	53.1	53.1	52.8
10	93.8	93.8	100	100	100	100	100	100	100	100	98.8
11	50	59.4	68.8	84.4	75	81.2	78.1	84.4	81.2	81.2	74.4
12	75	90.6	100	100	100	100	100	100	100	100	96.6
13	100	100	100	100	100	100	100	100	100	100	100
14	37.5	50	65.6	81.2	81.2	87.5	87.5	90.6	96.9	93.8	77.2
15	90.6	96.9	100	100	100	100	100	100	100	100	98.8
16	78.1	93.8	96.9	100	100	100	100	100	100	100	96.9
Mean	**82.6**	**89.3**	**92.4**	**94.7**	**94.3**	**95.3**	**94.9**	**95.5**	**95.7**	**95.5**	**93.0**

adapted EEG-Inception for multi-class classification of VEPs elicited by stimuli with different shades of gray. This approach was evaluated in a c-VEP-based BCI, reaching high command decoding accuracy.

The stimulation paradigm that has been used in this work has 2 key features designed to improve user-friendliness of c-VEP-based BCIs. First, we used a presentation rate of 120 Hz, in contrast to 60 Hz as in the majority of c-VEP studies [7]. In this respect, there is a consensus in the literature stating that higher presentation rates result in less visual fatigue while reducing the selection time [4]. In addition, non-binary stimulation patterns, which present stimulus with lower contrast due to the utilization of intermediate levels of gray, are presumably less prone to provoke visual fatigue and eyestrain [9]. In this study, we have tested stimulation patterns with 3, 5, 7 and 11 shades of gray, showing the feasibility of these non-binary sequences [6]. Given the advantages of our system, it could significantly enhance the user-friendliness of applications that demand continuous control, such as assistive systems designed for severely disabled people.

The proposed method for bit-wise reconstruction of non-binary stimulation patterns has the potential to enhance the development of c-VEP-based BCIs by providing a more general classification framework that can be used in a wide range of applications. In contrast to more extended methods based on CCA, which require EEG recordings for all the stimulation sequences used in the system, bit-wise reconstruction methods showed that they are able to cope with arbitrary binary stimulation patterns once calibrated [9]. The successful bit-wise decoding of non-binary sequences achieved in this work, with up to 11 levels of gray, is an indication of the potential for our strategy to handle more complex and diverse patterns of neural activity. This can help to improve the usability of c-VEP-based BCIs, making them more user-friendly while maintaining their reliability and efficiency. Additionally, this approach can provide new insights into the neural mechanisms underlying visual perception, which can be further explored in future studies. Overall, this study highlights the promising potential of bit-wise reconstruction methods for advancing c-VEP-based BCIs and other neurotechnologies.

Regarding the command decoding results of our evaluation experiments, there are several points that are worth discussing. All the tested m-sequences reached accuracies above 95% with high selection speed. However, there are significant differences between the command decoding accuracy achieved with the binary m-sequence and the rest ($p - value < 0.05$, Wilcoxon Signed Rank Test). In this regard, it would be desirable to reduce these differences as much as possible. Further improvements in the architecture of EEG-Inception could enhance multi-class classification. Interestingly, the system's performance did not decrease more for the sequences with more classes (e.g., $p = 7$ or $p = 11$), despite that these patterns provide less training examples per-class. Therefore, stimulation sequences with more levels of gray, which could be more comfortable for the user, would improve the system's usability without affecting performance. Similarly, the length of the m-sequences (see Table 1) did not have as much influence as it could be expected in advance, given that longer sequences have more stimuli per cycle. Thus, the length of the stimulation sequences could be reduced in future designs to increase selection speed. In this case, random stimulation patterns, instead of m-sequences with predefined length, could be more practical [2].

Despite the promising results of this work, we acknowledge several limitations. We focused on the adaptation of our bit-wise reconstruction strategy for non-binary sequences. Thus, we used well-tested stimulation patterns, such as the m-sequences, which show very low correlation between shifted versions [7]. Although this is a desirable characteristic to avoid selection errors, the use of m-sequences limits the total number of commands that can be encoded in the system. In future studies, our strategy should be tested with other types of stimulation patterns to enable an arbitrary number of commands, making an in-depth comparison with the circular shifting paradigm. The proposed deep learning framework could also be enhanced by applying transfer learning to increase the performance, robustness and calibration time of the system [13]. In this regard, a pretraining stage with data from other subjects could be beneficial and should be tested in the future. Finally, the proposed system has been tested with healthy subjects. Although this is an important step to show the feasibility of our approach, further validation in more practical settings with severely disabled subjects must be addressed in future works.

5 Conclusion

This study evaluated for the first time the feasibility of a bit-wise reconstruction approach of non-binary stimulation patterns for c-VEP-based BCIs using deep learning. Non-binary sequences, which use low-contrast stimuli, could reduce visual fatigue and eyestrain in these systems. The results showed that the proposed approach based on EEG-Inception achieved high command decoding accuracy regardless of the number of levels of gray used to encode the different commands. Therefore, the stimulation paradigm used in this work, which combines a high stimulus presentation rate at 120 Hz with non-binary sequences, together with our bit-wise reconstruction method, increase the usability of c-VEP-based BCIs while maintaining their characteristic reliability and efficiency.

Acknowledgements. This study was partially funded by "Ministerio de Ciencia e Innovación/Agencia Estatal de Investigación" and European Regional Development Fund (ERDF) [TED2021-129915B-I00, RTC2019-007350-1, and PID2020-115468RB-I00]; and "CIBER en Bioingeniería, Biomateriales y Nanomedicina (CIBER-BBN)" through "Instituto de Salud Carlos III". E. Santamaría-Vázquez, was in receipt of a PIF grant by the "Consejería de Educación de la Junta de Castilla y León".

References

1. Abiri, R., Borhani, S., Sellers, E.W., Jiang, Y., Zhao, X.: A comprehensive review of EEG-based brain-computer interface paradigms. J. Neural Eng. **16**, 011001 (2019). https://doi.org/10.1088/1741-2552/aaf12e, https://iopscience.iop.org/article/10.1088/1741-2552/aaf12e

2. Buračas, G.T., Boynton, G.M.: Efficient design of event-related FMRI experiments using m-sequences. NeuroImage **16**, 801–813 (2002). https://doi.org/10.1006/nimg.2002.1116
3. Chollet, F.: Keras (2015). https://keras.io
4. Gembler, F.W., Rezeika, A., Benda, M., Volosyak, I.: Five shades of grey: exploring quintary m -sequences for more user-friendly c-VEP-based BCIS. Comput. Intell. Neurosci. **2020**, 1–11 (2020). https://doi.org/10.1155/2020/7985010
5. Lin, Z., Zhang, C., Wu, W., Gao, X.: Frequency recognition based on canonical correlation analysis for SSVEP-based BCIS. IEEE Trans. Biomed. Eng. **53**, 2610–2614 (2006)
6. Martínez-Cagigal, V., Santamaría-Vázquez, E., Pérez-Velasco, S., Marcos-Martínez, D., Moreno-Calderón, S., Hornero, R.: Non-binary m-sequences for more comfortable brain-computer interfaces based on c-VEPs. Expert Syst. Appl. (2023, under review)
7. Martínez-Cagigal, V., Thielen, J., Santamaría-Vázquez, E., Pérez-Velasco, S., Desain, P., Hornero, R.: Brain-computer interfaces based on code-modulated visual evoked potentials (c-VEP): a literature review. J. Neural Eng. **18**, 1–21 (2021)
8. Nagel, S., Spüler, M.: Modelling the brain response to arbitrary visual stimulation patterns for a flexible high-speed brain-computer interface. PLOS ONE **13** (2018). https://doi.org/10.1371/journal.pone.0206107
9. Nagel, S., Spüler, M.: World's fastest brain-computer interface: combining EEG2code with deep learning. bioRxiv (2019)
10. Santamaría-Vázquez, E., Martinez-Cagigal, V., Vaquerizo-Villar, F., Hornero, R.: EEG-inception: a novel deep convolutional neural network for assistive ERP-based brain-computer interfaces. IEEE Trans. Neural Syst. Rehabil. Eng. **28**, 2773–2782 (2020)
11. Santamaría-Vázquez, E., Martínez-Cagigal, V., Gomez-Pilar, J., Hornero, R.: Asynchronous control of ERP-based BCI spellers using steady-state visual evoked potentials elicited by peripheral stimuli. IEEE Trans. Neural Syst. Rehabil. Eng. **27**, 1883–1892 (2019)
12. Santamaría-Vázquez, E., et al.: Medusa: a novel Python-based software ecosystem to accelerate brain-computer interface and cognitive neuroscience research. Comput. Methods Programs Biomed. **230**, 107357 (2023). https://doi.org/10.1016/j.cmpb.2023.107357
13. Santamaría-Vázquez, E., Martínez-Cagigal, V., Pérez-Velasco, S., Marcos-Martínez, D., Hornero, R.: Robust asynchronous control of ERP-based brain-computer interfaces using deep learning. Comput. Methods Programs Biomed. **215**, 1–10 (2022)
14. Wolpaw, J., Wolpaw, E.W.: Brain-Computer Interfaces: Principles and Practice. OUP USA (2012)

Spiking Neural Networks: Applications and Algorithms

Event-Based Regression with Spiking Networks

Elisa Guerrero$^{(\boxtimes)}$ (iD), Fernando M. Quintana(iD),
and Maria P. Guerrero-Lebrero(iD)

University of Cadiz, Cadiz, Spain
{elisa.guerrero,fernando.quintana,maria.guerrero}@uca.es

Abstract. Spiking Neuron Networks (SNNs), also known as the third generation of neural networks, are inspired from natural computing in the brain and recent advances in neuroscience. SNNs can overcome the computational power of neural networks made of threshold or sigmoidal units. Recent advances on event-based devices along with their great power, considering the time factor, make SNNs a cutting-edge priority research objective. SNNs have been used mainly for classification problems, but their application to regression tasks remains challenging due to the complexity of training with continuous output data. In the literature we can find some first approximations in regression, specifically, for problems of a single variable of continuous values. This work deals with the analysis of the behavior of SNNs as predictors of multivariable continuous values. For this, a data set based on events has been generated from a bouncing ball and an event-based camera. The goal is to predict the next position of the ball over time.

Keywords: Regression · Spiking Neural Networks · Neuromorphic Software · DVS

1 Introduction

Wolfgang Maass [1] classified past and current Artificial Neural Networks ANNs research into three generations. First generation of ANNs is characterised by a single layer of units using a binary activation function and Hebb learning rule, allowing networks to simulate any Boolean function [2]. Second generation is organised in multilayer architectures, using non-linear activation functions, that make them representationally meaningful to stack more than one layer, and the existence of their derivatives makes it possible to use gradient-based optimization methods for training [3]. Second generation networks with one hidden layer, are universal approximators, that is, they can approximate any continuous, analog function arbitrarily well [4]. With recent advances in availability of large labelled data sets, computing power in the form of general purpose GPU computing, and advanced regularisation methods, Deep Learning (DL) [5] have emerged in the last decade. Although they have allowed us to make breakthrough progress in

© The Author(s), under exclusive license to Springer Nature Switzerland AG 2023
I. Rojas et al. (Eds.): IWANN 2023, LNCS 14135, pp. 617–628, 2023.
https://doi.org/10.1007/978-3-031-43078-7_50

many fields, they are biologically inaccurate and do not actually mimic the actual mechanisms of our brain's neurons.

In the third generation, SNNs include the concept of time into their operating model, the idea is that neurons in the SNN do not fire at each propagation cycle (as it happens with traditional neural networks), but rather they fire only when a membrane potential reaches a certain threshold. This temporal dependence makes the dynamics of this model closely resemble that observed in primate brain activity [6].

Besides the biologically motivated definition of SNNs, there is a more pragmatic application-oriented view coming from the field of neuromorphic engineering, where SNNs are often called event-based instead of spiking [7]. Event cameras produce asynchronous events for each pixel instead of intensity images. These events are generated by changes in brightness, obtaining a high time resolution of the order of microseconds, low power consumption and reduced bandwidth, by emitting only independent events instead of complete images. The resulting spatiotemporal event patterns can then be processed through networks of spiky neurons [8].

This processing exhibits interesting properties that make it particularly suitable for applications that require fast and efficient computation and where the timing information of the input/output signals is crucial to make predictions correctly [9]. In general, most of the works addressed so far with SNN and event cameras have focused mainly on multiclass classification problems, and only a few have addressed regression tasks. In [10], authors propose the use of the membrane potential as the estimated output instead of spikes when using SNN. However, there is currently no unanimity within the scientific community as to whether these values of electric potential could be good estimators of continuous values.

In this paper we follow, and extend, the mentioned approach based on the membrane potential and a single output regression task, to apply SNN for a regression task of two continuous output variables, the prediction of the position of a bouncing ball. We have selected two specific SNN models in order to assess their behaviour when membrane potential is taken into account in the learning process.

To address this task, different considerations must be taken into account about the spikes treatment both at the input of the SNNs as well as about the specific SNN models.

The remainder of this paper is organized as follows: Sect. 2 reviews recent literature surrounding SNN applications, spike coding and regression tasks. Section 3 describes the methodology we follow to generate the dataset, performance metrics adopted, the neuron models and the specific architectures based on these models. Section 4 details the results obtained through the experiments and finally we present in Sect. 5 some concluding remarks and future research.

2   Related Work

Although SNNs still lag behind DL networks in terms of their performance, the gap is vanishing on some tasks, while SNNs typically require much lower energy for the operation. SNNs have largely been applied to image classification on large datasets such as ImageNet [11,12], other works have been focused on object detection tasks [13,14], pose estimation, action identification, etc. [15–19]. When computational resources or temporal information are crucial, SNNs could be the desired alternative. However most applications of SNNs are still limited to less complex datasets such as MNIST, N-MNIST, and N-Caltech101 [20], due to the complex dynamics of neurons and the non-differentiable nature of spike operations [21].

In regression tasks just a few research can be found based on event data or SNNs as learning models. In [22] they predicted a car steering angle using a Deep Learning approach specifically designed to work with the output of event sensors. The first, significant work, that considered SNNs as part of the learning architecture, was in [23], where authors dealt with the prediction of angular velocities of a rotating event camera in continuous time, they used a convolutional spiking neural network architecture with three, non-spiking, output neurons. More recently, [10] presented an interesting framework for nonlinear regression using SNNs. As they pointed out the use of SNN in regression it is determined by the treatment of events, not only at the input layers but the transformation into real numbers or continuous values of the output spikes.

At the input layers event data representations encode the event information related to a time-interval or temporal-window extracted from an event-stream.

In some works, events have been transformed into image-like representations compatible with natural images. The simplest way to use event data for supervised learning is to accumulate events pixel-wise over a period of time, either by counting them or by accumulating their polarities [22]. Other approaches include: 2D time surfaces or maps of most recent timestamps [24] or interpolated voxel grid [25]. This representation better preserves temporal information but requires more memory and more computations.

At the output layer, different spikes decoding strategies exist, based on rate decoding or latency decoding [26], however we follow the proposal of [10] where they designed a network topology based on different SNN layers and the output of the last spiking layer were transformed into a decoding layer taking the membrane potential of every time step as input and providing real numbers as output.

3   Methodology

3.1   Event-Based Dataset

Events produced by a bouncing ball were recorded using a DVXplorer event camera [27], which provides a high dynamic range, high temporal resolution, low latency, and low power consumption. This sensor is able to capture pixel-level brightness changes instead of standard intensity frames, consisting of a continuous flow of pixel events which represent the moving objects in the scene. At each pixel position $u_i = (x_i, y_i)$, the event-based camera produces events $e_i = (u_i, t_i, p_i)$ with polarity $p_i = (-1, 1)$ when the intensity changes above a threshold value at timestamp t_i [26].

Once the events have been recorded, we need to apply a representation method that converts asynchronous streams of events from event-based cameras to a sequence of sparse and expressive event frames [28]. Inspired by [22] the events recorded were binned over a time interval T (T=1000 microseconds) in a pixel-wise manner, obtaining 2D histograms of events. We used Tonic package [29] to transform events. The resolution of the camera were resized to 176×144 and time steps fixed to 100 miliseconds, providing a total of 166×100 frames.

a)   b)

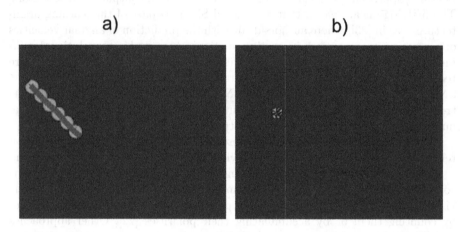

Fig. 1. Integrated events into 1000 microseconds bins, these events were captured by an event-based camera, the sensor were resized to 176×144 pixels. Each pixel (x,y) reports when it detects a relative change in the illumination intensity that is above or below a defined percentage of the previous intensity. In a) A 2D event frame (of size 176×144) accumulated over 100 time steps, red cross indicates the interpolated centers of the ball b) A 2D event frame (size 176×144) of a single time step and the interpolated center for this event set.

To obtain a labelled dataset, events must be associated with the exact position of the ball, so we applied the E2VID software [30] to generate a series of intensity frames from the event flow, in which the centers of the ball are calculated applying computer vision techniques. Once the position in each frame is obtained, a cubic interpolation is performed to calculate the position in each desired instant of time. Due to the nature of the data, they are asynchronous events with a temporal resolution of 1ms, so we obtained the position of the object at that resolution and averaged over the fixed time step. Figure 1a) shows

100 accumulated frames and the interpolated centers of the ball. Figure 1b) shows a single frame from the previous series of 100 binned frames and the corresponding center of the ball.

For learning purposes this dataset was split into two groups, the training set (70%) and the test set (30%).

3.2 Performance Metrics

To evaluate the performance of our network we used the root-mean-square error (RMSE), that measures the average difference between values predicted by a model and the actual values. It provides an estimation of how well the model is able to predict the target value (accuracy).

As a variance, RMSE can be interpreted as the standard deviation of the unexplained variance, and it has the useful property of being in the same units as the response variable. The lower the value of the Root Mean Squared Error, the better the model is.

The RMSE is particularly useful for comparing the generalization capacity of different regression models.

$$RMSE = \sqrt{\frac{\Sigma_{i=1}^{N}(target_i - predicted_i)^2}{N}} \qquad (1)$$

3.3 Neuron Model

In this work we used the Leaky Integrated-and-Fire (LIF) model given its computational efficiency and capability of capturing the essential features of information processing in the nervous system [31].

The LIF model is represented by a RC circuit with a certain threshold. The state of a neuron is described in terms of its membrane potential, which is determined by the synaptic inputs and the injected current that the neuron receives. When the membrane potential reaches the fixed threshold, an action potential (spike) is generated and the voltage is reset [32,33] (see Fig. 2).

This neuron is modelled by the following equation:

$$\tau_m \frac{du}{dt} = -(U - U_r) + RI_{i_n}(t) \qquad (2)$$

We refer to U as the membrane potential and to τ_m as the membrane time constant of the neuron, I_{i_n} is the input current and R the membrane resistance.

The standard LIF is a feed-forward neuron, then in order to take into account the changes in the recurrent connections and the dynamic strength of neuron connections over time, we also considered in our experiments the Recurrent LIF (RLIF) network.

RLIF assumes feedback connections where all spikes from a given layer are first weighted by a feedback layer before being passed to the input of all neurons [10]. This enables the network to use relationships along several time steps for the prediction of the current time step [26,33].

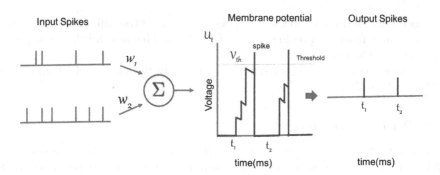

Fig. 2. The LIF model performs spatial and temporal integration of synaptic inputs, has a leaky membrane, generates a spike when the voltage reaches a certain threshold and goes refractory during the action potential.

3.4   Neural Architecture

When using event-based data the key question in regression is how to interpret the binary spike information in the output layer to obtain real numbers. In [10] authors proposed training membrane potential to regress a single output variable, following this work we trained the membrane potential of LIF and RLIF-based SNNs consisting of two output units of continous values, in order to be able to predict the 2D position of a bouncing ball over time.

To this end, a three-layer SNN architecture was adopted and tested on the event-based dataset using LIF and RLIF principles. In every case the architecture consisted of 3 fully connected layers, the input layer of 176×144 units, two hidden layers of H and H/2 units and 2-units output layer. Figure 3 illustrates the general topoloy used in this work.

The inputs of the first, second and third layer are passed through the fully connected activation function before entering the LIF/RLIF neurons. Therefore, information flows in the form of accumulated event frames into the input layer, then is transformed into binary spikes in the spiking layers and taking back as real numbers in the output layer.

The SNNs tested in this paper was developed in snnTorch [34], a Python package for performing gradient-based learning with spiking neural networks, extending the capabilities of PyTorch [35].

Networks were trained using the fast sigmoid as surrogate gradient function. The loss function was the mean-squared error (MSE) between the output and the target and the optimization method was ADAM with a learning rate of 0.001. Exponential decay rates for the first and second moment estimates was 0.95.

We assessed the performance of SNN to investigate the following questions:

a) Are LIF and RLIF models capable of predicting continuous event-based values with a sufficient accuracy?

b) How can the number of hidden units influence the accuracy of the results?

First, in order to check the predicting capability of LIF and RLIF models, we trained both, LIF and RLIF SNNs during 40 epochs and using 50 hidden units,

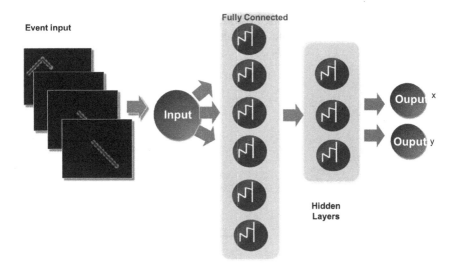

Fig. 3. LIF network scheme to illustrate the topology used in this work: a three-layer LIF network with H and H/2 hidden units at first and second hidden layers and 2 output units to predict the XY coordinates of the center position of the bouncing ball.

we considered the continuous values obtained from the membrane potential of the SNN output units as final response of the network. The purpose was to study if membrane potential values could be good estimates of the multivariate continuous target values instead of using some kind of codification of the output spikes.

Once this issue was checked, we designed a more complete battery of experiments in order to assess the generalization capability of the different models and architectures and the influence of the number of hidden units.

We compared results between LIF and RLIF SNNs with three different number of hidden units H, namely, H = 50, H = 200 and H = 500 for every first hidden layer, as well as the number of hidden units in the second layer was always assigned to half the number of neurons of the previous layer (H/2).

4   Results and Discussion

Every experiment was repeated 500 times and average values were calculated. We use box plots in order to visually show the distribution of numerical averaged RMSE and skewness by displaying the data quartiles (or percentiles) and averages.

For LIF based architectures of 50, 200 and 500 hidden units, Fig. 4 shows the RMSE error obtained after training and testing. From these results, it is straightforward to observe that the more hidden units, the more accurate results. As the training progresses along the epochs all the networks are capable of improving the results. But always 500 hidden units provide a better accuracy, maybe if less complex networks were trained during more epochs we could get acceptable results with all the networks, but at the cost of increasing the computation time.

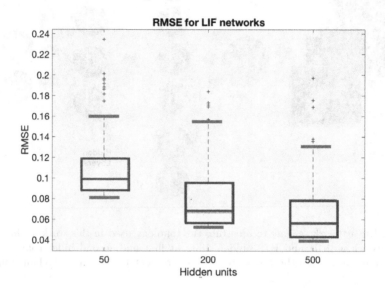

Fig. 4. Boxplot of RMSE test measurements for three different LIF SNNs.

For RLIF models the results show a different behaviour concerning the number of hidden units. In this case, the three architectures presented more similar averaged RMSE values as we can observed from the boxplots of Fig. 5.

We also show in Fig. 6 the line plots of the three averaged test RMSE measurements calculated over 100 epochs. On one hand, RMSE values for 50 and 500 hidden units are very similar during the first 20 epochs and then there exists a slight difference in favor of the 500 hidden units, since the RMSE measurement goes down during some more epochs to rise again from 27 epochs keeping variable during the following epochs.

The difference between 50 and 500 hidden units is not significant enough to opt for the most complex network. Thus, by analyzing the number of hidden units of the RLIF SNNs, although 50 hidden units could require more epochs to train the model, its simplicity against the 500 hidden units compensated the generalization capability, and did not suppose a significant increase in computation time. In general, it can be observed that 500 units did not improve the accuracy of the SNNs.

Table 1 reports the accuracy of each architecture in the form of the RMSE measurement, that allows us to quantitatively identify the models and the architectures that better fit the data. Last column indicates the number of epochs that each network reached the minimum test RMSE.

In all the cases we can observe that LIF models outperformed RLIF counterparts, which tell us that they could obtain a better generalization capability

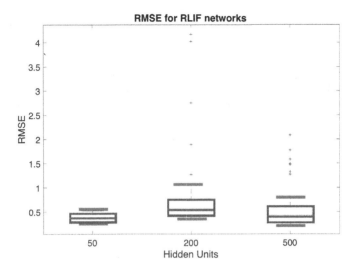

Fig. 5. Boxplot of RMSE test measurements for three different RLIF SNNs.

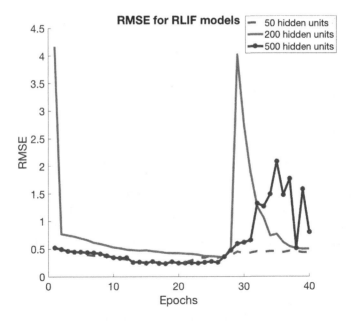

Fig. 6. Line plots of RMSE test measurements for three different RLIF SNNs.

than RLIF based networks. Recurrent LIF networks enable the network to use relationships along several time steps for the prediction of the current time step, so it could be expected that these networks performed better than simple feedforward LIF models, but with this specific task, better results were obtained with non-recurrent models. Thus, in this regression task, a feedforward spiking neural network based on LIF model can predict relatively well the position of the ball, and the more complex network the more accurate results. However it is important to bear in mind that training neuromorphic data is expensive as it requires sequentially iterating through many time steps, and this cost is incremented as the complexity of the network.

Table 1. Generalization Error.

Model	Hidden Units	RMSE	Epochs
LIF	50	0.0815	99
LIF	200	0.0518	98
LIF	500	0.0392	99
RLIF	50	0.2521	18
RLIF	200	0.3460	27
RLIF	500	0.2340	18

5   Conclusions and Future Work

The goal of this work has been to predict the position (XY coordinates) of a bouncing ball from 2D frames of events and using Spiking Neural Networks as predictors of these continuous values, using the membrane potential of LIF and Recursive LIF SNN models as the estimated outputs instead of the corresponding spikes. Results show how membrane potential on the output of SNNs models can be used as estimators of these continuous values with sufficient accuracy and encourage the research towards the comparison with other SNNs models and architectures. In addition we consider, as future work, the inclusion of convolutional layers to extract important features from the 2D event frames as well as we will investigate more complex datasets with different architectures of Deep SNNs for regression tasks, which will allow its application in robotic systems to object tracking.

Acknowledgements. F. M. Quintana would like to acknowledge the Spanish *Ministerio de Ciencia, Innovación y Universidades* for the support through FPU grant (FPU18/04321). This work was also supported by the project NEMO-VISION from the *Ministerio de Ciencia e Innovación*, PID2019-109465RB-I00/ AEI/10.13039/501100011033.

References

1. Maass, W.: Networks of spiking neurons: the third generation of neural net models. Neural Netw. **10**(9), 1659–1671 (1997)
2. McCulloch, W.S., Pitts, W.: A logical calculus of the ideas immanent in nervous activity. Bull. Math. Biophys. **5**(4), 115–133 (1943)
3. Rumelhart, D.E., Mcclelland, J.L.: Parallel Distributed Processing: Explorations in the Microstructure of Cognition. MIT Press (1986)
4. Cybenko, G.: Approximation by Superpositions of a Sigmoidal Function. Math. Control Signal. Syst. **2**, 303–314 (1989)
5. LeCun, Y., Bengio, Y., Hinton, G.: Deep learning. Nature **521**, 436–444 (2015)
6. Gerstner, W., Kistler, W.M., Naud, R., Paninski, L.: Neuronal Dynamics: From Single Neurons to Networks and Models of Cognition. Cambridge University Press (2014)
7. Human Brain Project. https://www.humanbrainproject.eu/en/. Accessed 20 Mar 2023
8. Hopkins, M., Pineda-García G., Bogdan P.A., Furber S.B.: Spiking neural networks for computer vision. Interface Focus **8**, 20180007 (2018)
9. Gallego, G., Rebecq, H., Scaramuzza, D.A.: Unifying contrast maximization framework for event cameras, with applications to motion, depth. IEEE/CVF Conf. Comp. Vis. Pattern. Recogn. **3867** (2018)
10. Henkes, A., Eshraghian, J.K.: Spiking neural networks for nonlinear regression. IEEE Trans. Neural Netw. Learn. Syst. (2022)
11. Rueckauer, B., Lungu, I.A., Hu, Y., Pfeiffer, M., Liu, S.C.: Conversion of continuous-valued deep networks to efficient event-driven networks for image classification. Front Neurosci. **11**, 682 (2017)
12. Sengupta, A., Ye, Y., Wang, R., Liu, C., Roy, K.: Going deeper in spiking neural networks: VGG and residual architectures. Front. Neurosci. **13**, 95 (2019). https://doi.org/10.3389/fnins.2019.00095
13. Kheradpisheh, S.R., Ganjtabesh, M., Thorpe, S.J., Masquelier, T.: STDP-based spiking deep convolutional neural networks for object recognition. Neural Netw. **99**, 56–67 (2018)
14. Zhou, S., Chen, Y., Li, X., Sanyal, A.: Deep SCNN-based real-time object detection for self-driving vehicles using LiDAR temporal data. IEEE Access **8**, 76903–76912 (2020)
15. Tavanaei, A., Ghodrati, M., Kheradpisheh, S.R., Masquelier, T., Maida, A.: Deep learning in spiking neural networks. Neural Netw. **111**, 47–63 (2019)
16. Mueggler, E., Rebecq, H., Gallego, G., Delbruck, T., Scaramuzza, D.: The event-camera dataset and simulator: event-based data for pose estimation, visual odometry, and SLAM. Int. J. Robot. Res. **36**(2), 142–149 (2017)
17. Rebecq, H., Horstschaefer, T., Scaramuzza, D.: Real-time visual-inertial odometry for event cameras using keyframe-based nonlinear optimization. In: Proceedings of the British Machine Vision Conference 2017, BMVC 2017. BMVA Press, London (2017)
18. Kirkland, P., Di Caterina, G., Soraghan, J., Andreopoulos, Y., Matich, G.: UAV detection: a STDP trained deep convolutional spiking neural network retina-neuromorphic approach. In: Tetko, I., et al. (eds.) Artificial Neural Networks and Machine Learning – ICANN 2019: Theoretical Neural Computation. ICANN 2019. LNCS, vol. 11727, pp 724–736. Springer, Cham (2019). https://doi.org/10.1007/978-3-030-30487-4_56

19. Orchard, G., et al.: A temporal approach to object recognition. IEEE Trans. Pattern Anal. Mach. Intell. **37**, 2028 (2015)
20. Orchard, G., Jayawant, A., Cohen, G.K., Thakor, N.: Converting static image datasets to spiking neuromorphic datasets using saccades. Front. Neurosci. **9**, 437 (2015)
21. Yamazaki, K., Vo-Ho, V.-K., Bulsara, D., Le, N.: Spiking neural networks and their applications: a review. Brain Sci. **12**(7), 863 (2022)
22. Maqueda, A.I., Loquercio, A., Gallego, G., García, N., Scaramuzza, D.: Event-based vision meets deep learning on steering prediction for self-driving cars. In: Proceedings of the 2018 IEEE Conference on Computer Vision and Pattern Recognition, pp. 5419–5427. IEEE Computer Society, Salt Lake (2018)
23. Gehrig, M., Bam Shrestha, S., Mouritzen, D., Scaramuzza, D.: Event-based angular velocity regression with spiking networks. In: IEEE International Conference on Robotics and Automation (ICRA), Paris (2020)
24. Lagorce, X., Orchard, G., Galluppi, F., Shi, B.E., Benosman, R.B.: HOTS: a hierarchy of event-based time-surfaces for pattern recognition. IEEE Trans. Pattern Anal. Mach. Intell. **39**(7), 1346–1359 (2017)
25. Shrestha, B., Garrick Orchard, G.: SLAYER: spike layer error reassignment in time. In: Advances in Neural Information Processing Systems, pp. 1417–1426 (2018)
26. Eshraghian J.K., et al.: Training spiking neural networks using lessons from deep learning. Neural Evolution. Comput. (2022)
27. Inivation. Understanding the performance of neuromorphic event-based vision sensors (2020). https://inivation.com/dvp/white-papers/
28. Liu, Q., Pineda-García, G., Stromatias, E., Serrano-Gotarredona, T., Furber, S.B.: Benchmarking spike-based visual recognition: a dataset and evaluation. Front. Neurosci. **10**, 496 (2016)
29. Tonic 1.2.6. https://tonic.readthedocs.io/en/latest/. Accessed 20 Mar 2023
30. Rebecq, H., Ranftl R., Koltun, V., Scaramuzza, D.: High speed and high dynamic range video with an event camera. IEEE Trans. Pattern Anal. Mach. Intell. (T-PAMI) (2019)
31. Nunes, J.D., Carvalho, M., Carneiro, D., Cardoso, J.S.: Spiking neural networks: a survey. IEEE Access **10**, 60738–60764 (2022)
32. Burkitt, A.N.: A review of the integrate-and-fire neuron model: I. Homogeneous synaptic input. Biol. Cybern. **95**, 1–19. (2006). https://doi.org/10.1007/s00422-006-0068-6
33. Wang, Z., Zhang, Y., Shi, H., Cao, L., Yan, C., Xu, G.: Recurrent spiking neural network with dynamic presynaptic currents based on backpropagation. Int. J. Intell. Syst. **37**–3, 2242–2265 (2022)
34. Snntorch Package. https://snntorch.readthedocs.io/en/latest/. Accessed 30 Mar 2023
35. Pytorch. https://pytorch.org. Accessed 20 Mar 2023

Event-Based Classification of Defects in Civil Infrastructures with Artificial and Spiking Neural Networks

Udayanga K.N.G.W. Gamage[1(✉)], Luca Zanatta[2], Matteo Fumagalli[1],
Cesar Cadena[3], and Silvia Tolu[1]

[1] Department of Electrical and Photonics Engineering,
Technical University of Denmark, Lyngby, Denmark
{kniud,mafum,stolu}@dtu.dk
[2] DEI Department, University of Bologna, Bologna, Italy
luca.zanatta3@unibo.it
[3] Autonomous Systems Lab, ETH Zurich, Zurich, Switzerland
cesarc@ethz.ch

Abstract. Small Multirotor Autonomous Vehicles (MAVs) can be used to inspect civil infrastructure at height, improving safety and cost savings. However, there are challenges to be addressed, such as accurate visual inspection in high-contrast lighting and power efficiency for longer deployment times. Event cameras and Spiking Neural Networks (SNNs) can help solve these challenges, as event cameras are more robust to varying lighting conditions, and SNNs promise to be more power efficient on neuromorphic hardware. This work presents an initial investigation of the benefits of combining event cameras and SNNs for the onboard and real-time classification of civil structural defects. Results showed that event cameras allow higher defect classification accuracy than image-based methods under dynamic lighting conditions. Moreover, SNNs deployed into neuromorphic boards are 65–135 times more energy efficient than Artificial Neural Networks (ANNs) deployed into traditional hardware accelerators. This approach shows promise for reliable long-lasting drone-based visual inspections.

Keywords: Event cameras · Spiking Neural Networks · Artificial Neural Networks · Civil Structural Defects

1 Introduction

Computer vision-based inspection and monitoring of civil infrastructures enable more time-efficient, cost-effective damage classification than current human manual inspection methods with the aim of improving infrastructure maintenance and safety. In recent years, much focus has been given to partially automating visual inspection and maintenance by using drones [7,8]. Drone-based inspection and maintenance help to minimize the inspection time and cost by increasing

I. Rojas et al. (Eds.): IWANN 2023, LNCS 14135, pp. 629–640, 2023.
https://doi.org/10.1007/978-3-031-43078-7_51

the availability of reliable data and minimizing the hazards and increasing the safety of the employees. With an onboard real-time defect detection system, low computational energy consumption is required without compromising the defect detection latency to increase the autonomous mission time of the drone.

Deep learning (DL)-based methods that rely on artificial neural networks (ANNs) and frame-based images have been implemented by the academia and industry [1] to pursue this goal. In [6] is presented a comprehensive survey on State-Of-The-Art (SOTA) DL architectures used to detect various types of defects in civil infrastructures e.g., bridges, highways, railways, tunnels, concrete buildings, and steel buildings. In [2], a fine-tuned DL model (VGG16) has been proposed for spalling and crack detection in concrete structures. In [3], a hierarchical multi-class classifier was successfully implemented with the inception-V3 architecture to classify cracks, corrosion, spalling, efflorescence, and nondefective surfaces in concrete bridges. In [4], ResNet-50, DenseNet, and HRNet architectures were separately trained to classify different levels of corrosion defect. PSPNet, DeepLab, and SegNet have been evaluated for pixel-wise segmentation of cracks and corrosion [5].

The previous works have obtained high classification performance; however, they did not study the robustness of SOTA methods under changing lighting conditions. As well known, vision-based methods are highly susceptible to environmental changes such as poor lighting [11]. Under low and dynamic lighting conditions, frame-based cameras generate low bright and low contrast images leading to increase misdetection and over-detection of defects with DL methods [2]. In addition, drone-captured images might be blurred. In this paper, we propose a novel framework that aims to improve defect classification under low or dynamic lighting conditions. Our framework relies on a biologically inspired vision sensor called event-based or neuromorphic camera [12], and on a new generation of ANNs called Spiking Neural Networks (SNN) [13] which promise higher expressiveness and lower energy consumption [12,13]. In literature, a crack vs non-crack classification with a VGG7-based SNN showed a 93% accuracy [14]. In their work, an event stream was generated based on the pixel's intensity value using Poisson coding, which is not the way that the real event-based cameras encode the scenes.

Many researchers have focused on developing onboard DL-based autonomous defect detection with drones [9]. As drones perform real-time operations in challenging environments such as in underground tunnels or top-of-the-sky scrappers, video streams cannot be transmitted due to weak communication with the ground station. Hence, having onboard detectors inside the drones is important. The use of onboard detectors may help to reduce the latency to identify a defect resulting in a faster inspection. Furthermore, thanks to the reduced communication overhead, the transmission power reduces consequently. In [10], authors have evaluated the detection accuracy on different single-board devices such as Nvidia Jetson Nano, Nvidia Jetson TX2, and Nvidia Jetson AGX Xavier that can be used for onboard defect detection. Here the authors have implemented the yolov4-tiny architecture-based DL algorithm on each of those devices to develop a fully automatic real-time onboard visual power line inspection scenario. Their work was based on frame-based images and ANNs. Also, they did not consider taking energy measurements or evaluating any neuromorphic hardware such as the Loihi chip [15].

In this work, we evaluate ANN and SNN implementation of four SOTA VGG architectures [1]. Besides, we provide an initial study on the advantage offered by the deployment in a neuromorphic processor (Loihi chip) to leverage the computational energy and latency of the classification system by exploiting sparsity [12,13].

In a nutshell, the main contributions of this paper are:

- Performance comparison of the event-based defect classification over a dataset of synthetic events vs the image-based defect classification with ANNs and SNNs.
- Performance comparison of the event-based defect classification over a dataset of real events vs the image-based defect classification with ANNs and SNNs.
- Energy efficiency comparison in SNN-based event-data classification vs ANN-based event-data and image-data classification.

The remaining sections of the paper are organized as follows. Section 2 provides a brief overview of event cameras and SNNs. Sections 3 and 4 introduce the methodology and the experimental setup, respectively. Section 5 presents the obtained results followed by the discussion in Sect. 6. Finally, Sect. 7 provides a conclusion and future work.

2   Background

Event-based cameras are more robust to dynamic lighting conditions than frame-based cameras, thanks to their high dynamic range [12]. Hence, event-based cameras have the potential to detect civil structural defects under extremely low or extremely high lighting conditions where the traditional frame-based cameras fail [12].

The spike stream in the output of an events camera is provided as input to a Spiking Neural Network (SNN) which computation unit, called a spiking neuron, is described as an ODE system. Two of the most used spiking neurons are LIF [16] and PLIF [17] which are described by the following equations:

$$\tau \frac{dv(t)}{dt} = -v(t) + I(t) \tag{1}$$

$$z(t+1) = \begin{cases} 1 & \text{if } v(t) \geq v_{th} \\ 0 & \text{otherwise} \end{cases} \tag{2}$$

where τ is the time constant, $v(t)$ is membrane potential, $I(t)$ is the pre-synaptic current developed by the accumulation of weighted inputs spikes coming from the neurons in the previous layer; $z(t+1)$ is the output of the spiking neuron at the next time step that depends on the threshold voltage v_{th}. Unlike

LIF, PLIF can learn the time constant during the training (τ). However, the computing drawback of spiking neurons lies in the threshold function which is non-differentiable. The scientific community has overcome this problem through different algorithms e.g., BPTT [18] and SLAYER [19] that implement surrogate function approximators of the gradient [20].

SNNs can be implemented in neuromorphic hardware (e.g., Intel Loihi [15]) that provides additional energy efficiency than traditional von-Neumann architectures [13].

3   Methodology

The overview of the main steps of the methodology for event-based defect classification is shown in Fig. 1. To overcome the lack of large event-based datasets, we utilized the v2e simulator [21] to generate a synthetic one. However, a real dataset generated with the DAVIS346 camera was also created to test and compare the ANNs/SNNs performance with synthetic and real data as described in Sect. 3.1.

In Sect. 3.2, the event encoding methods such as voxel grid and voxel cube are described. The encoded events are spatially resized to 128×128 resolution before being fed to the neural networks (Sect. 3.3).

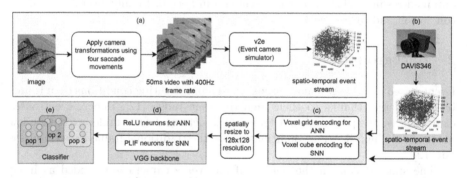

Fig. 1. Event-based classification methodology (a) Generation of events from a static image using event camera simulator (b) capture events from DAVIS346 real event-based camera (c) Encoding methods for spatio-temporal event streams (d) VGG backbone for ANNs consist of ReLU and the VGG backbones for SNNs consist of PLIF neurons (e) Classifier which output consists of three neuron populations.

3.1   Datasets

We generated a synthetic event-based dataset to deal with the lack of no publicly available event-based data related to structural defects. A dataset containing frame-based images of three civil structural defect classes was used [2]. The defect classes are: i) cracks, ii) spalling and iii) non-defective surfaces. Each class contains 5 000 samples that we split into three subsets (training, validation, and

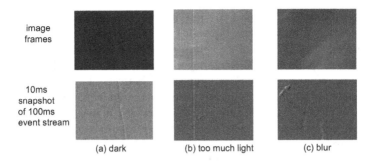

Fig. 2. Collected data with DAVIS346 under different lighting conditions (a) extremely low light (b) high light (c) blurred effect due to the low light.

test sets), which are composed of 4000, 500, and 500 samples, respectively. For each static image, a 50 ms video clip that consists of 20 intensity frames is created by applying camera transformations to simulate four saccade movements [22]. Clips are 400 Hz high-frame rate videos to avoid ghosting effects [23]. Each video clip is fed to v2e [21], which is the SOTA event camera simulator that models both event camera dynamics as well as the various types of noise associated with, to generate an event stream whose duration is 50 ms and the spatial resolution is 346×260. The resulting events have either positive or negative polarity. The positive polarity corresponds to the events for pixel intensity increments, while the negative polarity corresponds to pixel intensity decrements.

Moreover, by using a DAVIS346 event camera which is able to capture both RGB images and events with a spatial resolution of 346 × 260 pixel, we collected 35 samples under moving conditions, per class with different light conditions, i.e., low, normal, and high light in an underground tunnel (see Fig. 2). We captured each sample as an event stream of 100 ms temporal length as well as an RGB image. Samples show different light features, some of them are extremely illuminated while others are less illuminated and also blurred as shown in Fig. 2. In addition to intensity frames, Fig. 2 also shows the reconstructed frames by accumulating events for 10ms. In the end, we split this dataset collected with DAVIS346, into training and test set, which have respectively 10 and 25 samples per class.

3.2 Input Encoding for Deep Learning Architectures

The event stream needs to be preprocessed before feeding it into the neural networks. As the first step, we resized the images up to a 128×128 spatial resolution using a bi-linear interpolation and nearest neighbour interpolation for the RGB images and events data respectively. Then, for the ANNs, we downsampled the event streams in the temporal dimension to have ten temporal bins or channels by using the voxel grid encoding method explained in [24]. For the SNNs, we used a method named voxel cube explained in [25].

To determine the efficient hyperparameters, the number of time steps (T) and the number of micro bins, we evaluate the accuracy of our models for different combinations of T and the number of micro bins as shown in Fig. 3. In resolving the tie points for the accuracy of different T and number of micro bins combinations, we came up with a compromise between T and micro bins based on the fact that the number of time steps increased which increases inference latency. And as the number of micro bins increased it increases the neural network parameters causing to increase in memory requirement. In our analysis, T and the micro bins are chosen so that the product of T and the number of micro bins are always equal to almost 20 for the 50 ms synthesized event stream. This is because the minimum temporal distance between any two spikes in the synthesized events stream is 2.5 ms as we use 400 Hz video to generate the events. Based on the evaluation, we have chosen 5 time steps (T) and 2 micro bins since that combination outperforms other combinations in almost all cases.

3.3   Deep Learning Architectures

In this work, we investigated four state-of-the-art architectures: VGG9, VGG11, VGG13, and VGG16 as the backbones of our defect classifiers. The VGG block used is characterized by a 2D-convolutional 3×3 kernel followed by a 2D batch-normalization layer and a non-linear function that is a ReLU in the ANNs and a PLIF in the SNNs. The PLIF neuron provides higher accuracy in the classification tasks compared to LIF [25]. The output of the VGG9 backbone consists of 256 feature maps while the output of VGG11, VGG13, and VGG16 consists of 512 feature maps. These feature maps are fed into the classifier which is designed to refer to the work [25]. The classifier consists of a 1×1 convolution layer which outputs three channels, one channel for each class, each channel consists of three populations of neurons. Since we use 1D convolution, the classifier can process features maps of any size. A predicted logit value for each class is computed by summing up the outputs of all the neurons belonging to the relevant population. Finally, the predicted class is obtained using a softmax function.

After training the networks with the synthesised dataset, we fine-tuned the last three layers of each network to classify the dataset collected with DAVIS346 event camera until convergence.

4   Experimental Setup

To implement our neural networks, we used PyTorch [26] and SpikingJelly [27] to implement the PLIF activation function. Considering the work done in [25], where the authors solved a similar task, we used the same hyperparameters to train both ANNs and SNNs which are: Adam, CosineAnnealingLR, and categorical cross entropy for the optimizer, scheduler, and the loss function, respectively. The weights were initialized with a Xavier distribution and the bias to 0. As already mentioned in Sect. 3.2, both the image events are spatially reshaped to a resolution of 128×128. Furthermore, the images are normalized using the mean

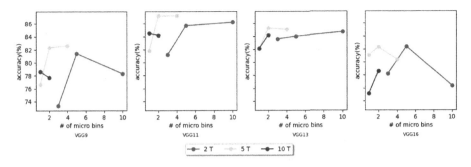

Fig. 3. Accuracy(%) variation with a different number of time steps (T) and micro bins.

and the standard deviation computed with the well-known ImageNet dataset [28] while the events are encoded either with a voxel grid or voxel cube as mentioned in Sect. 3.2. To use voxel cube encoding, we need to set two hyperparameters named time steps and micro bins. Hence, we performed a study in which we evaluated the performance of the networks varying both the time steps and the micro bins. The results of this study are shown in Fig. 3, in which the best performance is reached with time steps set to 5 and micro bins set to 2. Finally, we set the following learning rates: i) 5e-3 for the ANNs fed with the RGB data, ii) 5e-4 for the ANNs fed with the events, and iii) 1e-3 for the SNNs.

5 Results

We separated our analysis into three sections. In Sect. 5.1, we show the classification performance using the images and the events generated from the event-camera simulator. In Sect. 5.2, we provide the results for the classification performance under dynamic lighting conditions based on the event dataset collected using DAVIS346 real event camera. In both Sects. 5.1 and 5.2, we use accuracy (acc), sensitivity (sen), and specificity (spc) as the metrics for classification performance evaluation. Section 5.3 provides a comparison of energy consumption between ANNs and SNNs when classifying defects using intensity images and event-based data. The rough energy consumption of both ANNs and SNNs is computed by averaging the energy of the whole test set. In particular, for the SNNs, we consider all the spikes that trigger a Synaptic OPeration (SOP) and multiply that value by the SOP energy in the Loihi processor [15]. Similarly, for the ANNs, we compute the total number of MAC and multiply that value by the energy required by a single MAC on a SOTA DNN accelerator named Eyeriss v2 [29].

5.1 Event Camera Simulated Data Classification

Table 1 summarizes the performance of the neural networks used with the synthetic dataset. The networks fed with the RGB images outperform the ones fed

Table 1. Classification performance of ANNs and SNNs for different types of input.

	VGG9			VGG11			VGG13			VGG16		
	acc	sen	spc	acc	sen	spc	acc	sen	spc	acc	sen	spc
ANNs & RGB data	91.9	91.9	91.9	94.6	94.6	97.3	94.1	94.1	96.9	**95.0**	95.1	97.5
ANNs & events data	88.7	88.7	94.2	**93.3**	93.3	96.5	91.4	91.4	95.5	91.0	91.0	95.3
SNNs & events data	80.5	80.5	89.4	**87.2**	87.2	92.8	84.1	84.3	91.4	82.3	82.3	90.3

Table 2. ANN and SNN classification performance of images and event streams captured with DAVIS346 event camera.

	VGG9			VGG11			VGG13			VGG16		
	acc	sen	spc	acc	sen	spc	acc	sen	spc	acc	sen	spc
ANNs & RGB data	72.1	72.1	75.4	75.3	75.3	81.3	77.7	77.7	84.3	**78.0**	78.0	86.1
ANNs & events data	81.2	81.1	88.0	**88.2**	88.2	93.4	86.4	86.4	92.1	84.0	84.0	91.3
SNNs & events data	82.0	82.0	89.1	**86.4**	86.4	93.2	85.3	85.3	92.4	85.1	85.1	92.0

with the events data: 1–4% with respect to the ANNs and 10–13% with respect to the SNNs. Moreover, we can see that all ANNs outperform SNNs because the backpropagation algorithm for SNNs uses gradient approximators for the non-differential thresholding function of the spiking neurons. Though VGG16 shows the highest performance for image-based classification with ANNs, in event-based data classification, VGG11 outperforms all the other VGG backbones regardless of ANN or SNN. According to our belief, the reason for this observation can be explained as follows. As the VGG architecture becomes deeper and deeper, the tendency for over-fitting increases when the architectures are trained with event-based data compared to image-based data since the total number of features that need to be learned with binary event-based data is less than the features that need to be learned with image-data explain with 255 intensity values rather being binary values like positive or negative spikes.

5.2   Real Event Camera Data Classification Under Dynamic Lighting Conditions

Table 2 shows the classification results of the DAVIS346 dataset. The results show that the ANNs fed with the events dataset are more robust to the changing lights compared to the ANNs fed with RGB images as the former have an accuracy drop of 6% on average while the latter of 18%. Moreover, the SNNs are more robust than the ANNs since the accuracy drop between the synthetic dataset and the real one is around 1% in VGG11 while they increase accuracy by 1–2% with the other architectures. Besides, in this case, the performance of ANNs and SNNs in classifying event-based data is almost similar.

5.3 Energy Efficiency of ANNs and SNNs

In Fig. 4, we show the trade-off between the energy consumption and the accuracy of ANNs and SNNs. In particular, Fig. 4(a) shows the energy consumption vs the accuracy obtained with the synthetic dataset. Notice that the ANNs fed with the two datasets have the same energy consumption, while the SNNs consume 65–90 times less, paying the accuracy price. Moreover, Fig. 4(b) shows the results with the DAVIS346 dataset in which SNNs are 100–135 times energy efficient compared to their ANN counterparts. The SNNs' high power efficiency is majorly due to the high sparsity of the generated spikes. In our experiments, out of all the neurons present in VGG models, only about 14–17% of the neurons generated spikes.

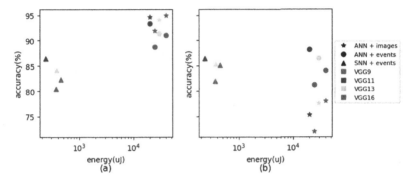

Fig. 4. Accuracy vs Energy consumption of ANNs and SNNs in (a) classifying images and events generated from event camera simulator without variation in lighting (b) classifying images and events from a DAVIS346 real event camera under dynamic lighting conditions.

6 Discussion

In this work, we studied a civil structural defect classification problem using event-based data, ANNs, and SNNs. The SOTA results for the defect classification are obtained with deep neural networks with intensity images [2] and the accuracy performance is around 95%. With the event-based classification of the same defects presented in [2] a 93% accuracy value was obtained with ANNs. However, based on our study we observed that the classification accuracy of SOTA image-based classification drops from 95% to 78% under the dynamic lighting conditions while event-based classification with ANNs shows an accuracy of 88% and event-based classification with SNNs shows an accuracy of 86%.

As per our knowledge, [14] is the only work that has studied an event-based defect detection problem. However, they did not consider the data of sensors to actually generate event data. In our case, we have considered the data of real sensors which are designed to encode the scene information as a sequence of spatio-temporal events. Also, the authors did not perform any comparative study of energy consumption of ANNs and SNNs in classifying defects. Unlike their study, in which they considered only one defect type, in our study, we considered two defect types together with non-defect category while performing a comparative study on energy consumption.

7   Conclusion

In this paper, we performed an initial study on separately identifying non-defective surfaces and types of structural defects using event camera data and with ANNs and SNNs. The results of the event-based classification were compared with SOTA image-based classification results. Though the image-based classifier still shows the best performance, it can be seen ANNs event-based classifier has the potential of classifying the defects with only a ca. 3–5% drop in evaluation metrics. However, under low-light, high-light, or blur conditions, the results of the event-based classification outperform the image-based classification. Compared to the event-based ANNs classifier, the event-based SNNs classifier shows a 6–10% drop in the evaluation metrics. However, SNNs can provide 65-135 times energy efficiency without compromising the inference latency on a neuromorphic processor. Based on our results, we observed that even though the latency for an SOP operation is high, still the high sparsity of SNNs results in a less number of SOP operations consequently without any latency increment. Hence the SNNs, event cameras and neuromorphic processors are a suitable combination for onboard online civil structural defect detection with drones. Some illumination effects in images can be created artificially and corrected with artificial neural networks. But extreme sunlight in the background of an e.g., wind turbine photo, which makes the turbine too dark and the background too bright, cannot be created artificially and cannot be fixed with neural networks. In contrast, since the pixels in the event camera capture the intensity differences, the events generated by event cameras still contain useful information irrespective of those kinds of extreme lighting effects which still helps to defect structural defects. In future work, a real event-camera-based dataset will be created to capture also extreme illumination effects. Further, the SNNs will be tested in a real embedded neuromorphic device in event-based defect detection applications.

Acknowledgements.

 This project has received funding from the European Union's Horizon 2020 research and innovation programme under the Marie Sklodowska-Curie grant agreement No: 953454.

References

1. Flah, M., Nunez, I., Ben Chaabene, W., et al.: Machine learning algorithms in civil structural health monitoring: a systematic review. Arch. Comput. Methods Eng. **28**, 2621–2643 (2021). https://doi.org/10.1007/s11831-020-09471-9

2. Yang, L., Li, B., Li, W., Liu, Z., Yang, G., Xiao, J.: Deep concrete inspection using unmanned aerial vehicle towards CSSC database. In: 2017 IEEE/RSJ International Conference on Intelligent Robots and Systems(IROS) (2017)
3. Hüthwohl, P., et al.: Multi-classifier for reinforced concrete bridge defects. Automat. Construct. **105**, 102824 (2019)
4. Synthetic Corrosion Synthetic Corrosion Dataset Dataset. Roboflow Universe (2022). https://universe.roboflow.com/synthetic-corrosion/synthetic-corrosion-dataset. Accessed 22 Nov 2022
5. Munawar, H., Ullah, F., Shahzad, D., Heravi, A., Qayyum, S., Akram, J.: Civil infrastructure damage and corrosion detection: an application of machine learning. Buildings **12** (2022). https://www.mdpi.com/2075-5309/12/2/156
6. Ye, X.W., Jin, T., Yun, C.B.: A review on deep learning-based structural health monitoring of civil infrastructures. Smart Struct. Syst. **24**(5), 567–585 (2019)
7. Nooralishahi, P., et al..: Drone-based non-destructive inspection of industrial sites: a review and case studies. Drones **5**, 106 (2021). https://doi.org/10.3390/drones5040106
8. (2022). https://reliabilityweb.com/articles/entry/the-use-of-drones-in-the-future-facility-maintenance-and-inspection-industr
9. Al-Kaff, A., Martín, D., García, F., Escalera, A., María Armingol, J.: Survey of computer vision algorithms and applications for unmanned aerial vehicles. Exp. Syst. Appl. 92, 447–463 (2018) https://www.sciencedirect.com/science/article/pii/S0957417417306395
10. Ayoub, N., Schneider-Kamp, P.: Real-time on-board deep learning fault detection for autonomous UAV inspections. Electronics **10** (2021). https://www.mdpi.com/2079-9292/10/9/1091
11. Ayoub, N., Schneider-Kamp, P.: Real-time on-board deep learning fault detection for autonomous UAV inspections. Electronics. **10** (2021). https://www.mdpi.com/2079-9292/10/9/1091
12. Gallego, G., et al.: Event-based vision: a survey. IEEE Trans. Pattern Anal. Mach. Intell. **44**(1), 154–180 (2022). https://doi.org/10.1109/TPAMI.2020.3008413
13. Roy, K., Jaiswal, A., Panda, P.: Towards spike-based machine intelligence with neuromorphic computing. Nature **575**, 607–617 (2019). https://doi.org/10.1038/s41586-019-1677-2
14. Xiang, S., Jiang, S., Liu, X., Zhang, T., Yu, L.: Spiking VGG7: deep convolutional spiking neural network with direct training for object recognition. Electronics **11**, 2097 (2022). https://doi.org/10.3390/electronics11132097
15. Davies, M., et al.: Loihi: a neuromorphic manycore processor with on-chip learning. IEEE Micro **38**(1), 82–99 (2018). https://doi.org/10.1109/MM.2018.112130359
16. Dayan, P., Abbott, L.F., et al.: Theoretical Neuro-science, vol. 806 (2001)
17. Fang, W., Yu, Z., Chen, Y., Masquelier, T., Huang, T., Tian, Y.: Incorporating learnable membrane time constant to enhance learning of spiking neural networks. In: International Conference on Computer Vision (2021)
18. Neftci, E.O., Mostafa, H., Zenke, F.: Surrogate gradient learning in spiking neural networks. IEEE Signal. Process. Mag. **36**, 61–63 (2019). https://doi.org/10.1109/MSP.2019.2931595
19. Shrestha, S.B., Orchard, G.: SLAYER: spike layer error reassignment in time. Adv. Neural Inf. Process. Syst. (2018)
20. Neftci, E.O., Mostafa, H., Zenke, F.: Surrogate gradient learning in spiking neural networks: bringing the power of gradient-based optimization to spiking neural networks. IEEE Signal Process. Magaz. **36**(6), 51–63 (2019). https://doi.org/10.1109/MSP.2019.2931595

21. Hu, Y., Liu, S.C., Delbruck, T.: v2e: from video frames to realistic DVS event camera streams. In: 2021 IEEE/CVF Conference on Computer Vision and Pattern Recognition Workshops (CVPRW) (2021). http://arxiv.org/abs/2006.07722. Automation

22. Orchard, G., Jayawant, A., Cohen, G., Thakor, N.: Converting static image datasets to spiking neuromorphic datasets using saccades. Frontiers (2015). https://www.frontiersin.org/articles/10.3389/fnins.2015.00437/full

23. Gehrig, D., Gehrig, M., Hidalgo-Carrió, J., Scaramuzza, D.: Video to Events: Recycling Video Datasets for Event Cameras. arXiv:1912.03095 (2019)

24. Bardow, P., Davison, A.J., Leutenegger, S.: Simultaneous optical flow and intensity estimation from an event camera. In: IEEE Conference on Computer Vision and Pattern Recognition (2016)

25. Cordone, L., Miramond, B., Thierion, P.: Object detection with spiking neural networks on automotive event data. In: International Joint Conference on Neural Networks (2022)

26. Paszke, A., et al.: PyTorch: An Imperative Style, High-Performance Deep Learning Library. arXiv preprint arXiv:1912.01703 (2019)

27. Fang, W., et al.: Spikingjelly (2020). https://github.com/fangwei123456/spikingjelly

28. He, K., Zhang, X., Ren, S., Sun, J.: Deep Residual Learning for Image Recognition. arXiv preprint arXiv:1512.03385 (2015)

29. Chen, Y.-H., Yang, T.-J., Emer, J., Sze, V.: Eyeriss v2: a flexible accelerator for emerging deep neural networks on mobile devices. IEEE J. Emerg. Select. Topic. Circuit. Syst. 9(2), 292–308 (2019). https://doi.org/10.1109/JETCAS.2019.2910232

SpikeBALL: Neuromorphic Dataset for Object Tracking

Maria P. Guerrero-Lebrero[(⊠)] [iD], Fernando M. Quintana[iD],
and Elisa Guerrero[iD]

University of Cadiz, Cádiz, Spain
{maria.guerrero,fernando.quintana,elisa.guerrero}@uca.es

Abstract. Most of widely used datasets are not suitable for Spiking Neural Networks (SNNs) due to the need to encode the static data into spike trains and then put them into the network. In addition, the majority of these datasets have been generated to classify objects and can not be used to solve object tracking problems. Therefore, we propose a new neuromorphic dataset, *SpikeBALL*, for object tracking that contributes to improve the development of the SNN algorithm for these type of problems.

Keywords: spiking neural networks · event cameras · object tracking · event-based dataset

1 Introduction

During the last years, deep learning has reached a level of human-like performance in many areas [1,2] owing to the combination of large scale datasets and increased computing power. Nevertheless, for gradient-based algorithms and the floating-point-based computation, artificial neural networks (ANNs) lack biological features and interpretability [3]. The current deep learning technology can be improved by blending computational neuroscience related knowledge and computer technology. The so-called third generation of artificial neural networks [4], Spiking Neural Networks (SNNs), simulate the human brain in terms of calculations and their representations, which shows strong biological interpretability. Neuromorphic engineering is an emerging paradigm in which elements of a computer are modeled after the human brain and nervous system. Mathematical models and their implementation in physical circuits are required for that purpose. Neuromorphic architectures are most often modelled from neurons and synapses. Neurons use electronic and chemical impulses to send information between different regions of the brain and the rest of the nervous system. Neurons use synapses to connect to one another and both are far more flexible, adaptable and energy-saving information processors than traditional computer systems.

© The Author(s), under exclusive license to Springer Nature Switzerland AG 2023
I. Rojas et al. (Eds.): IWANN 2023, LNCS 14135, pp. 641–652, 2023.
https://doi.org/10.1007/978-3-031-43078-7_52

Due to the natural characteristics of SNNs, their hardware implementation (analog or digital) generate advantages like real time processing and low consumption [5]. SNNs operate using spikes, which are discrete events that take place at points in time, rather than continuous values. The occurrence of a spike is determined by differential equations that represent various biological processes, the most important of which is the membrane potential of the neuron. Essentially, once a neuron reaches a certain potential, it spikes, and the potential of that neuron is reset.

In this work, we propose a new neuromorphic dataset, *SpikeBALL*, for object tracking that contributes to improve the development of the SNN algorithm for this type of problem in which, given the initialization of a specific objective (position of the object (x, y)), the trajectory of the objective will be followed in the recording set obtained by the event camera. The dataset is formed by trajectories of a ball generated by the players of a table football and belong to a single ball The dataset is made up of trajectories of a ball generated by the players of a table football and belong to a single ball hit by the players, following the rules of this game. These trajectories have been obtained by the event camera and their objective is to train a SNN to follow the ball and predict its trajectory.

For that purpose, the background of this problem is presented in Sect. 2. The methodology that has been followed to generate the dataset is presented in Sect. 3. Then, we describe the dataset in Sect. 4. Finally, the conclusions and future works are shown in Sect. 5.

2   Background

In recent years, many researchers have shown much interest in object tracking in video sequences [6]. In the field of image processing and computer vision, detecting the objects in the video and tracking its motion to identify its characteristics has been emerging as a demanding research area. Object tracking have been developed very fast in the past few years, most of these works have been developed using Convolutional Neural Networks (CNN) and Recurrent Neural Networks (RNN), however, few of them have used SNN to solve this type of problem [17]. Object tracking is a nontrivial problem in computer vision, and is widely used in sports events broadcasting, robotic, security monitoring and other fields. The use of Siamese networks [7] and Transformers have achieved to become very mature the object tracking with traditional cameras. However, when the objects are moving under extreme conditions of high speed and high dynamic range, traditional cameras have difficulty capturing their movements.

Recent developments in neuromorphic computing systems have focused on the creation of new hardware based on biological characteristics. An example of such hardware are event cameras, such as the Dynamic Vision Sensor (DVS), in which pixels of the DVS work independently (i.e., asynchronously). Each pixel announces when it discloses a relative change in the illumination intensity that is above or below a defined intensity threshold. A further advantage of this asynchronous approach is its enhanced dynamic range, since the pixel sensitivity does not have to be set globally in the event camera. In such a way, the unnecessary static information, such as background landscapes, is not recorded, only the dynamic information is registered. Consequently, event cameras are robust and accurate motion detectors and automatically filter out any temporarily redundant information [8]. This makes them extremely useful for scenes with motion like high-speed counting, or driving safety systems.

The currently widely used datasets, such as ImageNet [9] and COCO [10], have contributed significantly to the success of deep learning. Nevertheless, these type of datasets are not suitable for SNNs due to the need to encode the static data into spike trains and then put them into the network [11]. SNNs need datasets generated by neuromorphic camera DVS avoiding the missing information and can be fair to compare with the artificial neural networks. Researchers have proposed many neuromorphic datasets using DVS such as N-MNIST [12] or DVS-CIFAR10 [13] which are traditional classification datasets that follow predetermined or random trajectories of motion. On the other hand, other datasets has been obtained recording activities in natural environments using neuromorphic cameras, such as DVS-Gesture [14] and N-Cars [15]. However, the majority of these datasets have been generated to classify object and can not be used to solve object tracking problems. For this reason, with the creation of this dataset, we intend to contribute to the development of neuromorphic databases in which regression problems can be solved.

3   Methodology

In this work, a methodology have been developed to keep track of a ball inside a table football. The procedure that has been followed begins with the automated capture data. For that purpose, we use the DVS acquisition platform to shoot videos that are played on a monitor, and use the DV software [16] to collect the data automatically. The second stage consists of post-processing data in which the noisy data are removed. Finally, the data is labeled. Figure 1 shows the entire construction process of the dataset.

3.1   Automated Capture Data

The acquisition device used for generating the dataset have been DVXplorer camera of Innovation Company [16] which allows videos with a 640×480 resolution. DVS camera prevents external light changes from interfering with the experimental data collection. To display the trajectory of the ball, it has been necessary to design a bracket to hang the camera from the ceiling, this bracket has been created on a 3D printer. Figure 2 shows the equipped laboratory with the football table and the position of the event camera, hanging from the ceiling in the vertical line, perpendicular to the table, in order to capture the trajectories of the ball.

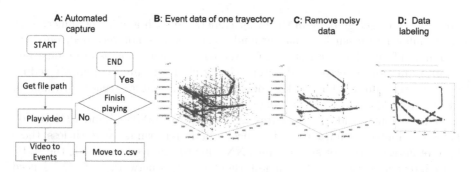

A: Automated capture   **B: Event data of one trayectory**   **C: Remove noisy data**   **D: Data labeling**

Fig. 1. Complete process of data generation. Phase A is the capture stage, including the building of the equipment environment, and the recording with the DV software. Phase B is the data acquisition as event data format. Phase C performs post-processing removing noisy data. And Phase D carry out the data labeling.

Fig. 2. The equipped laboratory to capture the trajectories of the ball with the event camera hanging from the ceiling.

In the experiment, the DV software is used to process the captured event data. When a player kicks the ball, the automatic capture process described in figure 1A is activated. A video is obtained for each hitting the ball. This process is finished when no player kicks the ball.

3.2   Post-processing of Data

The obtained videos contain information about the track of the ball, but also about the rods which have figures attached. For this reason, an algorithm has been developed in order to remove data that do not take in part of track of the ball. This algorithm has been designed exclusively for these dataset in order to obtain the cleanest tracks possible. It is shown, broad terms, in Algorithm 1 and it consists of determining a region which allowing computation of points that are part of the track.

Algorithm 1. Remove noisy data

1: Determine a optimal size region
2: **for** Every point i in trajectory **do**
3:     Calculate the region for i
4:     Save the points inside the region
5:     Add these points to the set of points belonging to the trajectory of the ball
6: **end for**

3.3   Data Labeling

Data labeling is carried out from the cleaned data. The track of the ball is divided into time frames of 10ms. The centre (cx, cy) of all the points that belong to a time frame t is calculated and (cx, cy) is the label for all points of the time frame t. This process is repeated for each time frame that have been divided the track of the ball. It is shown in Algorithm 2.

Algorithm 2. Data labeling

1: Divide the track in time frames of 10ms
2: **for** Every time frame t **do**
3:     Select the points that belong to t
4:     Calculate the centre (cx, cy) of these points
5:     Tag the points with the centre (cx, cy)
6: **end for**

---

4   Data Base Description

In order to store the dataset, we save the event data with the form of (x, y, t, p), where the first two items x, y are the pixel coordinates of the event, the third item t is the timestamp of the event, and the fourth item p is the polarity with value 1 and 0 indicating the increase or decrease of brightness separately. The two polarities are represented by two channels, and the pixels without events are filled with 0. In addition, the labels corresponding to the centers (cx, cy), generated as mentioned in Sect. 3.3 of this work, are stored. For each trajectory a *.csv* file is saved with the events write in previous format. The dataset is comprised of 10 different trajectories with between 80,000 and 100,000 events each one. One trajectory is shown in Fig. 3, in blue dots events positions (x, y) and green stars represent the target of these events.

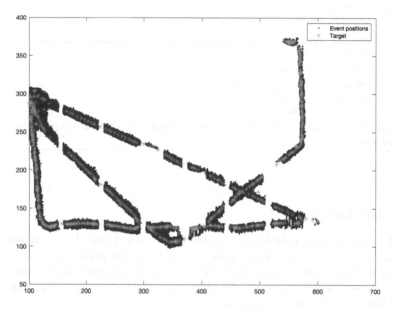

Fig. 3. One of the ten trajectories that are part of dataset. It comprises 83,580 events that are plotted in (x, y) format, where x, y are the pixel coordinates of the event in blue dots. Green stars are the targets of these events. (Color figure online)

5   Preliminary Results

5.1   Neural Arquitecture

With the intention of testing this dataset, *SpikeBALL*, a network has been used that avail of both the benefits of CNN and SNN.

On one hand, CNNs offer several distinct advantages that make them highly effective for computer vision tasks. These advantages include:

1. Local Connectivity and Shared Weights: CNNs leverage the concept of local connectivity, where neurons are connected to a small receptive field, resembling the receptive fields of neurons in the visual cortex. This local connectivity allows CNNs to capture local patterns and spatial dependencies efficiently. Additionally, CNNs employ weight sharing, where the same set of weights is shared across different spatial locations, enhancing parameter efficiency and generalization capability [18].
2. Translation Invariance: CNNs exhibit translation invariance, meaning they can detect and recognize patterns regardless of their location within an input image. This property is achieved through the combination of shared weights and convolutional operations, enabling CNNs to learn spatial hierarchies of features that are robust to translation [18].
3. Hierarchical Feature Extraction: CNNs automatically learn hierarchical representations of features from raw input data. Multiple convolutional layers allow CNNs to learn low-level features (e.g., edges, textures) in early layers and progressively extract higher-level features (e.g., shapes, objects) in deeper layers. This hierarchical feature extraction enables CNNs to capture complex patterns and learn rich representations [18].
4. Parameter Sharing and Model Size: CNNs significantly reduce the number of parameters compared to fully connected networks through weight sharing. By reusing the same set of weights across different spatial locations, CNNs achieve parameter efficiency, making them easier to train and less prone to overfitting, particularly with limited training data [18].
5. Spatial Invariance and Robustness: CNNs exhibit spatial invariance, enabling them to recognize patterns even with slight transformations, such as rotations or scale changes. This spatial invariance and robustness make CNNs well-suited for tasks where precise spatial location or scale is not critical, such as object recognition and image classification [18].

These advantages have contributed to the remarkable success of CNNs in various computer vision tasks, including image classification, object detection, semantic segmentation, and image generation.

On the other hand, SNNs offer several advantages over traditional neural networks. Firstly, SNNs capture the temporal dynamics of neural information processing by encoding information in the timing and order of spikes [4]. This temporal coding enables SNNs to represent time-dependent patterns and process dynamic inputs. Secondly, SNNs exhibit high energy efficiency due to their sparse and event-driven computations [23]. By generating spikes only when necessary, SNNs reduce overall power consumption and are well-suited for resource-constrained environments or low-power devices. Furthermore, SNNs have enhanced biological plausibility as they incorporate concepts such as refractory periods, spike timing, and synaptic plasticity mechanisms like Spike-Time-Dependent Plasticity (STDP) [4]. This biological fidelity makes SNNs valuable for studying and understanding neural information processing in biological systems. SNNs also demonstrate robustness to noise through their ability to filter out noisy input and focus on relevant spike timings [24]. This feature enables

SNNs to perform well in noisy or uncertain environments, enhancing their reliability in practical applications. In addition, SNNs operate in an event-driven manner, generating spikes only when significant changes occur in the input. This event-driven processing leads to efficient and real-time computations, as computational resources are allocated only when required.

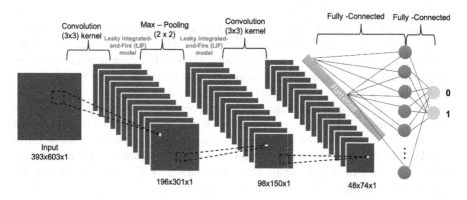

Fig. 4. Network architecture used that includes a CNN and an SNN. The SSN is implemented using a Leaky Integrate-and-Fire (LIF) model.

Figure 4 shows the architecture of the network used to validate the set of data proposed in this work. The SSN is implemented using a Leaky Integrate-and-Fire (LIF) model. LIF model is a commonly used neuron model in computational neuroscience and neural network simulations. It provides a simplified representation of the behavior of biological neurons, capturing essential dynamics while maintaining computational efficiency [25]. In the LIF model, the membrane potential of the neuron is modeled as an electrical potential that evolves over time. It integrates incoming synaptic inputs and generates output spikes based on a threshold mechanism. Notably, the LIF model incorporates a leak term, which accounts for the gradual decay of the membrane potential over time. The LIF model is computationally efficient and allows for analytical analysis, making it a valuable tool for investigating neural dynamics and simulating SNN. Although the LIF model simplifies the intricate dynamics of biological neurons, it serves as a foundational framework for understanding basic principles of neural information processing.

5.2   Performance Metrics

We propose root-mean-square error (RMSE) for evaluating the accuracy of the predictive model. It measures the average magnitude of the differences between the predicted values and the actual values. The RMSE metric provides a single value that represents the typical error or deviation of the predicted values from the actual values. It is especially useful in regression tasks, where the goal is to estimate a continuous target variable.

A lower RMSE indicates that the predicted values are closer to the actual values, implying higher accuracy and better model performance. On the other hand, a higher RMSE suggests larger errors between the predicted and actual values, indicating lower accuracy and poorer model performance.

$$RMSE = \sqrt{\frac{\Sigma_{i=1}^{N}(actual_i - predicted_i)^2}{N}} \tag{1}$$

5.3   Results and Discussion

The process has been repeated 300 times and each time 60 epochs have been performed. Figure 5 shows the RMSE calculated for each epoch. It can be seen that after 20 epochs the error stabilizes at around 1%.

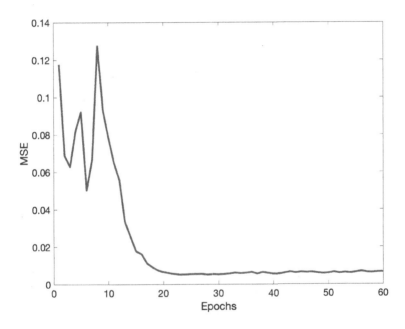

Fig. 5. RMSE depending on the epochs. After 20 epochs the error stabilizes at around 1%.

To qualitatively measure how good the prediction is, Fig. 6 shows one of the ten trajectories in which the positions of the events appear with blue dots, the centers calculated with the previously explained methodology with green stars, and the network prediction with red stars. It can be seen that the prediction generates values very close to the calculated centers, which corroborates the RMSE values obtained.

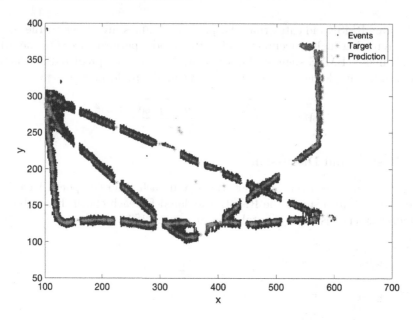

Fig. 6. One of the ten trajectories in which the positions of the events appear with blue dots, the centers calculated with the previously explained methodology with green stars, and the network prediction with red stars. (Color figure online)

6   Conclusions and Future Works

In this work, we propose a new neuromorphic dataset, *SpikeBALL*, for object tracking that contributes to improve the development of the SNN algorithm for these type of problems. In this way, a methodology has been developed in order to obtain a huge dataset to determine the trajectory of the ball in a table football. The procedure that has been followed begins with the automated capture data, post-processing data and, finally, the data labeling. Following this methodology, a dataset, comprised of 10 different trajectories with between 80,000 and 100,000 events each one, has been generated. Thanks to developed methodology the dataset is highly scalable, being able to generate a large volume of data in a short time.

In order to test the effectiveness of *SpikeBall* and the potential to provide new challenges for training SNN algorithms, experiments have been conducted on an architecture that combines the advantages of a CNN and an SNN. The results obtained show that *SpikeBall* can be useful to improve the prediction of SNNs in regression problems.

As future work, we think it would be good to use different architectures other than the one proposed together with classical regression methods to compare the results.

Acknowledgements. Fernando M. Quintana would like to acknowledge the Spanish *Ministerio de Ciencia, Innovación y Universidades* for the support through FPU grant (FPU18/04321). This work was also supported by the project NEMOVISION from the *Ministerio de Ciencia e Innovación*, PID2019-109465RB-I00/AEI/10.13039/501100011033.

References

1. Hirschberg, J., Manning, C.D.: Advances in natural language processing. Science **349**, 261–266 (2015)
2. Noda, K., Yamaguchi, Y., Nakadai, K., Okuno, H.G., Ogata, T.: Audio-visual speech recognition using deep learning. Appl. Intell. **42**, 722–737 (2015)
3. Rumelhart, D.E., Hinton, G.E., Williams, R.J.: Learning representations by back-propagating errors. Nature **323**, 533–536 (1986)
4. Maass, W.: Networks of spiking neurons: the third generation of neural network models. Neural Netw. **10**, 1659–1671 (1997)
5. Shen, G., Zhao, D., Zeng, Y.: Backpropagation with biologically plausible spatiotemporal adjustment for training deep spiking neural networks. Patterns **100522** (2022)
6. Barga, D., Thounaojam, D.M.: A survey on moving object tracking in video. Int. J. Inf. Theory **3**, 31–46 (2014)
7. Zhang, Z., Liu, Y., Wang, X., Li, B., Hu, W.: Learn to match: automatic matching network design for visual tracking. In: Proceedings of the IEEE/CVF International Conference on Computer Vision (IEEE), pp. 13339–1334 (2021)
8. Inivation. Understanding the performance of neuromorphic event-based vision sensors (2020). https://inivation.com/dvp/white-papers/
9. Deng, J., et al.: Imagenet: a large-scale hierarchical image database. In: 2009 IEEE Conference on Computer Vision and Pattern Recognition (IEEE), pp. 248–255 (2009)
10. Lin, T.-Y., et al.: Microsoft COCO: common objects in context. In: Fleet, D., Pajdla, T., Schiele, B., Tuytelaars, T. (eds.) ECCV 2014. LNCS, vol. 8693, pp. 740–755. Springer, Cham (2014). https://doi.org/10.1007/978-3-319-10602-1_48
11. Zhang, T., et al.: Self-backpropagation of synaptic modifications elevates the efficiency of spiking and artificial neural networks. Sci. Adv. **7**, eabh0146 (2021)
12. Orchard, G., Jayawant, A., Cohen, G.K., Thakor, N.: Converting static image datasets to spiking neuromorphic datasets using saccades. Front. Neurosci. **9**, 437 (2015)
13. Li, H., Liu, H., Ji, X., Li, G., Shi, L.: CIFAR10-DVS: an event-stream dataset for object classification. Front. Neurosci. **11**, 309 (2017)
14. Amir, A., et al.: A low power, fully event-based gesture recognition system. In: Proceedings of the IEEE Conference on Computer Vision and Pattern Recognition (IEEE), pp. 7243–7252 (2017)
15. Sironi, A., Brambilla, M., Bourdis, N., Lagorce, X., Benosman, R.: HATS: histograms of averaged time surfaces for robust event-based object classification. In: Proceedings of the IEEE Conference on Computer Vision and Pattern Recognition, pp. 1731–1740 (2018)
16. Inivation Hompage. https://inivation.com. Accessed 20 Mar 2023
17. Yongqiang, C., Yang, C., Deepak, K.: Spiking deep convolutional neural networks for energy-efficient object recognition. Int. J. Comput. Vision **113**, 54–66 (2015)

18. LeCun, Y., Bengio, Y., Hinton, G.: Deep learning. Nature **521**(7553), 436–444 (2015)
19. Simonyan, K., Zisserman, A.: Very deep convolutional networks for large-scale image recognition. arXiv preprint arXiv:1409.1556 (2014)
20. Krizhevsky, A., Sutskever, I., Hinton, G.E.: ImageNet classification with deep convolutional neural networks. Adv. Neural. Inf. Process. Syst. **25**, 1097–1105 (2012)
21. Long, J., Shelhamer, E., Darrell, T.: Fully convolutional networks for semantic segmentation. In: Proceedings of the IEEE Conference on Computer Vision and Pattern Recognition, pp. 3431–3440 (2015)
22. Szegedy, C., et al.: Going deeper with convolutions. In: Proceedings of the IEEE Conference on Computer Vision and Pattern Recognition, pp. 1–9 (2015)
23. Merolla, P.A., et al.: A million spiking-neuron integrated circuit with a scalable communication network and interface. Science **345**(6197), 668–673 (2014)
24. Sengupta, A., Ye, Y., Wang, R., Liu, Y.: Going deeper in spiking neural networks: VGG and residual architectures. Front. Neurosci. **13**, 95 (2019)
25. Izhikevich, E.M.: Simple model of spiking neurons. IEEE Trans. Neural Networks **14**(6), 1569–1572 (2003)

TM-SNN: Threshold Modulated Spiking Neural Network for Multi-task Learning

Paolo G. Cachi[1]([⊠]) (iD), Sebastián Ventura Soto[2](iD), and Krzysztof J. Cios[1,3](iD)

[1] Virginia Commonwealth University, Richmond, VA 23220, USA
{pcachi,kcios}@vcu.edu
[2] Universidad de Córdoba, Córdoba, Spain
sventura@uco.es
[3] University of Information, Technology and Management, Rzeszow, Poland

Abstract. This paper introduces a spiking neural network able to learn multiple tasks using their unique characteristic, namely, that their behavior can be changed based on the modulation of the firing threshold of spiking neurons. We designed and tested a threshold-modulated spiking neural network (TM-SNN) to solve multiple classification tasks using the approach of learning only one task at a time. The task to be performed is determined by a firing threshold: with one threshold the network learns one task, with the second threshold another task, etc. TM-SNN was implemented on Intel's Loihi2 neuromorphic computer and tested on neuromorphic NMNIST data. The results show that TM-SNN can actually learn different tasks through modifying its dynamics via modulation of the neurons' firing threshold. It is the first application of spiking neural networks to multi-task classification problems.

Keywords: Multi-task learning · Spiking neural networks · Loihi2 neuromorphic computer · Neuromodulation

1 Introduction

Multi-task learning (MTL) is a machine learning problem in which a model is trained to solve more than one task [3,16]. The goal is to improve the model's generalization ability by learning tasks in a shared feature space. This is useful when there is a significant amount of shared information between the tasks, which allows to learn shared features rather than learning them for each task separately. This leads to building a more efficient model, as it reuses features learned for one task to improve performance on another task [17,20]. Learning multiple tasks, however, encounters problems such as negative transfer, which happens when different tasks have conflicting goals such as when increasing performance for one task decreases performance for the other(s) [15,22].

Several solutions were proposed to deal with the negative learning problem [1,8,10,14]. The solution of interest here is [10], where the authors solved the multitask learning problem using an approach called single tasking of multiple tasks where the network is trained to solve multiple tasks doing only one

© The Author(s), under exclusive license to Springer Nature Switzerland AG 2023
I. Rojas et al. (Eds.): IWANN 2023, LNCS 14135, pp. 653–663, 2023.
https://doi.org/10.1007/978-3-031-43078-7_53

task at a time. They used attention-like mechanisms with an adversarial loss to allow for learning task-specific features. In other words, the network is forced to find different internal pathways for processing each task, which mitigates the negative transfer problem.

The above-described solution was implemented using classical artificial neural networks (ANN), which suffer from high energy consumption. A solution to the latter problem is to use spiking neural networks (SNN) that are much more energy efficient when run on neuromorphic computers, such as on Intel's Loihi2 [4,12]. SNN closely mimic the way the biological neural circuits operate, and which are known to consume very little energy [6,9]. Unlike ANN, which process data in a sequential manner, SNN process data in a dynamic event-driven fashion, with each neuron emitting a series of spikes in response to its input over time. This complex operation of SNN allows for their more powerful computations that are, however, still not fully understood. Although using SNN to solve simple learning tasks have been reported [2,18,21], they were not used for solving more complex tasks, such as multi-task learning.

The focus of this paper is to use SNN for solving multi-task classification problems. This will also provide for their better understanding and thus allow for their wider use. Specifically, we introduce SNN that changes its behavior, in terms of which classification task to perform, using single tasking of multiple task approach. To control the behavior of the network we modify the spiking neurons' firing threshold. This idea was inspired by the neuromodulation property of biological neurons, which change their internal dynamics based on external stimuli [11]. We named our network as threshold modulated spiking neural network (TM-SNN). Its architecture is shown in Fig. 1. It consists of three blocks. Each block (described in detail later) is built of one or more spiking neuron layers connected in a feed-forward fashion. For training TM-SNN we use SLAYER backpropagation algorithm that was developed to work with SNN [19]. TM-SNN is implemented in Intel's Lava neuromorphic framework that allows for its direct deployment on the Loihi2 computer and it was tested using different multi-task classification settings on NMNIST data [13].

The rest of the paper is structured as follows. Section 2 describes the TM-SNN. Section 3 describes experimental settings and results. The paper ends with conclusions.

2 Method

2.1 Problem Definition

In the setting of single tasking of a multi-task problem, we assume an input space X, where $X \in \mathbb{R}^n$ and a set of two (or more) classification labels $Y^{(1)}$ and $Y^{(2)}$, where $Y^{(1)} = \{y_1^{(1)}, y_2^{(1)}, ..., y_m^{(1)}\}$ and $Y^{(2)} = \{y_1^{(2)}, y_2^{(2)}, ..., y_p^{(2)}\}$. We plan to construct a SNN, F, with weights W and an internal parameter (in our case the firing threshold) φ that learns the transformations: $y_i^{(1)} = F(x_i \mid W, \varphi = \varphi_1)$ and $y_i^{(2)} = F(x_i \mid W, \varphi = \varphi_2)$.

2.2   Architecture

TM-SNN architecture is shown in Fig. 1. It consists of three spiking neuron blocks connected in a feed-forward fashion [5]. The spiking input signal is processed by the first block - the feature extraction block - into a latent p-dimensional spiking feature vector, which is then used to assign the multi-task labels using a label classifier block. A task classifier block is used for learning the specific task that is being performed. The idea behind this three-block architecture is to allow the feature extraction block receive feedback during training not only from the label classifier block (classification loss) but also from additional task classifier block (task loss). In this way, the task classifier block acts as a regularization mechanism for the feature extraction block. The task classifier block is not used during testing. Note that in contrast to the architecture proposed in [5], TM-SNN does not use a gradient reversal layer before the task classifier block. This is because we want the feature extraction block to learn specific feature vectors for each classification task rather than a common feature vector as done in [5].

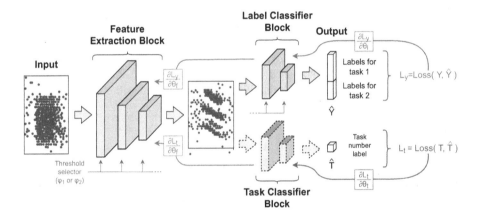

Fig. 1. TM-SNN architecture. It consists of three processing blocks connected in a feed-forward fashion: a feature extraction block and two classifier blocks. The label classifier outputs the labels for task 1 or task 2 (or more). The task classifier is used as a regularization mechanism to aid the feature extraction block learn a set of independent features for each task.

As the spiking neuron model to be used in TM-SNN, we use discrete-time approximation of the integrate and fire neuron model [7]. The equations for the membrane potential, $U_i^{(l)}[n]$, and synaptic current, $I_i^{(l)}[n]$, of neuron i in layer l are given by:

$$U_i^{(l)}[n+1] = \alpha U_i^{(l)}[n] + I_i^{(l)}[n] - S_i^{(l)}[n] \tag{1}$$

$$I_i^{(l)}[n+1] = \beta I_i^{(l)}[n] + \sum_j W_{ij}^{(l)} S_j^{(l)}[n] + \sum_j V_{ij}^{(l)} S_j^{(l)}[n] \tag{2}$$

where α and β are decay constants equal to $\alpha \equiv exp(-\frac{\Delta_t}{\tau_{mem}})$ and $\beta \equiv exp(-\frac{\Delta_t}{\tau_{syn}})$ with a small simulation time step $\Delta_t > 0$ and membrane and synaptic time constants τ_{mem} and τ_{syn}; W_{ij} are synaptic weights of the postsynaptic neuron i and presynaptic neurons j; V_{ij} are recurrent synaptic weights of neurons i and j within the same layer l; and $S_j^{(l)}[n]$ is the output spike train of neuron j in layer l at time step $[n]$. The output spike train is expressed as the Heaviside step function of the difference between the membrane voltage and the firing threshold φ as follows:

$$S_j^{(l)}[n] = \Theta(U_j^{(l)}[n] - \varphi) \tag{3}$$

2.3 Training

The goal of training is to learn weights, W, that predicts task 1 when using firing threshold $\varphi = \varphi_1$ and task 2 (or in general, more tasks) when $\varphi = \varphi_2$. To achieve this, TM-SNN is trained for both tasks concurrently using a per-batch task selection process. Before each batch sample presentation, a task to train TM-SNN for is randomly selected. If task 1 is selected, then the firing threshold of the feature extraction block and the label classifier block is set to $\varphi = \varphi_1$ and for task 2 it is $\varphi = \varphi_2$. After setting the firing threshold, the input samples are presented. Spike-based backpropagation SLAYER algorithm [19] is used to minimize both the label classifier and the task classifier loss functions. The combined loss, L, is calculated as:

$$L = (1 - \gamma) * L_y + \gamma * L_t \tag{4}$$

where L_y is the loss for the label classifier block given by $L_y = Loss(Y, \hat{Y})$; L_t is the loss for the task classifier block given by $L_t = Loss(T, \hat{T})$; and γ is a loss rate constant that controls the rate between the label and task classifier losses. The true labels for the label classifier block, Y, are constructed as a concatenation of Y_1 and $Y_2 = 0$ or $Y_1 = 0$ and Y_2, depending on whether task 1 or task 2 was selected. The task classifier block predicts 0 when trained for task 1 or 1 when trained on task 2 data. Note that the firing threshold is not changed for the task classifier block. This is because the goal is for the task classifier block to backpropagate the same information to the feature extraction block regardless of which task is being learned.

2.4 Testing

During testing, the firing threshold φ_1 or φ_2 is chosen first, depending on which task the testing is being performed. Second, test samples are input only to the feature extraction block and to the label classifier block. The task classifier block, as explained above, is not used.

2.5   Implementation

The network is implemented using Intel's Lava framework which consists of a set of libraries for development of neuromorphic simulation[1] Lava framework has been designed to allow for deployment on the Loihi2 neuromorphic computer. The code for the network implementation as well as all experiments and results are posted at GitHub.[2]

3   Experiments and Results

The performance of TM-SNN is tested on the neuromorphic NMNIST data (60K training and 10K testing samples) [13]. Four types of experiments are performed. First, training and testing performance of TM-SNN using different threshold values is reported see Fig. 2 and Table 1. The task classifier block is not used to assess the effects of the selected threshold values. Second, the influence of including the task classifier block in training is analyzed, see Table 2. Third, the results of TM-SNN operating as described above are compared with TM-SNN that uses the external input current (not the threshold) to control its behavior, see Table 3. Fourth, the ability of TM-SNN to learn more than two tasks is assessed, see Table 4.

3.1   Varying Threshold

Figure 2 shows accuracy of TM-SNN for two-task classification problem on the NMNIST data using different thresholds. Task 1 is the digit classification with 10 labels, and task 2 is the odd/even digit classification with 2 labels. Figure 2 also shows the results for a single-task SNN called ST-SNN, which was separately trained only on task 1 or only on task 2, to establish a base case. The network architecture for both TM-SNN and ST-SNN is essentially the same. It consists of two layers of 512 spiking neurons in the feature extraction block and two layers of 128 and 12 spiking neurons in the label classifier block. Note that φ_1 is set to 1.25 in all tests while φ_2 varies from 1.5 to 10. The constant φ_1 value is used to tune spiking neurons to operate in a normal operation mode (single tasking).

Notice in Fig. 2 that using $\varphi_1 = 1.25$ and $\varphi_2 = 5.0$ results in performance close to the base case scenario (when ST-SNN is trained on task 1 only). Using values for φ_1 and φ_2 close to each other ($\varphi_1 = 1.25$ and $\varphi_2 = 1.5$) achieves results in lower accuracies than the base case. On the other hand, using values that are too far apart (like $\varphi_1 = 1.25$ and $\varphi_2 = 10$) causes longer training times for TM-SNN (see blue line in Fig. 2).

Table 1 compares testing accuracy for both tasks for different firing threshold pairs, after 100 epochs. It also shows accuracy of the ST-SNN (base case).

Two conclusions can be drawn from the above results. First, similar to the training performance (shown in Fig. 2) TM-SNN performs better when the difference between φ_1 and φ_2 increases. Second, the best accuracies on both tasks

---

[1] The lava and lava-dl library are available at https://lava-nc.org.

[2] https://github.com/PaoloGCD/MultiTask-SNN.

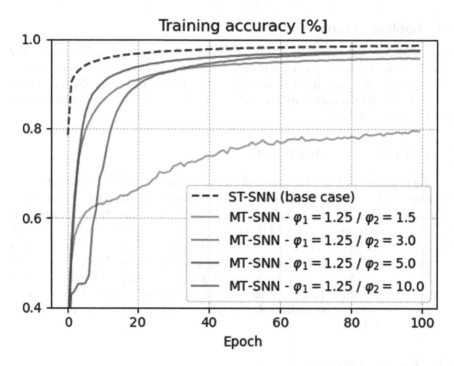

Fig. 2. Training accuracy of ST-SNN (base case) and TM-SNN using different threshold values; φ_1 is set to 1.25 while φ_2 changes from 1.5 to 10.

Table 1. Testing results of TM-SNN using different firing threshold values.

Model	Testing accuracy (%)	
	Task 1	Task 2
ST-SNN (base case)	*98.93*	*99.34*
TM-SNN - $\varphi_1 = 1.25$, $\varphi_2 = 1.5$	91.98	96.09
TM-SNN - $\varphi_1 = 1.25$, $\varphi_2 = 2.0$	95.50	98.51
TM-SNN - $\varphi_1 = 1.25$, $\varphi_2 = 3.0$	96.59	98.90
TM-SNN - $\varphi_1 = 1.25$, $\varphi_2 = 5.0$	97.80	**99.11**
TM-SNN - $\varphi_1 = 1.25$, $\varphi_2 = 10.0$	**97.85**	99.01

are lower than the accuracies of the base case, which is typical when solving multi-task problems. However, in order to improve this performance, we use the task classifier block, which results are described in the next section.

Importantly, it is also good to compare TM-SNN spiking outputs for each firing threshold, which is shown in Fig. 3 for thresholds $\varphi_1 = 1.25$ and $\varphi_2 = 5$ values.

Observe a drastic change in the output when the firing threshold is changed. Specifically, when $\varphi_1 = 1.25$, neuron 4 (corresponding to digit class 4) exhibits

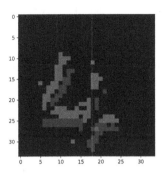

(a) Input of class/digit 4

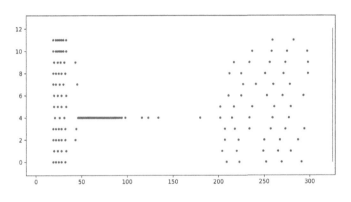

(b) Output with $\varphi_1 = 1.25$

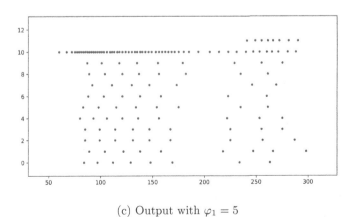

(c) Output with $\varphi_1 = 5$

Fig. 3. Example spiking output when TM-SNN is presented with input representing digit 4 (a) with $\varphi_1 = 1.25$ threshold (b) and with $\varphi_2 = 5$ (c).

the highest activity, while when $\varphi_1 = 5$, neuron 10 (corresponding to even-numbered class) shows the highest activity. Also notice that the overall firing rate of the output neurons for $\varphi_1 = 1.25$ is higher than for $\varphi_2 = 5$.

3.2   Using the Task Classifier Block in Training

The task classifier block is used to decrease the loss function value, which is inherent in multi-task problems. Table 2 compares accuracy when using task classifier block during training. Results are shown for the loss constant γ (Eq. 4) values equal to 0.5, 0.4, 0.3, 0.2 and 0.1. All tests are done using $\varphi_1 = 1.25$ and $\varphi_2 = 5$ values. For convenience of the reader, the results for ST-SNN and TM-SNN (repeated from Table 1) are also shown (two first rows).

Table 2. Testing accuracy of TM-SNN when using task classifier block.

Model	Testing accuracy (%)	
	Task 1	Task 2
ST-SNN - Base case	*98.93*	*99.34*
TM-SNN (without task classifier)	97.80	99.11
TM-SNN/$\gamma = 0.1$	97.90	99.29
TM-SNN/$\gamma = 0.2$	97.86	99.23
TM-SNN/$\gamma = 0.3$	**97.98**	**99.31**
TM-SNN/$\gamma = 0.4$	97.70	99.18
TM-SNN/$\gamma = 0.5$	97.77	99.20

We see that the addition of the task classifier block slightly increased accuracy of task 1 by 0.18% and by 0.20% on task 2, both for $\gamma = 0.3$. This small increase can be attributed to the fact that the task classifier is very simple (only two labels). Note that the task classifier block reaches a plateau very close to 100% accuracy after training for only 20 epochs.

3.3   Use of a Firing Threshold vs Using an External Input Current

For another comparison, Table 3, shows accuracy of a SNN that uses the external input current, called EC-SNN, instead of the firing threshold. The architecture of EC-SNN is essentially the same as TM-SNN. The training was done for 100 epochs using $I_{ext1} = 0$ for task 1 and I_{ext2} equal to 0.05, 0.1, 0.5, 1 and 5 for task 2.

Notice that while controlling I_{ext} the results are lower than when modifying the firing threshold of neurons. This result aligns with how biological circuits perform neuromodulation.

Table 3. Testing accuracy of TM-SNN when an external current I_{ext} is used to control the network operation.

Model	Testing accuracy (%)	
	Task 1	Task 2
ST-SNN (base case)	***98.93***	***99.34***
TM-SNN/$\gamma = 0.3$	**97.98**	**99.31**
EC-SNN/$I_{ext2} = 0.05$	95.63	98.06
EC-SNN/$I_{ext2} = 0.1$	96.05	97.86
EC-SNN/$I_{ext2} = 0.5$	96.07	97.66
EC-SNN/$I_{ext2} = 1.0$	95.78	97.62
EC-SNN/$I_{ext2} = 5.0$	92.20	97.95

3.4   Learning of Several Classification Tasks

Here we test the ability of TM-SNN to learn more than two tasks at the same time. Table 4 shows testing accuracies of TM-SNN trained with two, three, and four tasks. The firing threshold for the first tasks is set at 1.25 and for the other tasks are 5, 10 and 15. Task 1 and task 2 are the same classification tasks from the previous experiments (10 digit label classification and odd/even digit classification). Task 3 is the greater/less than 5 classification (2 labels) and task 4 is the module of 3 label classification (3 labels). The network architecture is the same as in the previous case with the exception that the number of output neurons are changed accordingly. Table 4 also includes the single task SNN (ST-SNN) trained with each task independently as reference.

Table 4. Testing accuracy of MT-SNN trained for four tasks.

Model	Testing accuracy (%)			
	Task 1	Task 2	Task 3	Task 4
ST-SNN (base case)	***98.93***	***99.34***	***99.01***	***98.97***
TM-SNN (2 tasks)	97.98	**99.34**	-	-
TM-SNN (3 tasks)	**98.24**	99.17	**98.84**	-
TM-SNN (4 tasks)	97.05	98.83	98.33	98.11

Results show that threshold modulation also works for cases involving more than two classification tasks. Interesting is the result of training TM-SNN for three tasks that resulted in higher accuracy than training for two tasks. However, the accuracy decreased when training for four tasks.

4   Conclusions

This paper proposed the first ever spiking neural network (SNN) for solving multi-task learning problems. It is called threshold modulated SNN as it uses modulation of a firing threshold of neurons to achieve multi-task learning. A different firing threshold is used for tuning the network to perform each specific classification task separately, while keeping only one set of network weights. For example, for solving a three-task classification problem TM-SNN uses three firing thresholds. TM-SNN was tested on two, three, and four multi-task classification problems using the NMNIST neuromorphic data set. We compared the results of TM-SNN with the results of SNN that was trained for solving each task independently (called ST-SNN). TM-SNN architecture (including one for ST-SNN) used in all experiments was basically the same and consists of three processing blocks. One block is for feature extraction, the second for label classification, and the third for task classification. The latter block performs regularization function and helps to mitigate negative transfer problem inherent to all multi-task learning problems. The results show that performance of TM-SNN is close to that of ST-SNN; in other words, one TM-SNN performs nearly as well as three independently trained ST-SNN networks for three separate classifications. We also compared the results of TM-SNN with the results of using the same TM-SNN architecture but that uses external input current for controlling which specific multi-task classification task to perform. The results show that modifying the firing threshold is better than controlling the network with external input current. TM-SNN was implemented and tested on the Lava platform associated with the Intel's Loihi2 neuromorphic computer.

References

1. Bilen, H., Vedaldi, A.: Universal representations: the missing link between faces, text, planktons, and cat breeds. CoRR abs/1701.07275 (2017). http://arxiv.org/abs/1701.07275
2. Cios, K.J., Shin, I.: Image recognition neural network: IRNN. Neurocomputing **7**(2), 159–185 (1995). https://doi.org/10.1016/0925-2312(93)E0062-I
3. Crawshaw, M.: Multi-task learning with deep neural networks: a survey. CoRR abs/2009.09796 (2020). https://arxiv.org/abs/2009.09796
4. Davies, M., et al.: Advancing neuromorphic computing with Loihi: a survey of results and outlook. Proc. IEEE **109**(5), 911–934 (2021). https://doi.org/10.1109/JPROC.2021.3067593
5. Ganin, Y., Lempitsky, V.: Unsupervised domain adaptation by backpropagation. In: Proceedings of the 32nd International Conference on International Conference on Machine Learning, ICML 2015, vol. 37, pp. 1180–1189. JMLR.org (2015)
6. Gerstner, W., Kistler, W.M., Naud, R., Paninski, L.: Neuronal Dynamics: From Single Neurons to Networks and Models of Cognition. Cambridge University Press, New York (2014)
7. Kaiser, J., Mostafa, H., Neftci, E.: Synaptic plasticity dynamics for deep continuous local learning (DECOLLE). Front. Neurosci. **14** (2020). https://doi.org/10.3389/fnins.2020.00424

8. Liu, P., Qiu, X., Huang, X.: Adversarial multi-task learning for text classification. CoRR abs/1704.05742 (2017). http://arxiv.org/abs/1704.05742

9. Maass, W.: Networks of spiking neurons: the third generation of neural network models. Neural Netw. **10**(9), 1659–1671 (1997). https://doi.org/10. 1016/S0893-6080(97)00011-7. https://www.sciencedirect.com/science/article/pii/ S0893608097000117

10. Maninis, K., Radosavovic, I., Kokkinos, I.: Attentive single-tasking of multiple tasks. CoRR abs/1904.08918 (2019). http://arxiv.org/abs/1904.08918

11. Marder, E.: Neuromodulation of neuronal circuits: back to the future. Neuron **76**(1), 1–11 (2012). https://doi.org/10.1016/j.neuron.2012.09.010

12. Orchard, G., et al.: Efficient neuromorphic signal processing with Loihi 2. CoRR abs/2111.03746 (2021). https://arxiv.org/abs/2111.03746

13. Orchard, G., Jayawant, A., Cohen, G.K., Thakor, N.: Converting static image datasets to spiking neuromorphic datasets using saccades. Front. Neurosci. **9** (2015). https://doi.org/10.3389/fnins.2015.00437

14. Rebuffi, S., Bilen, H., Vedaldi, A.: Efficient parametrization of multi-domain deep neural networks. CoRR abs/1803.10082 (2018). http://arxiv.org/abs/1803.10082

15. Rosenstein, M.T., Marx, Z., Kaelbling, L.P., Dietterich, T.G.: To transfer or not to transfer. In: NIPS 2005 Workshop on Transfer Learning, vol. 898 (2005)

16. Ruder, S.: An overview of multi-task learning in deep neural networks (2017). https://doi.org/10.48550/ARXIV.1706.05098. https://arxiv.org/abs/1706.05098

17. Schröder, F., Biemann, C.: Estimating the influence of auxiliary tasks for multi-task learning of sequence tagging tasks. In: Proceedings of the 58th Annual Meeting of the Association for Computational Linguistics, pp. 2971–2985. Association for Computational Linguistics, Online (2020). https://doi.org/10.18653/v1/2020.acl-main.268. https://aclanthology.org/2020.acl-main.268

18. Shin, J., et al.: Recognition of partially occluded and rotated images with a network of spiking neurons. IEEE Trans. Neural Netw. **21**(11), 1697–1709 (2010). https:// doi.org/10.1109/TNN.2010.2050600

19. Shrestha, S.B., Orchard, G.: SLAYER: spike layer error reassignment in time. CoRR abs/1810.08646 (2018). http://arxiv.org/abs/1810.08646

20. Standley, T., Zamir, A.R., Chen, D., Guibas, L., Malik, J., Savarese, S.: Which tasks should be learned together in multi-task learning? (2019). https://doi.org/ 10.48550/ARXIV.1905.07553. https://arxiv.org/abs/1905.07553

21. Tavanaei, A., Ghodrati, M., Kheradpisheh, S.R., Masquelier, T., Maida, A.: Deep learning in spiking neural networks. Neural Netw. **111**, 47–63 (2019). https://doi.org/10.1016/j.neunet.2018.12.002. https://www.sciencedirect. com/science/article/pii/S0893608018303332

22. Wang, Z., Dai, Z., Póczos, B., Carbonell, J.: Characterizing and avoiding negative transfer (2018). https://doi.org/10.48550/ARXIV.1811.09751. https://arxiv.org/ abs/1811.09751

Author Index

I. Rojas et al. (Eds.): IWANN 2023, LNCS 14135, pp. 665–669, 2023.
https://doi.org/10.1007/978-3-031-43078-7

Printed in the United States
by Baker & Taylor Publisher Services